BURNING PLASMA DIAGNOSTICS

BURNING PLASMA DIAGNOSTICS

An International Conference

Varenna, Italy 24 – 28 September 2007

EDITORS

Francesco P. Orsitto
ENEA, Frascati Research Center, Frascati, Italy

Giuseppe Gorini
Milano-Bicocca University, Milan, Italy

Elio Sindoni
Milano-Bicocca University, Milan, Italy

Marco Tardocchi
Milano-Bicocca University, Milan, Italy

SPONSORING ORGANIZATIONS
"Piero Caldirola" International Centre for the Promotion of Science and
International School of Plasma Physics
Plasma Physics Institute of the National Research Council
ENEA, Italy

AMERICAN INSTITUTE OF PHYSICS

Melville, New York, 2008
AIP CONFERENCE PROCEEDINGS ■ VOLUME 988

Editors:

Francesco P. Orsitto
ENEA, Frascati Research Centre
FPN Department
via E. Fermi 45
I-00044 Frascati
Italy
E-mail: Orsitto@frascati.enea.it

Elio Sindoni
Department of Environmental Science
Milano-Bicocca University
piazza della Scienza 3
I-20126 Milano
Italy
E-mail: Elio.Sindoni@unimib.it

Giuseppe Gorini
Physics Department
Milano-Bicocca University
piazza della Scienza 3
I-20126 Milano
Italy
E-mail: Giuseppe.Gorini@unimib.it

Marco Tardocchi
CNISM, Physics Department
Milano-Bicocca University
piazza della Scienza 3
I-20126 Milano
Italy
E-mail: Marco.Tardocchi@unimib.it

L.C. Catalog Card No. 2008921603
ISBN 978-0-7354-0507-3
ISSN 0094-243X
Printed in the United States of America

CONTENTS

FUSION PRODUCTS

PLASMA FACING COMPONENTS, RADIATION EFFECTS AND DUST

MAGNETICS, EQUILIBRIUM AND PLASMA CONTROL

PREFACE

This book of proceedings collects the papers presented at the International Conference on Burning Plasma Diagnostics held at Villa Monastero, Varenna (Italy) from 24 to 28 September 2007. The Workshop was organized by the "Piero Caldirola" International Centre For the Promotion of Science and International School of Plasma Physics (ISPP). Established in 1971, the ISPP has organized over seventy advanced courses and workshops on topics mainly related to plasma physics. In particular, courses and workshops on plasma diagnostics (previously held in 1975, 1978 1982, 1986, 1991, 1995, 1997 and 2001) can be considered milestones in the history of this Institution. Looking back at the proceedings of the previous meetings in Varenna, one can appreciate the progress in the field of plasma diagnostics over the past 30 years. The 2007 conference was co-organized by the Istituto di Fisica del Plasma of the National Research Council (CNR) and by the Centro Ricerche Frascati of the ENEA.

Previous Varenna meetings on diagnostics had covered diagnostics in present-day magnetically confined plasmas, but also the problems and challenges of ITER diagnostics. ITER (the International Thermonuclear Experimental Reactor, a joint venture between Europe, Japan, Russia, USA, China, India and South Korea) will need to measure a wide range of plasma parameters in order to reach and sustain high levels of fusion power and provide input to control systems with adequate reliability and long-term stability. The conference was the first appointment of the diagnostic community after the approval of ITER, and was therefore an opportunity to review the status of ITER diagnostics with particular reference to the capabilities of the present proposed systems to meet the requirements on the measurements. Critical issues related with the measurements of fast particles, fusion products, plasma facing components and radiation effects were addressed as well as new areas of diagnostic developments related with DEMO, the next step fusion reactor that will deliver electrical power from fusion.

The many excellent papers presented at the conference showed the strong interest in diagnostics throughout the world fusion community. The meeting was attended by over 130 participants, which is a record for this kind of meetings in Varenna. There were tutorial, invited and contributed papers presented in a number of oral sessions and in three poster sessions. The tutorial sessions were much appreciated and restore a Varenna tradition that had been interrupted after the 1991 course and workshop.

In the interest of speed of publication, these proceedings are reproduced directly from the camera-ready copy provided by the authors without refereeing or editing.

Thanks are due to the institutions that have contributed financially to the organization of this conference, in particular the Istituto di Fisica del Plasma of CNR, the ENEA and the University of Milano-Bicocca.

G Gorini, F P Orsitto, E Sindoni, M Tardocchi

October 2007

BURNING PLASMA
DIAGNOSTICS AND REQUIREMENTS

Experience On Plasma Diagnostics On JET

Peter E Stott

Campagne Palerme, chemin de Font Glory, 04100 Manosque, France

Abstract. This year marks the 30th anniversary of the decision on 17 October 1977 by the European Council of Ministers to proceed with the JET project. Very little thought had been given to the diagnostic requirements during the design phase – indeed many people assumed that JET would not do detailed physics studies and would need only a few basic diagnostics. Of course this view turned out to be incorrect – JET needed a comprehensive set of diagnostics and has conducted a very extensive physics programme. The large physical dimensions of JET, the radiation environment and the stringent interface standards radically changed our concept of diagnostics compared to the smaller fusion experiments that had gone before. In this paper I will review some of the experience on burning plasma diagnostics on JET.

Keywords: plasma, diagnostics, fusion, tokamak
PACS: 52.70.-m

INTRODUCTION

This year marks the 30th anniversary of the decision on 17 October 1977 by the European Council of Ministers to construct the large tokamak experiment known as the Joint European Torus (JET) on the Culham site in the UK. The design phase of JET had started in 1973 and was completed in 1975 but this was followed by a delay of two years before the project received approval in 1977. At the time this delay seemed catastrophic for fusion – though with hindsight the time taken to reach a decision to build JET now seems remarkable short compared that for a decision on ITER. Construction of JET started in 1978 and first plasma came five years later in June 1983. The only diagnostic at start up was a Rogowski coil and the first plasma current (figure 1) was displayed on an oscilloscope that had been hurriedly smuggled into the control room – though such instruments with hard wire connections to diagnostics attached to the tokamak were strictly prohibited.

JET had been designed as a limiter tokamak with vertically-elongated D-shaped toroidal field coils that allowed a D-shaped plasma cross-section. The original mandate had been for a plasma current of 3MA with circular poloidal cross-section but during the design stage the elongated poloidal cross-section allowed the specification to be extended to a plasma current of 4.8MA and later machine enhancements allowed JET to reach 7MA in the limiter mode.

CP988, *Burning Plasma Diagnostics*, edited by F. P. Orsitto, G. Gorini, E. Sindoni, and M. Tardocchi

While JET was under construction, the H-mode had been discovered in ASDEX and by 1986 JET started to operate with an improvised divertor-like configuration that gave improved plasma energy confinement. A new purpose-built pumped divertor with coils inside the vacuum vessel was installed in 1995. JET had been planned from the outset to operate with D-T – the first tokamak to do so – and the first operation with D-T in 1991 was followed by a more extensive campaign in 1997 that produced a record 16MW of fusion power for a few seconds. A further D-T campaign took place in 2003. JET was built and operated from 1978 until 1999 by the JET Joint Undertaking team. Since 2000 JET has been operated by the UKAEA for the European Fusion Associations.

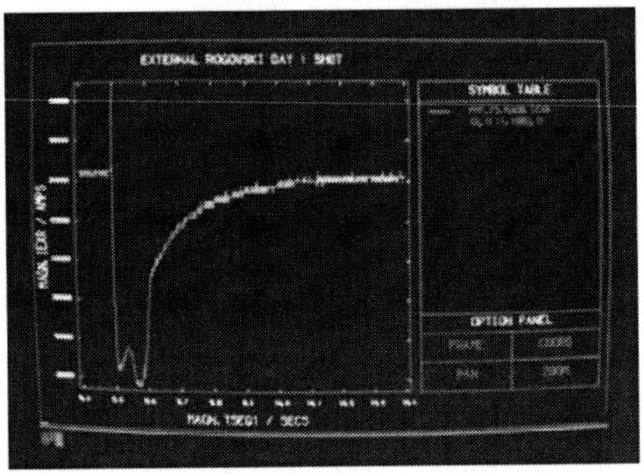

FIGURE 1. The plasma current from JET's first shot was displayed on an oscilloscope in the Control Room.

DIAGNOSTICS

Very little thought had been given to the diagnostic requirements during the JET design phase and serious work on diagnostics started only in 1979 when construction of the tokamak was already well advanced. Many people had assumed that JET would not do any detailed physics studies and would need only a few basic diagnostics. There was a widespread view that all diagnostics were unreliable and that sophisticated diagnostics would be impossible on a big experiment like JET. It was also assumed that changes to diagnostics would not be possible once JET started to operate and that systems, once installed, would have to last throughout the life of JET. Most of these assumptions turned out to be incorrect.

The assessment in 1979 identified a list of about 20 diagnostics systems. Most of these diagnostics were built in collaboration with teams in the other European fusion laboratories – the JET team was responsible for the overall performance specification and for the integration and interface of the diagnostic systems with the tokamak. This working arrangement proved to be very successful – bringing into play the diagnostic expertise of the whole European fusion programme as well as establishing the way for staffing the diagnostics and physics programmes during the JET operation phase. The fusion labs also made substantial contributions to the costs of building the diagnostic systems for JET. The budget that had been set aside for diagnostics in the initial JET costs estimate was seriously inadequate and had to be increased by at least an order of magnitude. But even within the increased budget, priorities had to be set and some systems deferred until a later stage of the project. The initial set of diagnostics systems has been upgraded and enhanced many times. Many new systems have been added to take advantage of new diagnostic developments and to meet new operational requirements so that by 1999 JET had more than 80 diagnostics in operation (see figure 2). I will not attempt in the space of this paper to describe any of these systems in detail as they have been extensively covered elsewhere – and here I will discuss just a few points in general.

A frequently-asked question was "why do you need so many diagnostics?" The term comes from the word "diagnosis" – derived from the Greek words *dia* meaning "through" and *gnosis* meaning "knowledge" – so diagnostics are the means to acquire a thorough knowledge of fusion plasma. Good diagnostics are essential for the progress of fusion research both to understand the physics and to optimize the operation. Diagnostics are increasingly important to control the operation of fusion plasma in order to optimize performance and to access regimes of improved confinement and stability. In order to characterize and understand fusion plasma, we need to measure many plasma quantities or parameters. Different diagnostics techniques are needed to measure the same parameter in different regions of plasma because of the very wide parameter ranges in fusion plasma. For example plasma temperature covers a range of five orders of magnitude – from core plasma with temperatures in the range 30keV to 1keV to the edge plasma with temperatures in the range 1 keV to 100eV – and the divertor plasma has temperatures down to 1eV. There are several different methods of measuring plasma temperature, but no single technique covers the whole range – so we need different instruments employing different techniques to cover the different plasma regions. Overlapping measurement methods are also necessary for cross-calibration and to ensure reliability and to provide confidence in interpretation.

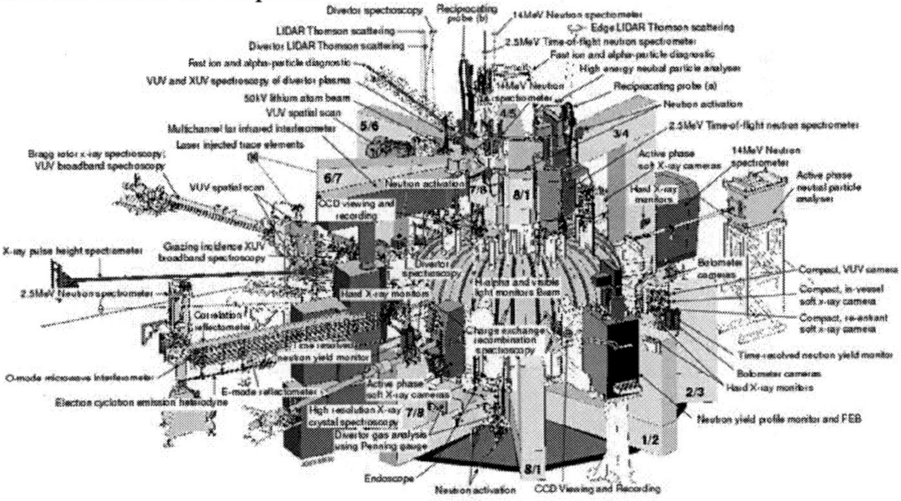

FIGURE 2. A schematic of the diagnostics on JET in 1999.

JET held many new challenges for diagnostics including the large physical dimensions, the stringent interface standards, the radiation environment and the requirements for remote handling & control. These requirements changed our concept of diagnostics and introduced diagnostic engineering. Diagnostics on earlier smaller tokamaks had generally been relatively small instruments that could be installed and moved around by hand. Access was easy and diagnostics could be modified, repaired and calibrated hands on – it easy to go into the experimental hall and make changes between shots if necessary. The very large size of JET required much larger diagnostic systems that required serious engineering. The JET interface standards – especially for diagnostics that were connected directly to the JET vacuum system – were stringent and diagnostics had to be designed from the outset for remote handling and remote control. Windows for diagnostics presented a serious challenge and there was great concern about the consequence of a window on the torus failing – especially during tritium operation – so double windows with separately-pumped interspaces were required.

The most significant challenge was due to the radiation environment. The JET tokamak is housed inside a large biological shield building – known as the torus hall – with high-density concrete walls that are 3m thick and with a roof that is 2m thick. Neutron and radiation levels inside the torus hall are many orders of magnitude higher than around smaller tokamaks. The preference for JET diagnostics was to locate all sensitive diagnostic components outside the torus hall but in cases where this was not possible we had to develop radiation-hardened detectors to use close inside the torus hall and to think about local shielding.

The Thomson scattering systems provide a good example of the benefits of robust diagnostic engineering with sensitive components located outside the torus hall so that they can be easily accessed and maintained and even upgraded when required. The first Thomson scattering diagnostic on JET was a single-point system with a ruby laser and detectors shielded from radiation by locating them above the roof of the torus hall. The laser beam entered the torus hall via a small labyrinth in the roof and passed vertically through the torus. The scattered light was collected with a large mirror telescope outside a cluster of windows on a horizontal port and relayed back through the roof. With this arrangement the electron density and temperature could be measured at only one spatial point in the plasma mid-plane whereas many smaller tokamaks of the day already had the capability to measure temperature profiles. The JET diagnostic was converted into the novel LIDAR system in 1987 in order to measure spatial profiles. The LIDAR technique uses a very short laser pulse which travels horizontally across the plasma mid-plane and spatial resolution is obtained from the time of flight using very fast detectors. Dedicated edge and divertor Thomson scattering diagnostics have been added more recently.

Systems with detectors located close to the plasma include the magnetic and bolometric diagnostics. The original set of magnetic diagnostics had saddle loops outside the torus and small B_θ coils inside the vacuum vessel to measure the shape and position of the plasma. The B_θ coils were housed inside inconel tubes that were welded onto the inside wall of the torus vacuum vessel but with their own vacuum systems that could be independently pumped in the event of a leak. Approval of this system took a long time and requirements for quality control during manufacture were a new experience for diagnosticians. But the efforts paid off and the original components of this system are still working reliably after 30 years – although the original set of coils has been enhanced by extra coils to measure the fine details of the X-point and divertor more accurately.

NEUTRON DIAGNOSTICS

Although neutron diagnostics had been used on many earlier experiments, JET and TFTR were the first tokamaks to produce sufficient neutrons for good measurements and therefore considerable effort was invested into building neutron diagnostics and interpreting their data. JET was equipped with various systems to measure the time-resolved neutron yield, the neutron emission profile and the neutron energy spectrum. Accurate neutron yield measurements required careful calibration and this was achieved with in-situ calibration using sources introduced into the torus when it was not under vacuum. Calibration also relied on a sample activation system where small samples of suitable materials contained inside polythene capsules are exposed to neutrons close to the plasma edge and transferred pneumatically to a remote and well shielded counting station to measure the residual activity of the radionuclide of interest. The JET activation system was designed primarily for calibration purposes but also it has been extensively exploited for a wider range of measurements including the time-integrated 2.5 and 14 MeV neutron yields.

The neutron profile cameras consist of two massive radiation shields in which are embedded multiple collimator channels through which the neutron detectors view the plasma along 19 lines-of-sight in orthogonal directions – one camera views vertically and the other camera views horizontally. Detectors located behind the collimators give time and space-resolved tomographic views of the neutron emission. These detectors have to be able to resolve the neutron energy spectrum in order to distinguish between the neutrons that reach the detector directly from the plasma and the neutrons that have been scattered (and thus lost energy) from the machine structure. The profile cameras have been very versatile instruments – fitted with other detectors they have been used for measurements of the bremsstrahlung radiation emitted by fast electrons generated during Lower Hybrid Current Drive and more recently there have been impressive results using the cameras for gamma ray spectroscopy.

The high resolution neutron spectrometers also require collimators and shielding – so some instruments have been placed outside the torus hall with the collimator built into the thickness of the wall or roof, while other spectrometers have been built closer to the torus within their own massive shields. A time-of-flight neutron spectrometer was used to measure the energy spectrum of 2.4 MeV neutrons from D-D plasmas. In this instrument, the neutrons scattered from a scintillator foil and were detected by a bank of neutron detectors placed at a distance of about 2 m. The neutron energy spectrum is determined from the distribution of the times of flight between the scintillator and the detectors. The time-of-flight technique was extended to 14 MeV neutrons using the associated-particle time-of-flight spectrometer where the neutrons are scattered from a passive foil. In this system the energy of the knock-on protons is measured with silicon diode detectors and the energy of the recoiling neutrons is measured by conventional time-of-flight methods. These instruments were installed in the roof lab and viewed the plasma through a vertical line of sight. In the mid 1990s, a new spectrometer with very high energy resolution based on the magnetic analysis of recoil protons was installed within a massive collimator and shield inside the torus hall and has given valuable results.

CONCLUSION

Diagnostics progress on JET has been driven by ever increasing demands to understand physics and by ever increasing requirements for plasma control. An important factor has always been the enthusiasm and ingenuity of diagnosticians in developing new methods and pushing back the limits on established ones. During the three decades that has passed since the first specifications were drafted for JET diagnostics in the late 1970s, there have been substantial advances in the underlying technology – for example lasers, detectors and microwave components have all improved significantly. In particular data acquisition and computer technologies have advanced substantially – vastly increasing the scope of the amount of data that can be stored and analyzed.

Diagnostics have been an essential contribution to the success of JET. It is interesting to see how the original set of around twenty diagnostics systems has been enhanced and extended so that JET now has more than eighty diagnostics. Implementing these diagnostics on JET was a quantum leap from the diagnostics that had been used on earlier fusion experiments. A similar step will be needed during the next few years to build and bring into operation the diagnostics that are required for ITER – and it is good to see that the ITER community is taking diagnostics very seriously. However it was disturbing to hear during the course of this meeting that studies for the DEMO fusion reactor have started by assuming that only a few basic diagnostics will be needed in a fusion reactor. This is all too reminiscent of the early days of JET. A fusion reactor will be like a high performance fighter plane – it will need sophisticated feedback control in order to keep flying – and this will require diagnostics to measure the many key parameters. It is no use deluding ourselves at this stage that diagnostics are not necessary in a fusion reactor. Undoubtedly the diagnostics for a reactor will be very difficult – but developing suitable diagnostic techniques and components that will work reliably in a reactor environment is a challenge that must be addressed and resolved if the goal of fusion energy is to be realized.

ACKNOWLEDGMENTS

I would like to acknowledge the work of my many colleagues at JET and other fusion laboratories who have contributed to this work and who have made such remarkable progress in the techniques of plasma diagnostics and in our understanding of magnetic confinement.

Some Lessons from the Burning Plasma Diagnostics in TFTR

Kenneth M. Young and David W. Johnson

Princeton Plasma Physics Laboratory, P.O. Box 451, Princeton, NJ 08543, USA

Abstract. A number of alpha-particle diagnostics were initiated successfully during the TFTR DT phase. These diagnostics were prototypical and were not able to operate under all conditions. Had TFTR continued to operate, the deficiencies should have been corrected. Effects of radiation on fiberoptics and other components, lack of sufficient power in collective scattering and the effects of neutral beams on pellet ablation impacting pellet charge-exchange were specific deficiencies in alpha – particle diagnostics. For understanding of the alpha source profile, the importance of very detailed calibrations of neutron diagnostics must be stressed. The interactions of fast ions with instabilities have been taken further in later devices but some aspects were already clear in the DT phase.

Keywords: Plasma diagnostics, Tokamaks, Burning Plasmas
PACS: 52.70.-m, 52.55.Fa

INTRODUCTION

Ten years ago, when TFTR was shut down it was very clear that a rich vein of physics had been struck in the D-T experimental campaign culminating in 10.7 MW of fusion power. Together with pioneering experiments into plasma confinement, the first serious effort at examining the physics of alpha-particles was a key element in the physics program. These physics results were well summarized in ref. 1, and in the references quoted therein.

A number of novel diagnostics were developed to permit the study of the plasma behavior and of the alpha-particles themselves during these D-T discharges and these instruments and the challenges associated with them form the primary topic of this paper. A particular issue was the relatively high radiation environment in which these diagnostics had to function. During the TFTR operation in deuterium, measurements of the dose levels at the site-boundary had indicated that the D-T experiments could be carried out without installation of a concrete shield closely surrounding the tokamak. A consequence was that many diagnostic instruments either had to perform in an unshielded location within the shielded Test Cell, be moved out beyond the shield walls or floor with extended signal transmission paths, or have local shielding constructed around them. (In the D-T experiments, the only complete loss of data was from the Si-diodes in the x-ray camera which was anticipated to saturate under the neutron flux, increased by about two orders of magnitude, and these diodes recovered for further operations with D-plasmas.)

This paper will address the lessons learned from the operation of diagnostics during the D-T program. Firstly it will consider the performance of the diagnostics

CP988, *Burning Plasma Diagnostics*, edited by F. P. Orsitto, G. Gorini, E. Sindoni, and M. Tardocchi

specifically developed for following the behavior of the alpha-particles. Then it will consider some aspects of instabilities and waves and their interactions with alpha-particles and fast-ions. Measurements of the retained tritium were also carried out and these may have significance for the inventory of tritium in ITER. Another specific issue for a burning plasma device is proper accounting of the neutron source, for reasons ranging from regulatory to defining the alpha-particle source; hence the detail of the calibration processes used at TFTR are described. Then two aspects of the radiation environment are considered, the first being the design of local shielding, which may have relevance for the port-plug design for ITER, and the second being compensation techniques for radiation-induced effects in fiberoptic cables.

PROTOTYPICAL ALPHA-PARTICLE DIAGNOSTICS

All the diagnostics applied to alpha-particle measurement have to be considered as prototypes, even though some testing on fast ions was done in D-plasmas with neutral beam or RF heating prior to the D-T operating phase. The scientific results from these instruments have been collected in ref. 2 and the potential for the techniques to be used in a next-step burning plasma device discussed in ref. 3. This paper sets out to look more closely at the issues involved in the measurements, and the lessons that could be carried forward to applying these techniques to a burning plasma device.

We consider first the measurements of the escaping particles. Scintillators behind slits and imaged through a refractive lens system onto fiberoptics outside the vacumm vessel allowed imaging of both the pitch angle and energy of the incoming particles [2]. The particles, on their loss orbits, entering the slit are moving mostly along the toroidal direction. Thus the slit has to be as close as possible to the plasma, while the scintillator is protected from the severe heat loads, and must have good accessibility along the field lines. The arrangement of three of the four detectors arrayed poloidally in TFTR is shown in Fig. 1a with the scintillators mounted behind narrow slits housed in graphite shielding [4,5]. They are close to one of the poloidal limiters placed around the tokamak. The relatively clear access in the toroidal direction is apparent. Such access could only be available in ITER, which will have very similar toroidal field strength to TFTR, with large gouges taken out of the blanket modules and excellent thermal shielding. There was another issue in TFTR where fluorescence in the optical fibers led to a plateau background in the scintillator image. An example of such an image is shown in Fig. 1b [5]. The fiber was shielded as well as possible up to very close to the vacuum window. The radiation dose rate at the fiber end was estimated to be ~ 30 Gy/s, perhaps three orders of magnitude larger than that presently expected outside a port plug in ITER.

Three different techniques were used to gather information about the confined alpha-particles. The first made use of the standard charge-exchange spectroscopy system (CHERS) using the heating beam to measure the rapid radial transport of thethermalized alpha-particles (helium ash) [6]. In TFTR the difference between

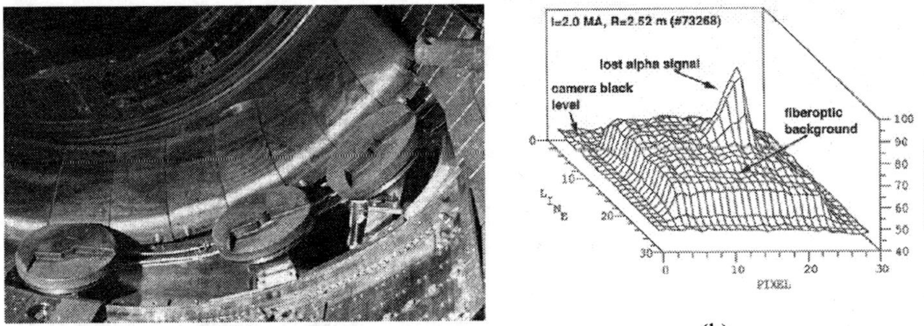

(a) (b)

FIGURE 1. (a) Photograph of three "mushroom" shields protecting the scintillator detectors of the escaping alpha-particle detectors. (b) The image of a scintillator during a D-T discharge, showing the peak due to the alphas and the plateau due to luminescence in the fiberoptics [5].

discharges in D-T and D plasmas, as nearly identical as possible, was used in the analysis. In ITER this method will not be necessary.

The second was a sophisticated adaptation of this CHERS technique to look at the slowing down alpha-particles up to energies of ~ 700 keV [7]. The light from these particles appear at the upper end of the dominant 468.6 nm He-line mostly coming from the thermal neutrals in the plasma edge. A sample spectrum is shown in Fig. 2. An important physics result from this instrument was that the particles are well confined with the diffusion coefficient, $0<D_\alpha<0.03\ m^2/s$. However, the measurement is difficult because the high-energy tail is only ~1% of the bremsstrahlung at 0.3 MeV, and the fluorescence in the fiberoptics for a D-T discharge generating ~10MW is ~10% of the bremsstrahlung. Hence the quoted data were taken after the neutral beams were turned off, made possible by the long slowing down time of the alpha-particles. Had there been a further D-T experimental campaign the fibers would have been better shielded and able to function more like those in ITER might be expected to [8].

The third technique made use of a fast neutral particle analyzer to observe the alpha-particles generated in the ablation cloud of a lithium pellet fired radially into the plasma [9]. The spatial resolution was obtained by time-of-flight of the pellet. By this technique it was shown that the alpha-particles slowed down classically. Again the measurements were made after the beams were turned off. In this instance, the pellets ablated too fast in the beam. This technique would not be possible in ITER because of its larger size and higher temperatures, but also because of its use of heating beams.

A detailed study of ion cyclotron emission from the plasma was made. The emission comes for the scrape-off region on the low field side, and while it showed strong signals for the beam particles, only very transient signals for the fusion product particles were seen [10]. Attempts to make use of collective Thomson scattering at too low an input power level [11] and of the high-energy neutrons created by knock-on reactions of the fusing particles in the plasma [12] were not successful.

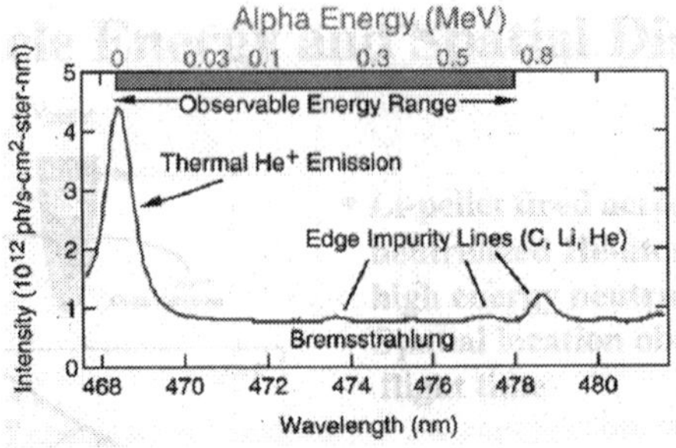

FIGURE 2. Helium spectrum from which the information on the slowing-down alpha-particles is extracted [7].

FAST ION LOSSES AND INSTABILITIES

Instabilities in the plasma can be expected to cause losses of fast ions, and equally these ions can be expected to drive instabilities. Thus it was important to study the relationship between instability and the fast ions in TFTR. All the measurement techniques for fluctuations i.e reflectometry, electron cyclotron emission, magnetics and neutrons, were unaffected by the radiation environment except x-ray imaging. A new reflectometry technique for measuring small-scale density fluctuations inside the plasma core was specifically developed [13]. The escaping-alpha diagnostics detected enhanced losses due to kinetic ballooning modes and disruptions as well as those due to neoclassical tearing modes (NTM) shown in Fig. 3a [14]. The strong rotation of the mode causes the strong peaking at an individual lost-alpha detector. Another example is shown in Fig. 3b [15] where fast ions are driven out by toroidal Alfvén eigenmodes (TAE) created by the heating systems. Theory had anticipated that alpha-particle driven TAE would be stable for normal shear D-T discharges and would only become unstable for weak central shear discharges after the beams were turned off. This was satisfactorily demonstrated experimentally. The mode amplitudes were small and no additional losses were detected. The situation will be much different in ITER where multiple AE modes across the profile could lead to significant loss of the energetic alpha-particles.

MEASUREMENTS OF TRITIUM IN THE PLASMA

Measurement of tritium in the plasma in the presence of a largely deuterium plasma is a challenge. A Fabry-Perot spectrometer was used on TFTR, but it provides only information about the constituents in the plasma edge [16]. Core-integrated ratios of the central densities of the two fuel elements were determined from the 14 MeV and 2.54 MeV neutron peaks [17]. This is a difficult measurement in full D-T operation because there is a difference of two orders of magnitude between the sizes of the two peaks and because of the higher energy neutrons being scattered down in energy. However, this technique can be very valuable in assessing the removal of tritium with

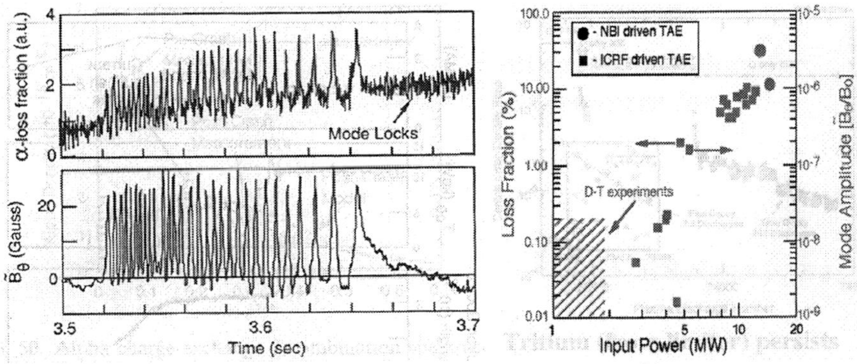

(a) **(b)**

FIGURE 3. (a) Alpha-particle loss driven by NTM modes [1]. (b) Fast-ion loss can be driven by TAE modes created by the heating sources [15].

time after D-T operation. In operation with pulses using deuterium beams after a sequence of D-T operation, there was a rapid initial decay of n_T/n_D ratio of the densities in the core to below 1×10^{-2} but then it required several hundred high power discharges to reduce the ratio by a further order of magnitude. Standard discharge cleaning techniques and ohmic discharges were not effective. This showed how difficult it is to remove tritium from a carbon first wall.

CALIBRATION OF NEUTRON MEASUREMENT SYSTEMS

Neutron measurements are key to demonstrating the quality of performance of the device, to managing the operation in terms of radiation doses and material activation, as well as partaking in the physics of the evolving experiments. One significant measurement is that of the region where the alpha-particles, the prime heating source, are generated in fusion reactions.

Because of the sensitivity of the calibration to the movement of material in the vicinity of detectors, the calibration was carried out when all significant installations were in place just prior to the D-T operation. A valuable asset in checking on a significant impact of any later material changes for a specific detector was an activation foil system placed very close to the plasma which gave very precise comparison of the integrated neutron fluxes between pulses.

The calibrations made use of a ^{252}Cf source and a small accelerator-based "bore-hole" source generating $\sim 1 \times 10^8$ n/s of 14 MeV neutrons [18]. These sources are very much weaker than the distributed source of $>1 \times 10^{17}$ n/s in the D-T discharges. The sources were moved in steps toroidally along the axis of the vessel and also radially across at the midplane. For the neutron collimator system (camera), which is below the vacuum vessel, the steps were very small to be able to pick up slight mechanical deviations. Some data is shown in Fig. 4a. In order to get sufficiently good counting statistics the calibration required about a week's time. It could not be repeated once the vacuum

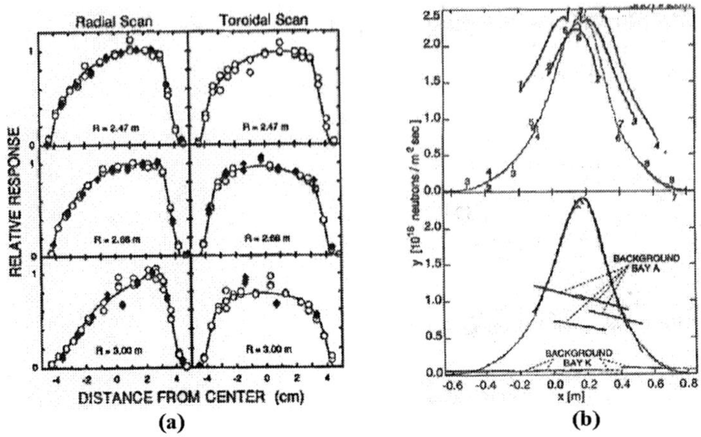

FIGURE 4. (a) Detailed structure visible in the collimator channels during neutron calibration: the source was moved in small steps across the collimator [18]. (b) The upper picture shows the signals on twelve individual channels in the collimator as the plasma was moved across the collimator. The lower picture shows the neutron source profile integrated from these curves [19].

vessel became activated, so a novel technique of making use of the plasmas by moving (jogging) it during the pulse was developed to check on calibrations and for calibrating newly installed channels [19]. Data for a set of 12 channels for a jogged plasma, together with the final integrated profile, is shown in Fig 4b.

Calibration of the ITER neutron systems will be very much harder because of the thick shielding structures, even if a small neutron source can be developed with several orders of magnitude greater flux. The TFTR experience, even with well-engineered mechanical components shows that modeling will not provide sufficiently good cross calibration between camera channels.

DIAGNOSTICS IN A RADIATION ENVIRONMENT

In addition to devising new alpha-particle and neutron diagnostics, the issue of a greatly increased radiation environment had to be addressed. The first of the two major activities were the design and fabrication of shielding for diagnostics positioned relatively close to the tokamak, mostly inside its shielded hall. In a few cases where a suitable shielding structure was not feasible, the diagnostics were moved behind penetrations in the hall's walls or floor. The second was setting up compensation for radiation effects in fiberoptics. Since radiation concurrently induces both fluorescence and absorption in the fibers, both spectrally dependent in the useful near UV to near IR range, it is not simple to provide compensation.

In order to optimize the shielding for any specific diagnostics a number of inputs are necessary for the modeling. 3-d MCNP modeling was used, though 1-d was usually sufficient [20]. Firstly a likely physical geometry was developed. Then the other inputs were the predicted source spectra, the detector response functions (to neutrons and gammas), and the preferred shielding materials. Generally the shields

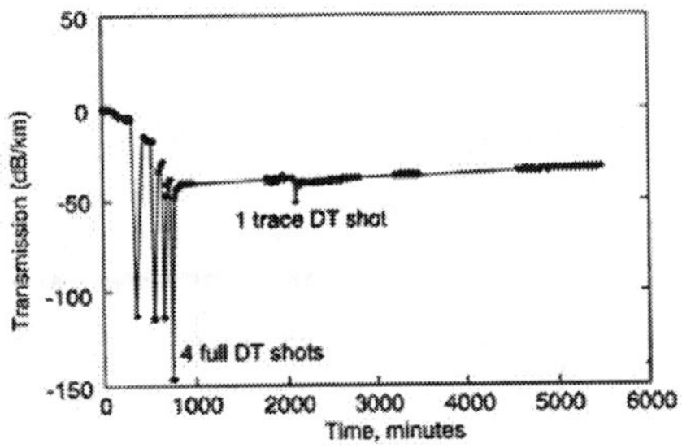

(a)

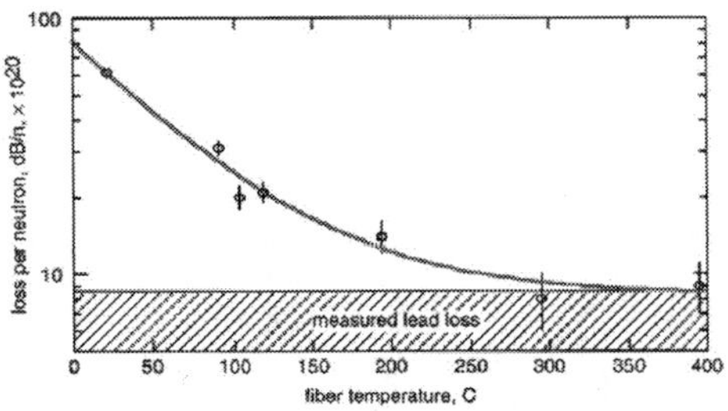

(b)

FIGURE 5. (a) The decay of the absorption in a 600 m-core PCS fiber after D-T discharges. (b) the effect of temperature in reducing the transmission loss [21].

were made of boronized polyethylene with a lead enclosure close to the detector. The resulting shields were very effective though some correction of data was still necessary. Noise would manifest itself as quasi-dc enhancement of the normal background level or with spikes corresponding to (n,p) or (n,α) reactions in the detectors.

Ramsey devised a scheme whereby a compensation fiber could be run alongside any data taking optical fiber [21]. The compensation fiber was equipped with stepped oscillating LED light sources, one passing down the fiber and one reference. Without the lights on, the luminescence is seen. With the source transmitting down the fiber, the sum of the luminescence and transmitted light through the fiber is seen. The reference source adds its light to the luminescence. In this way the transient luminescence and transmission loss are documented. The absorption increases with time and there is a relatively rapid recovery immediately after a shot, even at room temperature. But Fig. 5a shows the recovery of a fiber after four D-T shots showing the long-term recovery continuing over four days. There is a strong effect of heating the fibers (they must be metal-clad) and this is shown in Fig. 5b; loss is negligible above ~300°C. Heating has little impact on the luminescence. Hence for ITER, for which the dose per long pulse outside a vacuum window will be about the same as for a TFTR pulse, heated fibers and an equivalent compensation loop should be sufficient for most techniques using fiberoptics. A disquieting observation on TFTR was that some fibers drawn from the same boule showed markedly different absorption.

SUMMARY

Many important physics questions were left over from the D-T experiments on TFTR. This paper has only addressed some of the issues arising for measurement capabilities which are relevant for ITER. Alpha-particle diagnostics are extremely difficult and complex. Measurement of the escaping alpha-particles will be extraordinarily difficult because of the conformal first wall geometry and heat loads close to the plasma. Temperature measurement by infra-red camera may give some information on total loss but none of the energy and pitch-angle information one would like. Alpha-CHERS looks a feasible technique for ITER particularly if a good view of the heating beam can be found. Collective Thomson scattering with its recent developments at TEXTOR and ASDEX-U in the microwave range of frequencies [22] looks promising, but it is highly desirable that the technique be demonstrated under realistic alpha-particle like conditions.

Because good confinement of the alpha-particles is essential on ITER, a full panoply of fluctuation diagnostics with excellent spatial coverage must be available. Since TFTR the family of AE instability modes has increased and now instabilities up to very high frequencies (>1MHz) are thought to be hazardous.

The knowledge gained on TFTR about fiberoptic behavior under fusion-relevant radiation has already been of great value to the ITER R&D program. Luminescence should not be too significant for well-chosen fibers, and heating the fibers appears promising to anneal radiation-induced absorption. Perhaps the knowledge gained on designing shielding can be helpful in optimizing the shielding of the port-plugs as well as for the necessarily close-in diagnostics (e.g. x-ray spectroscopy).

ACKNOWLEDGEMENTS

This work could not have been done without the efforts of the full TFTR team. We are particularly grateful to Rich Hawryluk, Doug Darrow and Lane Roquemore for their more recent help. The writing of this paper was supported, in part, by US DOE Contract DE-AC02-76CH03073.

REFERENCES

1. R.J. Hawryluk, *Rev. Mod. Phys.*, **70**, 537 (1998).
2. S.J. Zweben et al., *Nucl. Fus,*, **40**, 91 (2000).
3. K.M. Young, "Alpha-particle measurements needed for burning plasma experiments" in *Advanced Diagnostics for Magnetic and Inertial Fusion*, edited by P.E. Stott and A.J. Wootton, Kluwer Academic/Plenum Press, New York 2002, p59.
5. D.S. Darrow et al., *Rev. Sci. Instrum.*, **66**, 476 (1995).
6. E.J. Synakowski et al., *Phys. Rev. Lett.*,**75**,3689 (1995).
7. G.R. McKee et al., *Rev. Sci. Instrum.*, **66**, 643 (1995).
8. M. von Hellerman et al., *Rev. Sci. Instrum.*, **77**, 10F516 (2006).
9. R.K. Fisher et al., *Rev. Sci. Instrum.*, **68**, 336 (1997).
10. S. Cauffman and R. Majeski., *Rev. Sci. Instrum.*, **66**, 817 (1995).
11. J.S. Machuzak et al., *Rev. Sci. Instrum.*, **68**, 458 (1997).
12. R.K. Fisher et al., *Rev. Sci. Instrum.*, **68**, 1103 (1997).
13. R. Nazikian and E. Mazzucato, *Rev. Sci. Instrum.*, **66**, 392 (1995).
14. K.M. McGuire et al., in *Fusion Energy: Proceedings of the 16th IAEA Fusion Energy Conference*, Vol. 1, p 19 (IAEA, Vienna, 1997).
15. R.J. Hawryluk et al., *Phys. Plasmas*, **1**, 1560 (1994).
16. C.H. Skinner et al., *Rev. Sci. Instrum.*, **66**, 646 (1995).
17. S.E. Kruger et al., *Nucl. Fus,*, **36**, 1053 (1996).
18. D.L. Jassby et al., *Rev. Sci. Instrum.*, **66**, 891 (1995).
19. S. von Goeler et al., *Rev. Sci. Instrum.*, **68**, 548 (1997).
20. S.L. Liew et al., *Fusion Technol.*, **8**, 1020 (1985).
21. A.T. Ramsey, *Rev. Sci. Instrum.*, **66**, 871 (1995).
22. S.B. Korsholm et al., *Rev. Sci. Instrum.*, **77**, 10E514 (2007).

Diagnostics Challenges for Burning Plasma Experiments

A.J.H. Donné[1] and A.E. Costley[2] for the ITPA Topical Group on Diagnostics

[1]*FOM-Institute for Plasma Physics Rijnhuizen, Association EURATOM-FOM, Partner in the Trilateral Euregio Cluster, Partner in ITER-NL, PO Box 1207, 3430 BE Nieuwegein, The Netherlands*
[2] *ITER International Organization, Cadarache Centre, 13108 St Paul-Les-Durance, France*

Abstract. Many challenges are encountered in the design of diagnostics for a burning plasma experiment commencing with the definition of the measurement requirements in a coherent way. The specific hostile environment of the burning plasma experiment gives rise to many phenomena that are new to the diagnostics design (radiation-induced effects, long pulses, high particle fluxes, strong relativistic effects, etc.). This presents an overview of the main diagnostics challenges and briefly outlines some of the solutions being developed to meet them.

Keywords: Plasma diagnostics, Fusion, ITER
PACS: 52.70.-m, 52.55.-s

INTRODUCTION

The development of diagnostics for a next step Burning Plasma eXperiment (BPX) such as ITER is a major challenge [1-4]. This is because the measurement requirements are in general the same or more stringent than on present day devices while the environment in which the diagnostics have to be implemented is much harsher. Moreover, in view of the key role of many measurements in the real-time control of the plasma (e.g. in the stabilization of neoclassical tearing and resistive wall modes, and active control of the presence and properties of internal transport barriers), a high level of reliability and robustness in the diagnostic systems is required.

Implementation of existing diagnostic techniques requires substantial design and in some cases dedicated R&D. Moreover, some plasma parameters cannot be measured by existing techniques under BPX conditions and new techniques are required. The starting point in the development of the diagnostics for a BPX (see Fig. 1) is to identify and develop the requirements and their justifications for the large range of measurements. After this the appropriate diagnostic techniques can be selected, and their implementation on BPXs can be developed, including integration with other machine and/or diagnostic components.

In this paper a brief overview of the diagnostic challenges for BPXs will be given, using Fig. 1 as a guideline. Firstly, the use of diagnostics for machine protection and basic plasma control, advanced plasma control and for physics evaluation will be illustrated along with the associated measurement requirements. Secondly, the specific

CP988, *Burning Plasma Diagnostics*, edited by F. P. Orsitto, G. Gorini, E. Sindoni, and M. Tardocchi

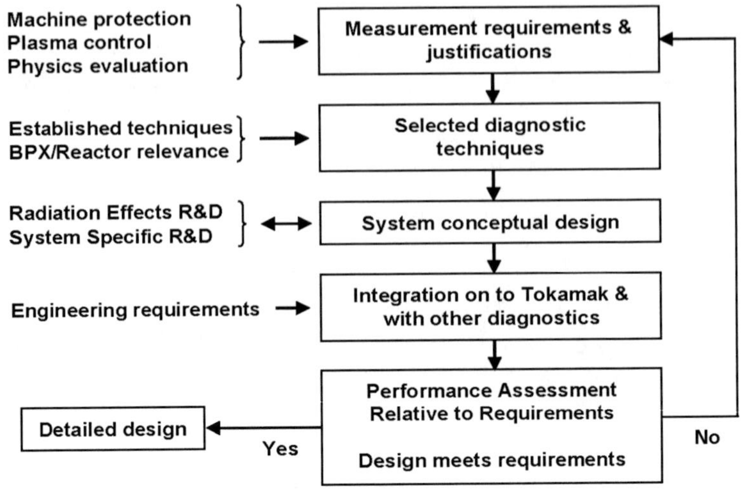

FIGURE 1. Flow diagram of the diagnostic design process [5].

environmental effects in a BPX will be introduced. These are amongst others related to:

- the high neutron- and gamma flux and fluences which lead to radiation induced effects in electrical and optical components as well as to nuclear heating;
- the high particle fluxes (e.g. from recombination) on front-end diagnostic components (in particular mirrors) which lead to erosion and re-deposition;
- the long pulse lengths, which demands high I have alof the diagnostics;
- the specific properties of the BPX being a nuclear device, which complicates maintenance. Diagnostics should be robust and designed to survive the full life cycle of the BPX device;
- the limited number of diagnostic ports, which necessitates integration of multiple diagnostics in a single port plug.

Although the abovementioned issues are generic for any BPX, the remainder of this paper is specifically focused on ITER.

CHALLENGES IN THE DIAGNOSTIC DESIGN PROCESS

Measurement Requirements And Justifications

ITER will require an extensive set of diagnostics to provide measurements for machine protection (e.g. gap between separatrix and first wall, first wall temperature), plasma control (e.g. plasma shape and position, plasma current) and for physics studies (e.g. confined alpha particles, alpha-driven modes, electric fields, fluctuations). In total about 45 parameters need to be measured. In Table 1 an overview is given of the measurements that are required for the various operational phases of ITER [1]. A distinction has been made between measurements that are needed for control purposes and measurements for physics evaluation.

TABLE 1. Overview of the measurements required in the various operational phases of ITER [1]

Operating Scenario	Required Measurements[1]	
	Control	**Evaluation**
H phase. Inductive. Ohmic L- mode. Limited H- Mode	Plasma shape and position, vertical speed, B_{tor}, I_p, V_{loop}, q(a), q(95%), locked modes, m = 2 modes, low m/n MHD modes, line-averaged density, runaway electrons, surface temperature of divertor plates and first wall	halo current, impurity identification and influx, $n_e(r)$ and $T_e(r)$ in core, T_i in core, P_{rad} from core, line-averaged Z_{eff}, H/L mode indicator, gas pressure and composition (divertor and duct), ELM occurrence and type
D phase. Inductive. ELMy H-mode	Plasma shape and position, vertical speed, B_{tor}, I_p, V_{loop}, q(a), q(95%), locked modes, m = 2 modes, low m/n MHD modes, line-averaged density, runaway electrons, surface temperature of divertor plates and first wall, **H/L mode indicator, ELM occurrence and type, divertor detachment**	halo current impurity identification and influx, $n_e(r)$ and $T_e(r)$ **profile**, T_i in core, $\mathbf{P_{rad}(r)}$, line-averaged Z_{eff}, gas pressure and composition (divertor and duct), **plasma shape and position (1000 s), β, $\mathbf{P_{fus}}$, heat deposition profile in divertor**
D/T phase. Inductive. ELMY H Mode	Plasma shape and position (**1000 s**), vertical speed, B_{tor}, I_p, V_{loop}, q(a), q(95%), locked modes, m = 2 modes, low m/n MHD modes, line-averaged density, runaway electrons, surface temperature of divertor plates and first wall, H/L mode indicator, ELM occurrence and type, divertor detachment, **$\mathbf{T_e(r)}$ in core, $\mathbf{T_i}$ in core, $\mathbf{P_{rad}}$ from core, $\mathbf{P_{fus}}$, $\mathbf{n_{He}(r)}$, $\mathbf{n_{He}}$ in divertor, $\mathbf{n_T/n_D}$ in core, divertor ionisation front position, $\mathbf{v_{tor}(r)}$ and $\mathbf{v_{pol}(r)}$**	halo current impurity identification and influx, $n_e(r)$ profile, $T_e(r)$ at edge, $P_{rad}(r)$, line-averaged Z_{eff}, gas pressure and composition (divertor and duct), β, heat deposition profile in divertor, **neutron and alpha source profiles, impurity profile, position of q = 1.5 and q = 2 surfaces, $\mathbf{T_i(r)}$ in core, $\mathbf{Z_{eff}(r)}$, D and T influx, neutral density (near wall), $\mathbf{n_e}$ and $\mathbf{T_e}$ in divertor, impurity and DT influxes in divertor with spatial resolution, alpha loss, neutron fluence, erosion of divertor tiles**
High power D/T Phase. Inductive ELMY H mode. High β	Plasma shape and position (1000 s), vertical speed, B_{tor}, I_p, V_{loop}, q(a), q(95%), locked modes, m = 2 modes, low m/n MHD modes, line-averaged density, runaway electrons, surface temperature of divertor plates and first wall, H/L mode indicator, ELM occurrence and type, divertor detachment, $T_e(r)$ in core, T_i in core, P_{rad} from core, P_{fus}, $n_{He}(r)$, n_{He} in divertor, n_T/n_D in core, divertor ionisation front position, $v_{tor}(r)$ and $v_{pol}(r)$, **β, position of q = 1.5 and q = 2 surfaces, high sensitivity measurements of $\mathbf{n_e}$ and $\mathbf{T_e}$, detection and measurement of NTMs.**	halo current impurity identification and influx, $n_e(r)$ profile, $T_e(r)$ at edge, $P_{rad}(r)$, line-averaged Z_{eff}, gas pressure and composition (divertor and duct), heat deposition profile in divertor, neutron and alpha source profiles, impurity profile, $T_i(r)$ in core, $Z_{eff}(r)$, D and T influx, neutral density (near wall), n_e and T_e in divertor, impurity and DT influxes in divertor with spatial resolution, alpha loss, neutron fluence, erosion of divertor tiles
Hybrid operation	Plasma shape and position (**1500s**), vertical speed, B_{tor}, I_p, V_{loop}, q(a), q(95%), locked modes, m = 2 modes, low m/n MHD modes, line-averaged density, runaway electrons, surface temperature of divertor plates and first wall, H/L mode indicator, ELM occurrence and type, divertor detachment, $T_e(r)$ in core, T_i in core, P_{rad} from core, P_{fus}, $n_{He}(r)$, n_{He} in divertor, n_T/n_D in core, divertor ionisation front position, $v_{tor}(r)$ and $v_{pol}(r)$, β, position of q = 1.5 and q = 2 surfaces, high sensitivity measurements of n_e and T_e, detection and measurement of NTMs.	**q(r)**, halo current impurity identification and influx, $n_e(r)$ profile, $T_e(r)$ at edge, $P_{rad}(r)$, line-averaged Z_{eff}, gas pressure and composition (divertor and duct), heat deposition profile in divertor, neutron and alpha source profiles, impurity profile, $T_i(r)$ in core, $Z_{eff}(r)$, D and T influx, neutral density (near wall), n_e and T_e in divertor, impurity and DT influxes in divertor with spatial resolution, alpha loss, neutron fluence, erosion of divertor tiles
Steady state operation	Plasma shape and position (**3600s**), vertical speed, B_{tor}, I_p, V_{loop}, q(a), q(95%), locked modes, m = 2 modes, low m/n MHD modes, line-averaged density, runaway electrons, surface temperature of divertor plates and first wall, H/L mode indicator, ELM occurrence and type, divertor detachment, $T_e(r)$ in core, **$\mathbf{T_i(r)}$ in core**, P_{rad} from core, P_{fus}, $n_{He}(r)$, n_{He} in divertor, n_T/n_D in core, divertor ionisation front position, $v_{tor}(r)$ and $v_{pol}(r)$, β, position of q = 1.5, q = 2 **and $\mathbf{q_{min}}$** surfaces, high sensitivity measurements of n_e and T_e, detection and measurement of NTMs, **high resolution measurements of the gradient of $\mathbf{T_e}$ and $\mathbf{T_i}$, measurement of RWMs**	q(r), halo current impurity identification and influx, $n_e(r)$ profile, $T_e(r)$ at edge, $P_{rad}(r)$, line-averaged Z_{eff}, gas pressure and composition (divertor and duct), heat deposition profile in divertor, neutron and alpha source profiles, impurity profile, $Z_{eff}(r)$, D and T influx, neutral density (near wall), n_e and T_e in divertor, impurity and DT influxes in divertor with spatial resolution, alpha loss, neutron fluence, erosion of divertor tiles

1 Parameters in bold font are new to the function (control or evaluation).

For each of the parameters listed in Table 1 detailed measurement requirements have been developed that specify the spatial and temporal resolution and accuracy that are required for the measurement of the plasma parameter under various conditions. As an example the requirements for the measurement of the electron temperature are shown in Table 2. For the full list of measurement requirements one is referred to the recently published *Progress in ITER Physics Basis* [1].

TABLE 2. Measurement requirements for the electron temperature

	Condition	**Range (keV)**	**Δt (ms)**	**Δx**	**Accuracy**
Core T_e	$r/a < 0.9$	0.5 - 40	10	a/30	10%
Edge T_e	$r/a > 0.9$	0.05 - 10	10	5 mm	10%

Each of the measurement requirements is accompanied by a detailed justification (see [1] for an example). The development of the measurement requirements has been an on-going process for more than a decade. ITER is the first magnetic confinement device for which the measurement requirements have been systematically developed during the machine design phase. This coherent approach has been taken in order to achieve an optimised, fully integrated, measurement system with a capability closely matched to the requirements. At all times the goal has been to develop the minimum requirements that still fully support the main aims of the measurement (machine protection and basic control, advanced control and physics evaluation). In case a specific requirement cannot be met for whatever reason, one can directly conclude from the justifications what part of the ITER research programme or machine operation will be affected. It is fair to state that the development and fine-tuning of the measurement requirements (to keep track of developments in the field) has been, and still is, a substantial challenge on its own. A large fraction of the measurement requirements is based on experience at present devices and extrapolations towards ITER. However, ITER will be the first device operating in regimes with dominant heating by alpha particles, and therefore new physics regimes. This implies that part of the measurement requirements have been developed on the basis of modelling calculations – where possible supported by experiments (e.g. the alpha-particle simulation experiments on JET [6]).

Selection Of Diagnostic Techniques

Once the measurement requirements have been defined the diagnostic techniques to perform the measurements (see Fig. 1) can be selected. Although many of the techniques used in present devices can also be used on ITER, it is necessary to keep in mind that they will be used in an environment that is much harsher. Further, some of the required diagnostic functionalities go the beyond the scope of those at present devices. It can be seen from Table 1 that in ITER many more measurements - even profiles of parameters such as plasma rotation, electron temperature, etc. - will be used for control purposes. It is necessary to carefully assess whether the diagnostic techniques established at present devices can be made more robust to operate in the harsh ITER environment and whether – if applicable – their scope can be expanded.

As already mentioned, ITER be the first device with dominant alpha heating. Because of the new physics associated with this, a range of new measurements is needed (e.g. the measurement of confined and escaping alpha particles) for which new diagnostic techniques need to be developed. Whenever possible these techniques should be tested on present devices. Specific examples are fast ion collective Thomson scattering for studying fast particles including alphas, fast-wave reflectometry for measuring the fuel ion ration, and techniques to measure escaping alpha particles.

The ITER Environment

The ITER environment is very harsh for diagnostics [1,2,7]. Compared to the discharge of JET with the highest fusion performance of 16 MW thus far [6] ITER will have:

- neutral particle fluxes emerging from Charge Exchange reactions that are about a factor of 5 higher;
- neutron and gamma fluxes that are both an order of magnitude higher;
- plasma discharges that are at least a factor of 100 longer;
- neutron fluences that are at least $>10^5$ times higher;
- nuclear heating up to 1 MW/m^3 (at JET essentially zero).

Consequentially a range of phenomena have to be considered that are new to the diagnostic design. Several of these phenomena can affect the proper function of electrical components: Radiation-induced conductivity (RIC); Radiation induced electrical degradation (RIED); Radiation-induced electromotive force (RIEMF); Radiation-induced thermo-electric sensitivity (RITES); Thermal-induced electromotive force (TIEMF; see Fig. 2). Some of the effects depend on the irradiation dose and some depend on the dose-rate. Even straightforward components such as mineral insulated cables used to transmit signals from in vessel detectors to data acquisition equipment outside the tokamak hall, and coils to measure magnetic fields in the vacuum vessel, suffer from these effects. The challenge is to develop cables, coils, etc. in which the combined action of the above mentioned effects does not deteriorate the function of the diagnostic (component).

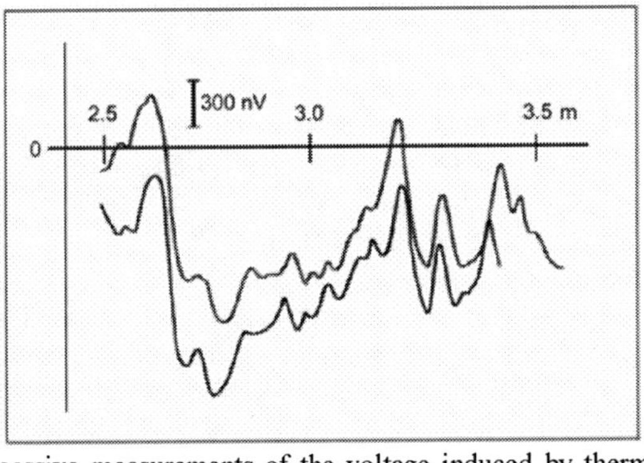

FIGURE 2. Two successive measurements of the voltage induced by thermal effects in MI cable, heated over 2 cm by airgun [8].

Refractive materials (lenses, fibers, windows) will suffer from Radiation induced absorption (RIA) and Radioluminescence or Radiation-Induced Emission (RL or RIE;

see Fig. 3). Although the effects can be reduced to some extend by annealing at elevated temperatures and hydrogen hardening, refractive materials can in general not be used at close proximity to the plasma. For most of the optical diagnostics this implies that refractive components can be used from the end of the port plug onwards. Inside the port plug only mirrors and/or waveguides can be used to transmit the radiation to and from the plasma.

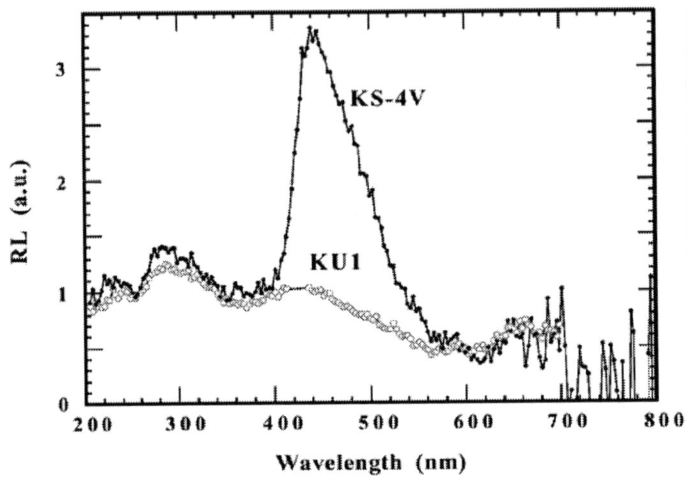

FIGURE 3. Radioluminescence (RL) of KU-1 and KS-4V quartz fibers exposed to a radiation field of 700 Gy/s [9].

The radiation leads further to nuclear heating and changes in other material properties such as activation, transmutation and swelling. All these effects can strongly affect the proper operation of diagnostics and diagnostic components. The challenge is to either use materials that are robust to these effects or – in case the first is not possible – to use alternative diagnostic techniques for measuring certain plasma parameters.

In addition to the material changes due to the high radiation levels, the nuclear environment sets stringent demands on the engineering of the diagnostic systems (10). The diagnostics should be designed in such a way that there is an adequate shielding of neutrons streaming out of the vessel. For optical and microwave diagnostic systems installed in ports, this is achieved by means of labyrinths in the port plugs. It is necessary to include a double tritium barrier, with a monitored volume between the two barriers. Components inside the vacuum vessel and inside the port plugs must, as far as possible, be capable of being repaired or replaced using remote handling tools.

Many diagnostics on ITER will use mirrors to transmit radiation to and from the plasma. Because of the high level of particle fluxes, the mirrors are subject to erosion and deposition. Both of these effects can strongly reduce the polarization dependent reflectivity of the mirrors, and thereby deteriorate the diagnostic. In erosion dominated regions it is possible to use mirrors of single crystalline material (tungsten, molybdenum, stainless steel) that are rather resistant to erosion. Deposition has in almost all cases a strong deleterious effect on the reflective properties of the mirrors.

Apart from research in mirror materials, the geometries of mirrors (e.g. at the end of a duct with baffles) much work done to develop means to protect mirrors (e.g. shutters, gas flow in front of the mirrors) or to mitigate deposition (e.g. operating the mirrors at elevated temperatures). In DIII-D it has been demonstrated that mirrors in the divertor heated to 200-300 °C suffer much less from deposits than mirrors at the same place, operated at room temperature [11].

The temperatures in ITER can reach values up to about 40 keV, which implies that there are strong relativistic effects (see Fig. 4). This can have a large effect on microwave diagnostics because of the downshifted ECE emission (e.g. in contrast to present devices the 2nd harmonic X-mode cannot be used for determining T_e in the plasma core. Instead the 1st harmonic O-mode needs to be utilized. The downshifted emission also strongly limits the operational ranges of reflectometers to measure the density profile. The core density profile can only be measured by reflectometry via the lower cutoff from the high field side.

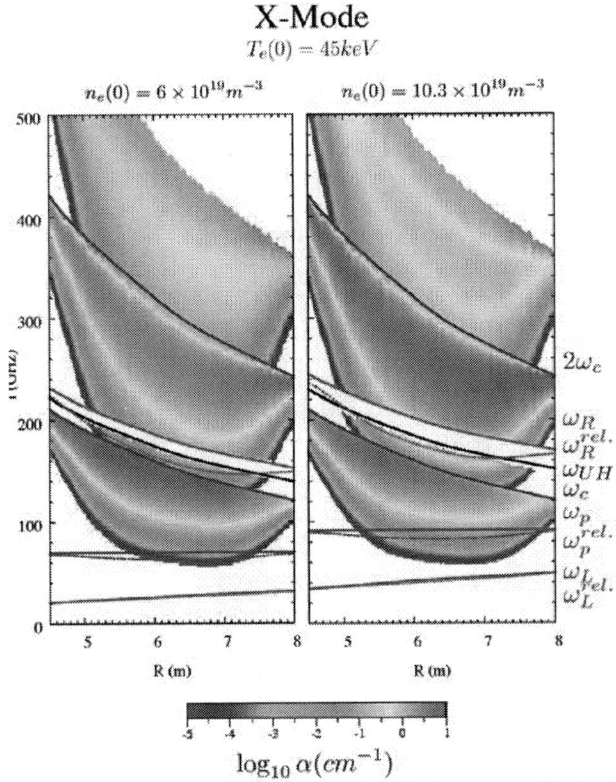

FIGURE 4. Local absorption contours for X-mode propagation at perpendicular incidence for two densities, at high temperature (T_e on axis is 45 keV) representing an extreme case of what might occur in an ITER steady-state scenario [12].

Finally, the need to maintain a measurement during the long pulses should be mentioned. These cause problems for traditional methods such as inductive magnetics used to measure magnetic fields. During the long pulse duration the integrators can be subject to drift. The combination of the long pulse and high levels of radiation constitutes a very difficult measurement problem. In addition to trying to make these techniques more robust and radiation hard, research is on-going to develop radiation-hard steady state magnetic field sensors based on the Hall effect [13], and to develop alternative techniques such as position reflectometry to measure the gap between the density at the separatrix and the first wall.

DIAGNOSTIC INTEGRATION

Diagnostics will be installed in five ports at the equatorial level and twelve at the upper level. In addition, there will be special instrumented cassettes in the divertor. To make the best possible use of this limited space, several diagnostics will be installed in each port [14] and this puts high demands on the integration (see Fig. 5). In most cases the various diagnostics will be developed by different teams in the ITER parties. Effects such as the neutron shielding, the total mechanical strength of the port plug and deformations under ITER operation have to be assessed for the full port plug, and these assessment must be redone if changes are made to the design of the individual diagnostics.

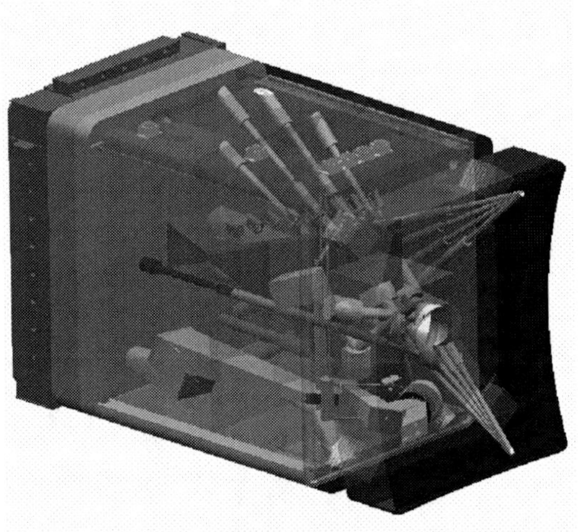

FIGURE 5. Cut-away view of equatorial port number 1 providing access for the radial neutron camera, high resolution neutron spectrometer, a visible IR-TV camera, gamma ray spectrometers, bolometry, one of the divertor impurity monitors, and one of the viewing systems for Motional Stark Effect diagnostic (courtesy C.I. Walker).

DIAGNOSTIC ASSESSMENT

The last step in the diagnostic design cycle (see Fig. 1) is the assessment of the diagnostic capabilities with respect to the measurement requirements. This is done frequently and Table 3 shows the current assessment. For the parameters printed in black it is expected that the requirements can be met. For the parameters in blue it is not yet clear whether the requirements can be fully met. For the parameters in red it is expected that the present requirements cannot be met. For the latter two sets of parameters it is necessary to go back to the measurement requirements (see Fig. 1) to judge whether they can be somewhat relaxed without harming the operation of the machine or the intended physics programme. If this is not possible it is necessary to try to improve the diagnostic design and/or to develop novel techniques in order to meet the requirements.

Table 3. Assessed measurement capability relative to requirements [1]

GROUP 1a Measurements for machine protection and basic control	GROUP 1b Additional measurements for control in specific scenarios	GROUP 2 Additional measurements for performance evaluating and physics
Expect to meet the measurement requirements		
Plasma shape and position, separatrix–wall gaps, gap between separatrixes Plasma current, $q(a)$, $q(15\%)$ Loop voltage Fusion power Line-average electron density 'Halo' currents Radiated power (main plasma, X-point & divertor) Divertor detachment indicator (J_{sat}, n_e, T_e at divertor plate) Disruption precursors (locked modes, $m=2$) H/L mode indicator Z_{eff} (line-averaged) n_T/n_D in plasma core ELMs Gas pressure (divertor and duct) Gas composition (divertor and duct)	Electron temperature profile (core) Electron density profile (core and edge) Ion temperature profile (core) Z_{eff} profile Ionization front position in divertor n_e of divertor plasma Low m/n MHD activity Sawteeth Neutron fluence	TAE Modes, fishbones T_e profile (edge) MHD activity in plasma core
Expect to partly meet the measurement requirements		
$\beta_N = \beta_{tor}(aB/I)$ Impurity and D,T influx (divertor and main plasma) Surface temp. (divertor and upper plates) Surface temperature (first wall) Runaway electrons Dust	Neutron and α-source profile Helium density profile (core) Plasma rotation (toroidal and poloidal) Current density profile (q-profile) Radiation power profile (core, X-point and divertor) Helium density (divertor) Heat deposition profile (divertor) Impurity density profiles Neutral density between plasma and first wall T_e of divertor plasma Net erosion (divertor plate)	Confined α-particles n_e, T_e profiles (X-point) $n_T/n_D/n_H$ (edge) $n_T/n_D/n_H$ (divertor) T_e fluctuations n_e fluctuations Radial electric field and field fluctuations Edge turbulence
Expect to not meet the measurement requirements		
	Alpha-particle loss	T_i in divertor Plasma flow (divertor)

SUMMARY AND CONCLUSION

Setting the requirements for measurements on ITER is a challenge in itself in which the needs of the machine operation and the intended research programme must be carefully considered. A method for doing this has been developed and detailed requirements established. The hostile ITER environment gives rise to a range of phenomena that are new to the diagnostic design, including many radiation-induced and temperature-induced effects. The long pulses and the high plasma temperatures impose additional challenges to some of the diagnostics. In order to make the best possible use of the limited available space, the integration of diagnostics must be optimized. The diagnostics and port plugs will be supplied by the different ITER parties and will be designed in detailed and constructed in locations remote from the tokamak. Taken together these challenges mean that the step to ITER diagnostics is the most substantial challenge ever encountered in the development of diagnostics for high temperature plasmas.

ACKNOWLEDGMENTS

We would like to dedicate this paper to the memory our former colleague Ber de Kock, who worked in the diagnostics staff of the ITER Joint Central Team during the EDA phase and who gave many contributions to the field of ITER diagnostics. Unfortunately he recently passed away after a long illness. We will always remember his positive attitude and his many contributions to the field.

This report was prepared as an account of work undertaken within the framework of ITER Transitional Arrangements (ITA). These are conducted by the Participants: the European Atomic Energy Community, Japan, India, the People's Republic of China, the Republic of Korea, the Russian Federation, and the United States of America, under the auspices of the International Atomic Energy Agency. The views and opinions expressed herein do not necessarily reflect those of the Participants to the ITA, the IAEA or any agency thereof. Dissemination of the information in this paper is governed by the applicable terms of the former ITER EDA Agreement. The work of the first author was carried out within the framework of the European Fusion Programme under the contract of Association between EURATOM and FOM and with financial support from NWO.

REFERENCES

1. A.J.H. Donné *et al., Nucl. Fusion* **47**, S337 – S384 (2007).
2. G. Vayakis *et al., accepted for publication in Fusion Sci. Techn.* (2008).
3. A.E. Costley *et al., 'Requirements and Issues in Diagnostics for Next Step Burning Plasma Experiments'* in Advanced Diagnostics for Magnetic and Inertial Fusion, edited by P E Stott et al (New York, Kluwer Academic/Plenum Press), p. 1 (2002).
4. A.E. Costley, T. Sugie, G Vayakis and C.I. Walker, *Fusion Eng. Design,* **74**, 109 – 119 (2005).
5. A.J.H. Donné and A.E. Costley, *IEEE Transact. on Plasma Science* **32,** 177 – 186 (2004).
6. P.R. Thomas *et al., Phys. Rev. Lett.* **80**, 5548 – 5551 (1998).
7. A.E. Costley *et al., Fusion Eng. Design* **55**, 331 – 346 (2001).
8. R. Vila and E.R. Hodgson, *J. Nucl. Mat.* **367,** 1044 – 1047 (2007).
9. A. Moroño and E.R. Hodgson, *J. Nucl. Mat.* **367,** 1107 – 1111 (2007).
10. C I Walker *et al., Fusion Eng. Design* **51-52**, 377 – 385 (2000).
11. D. Rudakov *et al. Rev. Sci. Instrum.* **77,** 10F126 (2006).
12. G. Vayakis *et al., Nucl. Fusion* **46**, S836 – S845 (2006).
13. I. Bolshakova *et al., submitted to Plasma Phys. Rep.*
14. C.I. Walker *et al., Rev. Sci. Instrum.* **75**, 4243 – 4246 (2004).

Requirements for Plasma and First Wall Measurements on ITER

A.E. Costley[1], P Andrew[1], R L Boivin[2], C Ingesson[3], D Johnson[4], G Vayakis[1] and K Young[4]

[1] *ITER International Organization, Cadarache Centre, 13108 St Paul-Les-Durance, France*
[2] *General Atomics, San Diego, California, 92186, USA*
[3] *EFDA Close Support Unit, Boltzmannstr. 2, D-85748 Garching, Germany*
[4] *Princeton Plasma Physics Laboratory, Princeton, NJ 08543, USA*

Abstract. In this paper we outline the process that is in use for developing the requirements for plasma and first wall measurements on ITER. These requirements are the starting point in the selection, design and integration of the individual measurement systems that together make up the ITER diagnostic system. We describe the issues that have arisen in the defining process and we give the current status. We take a brief forward look at some aspects of the measurement requirements for a next step (DEMO).

Keywords: Plasma diagnostics, Fusion, ITER
PACS: 52.70.-m, 52.55.-s

INTRODUCTION

In order to support the planned experimental programme on ITER, an extensive range of key plasma and first wall parameters should be measured. Some of the measurements will be used in real time to prevent the onset of conditions that could potentially damage the first wall and other in-vessel components (machine protection). Others will be used in real-time feedback control loops to control the value of key parameters at values required for specific plasma performance (plasma control); while yet others will be used to evaluate the plasma performance and to provide information on key phenomena fundamental to ITER performance (physics studies). The measurements of some parameters may contribute to all three roles although the specifications of the measurements (accuracies, resolutions etc.) may be different depending on the role.

An extensive diagnostic system is being developed to make the necessary measurements [1]. The starting point for the selection and design of the individual measurement systems included in the diagnostic system is the detailed requirements for measurements (Fig. 1). In this paper we address the questions why do we need to set requirements, how have they been set, and what is the current status? The issues that arise in such an endeavour are outlined and a brief forward look to the setting of

CP988, *Burning Plasma Diagnostics*, edited by F. P. Orsitto, G. Gorini, E. Sindoni, and M. Tardocchi
2008 American Institute of Physics 978-0-7354-0507-3/08/$23.00

requirements for measurements on the devices that will follow ITER is included. The full requirements have been published recently [2] and are not repeated here.

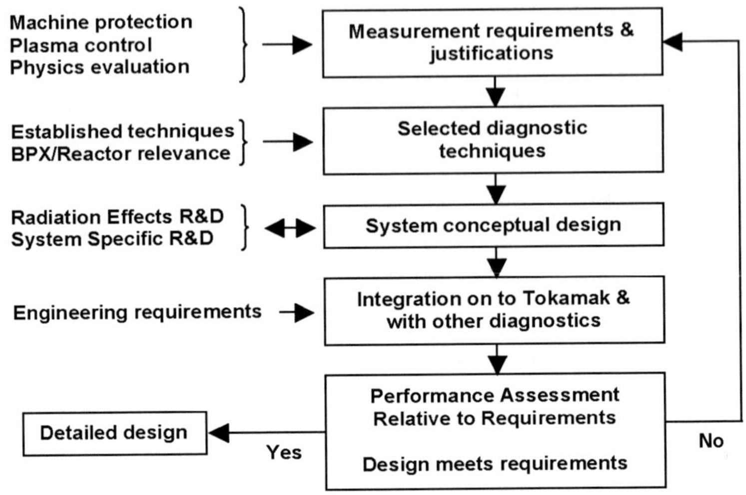

FIGURE 1. Flow diagram of the diagnostic design process.

WHY DO WE NEED TO SET REQUIREMENTS?

As on all other machines, ITER will have a limited number of ports that can be used for diagnostics, limited space around the machine, and, of course, a finite diagnostic budget. For the planned start up scenario, 12 upper ports and four equatorial ports are available for diagnostics. In addition, diagnostics can be installed in two equatorial ports providing they can be easily removed so that remote handling equipment can deployed into the vacuum vessel through these ports during shutdowns. In addition to housing diagnostic equipment these ports have to accommodate shielding blocks to prevent neutron streaming that would otherwise lead to activation of the vacuum vessel flange and heating of the toroidal field coils. At the divertor level, diagnostics (electrical and optical) can be installed at six locations. Diagnostic components – sensors, waveguides, and wiring – can also be installed in the vacuum vessel at specific locations. Within the ITER costing, about 160 kIUA, that is about 5.3% of the total machine construction cost, is provided for diagnostic hardware (including the Diagnostic Neutral Beam) (1 kIUA = $1M (1989)). Together these finite resources set a 'budget' that is to be used to supply the diagnostic system and, of course, a system well adapted to the needs is required. Essentially it is the existence of these boundary conditions that drives the need to set requirements.

In setting the requirements we need to determine what needs to be measured, how well the measurements have to be made, and what are the relative priorities in the measurements. The requirements must be pitched at the optimum level; that is demanding enough to satisfy the intended use of the measurements but not too demanding to require over complicated measurement systems that add cost and complexity and may require significant changes in the design of the interfacing tokamak systems. Complexity relates to reliability and maintainability, which are also important factors in this tradeoff.

HOW HAVE THE REQUIREMENTS BEEN SET?

The requirements have been set through an iterative process that has essentially four steps, and is shown schematically in Fig. 2.

Step 1. The first step was to identify the functions that the measurements have to perform. As mentioned above this is to provide measurements for machine protection, plasma control and physics evaluation. There may also be a need to provide local measurements for use in safety systems, that is to protect personnel. Examples of such measurements are the amount and type of dust in specific locations in the tokamak and the amount of tritium retained. It is still under discussion whether such measurements will be needed.

Step 2. On the basis of experience of the use of diagnostics on existing large tokamaks, and the judged likely need for measurements on ITER, a first set of parameters, such as plasma shape and position, plasma current, electron temperature and density, impurities, first wall temperature etc, that have to be measured in order to meet the functional requirements was identified. For each parameter target draft specifications quantifying the range, spatial and temporal resolutions, and accuracy needed in the measurement was set (Initial Set).

Step 3. The draft target specifications were reviewed by members of groups that might eventually use the measurements. The draft specifications were reviewed by the members of all the physics groups of the ITPA and by the members of the relevant groups in the ITER central team. More than 100 physicists were involved in this process. A revised draft set of requirements was developed by the ITER Expert Group on Diagnostics (which later became the ITPA Diagnostic Topical Group) working in collaboration with the ITER central team staff.

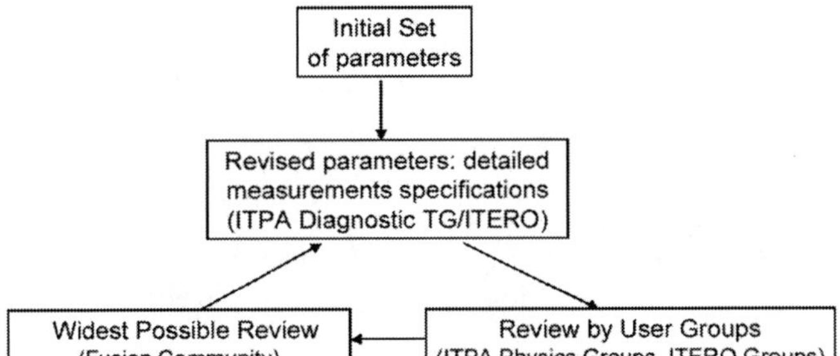

FIGURE 2. Flow diagram for setting of measurement requirements.

Step 4. The draft set of specifications were made available by publication, presentation at conferences and by installing them in the public area of the web page of the ITPA Diagnostic Topical Group [3] for the widest possible review in the plasma physics community. An example, of a detailed specification is shown in Table 1.

TABLE 1. Measurement requirements for the electron temperature

	Condition	Range (keV)	Δt (ms)	Δx	Accuracy
Core T_e		0.5 - 40	10	a/30	10%
Edge T_e	r/a > 0.9	- 10	10	5 mm	

In addition to the detailed specification, a short justification has been developed for each parameter. This aims to document the reasoning behind the value of each of the parameters set in the table: it shows a direct connection between the values in the table and the intended use. In addition to justifying the choices, it has another use. If a parameter cannot be measured as required, then a simple inversion of the justification gives an impact statement of what will not be available. An example of the justification is shown in Table 2.

TABLE 2. Example of Justification: Te measurements

The electron temperature profile is a major indicator of plasma performance and a key component of transport analyses. The profile is key information in instability analyses. Steep transport barriers are observed inside the plasma core and electron temperature pedestals at the edge play a role in analysis of the transport. A time resolution of 10 ms is short compared to times of interest and allows for study of MHD. For kinetic control of the stored energy or ITB gradient, 10 ms time resolution is expected to be sufficient since this is much faster than the typical core confinement times and of the same order as the actuator (heating) response time. The core temperature should be measured with 10% accuracy or better to enable a useful determination of the stored energy and to support useful analysis of plasma performance with transport codes. To resolve fine structure on the profile such as internal transport barriers it should be measured at 30 locations across the profile. The edge, with much steeper gradients should be measured with only 0.5 cm between measurements at about 20 locations.

The process is on-going and should continue. It is expected that the functional uses of the measurements will become better defined as the plans for the operation and research plan of ITER are developed, and the specifications of the measurements should be updated correspondingly.

CURRENT STATUS: DIFFERENT REFERENCE LEVELS

In the development process, we have found it convenient to define different levels of requirements owned and maintained by different groups and individuals. There are essentially three levels.

At the lowest level all comments and suggestions for changes are integrated into one file. This is called **Reference Level 2**. No judgment is exercised at this level. Routinely Reference Level 2 and the accompanying justifications are reviewed by the members of the ITPA Topical Group (TG) for Diagnostics and, after careful consideration of the justifications, **Reference Level 1** is formed and updated. As far as possible, consensus within the TG is sought in this process. According to the process agreed for this step, if there any disagreements the Chair and Co-Chair of the TG have the power to decide. Thus far it has never been necessary to exercise this power. At the highest level, the measurement requirements are included in the main project requirements documents of ITER. Periodically these are updated and at that time changes in the measurement requirements are proposed based on Reference level 1 and the developed justifications. If approved through the ITER change process, the revised requirements are adopted into the ITER project requirements documents. The requirements at this level are called **Reference Level 0**. Currently Reference levels 1 and 0 are the same.

ISSUES AND CHALLENGES

This is a complicated process involving several different aspects and some human judgment and so it is not surprising that some issues have arisen.

Issue 1. Setting of Optimised Requirements. Diagnostics consume budget, take space in and around the machine, can impact other systems through interfaces, and, through failures in their operation and/or their hardware, can endanger machine operation. The diagnostic system must, of course, meet the functional needs but it must also not be overly specified which could lead to an over complicated system. Getting the balance right is important and is a non-trivial task. The extensive and rather formal review and check process that we have developed is intended to achieve the best possible optimized set.

Issue 2. Setting of Priorities. It is essential that some requirements are met while for others it is only desirable. These priorities have to be taken through to the system selection, resource allocation and system design. The optimisation and prioritisation are done by examining the consequences of a requirement not being met, with order being first priority those that affect machine protection followed by plasma control and then physics studies. After this process it is necessary to check that a measurement crucial to the physics mission is not neglected or under represented. If safety class measurements are added to the diagnostic scope, then these will form the highest priority.

Issue 3. Definitions. Specifying precise values for parameters such as spatial resolution and accuracy immediately raises the question of the definitions of these parameters and different definitions are used in scientific work. For example, in one definition of spatial resolution two features are said to be resolved if the signal between them drops to ~ 0.8 of the signal in the feature(s), similar to a definition of resolution often used in spectroscopy. Another definition is based on the expected gradient. For accuracy, it is usually better to talk in terms of uncertainity and, as is well known, there are two types of uncertainity - systematic and random. Accuracy/uncertainty and spatial resolution are linked: if the uncertainty is ΔP then it can be argued that the best spatial resolution that can be achieved is $\delta r \sim \Delta P / \nabla P$ [4].

Some diagnostics provide localized information by inverting the signals measured along several different lines of sight. Others give direct localised measurements. This raises the question of how to compare these measurements and especially the spatial resolution. A definition of spatial resolution, ideally consistent for both local measurements and for profiles derived from non-local measurements is required. One way of circumventing this problem is to carry out model simulations. Starting with a simulated plasma with spatial variations the measurement is modeled. The results can then be directly compared [4].

Similarly with time resolution it is necessary to properly take into account sampling rate, integration time and synchronisation.

Issue 4. Measurement Requirement versus Performance Specification. As mentioned above, the measurement requirements are determined from a careful assessment of the use of the measurements in machine protection, plasma control etc. However, the measurements will be provided by a diagnostic system which will consist of about 40 individual diagnostic systems and these will be procured against a certain performance specification that can be tested without plasma; that is there is a step between the instrument performance and the required plasma/first wall measurement, and then another step from the individual system performance to the integrated performance of the diagnostic system. In some cases, the first step is straightforward, for example, measurement of the neutron flux using a calibrated neutron detector. In other cases it is much more complicated; for example the measurement of the electron temperature profile using electron cyclotron emission (ECE) measurements in which information on the magnetic field in the plasma is needed in the analysis process and this is provided by another diagnostic. The integrated capability is, of course, the integrated capability of the individual measurement systems that constitute the diagnostic system, and, as mentioned, in some cases the performances are inter-linked. The situation is shown schematically in Fig. 3 for the core temperature which is measured by the LIDAR, the ECE diagnostics and the soft X-ray diagnostic. Each has unique limitations on its capability and cannot fully meet the measurement requirements but the performance of the three diagnostics together should be close to that required. The performance limitations are reflected in the functional specifications used for procurement. Another important procurement specification is the testing needed to qualify the diagnostic without plasma. Definition of these procurement specifications is currently underway

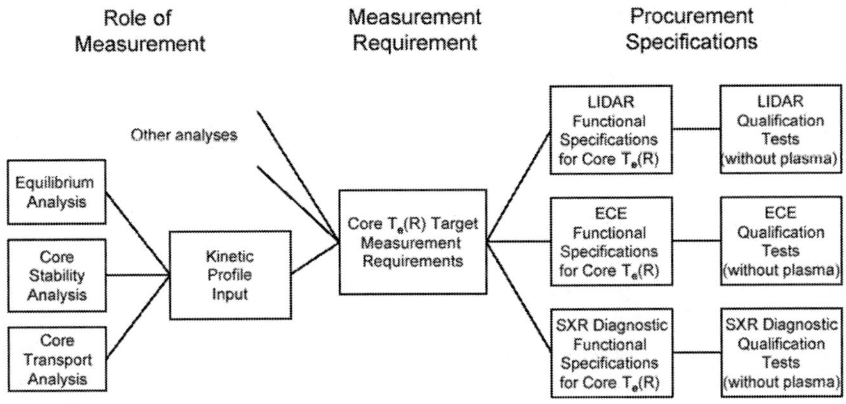

FIGURE 3. Flow diagram for setting of procurement specifications.

LOOKING FORWARD

As in many other areas, the investigations and developments for ITER will lay the foundations for the machines that will follow ITER. These will be demonstration reactors - frequently termed DEMO. The diagnostic system for a DEMO is, of course, far from being defined at this stage. However, since there will inevitably be a step from ITER to DEMO, there will still be an experimental aspect involved. Hence, possibly for DEMO there might be an extended set of diagnostics and corresponding measurement requirements to deal with the start up and determination of the optimum operating conditions. All the aspects mentioned herein in relation to the setting of requirements for measurements - optimisation, prioritisation, definitions etc - will be important, but because of the scale and cost of the undertaking, the issues mentioned above must have been fully resolved: the measurement specifications will have to be at the level of engineering requirements.

SUMMARY

The requirements for plasma and first wall measurements on ITER are highly developed, well documented and have supporting justifications. The process for making, reviewing and changing these requirements is also highly developed. It is expected that the detailed specifications will continue to develop as the needs for measurements on ITER become even better defined. The requirements are already being used to guide the system selection and detailed design. Issues such as optimisation, prioritisation, definitions etc are being addressed in current work. It is clear that we will learn on ITER the measurements needed to support reactor level operation in the next generation of fusion machines (DEMO) and how to specify them.

ACKNOWLEDGMENTS

This report was prepared as an account of work undertaken within the framework of ITER Transitional Arrangements (ITA). These are conducted by the Participants: the European Atomic Energy Community, Japan, India, the People's Republic of China, the Republic of Korea, the Russian Federation, and the United States of America, under the auspices of the International Atomic Energy Agency. The views and opinions expressed herein do not necessarily reflect those of the Participants to the ITA, the IAEA or any agency thereof. Dissemination of the information in this paper is governed by the applicable terms of the former ITER EDA Agreement.

REFERENCES

1. A.E. Costley, T. Sugie, G Vayakis and C.I. Walker, *Fusion Eng. Design,* **74**, 109 – 119 (2005).
2. A.J.H. Donné *et al., Nucl. Fusion* **47**, S337 – S384 (2007).
3. http://www.rijnh.nl/ITPA/
4. L.C. Ingesson, D.J. Campbell, M. Cecconello, E. Ciattaglia, P. Dirken, N. Hawkes, M. Walsh, M. von Hellermann, S. Kálvin, P. McCarthy, O. Neubauer and L. Petrizzi, *Rev. Sci. Instrum.* 77, 10F502-1–3 (2006).

Technology Issues of Burning Plasma Diagnostics

A S Kaye

JET-EFDA, Culham Science Centre, Abingdon OX12 3DB, U.K.

Abstract. The ITER Tokamak will require many diagnostics both for safe and reliable operation of the machine and for understanding of the physics underlying the performance. The design of these diagnostics raises many challenging technical issues not faced on smaller machines. These arise partly from the increase demands on established diagnostics arising from the increased size, higher magnetic field, large heating power, and in particular the dramatically longer pulse duration of ITER, which make issue such as power loading on first wall components more challenging. The demands on reliability and availability of the machine in order to achieve the objectives within the agreed time schedule also place severe additional demands on the design, quality assurance and maintainability of diagnostics. ITER will produce many orders of magnitude more neutrons than previous Tokamaks and will be a licensed nuclear facility. This has important implications for the traceability, quality assurance and availability of safety critical diagnostics, and for the control of the design and procurement of all diagnostics. The high neutron flux/fluence also constrains the design of diagnostics, which must offer shielding consistent with the allowable dose rates on critical components of the Tokamak, and themselves be tolerant of the radiation level at the diagnostic. This paper presents an overview of the more critical issues for ITER diagnostics.

Keywords: Burning plasmas, ITER, Diagnostics,

INTRODUCTION

With the construction of the ITER Tokamak, fusion engineering is moving strongly forward in the demonstration of key technologies for the construction of Tokamak power generating reactors. ITER and subsequent Tokamak reactors differ fundamentally from present large Tokamaks in that they will rely predominantly on plasma heating by the 3.5 MeV alpha particles generated in the deuterium-tritium fusion reaction. Such burning plasmas may also ignite, that is, rely solely on alpha particle heating with no external heating, but this is not necessarily the case and is not foreseen on ITER where the design value of Q (the ratio of fusion power to external heating power) is 10. Burning plasmas raise a range of issues for Tokamak diagnostics, partly from the increased size, magnetic field and plasma current required to achieve a burning plasma, and partly from the consequences of the plasma burn, key amongst these being the intense neutron flux at the first wall and the need to use tritium as a fuel.

The increased scale impacts mainly on diagnostics near the first wall, where they are subject to increased incident power and in particular energy density both during steady operation and during fault conditions. The high neutron flux leads to a wide range of requirements. The use of radiation resistant materials and designs in the first wall and the port plugs must take account of both prompt effects during irradiation and long term effects due to the accumulating damage in materials. The neutrons must be predominantly contained by the shield modules within the Tokamak vessel in order to protect external equipment and manage the activation levels and dose rates around the machine. The design of diagnostics, as with other equipment, must not allow significant breach of this containment. Neutrons activate many materials, which necessitates the use

CP988, *Burning Plasma Diagnostics*, edited by F. P. Orsitto, G. Gorini, E. Sindoni, and M. Tardocchi
2008 American Institute of Physics 978-0-7354-0507-3/08/$23.00

of remote handling for in-vessel maintenance and repair activities. Such interventions must be kept to a minimum, which requires that all diagnostic equipment must be designed and manufactured to achieve demanding levels of reliability, availability and maintainability. Burning plasma Tokamaks require a substantial inventory of tritium and unlike present generation machines, are licensed nuclear facilities. All equipment, including diagnostics, must comply with the conditions of the Safety Case supporting such a license, and must be demonstrated to the Regulator to comply. This has important ramifications not only for the technical quality of the design, but also for the documentation and traceability of equipment, and for the on-going maintenance and inspection after installation.

This paper discusses these various issues in relation to the design of diagnostics for burning plasma devices, using the JET facility to illustrate the scale of the issues.

THE BURNING PLASMA TOKOMAK ENVIRONMENT

Power production from fusion in large Tokamaks relies on the fusion of deuterium and tritium nuclei to form helium nuclei (alpha particles) and free neutrons. Each reaction releases around 17MeV of energy, of which some 3.5MeV is carried by the alpha particle. The neutron travels directly into the first wall where the bulk of the energy is absorbed in the first 100-200mm of the steel and water shield modules (and used to generate steam in an eventual reactor). The alpha particle is contained in the plasma by the magnetic field and loses energy by collision with other charged particles, heating the plasma in the process.

The key parameters of a burning plasma device are rather narrowly constrained. The cross section for the fusion reaction peaks at 10-20 keV ion temperature. The reaction rate depends on the square of the ion density, which is therefore to be maximised, but which is limited by the plasma pressure, which in turn is determined by the magnetic field. This is limited to around 6 Tesla for existing superconducting technology in devices of this scale - the corresponding density is then around $10^{20}m^{-3}$. The confinement improves strongly with plasma current, but this is also limited by the magnetic field and the requirements for MHD stability of the plasma column, which depends on the plasma size. Increasing the plasma volume increases the fusion power more rapidly than the power losses to the walls – at some critical volume these approach a balance and a burning plasma is achieved. Scaling from many Tokamaks shows the critical volume to be around $1000m^3$. The corresponding plasma current is then typically 20MA, and the fusion power around 1 GW. Whilst the ITER device was originally designed to ignite, it was subsequently scaled down to achieve Q=10. This modestly reduces the size and plasma current. The basic parameters of ITER are given in Table 1 [1].

Table 1. The basic parameters of ITER for full performance and quasi steady state operation.

	Full performance	**Steady state**
Major radius, m	6.2	6.2
Minor radius, m	2.0	2.0
Toroidal field on axis, Tesla	5.3	5.3
Plasma current, MA	15	9
External heating power, MW	50	60
Fusion power, MW	500	350
Alpha heating power, MW	100	70

Some important design parameters for burning plasma diagnostics also follow from this outline design. Thus, the neutron flux density at the first wall has to be around 0.5-1MW/m^2. About half of the alpha and additional heating power ultimately appears on the divertor at a power density of 10 MW/m^2 at the strike points. The balance is radiated by x-rays/vuv onto the FW, giving around 0.1 MW/m^2. The divertor and first wall subject to these power loadings is slowly eroded, and the material redeposited around the machine. This re-deposition is prone to trap tritium and must be controlled.

A critical issue for burning plasma devices is plasma purity, normally specified as Z_{eff}, a measure of the average charge of the ions. A plasma must be charge neutral, so this parameter is one if all ions are singly charged hydrogen isotopes. In practice, impurities arise from the influx of first wall materials, and in ITER the target Z_{eff} is 2. Heavy ions are particularly deleterious. Thus a single nickel ion in the plasma core will have lost by ionisation perhaps 25 electrons, but retain a few. The high net positive charge on the ion displaces 25 hydrogen isotopes at constant electron density, leading to a rapid dilution of the fuel. Furthermore, the remaining few electrons result in intense line radiation which has a strong cooling effect on the plasma. The design of plasma facing components must be consistent with very low impurity influx - the choice of materials for plasma facing components is strongly limited as a result. In ITER, only carbon, beryllium and tungsten are considered. The presence of oxygen is also highly deleterious. The Tokamak vessel and in-vessel equipment must maintain ultra-high vacuum conditions, and will normally be bakeable to ~ 200C to maintain cleanliness.

Transient phenomena are also important. Instabilities known as ELMs (Edge Localised Modes) occur in the edge plasma. Large ELMs in a burning scale Tokamak lead to the deposition of 10 MJ or more on the first wall/divertor components in less than one millisecond at a frequency of perhaps 10 Hz. Such events cannot be accommodated with available materials without surface melting/ablation. These large ELMs can be suppressed, but designs of first wall components must accommodate a small number of such events and smaller residual ELMs of order 1 MJ energy at ~ 100Hz. Disruptions, in which the plasma current disappears in a time of tens of milliseconds, can arise for various reasons, and may be associated with loss of vertical position control. These events induce large eddy currents which in turn induce large forces on in-vessel components. Furthermore, part of the plasma current may flow in the first wall and also induce large forces on the affected components. The stored energy in the plasma is 2-3 GJ and this is deposited onto the first wall in a few milliseconds. This can cause further erosion and re-deposition of first wall material.

This is a very challenging environment in which to implement diagnostics.

IMPLICATIONS FOR DIAGNOSTICS

Diagnostics on burning plasma Tokamaks must accommodate both the effects of scaling up the Tokamak to achieve burn, and the effects of the use of tritium and the production of neutrons during the burn. Key ITER parameters have been given in Table 1 above. The change relative to JET of these and other parameters is listed in table 2. A number of points are noted from this table. ITER is big with high magnetic field, and with correspondingly high plasma current. In particular, it has a much longer pulse length (500-3000 sec, of which typically 100 sec is required just to ramp up to full performance). On the other hand, the heating power density and first wall power loading are similar. The rate of decay of magnetic field in disruptions is rather less than

Table 2. Change in ITER parameters compared to JET (FW: first wall).

	Relative to JET
Volume	x 10
Toroidal field	x 1.5
Plasma current	x 3
Pulse length	x 25 ... 150
Tritium site inventory	x 1000
In-vessel mobilisable tritium inventory	x 100
Heating power per m^3, incl. Alphas	x 0.5
FW Power density (excl. neutrons)	x 1
FW Energy density (excl. neutrons)	x 25 ... 150
FW Erosion/redeposition per pulse	x 100 600
FW Neutron flux per m^2 per sec	x 8
FW Neutron fluence per m^2	x 100,000
dBv/dt in disruption, outer wall	x 0.5
dBr/dt in disruption, outer wall	x 0.25
Disruption heat blast, J/m^2	x 10
Vertical force on vessel	x 10
Sideways force on vessel	x 4
Sideways acceleration of vessel	x 0.15

in JET, although eddy currents may be more of an issue due to the need for massive structures for neutron shielding. In contrast, the heat blast from plasma stored energy is substantially higher. Whilst the forces on the vessel are large, the vessel is also massive and the resulting acceleration is rather reduced compared to JET. Even the peak neutron flux density (n/m^2/sec) is not so different. The dramatic changes arise from the pulse length, as a result of which the energy density at the first wall, the wall erosion, and in particular the neutron fluence are strongly increased. The large increase in tritium inventory also has important implications discussed below.

The main implications for diagnostics are then anticipated to be a need for active cooling and control of thermal stresses for all systems near the FW, a need for tolerance of FW erosion and re-deposition (which may take the form of provision for in-situ cleaning), and in particular the requirements arising from the enhanced tritium content and neutron fluence to be discussed below. In general, all in-vessel diagnostic systems will be subject to large thermal and mechanical stresses, limited selection of materials, tight constraints on surface power loading and impurity influxes particularly in the divertor, remote handling maintenance and inspection, and have to operate close to allowable limits on material properties. Furthermore the diagnostic must not degrade the adjacent first wall components, themselves designed close to engineering limits.

SPECIFIC DESIGN ISSUES

Disruptions

A disruption is a sudden collapse of the plasma current which can disappear in a time of typically a few tens of milliseconds in ITER. There are various causes, including operation beyond plasma stability limits in density or current, a sudden influx of impurities such as a FW fragment falling into the plasma, loss of vertical position control, loss of additional heating in vulnerable plasma scenarios,.... These events induce very large forces on in-vessel components and ultimately on the vacuum vessel itself, in many cases exceeding the endurance limit of materials and therefore leading to fatigue and risk of failure. These events are a major constraint on Tokamak operation and the number and severity of events must be strictly controlled.

There are three classes of disruptions to be considered in the design:

- Centred disruptions. There is a large change in poloidal magnetic field, and radially inwards net force on the vacuum vessel.
- Vertical displacement events (VDE) arise from a loss of vertical position control, allowing the plasma to drift vertically and contact the FW. These events produce both poloidal field change and a transient increase in field normal to the FW during the plasma displacement. Part of the plasma current may flow in the FW where it is impacted by the plasma. This 'halo' current can lead to very high local forces on FW components. There is a large net vertical force on the vessel.
- Asymmetric VDE's arise when the plasma column tilts during the vertical displacement and does not rotate. This leads in addition to a net horizontal force on the vacuum vessel and may lead to a peaking of the halo current profile.

In all cases, the rapid change in magnetic field induces eddy currents in FW and vacuum vessel components. These currents combine with any halo currents, and where they cross the magnetic field produce large jxB forces. The magnetic field needs to be calculated self consistently, as the eddy currents are sufficiently large to substantially modify the fields. The analysis of these transient, three dimensional currents crossing a time varying field, and the resulting forces and stresses is a critical part of the design of in-vessel systems. Sophisticated software is available for this analysis. In asymmetric disruptions, the sideways acceleration induces inertial loads which must also be taken into account. Whilst the rates of change of field in ITER are a little less than in JET, the field itself is higher and the structures are larger, which will give higher eddy currents. The size of the structures also give greater strength for supporting the resulting forces. Typically, the effects of scale length cancel, as do the changes in field, with the exception that the VDE events will be less severe in burning plasmas due to the slow vertical displacement of the enlarged plasma. Whilst specific systems must be analysed in detail, the stresses in burning plasma devices will be rather similar to equivalent systems in existing machines.

Disruptions also lead to a sudden collapse in stored energy in the plasma which gives rise to a burst of intense radiation onto the FW. In burning plasma devices, the deposited energy density is an order of magnitude higher (3 MJ/m^2 in ITER, with an additional 3 MJ/m^2 magnetic energy on a timescale of 10 millisecond) and may lead to surface melting. Under some conditions, runaway electrons are generated which intercept the outer wall. In ITER, these have an energy of around 10 MeV. At these extreme energies, runaway electrons will pass through light elements such as carbon or beryllium, to be dumped in the underlying copper/steel structure at very high power density.

In the case of ITER, various bounding disruption scenarios have been defined [2] and all diagnostics must remain within allowable limits for all scenarios.

Nuclear Safety Aspects

The major impact of burning plasmas on component design arises from the large tritium inventory and the associated neutron production, see Table 3 below. These impact on the design in two important respects. Firstly, the tritium inventory required to run a burning plasma experiment with high availability is some kilograms, 3kg in the case of ITER. Some of this tritium will remain trapped in the torus. In the JET DTE tritium campaign, some 10% of the 100gm of tritium cycled through the torus remained in-vessel at the end of the campaign. Of this 10%, some two thirds was subsequently recovered mainly by baking in damp air. The balance remained permanently trapped in vessel in co-deposited carbon flakes from re-deposition of FW material. This remains a critical issue for burning plasma devices. The burning plasma produces an intense flux of neutrons, and also activates the irradiated material which prevents manned access to the torus even after deuterium operation. As a result of these issues, a burning plasma device is necessarily a licensed nuclear facility. This requires that the management of safety and the environment is detailed in a Safety Case approved by the appropriate regulatory authorities, in the case of ITER, by the French Regulator. This safety case details all safety related aspects of the design and operation of the facility, including detailed analysis of a set of bounding accidents, and is approved and monitored by the Regulator. The facility must demonstrate compliance with the safety case, including design, manufacturing quality, inspection, testing, commissioning, operation, maintenance and de-commissioning. This applies to diagnostics as to other equipment. Some diagnostics will be Safety Related Equipment which has a function in ensuring the safety of operations. Such equipment will be subject to the full panoply of controls, and must be designed to a level of integrity which ensures compliance. Other diagnostics will not be SRE, but will nonetheless have to comply with conditions in the safety case, for example for control of the tritium boundary or limits on potential water or air leaks into the torus. It is important that specific requirements of diagnostics are included within the Safety Case as subsequent modification is very difficult. The Safety Case may affect the design of diagnostics in various respects, including;

- Level of integrity/redundancy (3 levels are defined within ITER CODAC)
- Earthquake tolerance
- Water ingress: maximum potential volume and flow rate (pipe size)
- Air ingress: maximum oxygen flow rate (window size)
- Integrity of tritium boundary: double containment windows and foils.
- Activation: materials selection
- Maintainability: dose to workforce ALARA/remote handling

Quality assurance, traceability and documentation generally are critical requirements and mandatory for all equipment intended for installation on the torus.

Design Issues with Neutrons

The high neutron flux and fluence at the first wall has major implications for the design of diagnostics, as with other systems. The neutron power density at the first wall is typically 0.5 MW/m^2. This power is dissipated in the shield modules with an e-folding

Table 3. Tritium and Neutron Data for JET and ITER

	JET	ITER
Site tritium Inventory, kg	0.1	3
In-vessel mobilisable tritium, kg	0.01	1
Fusion power, MW	15	500
Lifetime neutron production	2.10^{21}	10^{27}
Neutron flux density, MW/m^2	0.1	0.78
Neutron fluence, MW.years/m^2	0	0.3
DPA at FW	0	3-5

length of about 100mm. The resulting power density gives rise to severe thermal stresses which requires closely spaced cooling channels in this part of the shield or of any diagnostic equipment located close to the FW.

Neutrons also degrade materials. The fluence at the first wall in ITER is equivalent to 3-5 DPA in structural materials. This is about the level at which degradation of metals starts to become important. In the next generation of power producing reactors, the fluence will be such as to raise serious issues with FW materials. The IFMIF facility is planned to qualify materials for this environment. Mirrors at the FW may suffer a loss of reflectivity. It is not possible to locate ceramic materials for windows or leadthroughs in this environment. The flux density is much reduced at the rear of the shield/port plug. In particular, the energy spectrum of the neutrons is now similar to fission reactors and there is a large body of relevant data on the behavior of structural materials. The dose rate here is around 0.03 Gy/s and the total dose around 10^6 Gy. This is sufficient to produce significant degradation of the electrical and optical properties of ceramics typically used for windows, electrical leadthroughs and optical fibres in diagnostic systems. Thus darkening of fibres, increase of the loss tangent and conductivity of ceramics (REID, RIDD etc) need to be considered. These effects have been widely studied and reported [3]. The use of electronics and some organic materials may also be an issue in this area. In the ITER galleries, there will not be a problem with materials, apart possibly from prompt effects during transit of highly active flasks to the hot cell.

Neutrons activate materials which leads to restrictions and prohibition on manned access to major parts of the facility. In particular, no manned access in-vessel is possible after the start of operations in deuterium. All in-vessel and in-port equipment must therefore be designed for remote handling removal, repair and replacement. Some areas of the machine are very difficult to access remotely, notably the cryostat. It is therefore essential to provide sufficient shielding of the neutrons to allow limited manned access. In addition, there is important advantage in maintainability in having manned access behind the port plugs and the specification on ITER allows controlled access 12 days after ending operation (<100 microSieverts/hr). The biological shield on ITER allows controlled but unlimited manned access to the galleries (<10 microSieverts/hr). Neutrons also lead to power deposition within the superconducting coils. The heat capacity of liquid helium and the Carnot efficiency of helium refrigerators are very low, and such nuclear heating has to be kept within tight limits. These constraints require the provision of an effective neutron shield behind the first wall, which in the case of ITER, together with the vacuum vessel, comprises some 1.5 m of steel and water to provide around five orders of magnitude attenuation of the power flux. There are then severe constraints on the design of penetrations through the shield to maintain this attenuation. Penetrations must be kept to a minimum area, and labyrinths used within the port plugs to block neutron streaming. Further labyrinths are likely to be necessary at the biological shield to achieve the required dose rates in the galleries.

Reliability and Availability

Burning plasma devices are major facilities and represent a large investment of time and resources. Effective utilisation of this investment demands high levels of reliability and availability of the facility. In the case of ITER, during the final DT phase of operation, a reliability of 90% (the proportion of pulses achieving the objective of the pulse) and a 'high level' of availability (readiness for pulsing) are specified. It can be inferred from the programme and fluence requirements that the machine is pulsing at full performance and full duty cycle (25 minutes between pulses) for 25% of the total time over ten years. Operation under realtime, feedback control requires some 45 diagnostics to be available. These are extremely demanding requirements which will only be achieved with correct design, QA, testing, commissioning, maintenance...of all systems, including diagnostics. In particular, it is essential to design for benign failure which has low impact on operations, for example by providing redundancy in the provision of key measurements, and by ensuring ease of recovery of operation, for example by ease of maintenance or by provision of standby capacity. Formal procedures exist for the reliability assessment of equipment at the design stage, referred to as RAMI (Reliability, Availability, Maintainability and Inspectability). A key adjunct to this process is FMEA (Failure Modes and Effects Analysis), a process for systematic analysis of failure modes and the impact of such failures on the operation of the facility. Such processes will be an essential part of the design of diagnostics on burning plasma Tokamaks.

In practice, the reliability and availability issue will generate similar requirements to the Safety Case: design integrity, assessment and approval; documentation and traceability; QA; inspection and testing; maintenance... In this case, the work is not subject to control by the external Regulator, but in most other respects, the requirements are the same.

SUMMARY AND CONCLUSIONS

Burning plasma Tokamaks present substantial new issues in the design of diagnostics. Such devices are necessarily large compared to present generation Tokamaks, have high magnetic field and in particular long pulse length both to ramp up to burning conditions and to allow exploitation of the facility. The tritium inventory will be large and the neutron fluence will be increased by many orders of magnitude. These factors have many implications for diagnostics. In-vessel systems are of necessity remote handlable, have active cooling and can utilise only a very restricted range of radiation tolerant materials. Diagnostic penetrations in the shield must preserve the high levels of attenuation of the neutrons, using labyrinths to prevent neutron streaming. Dose rates at the port flange allow ceramic windows and leadthroughs, and optical fibres subject to careful design. Further labyrinths are required at the biological shield. Such devices will be licensed nuclear facilities, and diagnostics will be constrained by the provisions of the safety case and subject to external regulation. Effective utilisation of the facility requires high levels of reliability and availability imposing further constraints on the design and a need for redundancy and benign failure modes. All systems will require a high level of design assessment and approval, QA, documentation and traceability, testing and maintenance to deliver the required performance.

REFERENCES

1.ITER Project Integration Document (PID), ITER IDM D 2234RH
2.ITER Load Specification, ITER IDM D 222QGL
3.E Hodgson, these proceedings.

Reactor Implications For Burning Plasma Diagnostics

C. Gordon, D. Baker, C. Walker

ITER Organization, Cadarache JWS, 13108 St Paul-lez-Durance, France

Abstract. Reactors based on DT fusion will impose certain requirements and pose challenges for diagnostics. These relate, for example, to the need for such facilities to address the safety aspects of using tritium and high neutron fluxes leading to activation and radiation fields. There are also implications arising from the need for high machine availability and hence reliability of critical diagnostics. This paper will focus on the safety and licensing implications using design issues for typical diagnostic types to illustrate the above.

Keywords: Diagnostics, Plasma, Safety, Licensing.
PACS: 28.52.-s Fusion reactors 28.52.Nh Safety

INTRODUCTION

Reactors based on DT fusion will impose certain requirements and pose challenges for diagnostics. These relate, for example, to the need for such facilities to address the safety aspects of using tritium and producing high neutron fluxes leading to activation and radiation fields. There are also implications arising from the need for high machine availability and hence reliability of critical diagnostics. This paper focuses on safety and licensing implications. Safety aspects relevant to diagnostics include:

- confinement of radioactive materials;
- shielding for protection of personnel.

In both cases it is the diagnostic penetration through the bulk shielding and vacuum vessel (that acts as the primary confinement barrier for radioactive material) that is at issue.

The presence of quantities of tritium and activation products and radiation fields mean that a fusion reactor needs to be licensed by nuclear regulatory authorities. For this one needs to:

(a) take the necessary safety provisions in design and operation, and

(b) demonstrate these are adequate to the regulator.

It is the level of proof to be submitted to a regulator that leads to an additional challenge.

ITER is a joint international research and development project that aims to demonstrate the scientific and technical feasibility of fusion power. The partners in the project are the European Union, Japan, the People's Republic of China, the Republic of India, the Republic of Korea, the Russian Federation and the USA. ITER will be constructed at Cadarache in France. ITER will accomplish its aims by demonstrating high power amplification and extended burn of deuterium-tritium plasmas, with

CP988, *Burning Plasma Diagnostics*, edited by F. P. Orsitto, G. Gorini, E. Sindoni, and M. Tardocchi
2008 American Institute of Physics 978-0-7354-0507-3/08/$23.00

steady-state as an ultimate goal, by demonstrating technologies essential to a reactor in an integrated system, and by performing integrated testing of the high-heat-flux and nuclear components required to utilize fusion energy for practical purposes. In short, ITER will have a reactor-relevant burning plasma and prototypical fusion power technologies.

From a safety perspective, issues for ITER arise from potential hazards associated with a burning plasma:

- Tritium as fuel and bred in the blanket;
- Activated components (arising primarily from the 14 MeV neutron irradiation);
- Activated / tritiated dust produced by erosion of plasma facing components;
- Activated corrosion products in cooling water, and
- Other toxic materials e.g. beryllium.

This paper uses the safety and licensing issues for ITER to examine the reactor implications for burning plasma diagnostics. It describes:

- the safety approach for ITER;
- the confinement approach for ITER in-vessel;
- issues for penetrations and diagnostics, and
- licensing.

SAFETY APPROACH FOR ITER

The safety approach for ITER exploits fusion's favourable safety characteristics to the maximum extent feasible. Relevant characteristics are:

- plasma burn is terminated inherently when fuelling is stopped, is self-limiting with regard to power excursions, and is passively terminated by the ingress of impurities under abnormal conditions (e.g. by evaporation or gas release or by coolant leakage);
- the energy and power densities are low, the energy inventories are relatively low and large heat transfer surfaces and masses are available as heat sinks;
- the releasable inventories are limited: the bulk of the activity is bound in activated structural components;
- physical barriers exist inherent to the tokamak concept and must be leak-tight for operational reasons.

The main safety concern is the confinement of tritium and activated dust, and activated corrosion products in the coolant of the plasma-facing components. This is achieved in the design through multiple confinement barriers to implement the defense-in-depth approach. What this means is that providing and protecting confinement barriers is the main safety function for ITER for protection of the public and workers from radiation and from chemical and radioactive contamination. For direct radiation, assessments confirm that buildings and structures provide adequate shielding for the public. In addition, it is necessary to limit/prevent worker access to hazardous locations.

The confinement barriers need to perform well with certain orders of magnitude in releases, e.g.:

- tritium releases limited to ~gram quantities in normal operations or ~few grams in case of accident; 1000's g tritium possible on site;
- dust releases limited to ~gram quantities in normal operations or ~few 10's grams in case of accident;100's kg activated dust possible;

In implementing confinement, the approach is to limit the spread of contamination close to its source and to limit and control potential releases by providing:

- reliable barriers as part of the process systems;
- secondary confinements as necessary to limit the spread of contamination; and
- filter/detritiation systems to process exhausts.

REQUIREMENTS FOR CONFINEMENT SYSTEMS

Radioactive inventories at risk in the vacuum vessel are tritium and activated dust. These are normally confined within the vacuum vessel and its extensions, penetrations and connected systems. Penetrations through confinement systems (e.g. diagnostic systems) require analyses to verify an adequate level of reliability. In the event of failure, contamination is confined within the port cells which are also connected to the cooling systems vault. These areas are equipped with systems to control pressure and filter and detritiate releases. Figure 1 illustrates a simplified view of the implementation of the confinement scheme for the in-vessel inventories.

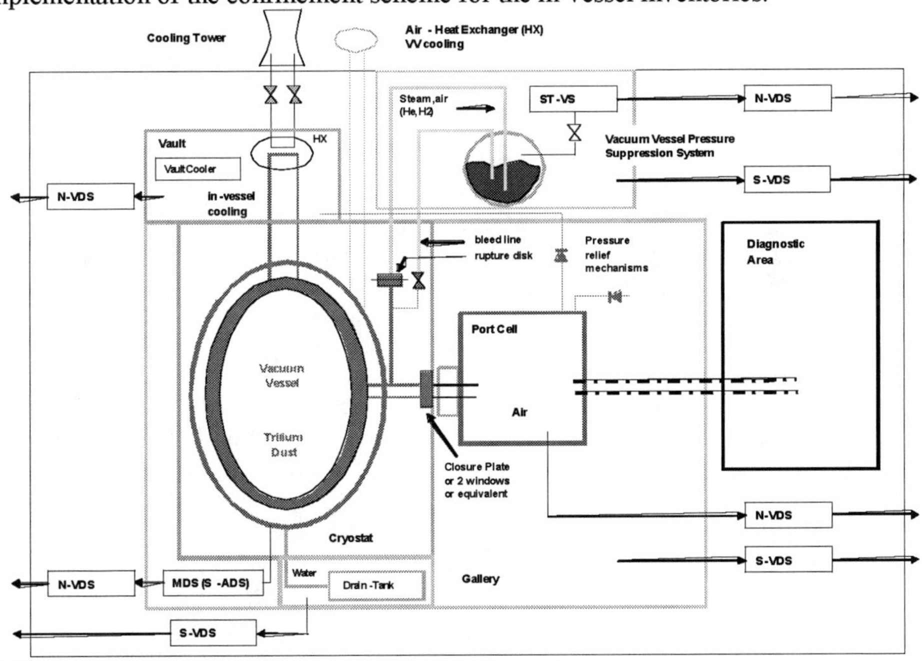

FIGURE 1. Simplified view of implementation of confinement

A set of general requirements have been developed for the systems implementing confinement to ensure they provide the required performance during and following accidents for which they are needed:

- leak-tight (less than leak rate assumed in the safety analysis),
- withstand pressures and environments from accident sequences,
- return volume to specified pressure values within a specified period following accidents,
- provide a detritiated, filtered, controlled and monitored pathway to specified release point following an accident until releases without their operation are acceptable,
- provide signals to actuate safety actions such as isolation of confinement,
- provide backup power to active systems and monitoring,
- valves implementing confinement operate within specified periods after detection and assume their safe position on loss of power,
- reliable - independent (no common parts) and physically separated as far as reasonably achievable,
- allow for initial and periodic assessment of performance,
- layout to avoid potential damage e.g. by movement of equipment during maintenance.

Diagnostics providing a confinement function must also follow these requirements.

DIAGNOSTIC ASSEMBLIES AND CONFINEMENT BOUNDARIES

Some diagnostic assemblies comprise part of the primary vacuum boundary and hence implement a confinement function. In ITER there will be $\geq$ 118 primary vacuum window assemblies. However, they are significantly different from the main, vacuum vessel, boundary. Compared with the vacuum vessel structure, diagnostic penetrations may require an optically transparent element that is brittle, and sometimes weak. Generally, special materials (compared with the 316LN (IG) stainless steel used in the vacuum vessel) are required to meet the diagnostic functional requirements. These can include fused silica, beryllium foil, quartz, sapphire, etc.

From a safety perspective, diagnostic (and heating) systems require particular attention because many diagnostic systems can penetrate not only the vacuum vessel structure, but also the port cells and gallery of the tokamak building into areas that may not be connected to a detritiation / filter system. This is illustrated in Figure 2. Confinement barriers are needed in such a system to ensure that there can be no direct connection from the vacuum vessel radioactive inventories to the environment.

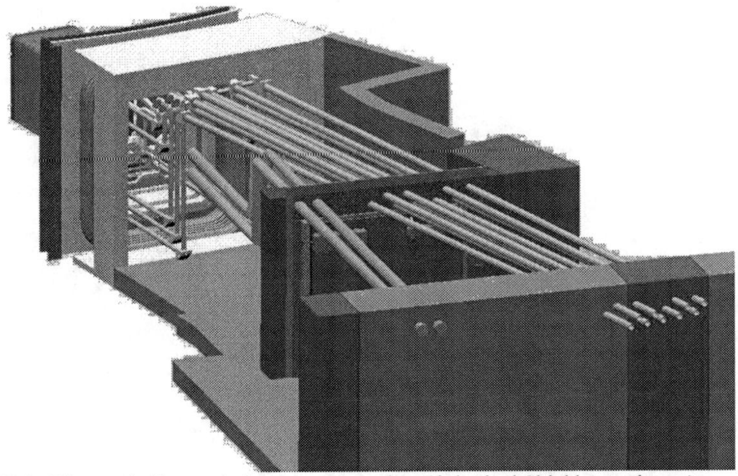

FIGURE 2. Diagnostic Transmission Lines Crossing Biological Shield (not shown), Port Cell Door Lintel and Tokamak Building Wall for ITER LIDAR and polarimetry;

To address these safety issues, the following strategy is applied to diagnostic confinement barriers:

1. minimize the likelihood of failure of a confinement barrier
2. mitigate the effect of failure of a confinement barrier
3. anticipate window failure as part of the safety case, i.e., rather than prove failures can never happen, show consequences of failure are acceptable.

Systems Design

Typical diagnostic penetration assemblies are illustrated in Figure 3. The objective is to both reduce the likelihood of failure of the assembly and mitigate the effects should it fail in order to allow continued machine operation (even if the diagnostic system can not function) and limit the spread of contamination. This is accomplished by:

- design provisions to the extent practical, e.g. by redundancy, separation, diversity of barriers as noted above for the general requirements for confinement barriers,
- developing and applying implementation and manufacturing rules for specific barriers to ensure reliable performance under normal and accident conditions, and
- developing and applying testing rules for specific barriers.

Window type barriers (valve barriers) are used between vacuum vessel and port cell, and those with two windows are designed to be as independent and diverse as practicable.

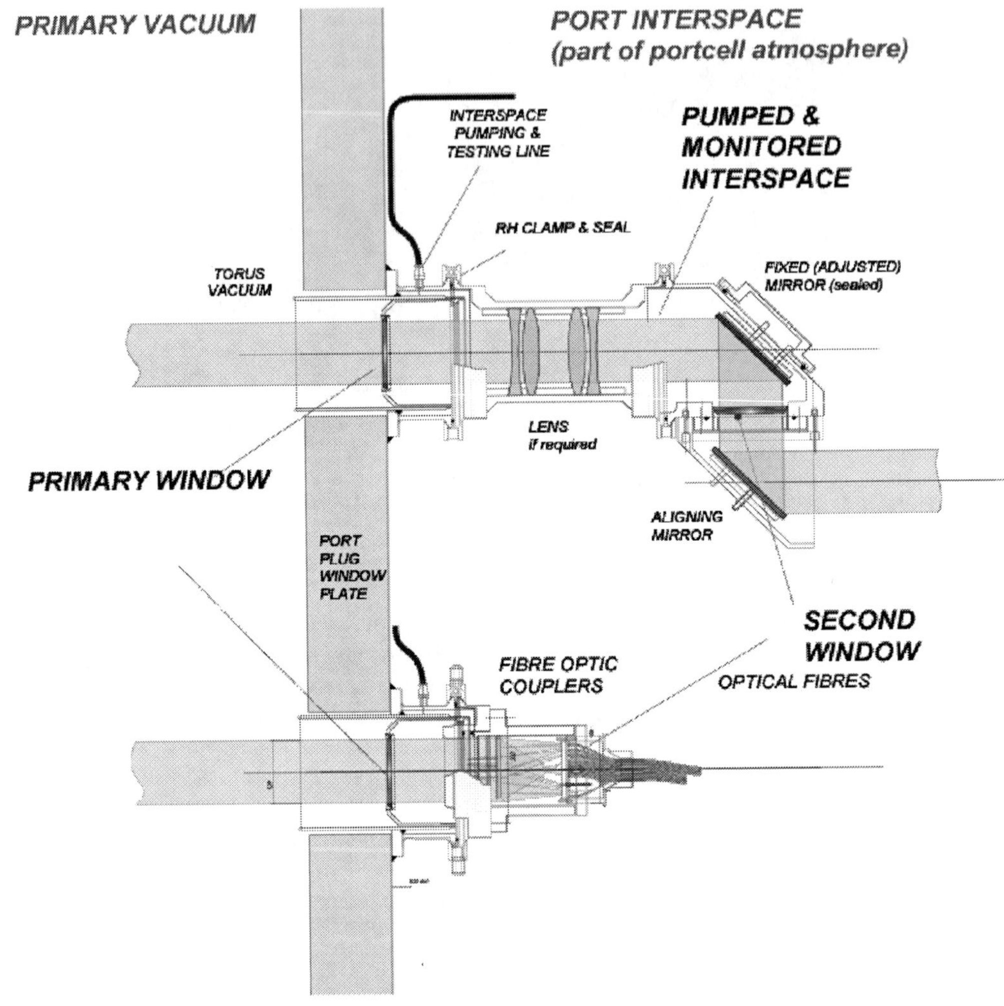

FIGURE 3. Typical Diagnostic Penetration Assemblies.

Technical Specifications

One aspect of minimizing the likelihood of failure of an individual element of a diagnostic penetration assembly is developing and applying implementation rules for its design, fabrication, and installation. These technical specifications take into account the operational experience and R&D for similar components to ensure they can withstand operational, maintenance and accidental conditions (e.g. pressure, temperature, radiation and magnetic fields, etc.). The technical specification addresses:

- o general provisions (including design rules and quality assurance);
- o window assembly duty (load, pressure, transmission requirements, etc);

- o material (general requirements and acceptance criteria for material procurement);
- o examinations;
- o fabrication;
- o tests (before, during, after manufacture, installation);
- o window assembly implementation (e.g. definition of vacuum boundaries, monitoring of interspaces), and
- o in-service inspection requirements (e.g. continual monitoring vacuum tightness and optical performance).

The technical specification will be specific for each window design, but will make use of standard rules where possible.

Type Tests

To provide additional assurance that the likelihood of failure of an individual element of a diagnostic penetration assembly is acceptably low, testing is foreseen. These will test windows of the same design and construction process as window assemblies that are to be supplied to ITER. In addition, some window types will require destructive tests to prove acceptability. An example of a typical window assembly acceptance test would include:

- o pressure test (e.g., 0.3 MPa internal, 1 hour without failure);
- o leak test (e.g. leakage $<10^{-10}$ $Pa.m^3\ s^{-1}$ after the pressure test);
- o thermal cycling (e.g. to 250°C, 5 times);
- o vacuum leak test (e.g. with internal / external vacuum/pressure);
- o dimensional control (to show all dimensions within tolerance).

Such acceptance tests provide assurance that the assembly will meet its operational and safety performance requirements prior to installation in the machine.

LICENSING IMPLICATIONS

One of the Regulators' concerns about ITER can be anticipated to be that if these diagnostic penetrations are not reliable, then the risk to workers and public could be unacceptable, and there are no well-established design standards to rely on to establish the expected reliability. For ITER, the justification of its position includes the following:

- o design that reduces the likelihood of failures (system design, technical specifications, testing as noted above);
- o safety case for the Rapport Préliminaire de Sûreté that shows consequences of failures are mitigated and limited, and
- o supplemental engineering justification for the acceptability of the diagnostic assemblies.

Supplemental Engineering Justification

The Supplemental Engineering Justification is intended to collect into a single dossier the underlying basis that provides the ITER Organization with confidence that the diagnostic assemblies will reliably function as intended. This includes summarizing the extensive worldwide experience on existing machines, the additional R&D planned, and the relevant industrial information. It will contain:

- Descriptions of the design and operational experience on existing machines, e.g. JET.
  - Design approaches
  - Causes of failures
  - Failure modes
- Descriptions of the design, testing, etc. to counter known failure modes.
  - Design analysis
  - Type testing
  - Material testing, including effects of irradiation.
  - Qualification testing for manufacturing
  - Acceptance testing during and after manufacturing
  - Installation testing
- Additional R&D Plan
  - Tests proposed to validate the defined requirements for ITER.
- Existing Industrial Specifications Document
  - Technical specifications or norms used and developed for other similar scopes.
  - Literature and industrial search of equivalent, applicable industrial specifications .
  - A general comparison between these specifications.
  - Considerations on the applicability of these specifications for ITER.

The objective is to provide evidence to the regulator to justify the design comparable to that which would be provided by using a recognized design code or standard.

SUMMARY

Diagnostics in fusion reactors will need to satisfy safety and licensing requirements in addition to their functional requirements. As part of barriers for radioactive materials, they will need to provide confinement reliably under normal and accident conditions. In addition to meeting these safety expectations, it will be necessary to justify the adopted approach to the regulator.

ITER has adopted a confinement approach taking into account the favourable safety characteristics of fusion with the aim of minimizing the likelihood of failure, limiting the spread of contamination in the event of failure and ensuring releases are well below the facility safety objective targets. For diagnostic assemblies this involves a combination of system design, technical specifications, and testing. This is supplemented by a compilation of additional information in an engineering justification. Finally, it includes the demonstration of limited consequences in the event of failure as part of the safety case.

ACKNOWLEDGEMENT

This report was prepared as an account of work by or for the ITER Organization. The Members of the Organization are the People's Republic of China, the European Atomic Energy Community, the Republic of India, Japan, the Republic of Korea, the Russian Federation, and the United States of America. The views and opinions expressed herein do not necessarily reflect those of the Members or any agency thereof. Dissemination of the information in this paper is governed by the applicable terms of the ITER Joint Implementation Agreement.

Plant Models for DEMO

David Maisonnier

EFDA CSU Garching
Boltzmannstrasse 2
D-75748 Garching, Germany

Abstract. The European Power Plant Conceptual Study (PPCS) has been a study of conceptual designs for commercial fusion power plants. It focused on five power plant models, named PPCS A, B, AB, C and D, which are illustrative of a wider spectrum of possibilities. The PPCS study highlighted the need for specific design and R&D activities as well as the need to clarify the concept of DEMO, the device that will bridge the gap between ITER and the first fusion power plant. An assessment of the PPCS models with limited extrapolations has led to the clarification of the objectives of DEMO. Many parameters will have to be controlled in DEMO in order (1) to control the machine, (2) to satisfy the testing requirements, (3) to satisfy regulatory requirements (primarily safety), and (4) to protect the investment. On the other, DEMO will ulilise one or two plasma scenarios only.

Keywords: DEMO, Reactor studies
PACS: 28.52-s, 28.52.Lf, 28.52.Nh

INTRODUCTION

The European fusion programme is "reactor oriented" and it is aimed at the successive demonstration of the scientific, the technological and the economic feasibility of fusion power. For a reactor-oriented fusion development programme, it is essential to have a clear idea of the ultimate goal of the programme, namely a series of models of fusion power plants, in order to define the correct strategy and to assess the pertinence of the on-going activities.

The European Power Plant Conceptual Study (PPCS) has been a study of conceptual designs for commercial fusion power plants. It focused on five power plant models, named PPCS A, B, AB, C and D, which are illustrative of a wider spectrum of possibilities [1, 2]. They are all based on the tokamak concept and they have approximately the same net electrical power output, 1500 MWe. These span a range from relatively near-term, based on limited technology and plasma physics extrapolations, to an advanced conception. All five PPCS plant models differ substantially from the models that formed the basis of earlier European studies. They also differ from one another, which lead to differences in economic performance and in the details of safety and environmental impacts.

The PPCS study highlighted the need for specific design and R&D activities, in addition to those already underway within the European long term R&D programme, as well as the need to clarify the concept of DEMO, the device that will bridge the gap between ITER and the first-of-a-kind fusion power plant (FPP). A detailed assessment

CP988, *Burning Plasma Diagnostics*, edited by F. P. Orsitto, G. Gorini, E. Sindoni, and M. Tardocchi
2008 American Institute of Physics 978-0-7354-0507-3/08/$23.00

of the PPCS models with limited extrapolations identified the physics issues that must be addressed to establish the DEMO physics basis. In parallel, a number of technological issues are being addressed to establish the basic features of DEMO: they are primarily related to the divertor and blanket concepts, to the maintenance scheme of the internal components and to structural materials.

In order to clarify the concept of DEMO it is also necessary to have a clear view of the fusion development strategy that is being pursued. The reference European fusion development scenario, sometimes called the "fast-track" [3], is being reassessed and developed in detail.

THE POWER PLANT CONCEPTUAL STUDY

PPCS Plant Models

The interrelationships of plasma performance, materials performance, engineering, economics and other factors were explored using the systems code PROCESS [4]. The PROCESS code employs a plasma physics module that, in its original form, was developed for the Conceptual Design Activity phase of ITER and used to explore the early ITER design. This was modified to reflect further developments and has been updated to incorporate modern scaling laws. The most important aspects of the plasma physics are the use of the recommended ITER H mode confinement scaling laws (IPB98y2); a divertor module based on a simplified divertor model benchmarked to 2D code runs; a synchrotron reflection coefficient based on experimental measurements; and a current drive efficiency calculated using Neutral Beam Injection efficiency based on a modified Mikkelsen-Singer calculation.

For the three near-term models, PPCS-A, PPCS-AB and PPCS-B, the plasma physics scenario represents, broadly, parameters about thirty percent better than the design basis of ITER: first stability and high current-drive power, exacerbated by divertor heat load constraints, which drive these devices to larger size and higher plasma currents. The main parameters of the PPCS models are given in tables 1 and 2.

The technology employed in models A, AB and B stems from the use of near-term solutions for the blanket. They are based, respectively, on the "water-cooled lithium-lead", the "helium-cooled lithium-lead" and the "helium-cooled pebble bed" which are being developed in the European fusion programme [5]. All of these concepts are based on the use of a low activation steel as the main structural material. Associated with these are water-cooled and helium-cooled divertors. The water-cooled divertor is an extrapolation of the ITER design. The helium cooled divertor has been chosen for models AB and B to prevent the risk of hydrogen formation resulting from the water-beryllium reaction in case of accident (model B) and to simplify the balance of plant by having the same coolant for all the internal components (models AB and B). The development of such a component represents a significant challenge. An ad-hoc R&D programme was initiated in 2003 [6] and preliminary results have demonstrated the feasibility of this concept.

TABLE 1: Main physics parameters of the PPCS Models					
Parameter	**Model A**	**Model AB**	**Model B**	**Model C**	**Model D**
Fusion Power (GW)	5.00	4.29	3.60	3.41	2.53
Aspect Ratio	3.0	3.0	3.0	3.0	3.0
Elongation (95% flux)	1.7	1.7	1.7	1.9	1.9
Triangularity (95% flux)	0.25	0.27	0.25	0.47	0.47
Major Radius (m)	9.55	9.56	8.6	7.5	6.1
TF on axis (T)	7.0	6.7	6.9	6.0	5.6
Plasma Current (MA)	30.5	30.0	28.0	20.1	14.1
β_N(total)	3.5	3.5	3.4	3. 4.0	4.5
H_H (IPB98y2)	1.2	1.2	1.2	1.3	1.2
Bootstrap Fraction	0.45	0.43	0.43	0.63	0.76
n/n_G	1.2	1.2	1.2	1.5	1.5
Q	20	16.5	13.5	30	35
Av. neutron wall load (MWm^{-2})	2.2	1.8	2.0	2.2	2.4
Divertor Peak load (MWm^{-2})	15	10	10	10	5
Z_{eff}	2.5	2.6	2.7	2.2	1.6

TABLE 2: Main thermodynamic parameters of the PPCS Models					
Parameter	**Model A**	**Model AB**	**Model B**	**Model C**	**Model D**
Fusion Power (MW)	5000	4290	3600	3410	2530
Blanket Gain	1.18	1.18	1.39	1.17	1.17
Blanket Power (MW)	4845	4470	4252	3408	2164
Divertor Power (MW)	894	981	685	583	607
LT Shield Power (MW)	-	-	67	-	-
Pumping Power (MW)	110	400	375	87	12
Heating Power (MW)	246	257	270	112	71
H&CD Efficiency	0.6	0.6	0.6	0.7	0.7
Gross Electric Power (MW)	2066	2385	2157	1696	1640
Net Electric Power (MW)	1546	1500	1332	1449	1527
Plant Efficiency[1]	0.31/0.33[2]	0.35	0.36	0.42	0.6

[1] The plant efficiency is defined as the ratio between the net electric power output and the fusion power.
[2] Depending on the divertor concept used, either an ITER-like conception with water outlet temperature of 150°C or a more advanced conception with water outlet temperature of 300°C.

PPCS C and D are based on successively more advanced concepts in plasma configuration and in materials technology. In both cases the objective is to achieve even higher operating temperatures and efficiencies. Their technology stems, respectively, from a "dual-coolant" blanket concept (helium and lithium-lead coolants with steel structures and silicon carbide insulators) and a "self-cooled" blanket concept (lithium-lead coolant with a silicon carbide structure).

For the balance of plant, model A is based on PWR technology, which is fully qualified. Models AB and B rely on the technology of helium cooling, the industrial development of which is starting now, in order to achieve a higher coolant temperature and a higher thermodynamic efficiency of the power conversion system.

In the PPCS models, the favourable inherent features of fusion have been exploited, by appropriate design and material choices, to provide safety and environmental advantages and to satisfy the basic requirements set out at the beginning of the study [1, 2].

Maintenance Scheme and Blanket Segmentation

In a fusion power plant the divertor is expected to be replaced every 2.5 full-power-years because of erosion, and the blanket every 5 full-power-years, corresponding to not more than 150 dpa of neutron damage in the steel of the first wall.

ITER uses a segmentation of the blanket into 421 modules. In a power plant, such a large number of modules would result in an availability barely above 50%, which is unacceptable. As an alternative, a segmentation of the blanket into the smallest possible number of "large modules" has been assessed. The maximum size of a module is determined by the size of the quasi-equatorial ports through which the modules must pass, which is limited by the magnet arrangements. The total number of modules is between 150 and 200. The feasibility of suitable blanket handling devices was investigated, and it was assessed that a plant availability of at least 75% could be achieved [7].

The larger the individual blanket element to be removed, the larger the electro-magnetic (EM) loads that will act upon that element. Already in ITER, the blanket modules are "slit" in order to reduce the loads in case of plasma disruption. Very rough evaluations indicated that, with a blanket segmentation in "large modules", it might not be possible to develop a mechanical attachment system able to support the loads. To accommodate these two conflicting requirements, the "multi-module" concept has been proposed and will be further assessed. The basic principle is to consider a "strong-back" structure onto which are attached a number of smaller blanket "modules". The large size of the strong-back would allow a limited number of pieces to be handled remotely, whilst the small size of the "modules" would reduce the EM loads to acceptable values.

EUROPEAN FUSION DEVELOPMENT STRATEGY

In order to clarify the concept of DEMO it is necessary to have a clear view of the fusion development strategy that is being pursued. The reference European fusion development scenario, sometimes called the "fast-track" [3], is being reassessed and developed in detail [8]. The name "fast track" was coined in 2001 to reflect the nature of a development strategy with one step only between ITER and the first FPP, as opposed to a two steps strategy with two big machines: DEMO and PROTO.

The scenario is outlined in figure 1. The development path can be summarised as "3×10": 10 years to build ITER, 10 years to design DEMO and 10 years to build DEMO. The materials irradiation facility IFMIF constitutes an integral part of the development scenario, as well as a number of test facilities and satellite machines (not shown). Indeed, the risk behind this scenario is linked in particular to the number and objectives of the facilities and satellite devices which will be built and operated during the next 30 years.

This development scenario is based on purely technical considerations, i.e. it is assumed that all key decisions are taken without delays, and it rests on two key hypotheses. Firstly, although the cost of the programme has not been evaluated in any detail, no significant budget variation is foreseen. Secondly, the construction of

DEMO is assumed to start after validation of the DEMO physics basis towards the end of ITER phase 1 operation.

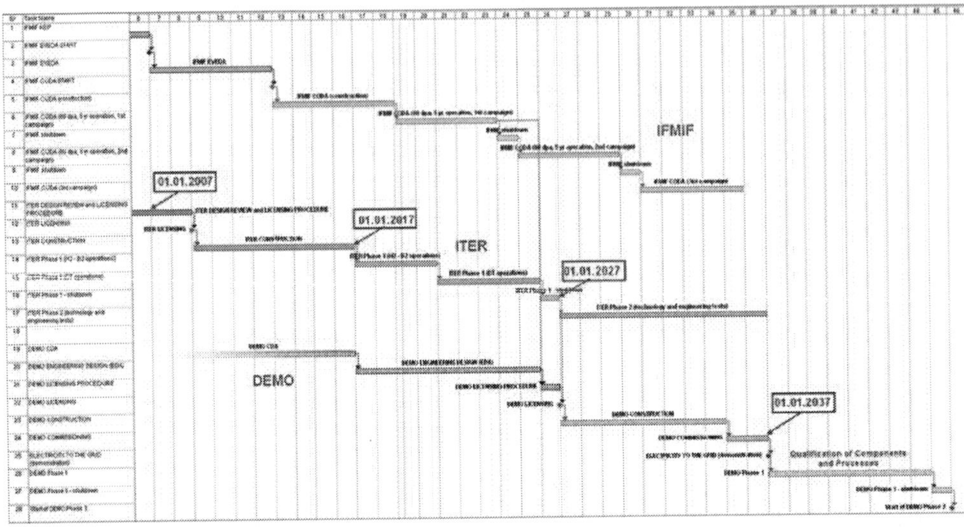

FIGURE 1. "Reference" European Fusion Development Scenario. The key assumption is that DEMO construction starts only after the establishment of theDEMO physics basis during ITER Phase 1 operation.

DEMO TECHNOLOGICAL OBJECTIVES

If there is a quasi-consensus today on a one-step fusion development strategy [8, 9], there are discrepancies on the objectives of DEMO. It is often stated that DEMO should be very similar to a first FPP and that it will differ mainly because of a lower availability. The first FPP will be funded, primarily, with private funds from a Utility and, to do so, the Utility will require enough evidence about the viability of the project. This means, inter alia, that all components and processes to be implemented in the FPP be qualified. A component or process will be qualified after having operated in reactor relevant conditions for a duration comparable to its expected lifetime. The lifetime of the blanket is assumed to be 5 full-power-year (FPY), corresponding to around 150dpa in the steel. If we assume, in DEMO, a neutron wall loading of $2MW/m^2$, 8 years are required to achieve approximately 50dpa[1] with an average availability of 33%.

One of the main objectives of DEMO could be to qualify components and processes in reactor relevant conditions and at least 10 years of DEMO operations would be required to provide sufficient confidence to start construction of a FPP. Moreover, if the purpose of DEMO is to produce neutrons for testing, no time will be available for physics research. The DEMO physics basis must be worked out in ITER

[1] A flux of 14MeV neutrons of $4.5\times10^{17}n/s{\cdot}m^2$ corresponds to a neutron wall load of $1MW/m^2$ and to 3×10^{-7}dpa/s (Fe) at the first wall. Therefore, a neutron fluence of $1MW/y{\cdot}m^2$ corresponds to 9.5dpa (Fe) at the first wall.

and any further plasma scenario optimisation will have to be developed in parallel to DEMO (e.g. during ITER phase 2 operation or in satellite devices).

Alternatively, it is worth considering whether the nuclear qualification of components and processes could be achieved, at least in part, in another device, e.g. a Component Test Facility (CTF). Should this be possible, a different fusion development scenario ought to be considered and the objectives of DEMO would differ from those proposed here-above.

DIAGNOSTICS FOR DEMO

General

The nuclear environment of DEMO will impose very severe limitations, in particular on any system installed inside the vessel. The effects of the nuclear environment on potential diagnostics are listed in detail in [10]: Radiation Induced Conductivity (RIC), Radiation Induced Electro-Motive Force (RIEMF), Radiation Induced Thermo-Electric Sensitivity, Radiation Induced Electrical Degradation (RIED), lifetime limitation etc.

Whatever the objectives of DEMO, diagnostics will be required to control long pulses (eg measurement of steady-state magnetic fields and of plasma potion) and high power operation (measurement of erosion of plasma facing components and of dust). Only limited validation of these can be achieved in existing machines, so that operation in ITER will be essential to assess and to qualify these DEMO diagnostics. On the other hand, a DEMO devoted to testing and qualification of components will utilize one or two plasma operating scenarios only, which should have been fully validated in ITER and during DEMO commissioning. In other words, DEMO will not be used for physics research, so that a limited number of diagnostics should be adequate. It will be another objective of ITER to identify the minimum set of diagnostics needed in DEMO.

Safety Requirements

DEMO, like ITER, will be a nuclear facility subject to the legal constrains of the country where it will be built. In any nuclear facility, a number of components are classified as “Safety Important Component” or SIC because they ensure a safety function. In fission plants there are different levels of SIC but, to date, ITER has elected to have only one SIC level, corresponding to the most demanding one. SIC components are essential for the implementation of the “defense-in-depth” safety approach: prevention, detection, and mitigation of incidents/accidents. Safety analyses allow to identify key accidents and to identify the parameters that must remain within a given range in order to prevent their occurrence. Similarly, should an accident occur, the safety analyses identify the parameters that allow its detection. Finally, following an accident, a number of systems will be activated to mitigate its consequences.

All components involved in the prevention, detection and mitigation of accidents are classified as SIC. All systems required to measure the parameters required for the prevention, detection and mitigation of accidents are classified as SIC. On the basis of the result of the safety analysis, the owner of the plant proposes a classification of components as SIC, which must be agreed by the safety authority. The measurement of all parameters involved must be duplicated with full "physical separation" (e.g. cables must be routed in different tunnels). If possible, the safety authority is also likely to request "technological duplication", i.e. the measurement of a given parameter using different technologies.

The Safety Authority will consider with equal attention all SIC relevant parameters, whether they are used to measure the flow rate in a cooling pipe or the power of the fusion reaction taking place in the vacuum vessel. There will be differences, e.g. some components must fulfill their function at all time (i.e. also during an earthquake, for instance the confinement barriers) whilst other must be operational only at certain times (plasma parameters will have to be measured only during the fusion reaction). Also, the level of reliability required may differ depending on the criticality of the parameter measured, but, in most cases, it will be considerably higher that what is achievable today.

SIC Diagnostics

SIC diagnostics related to plasma control for ITER are still under discussion. The final set will be proposed in the Preliminary Safety Report and it will have to be approved by the French Safety Authority. The SIC diagnostics related to plasma control for DEMO will probably be very similar. It is not very useful to try to guess which will be these parameters, but the followings are likely to be included in the list: fusion power, fuelling rate, plasma ion temperature, halo current, amount of dust in the primary vacuum enclosure.

CONCLUSION

The DEMO studies in Europe started 2 years ago. The identification and the analysis of DEMO critical issues constitute the first results of these studies. This has been possible in particular because we have focused the work around the development of a specific machine, not because we believe it to be the best one, but because it allows to address design integration aspects.

The selection of diagnostics for DEMO will depend on the exact objectives of DEMO and on the selected plasma parameters. The nuclear environment will impose sever limitations and a number of diagnostics used in ITER will be unsuitable.

A limited number of plasma parameters will be important for safety and the diagnostics used to measure them will be classified as "Safety Important Component", which will impose in very severe requirements.

ACKNOWLEDGMENTS

The author wishes to acknowledge the work of all persons having contributed to the PPCS and of all persons currently involved in the DEMO studies, with particular emphasis for the contribution of the EFDA DEMO Working Group. Also, the author wishes to acknowledge the inputs on safety issues from Mrs. J. Elbez-Uzan (CEA). As the PPCS, DEMO is a collaborative effort between several European Fusion Associations and the European industrial consortium EFET. The European DEMO studies are jointly funded by the European Communities and by government agencics of France, Germany, Hungary, Italy Poland and the United Kingdom. The views and opinions expressed herein do not necessarily reflect those of the European Commission.

REFERENCES

1. D. Maisonnier et al., Final Report of the European Power Plant Conceptual Study, 2005, EFDA-RP-RE-50, available online www.efdfa.org.
2. D. Maisonnier et al., Power Plant Conceptual Studies in Europe, *Proc. of 21st Fusion Energy Conference*, Chengdu (2006) IAEA-CN-149-FT/1-2.
3. European Council of Ministers, Conclusions of the fusion fast track experts meeting held on 27 November 2001 on the initiative of Mr. De Donnea (President of the Research Council) EUR (02) CCE-FU/FTC 10/4.1.1, Brussels 5 December 2001 (commonly called the "King Report").
4. I. Cook, R.L. Miller and D.J. Ward, Prospects for economic fusion electricity, *Fusion Engineering and Design* 63-64 (2002) 25-34.
5. M. Gasparotto and R. Laesser, The European Fusion Energy Research Programme Towards the Realisation of a Fusion Demonstration Reactor, *Proc. of Int. Conf. Nuclear Energy for New Europe, Slovenia*, 2006.
6. P. Norajitra et al., He-cooled Divertor Development: Technological Studies and HHF Experiments for Design Verification, *Proc. of 21st Fusion Energy Conference*, Chengdu (2006) IAEA-CN-149-FT/P5-14
7. D. Maisonnier, European DEMO Design and Maintenance Strategy, to be published in *Proc. of ISFNT*-8, Germany, 2008.
8. D. Maisonnier and J. Hayward, Technological and Engineering Challenges of Fusion, *Proc. of the 2nd IAEA Technical Meeting on First Generation of Fusion Power Plants*, Vienna, 2007.
9. National Policy on Future Nuclear Fusion Research and Development (tentative translation), Japan Atomic Energy Commission – Advisory Committee on Nuclear Fusion, 26.10.2005.
10. A.J.H Donné et al., Chapter 7: Diagnostics, *Nuclear Fusion* 47 (2007), 337-384

ECE, INTERFEROMETRY, POLARIMETRY AND THOMSON SCATTERING

Physics of ECE Temperature Measurements and Prospects for ITER

E. de la Luna* and JET-EFDA contributors†

**Asociación EURATOM-CIEMAT, CIEMAT, Madrid, Spain*
†See annex of M.L. Watkins et al, Overview of JET results, Proc. 21st IAEA Fusion Energy Conference, Chengdu, China (2006)

Abstract. The physics of the electron cyclotron emission (ECE) temperature measurements is reviewed. The current understanding of the expected ECE spectra in ITER is summarized, for perpendicular as well as oblique propagation. The relevance of the use of oblique ECE for investigating the shape of the electron distribution function at low energies is discussed.

Keywords: ECE, ITER, Oblique
PACS: 52.25.Tx, 52.70.Gw

INTRODUCTION

The capabilities of electron cyclotron emission (ECE) for electron temperature (T_e) measurements have been used for over four decades of magnetic fusion research. In this period the ECE diagnostics have evolved from proof-of-principle experiments [1] to becoming routinely available on all major fusion devices. In addition to the traditional time- and space- resolved electron temperature profile measurements, ECE diagnostics are employed for a wide variety of measurements in the plasma. This includes core MHD localization, perturbative transport experiments, real time control of ITBs and NTMs and electron distribution function measurements [2]. The good spatial and temporal resolution achieved with the available state-of-the-art multichannel radiometers has been a key feature in most of the measurements mentioned above. Moreover, ECE measurements have proved to be a rich source of new and, in some cases, puzzling phenomena, stimulating the development of more specific and ingenious diagnostic applications, such as oblique ECE [3] or sophisticated systems for 2D imaging [4].

However, as magnetic fusion experiments progress toward larger physical dimensions, higher temperatures and densities, the measurement of T_e from the ECE radiated power becomes more challenging. The purpose of this tutorial paper is to review the physics of evaluating T_e from ECE measurements (for other applications see [5] and references therein) and the prospects for the ECE diagnostics in ITER as it is seen from the perspective of today's experiments. A few examples of experimental data from JET (the closer fusion device to ITER parameters) are used to illustrate the limitations that the ECE diagnostics will encounter when applied to ITER. The main characteristics of the expected ECE spectra in ITER have been discussed in previous Varenna workshops [6][7]. In this paper we extend these earlier results by including the analysis of the emission using different observation angles with respect to the magnetic field.

CP988, *Burning Plasma Diagnostics*, edited by F. P. Orsitto, G. Gorini, E. Sindoni, and M. Tardocchi
2008 American Institute of Physics 978-0-7354-0507-3/08/$23.00

BASICS OF THE ECE THEORY

The source of the cyclotron emission is radiation from the orbital motion of the electrons along the magnetic field lines. A single electron of speed v radiates in an infinite set of harmonics s of the electron cyclotron angular frequency $\omega_{ce} = eB/m_e$, where B is the magnetic field and m_e is the electron rest mass, giving $f_{ce} = \omega_{ce}/2\pi \approx 28$ GHz/T. The polarization of the emitted wave depends on the direction of observation, relative to the magnetic field. For perpendicular observation, the radiant energy propagates in two characteristic modes called the eXtraordinary or X-mode and the Ordinary or O-mode, characterized by the direction of the wave electric field perpendicular or parallel to the confining magnetic field, respectively.

To calculate the observed emission spectrum two steps are required. The first step involves the evaluation of the emission (β) and absorption (α) coefficients [8]:

$$\beta(\omega, N_{\parallel}) \propto \sum_{s=-\infty}^{+\infty} \int \frac{du^3}{\gamma} \delta(\gamma - s\omega_{ce}/\omega - N_{\parallel}u_{\parallel}) \mid Pol \mid^2 f(u) \tag{1}$$

$$\alpha(\omega, N_{\parallel}) \propto \sum_{s=-\infty}^{+\infty} \int \frac{du^3}{\gamma} \delta(\gamma - s\omega_{ce}/\omega - N_{\parallel}u_{\parallel}) \mid Pol \mid^2 \left[\frac{s\omega_{ce}}{\omega} \frac{1}{u_{\perp}} \frac{\partial}{\partial u_{\perp}} + N_{\parallel} \frac{\partial}{\partial u_{\parallel}} \right] f(u) \tag{2}$$

where $\vec{u} = \gamma\vec{v}/c$ is the normalized relativistic momentum, $N_{\parallel} = k_{\parallel}c/\omega$ is the parallel component of the refractive index ($k_{\parallel}$ is the component of the wave vector parallel to the magnetic field and c is the speed of light), $\gamma = (1 - v^2/c^2)^{-1/2}$ is the relativistic factor, $\mid Pol \mid^2$ is a weighting function that depends on the polarization and the harmonic number, $f(u)$ is the electron distribution function and the subscripts $\parallel$ and $\perp$ refer to the parallel and perpendicular components with respect to the magnetic field. From these simplified expressions some general properties can be learned:

- The δ- function in the integrals of (1-2) implies that only electrons that satisfy the resonant condition:

  $$\omega = \frac{s\omega_{ce}}{\gamma} - k_{\parallel}v_{\parallel} \tag{3}$$

  can contribute to the emission and absorption processes. At each harmonic, three effects will lead to variation of the emission frequency: the radial variation of the magnetic field that changes the cyclotron frequency, a frequency shift due to the relativistic mass which lowers the emission frequency as the energy of the resonant electrons increases ($\gamma m_e c^2$) and the Doppler shift caused by the relative motion between the emitting electrons and the observer.
- The energy and pitch-angle dependence of the weighting function $\mid Pol \mid^2$ is different for the two different polarization modes and/or for the different harmonics (for a more detailed description see [8]).

In the second step, the radiation transport equation [9] is solved in order to determine how much of the emitted radiation emerges outside the plasma:

$$\frac{dI(\omega)}{dl} = \beta - \alpha I(\omega) \tag{4}$$

where $I(\omega)$ is the specific intensity at a frequency ω and l is the spatial coordinate along the ray trajectory (the ray refractive index is approximated by unity). Eq. (4) takes into account that part of the emitted radiation is absorbed as the wave propagates through the plasma toward the detection system. For the specific case of observation from the low-field side (LFS) of the plasma, the solution of Eq. (4) is of the form:

$$I(\omega,N_{\parallel}) = I_{in}(\omega,N_{\parallel})\exp\left(-\tau_{12}(\omega,N_{\parallel})\right) + \int_{l_1}^{l_2} \beta(l,\omega,N_{\parallel})\exp\left(-\tau(l,\omega,N_{\parallel})\right) dl \quad (5)$$

where $l_1 \leq l \leq l_2$, being l_1 and l_2 the initial and final points of the ray within the plasma boundaries, the exponential function in the integral is the transmission coefficient and τ is the optical depth defined by:

$$\tau(l,\omega,N_{\parallel}) \equiv \int_{l}^{l_2} \alpha(l',\omega,N_{\parallel})dl' \quad (6)$$

The first term in Eq. (5), which can formally be written as $I_{in}(\omega,N_{\parallel}) = R_w I(\omega,N_{\parallel})$, is the radiation that is reflected off the inner wall of the tokamak and/or the cut-offs, being $\tau_{12}(\omega,N_{\parallel}) \equiv \int_{l_1}^{l_2} \alpha(l,\omega,N_{\parallel})dl$ the total optical depth along the path that the ray followed in the plasma. Note that reflections generally also imply significant polarization scrambling and this effect is also included in $R_w(\leq 1)$.

In the case of a plasma in local thermodynamic equilibrium at a local temperature T_e, the emission and absorption coefficients are not independent but related by Kirchhoff's law, their ratio equal to the blackbody radiation (I_{BB}), which in the classical limit of the Plank's law can be written as:

$$\frac{\beta(\omega,N_{\parallel})}{\alpha(\omega,N_{\parallel})} = I_{BB}(\omega,N_{\parallel}) = \frac{\omega^2}{8\pi^3 c^2}T_e \quad (7)$$

Substituting (7) into (5), the radiation temperature $T_{rad}(\omega,N_{\parallel}) = (8\pi^3 c^2/\omega^2)I(\omega,N_{\parallel})$ can be conveniently written as:

$$T_{rad}(\omega,N_{\parallel}) \approx \langle T_e \rangle \times \frac{1-\exp\left(-\tau(l,\omega,N_{\parallel})\right)}{\left[1-R_w\exp\left(-\tau_{12}(\omega,N_{\parallel})\right)\right]} \quad (8)$$

where

$$\langle T_e \rangle = \frac{\int_{l_1}^{l_2} T_e(l)G(l,\omega,N_{\parallel})dl}{\int_{l_1}^{l_2} G(l,\omega,N_{\parallel})dl} \quad (9)$$

and

$$G(l,\omega,N_{\parallel}) = \alpha(l,\omega,N_{\parallel}) \times \exp\left(-\tau(l,\omega,N_{\parallel})\right) \quad (10)$$

defines the emitting region, with $\langle T_e \rangle$ being the average temperature within the emitting volume. This factor $G(l,\omega,N_{\parallel})$ has a nearly Gaussian profile and its width is determined by the reabsorption, as described by the transmission factor in Eq. (10). When the reabsorption is large, τ»1 (optically thick plasma), the emitting region is a narrow zone close to the resonance position and the emitted radiation is a measure of the local T_e (see Fig. 1). Note that for optically thick harmonics T_{rad} is independent of wall reflections.

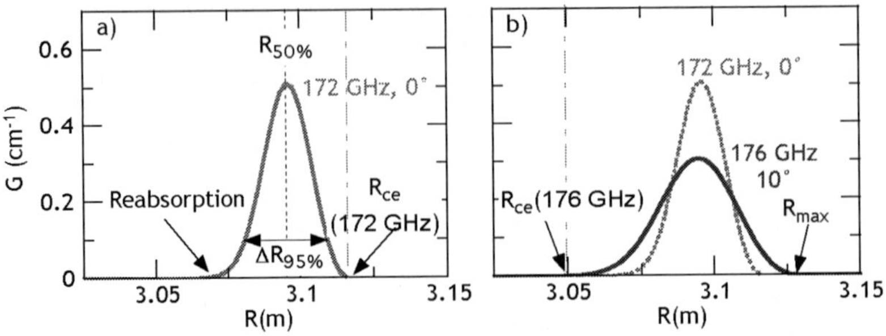

FIGURE 1. The function $G(\omega,R)$ (Eq. 10), determining the size of the emitting region, is shown for two observation angles: a) $\theta = 0^o$ and b) $\theta = 10^o$ for JET plasma parameters (T_{e0} =7 keV, $n_{e0} = 3.5\times10^{19}$ m^{-3}, B= 3.2 T). For $\theta = 0^o$ the emission comes from a thin layer on the left of the resonance position R_{ce} whose width is determined by the reabsorption, while for the oblique observation the electrons responsible of the observed radiation are located in a region $R_{ce} < R < R_{max}$, where R_{max} is defined by Eq. (11). The location of the emission is given by the center of gravity of the function $G(\omega,R)$ and the effective resolution by the radial extension of the region responsible for 95% ($\Delta R_{95\%}$) of the observed intensity.

A couple of comments are in order. First, the spatial localization of the emission discussed above is valid when there is only one harmonic contributing to the emitted radiation at a given frequency. Second, the observed emission is not only localized in real space (along the line of sight) but also in velocity space ($u_{\parallel}, u_{\perp}$). The correlation between the energy and location of the resonant electrons has immediate physical consequences for the interpretation of the ECE measurements. Let us consider first observation perpendicular to the magnetic field from the low-field side, which is the most common set-up for temperature measurements. For $N_{\parallel}$=0 the resonance condition is reduced to $\omega = s\omega_{ce}/\gamma$, consequently only electrons with $s\omega_{ce}/\omega > 1$ can contribute to the emission. Because in toroidal fusion devices the magnetic field decreases monotonously along the major radius, the emission at a given frequency is not only originated near the resonance but also from locations in the plasma with higher B and therefore generated by electrons with higher energy ($E = m_e c^2(\gamma - 1) = m_e c^2(s\omega_{ce}/\omega - 1)$). This emission is often called downshifted emission. To complete the picture we need to include the absorption along the propagation path. Since the absorption depends on the number of resonant electrons, when the distribution function is Maxwell-like, that is when the electron population decreases monotonically with increasing energy ($f(E) \propto \exp(-E/T_e)$), the low-energy electrons near $\omega \approx \omega_{ce}$ reabsorb the emission coming from the energetic electrons located further inside the plasma ($\omega > s\omega_{ce}$) and, as a result, the radiation that reaches the antenna comes from relatively low energy electrons ($E < T_e$) and is well localized in real space (see Fig. 1a).

Oblique ECE

We now consider the case of oblique observation, i.e. $N_{\parallel} \neq 0$ with $N_{\parallel} < 1$. In this case the resonant condition for oblique observation can only be fulfilled by electrons that satisfy:

$$\left[(s\omega_{ce}(l)/\omega)^2 - 1 + N_{\parallel}(l)^2\right] > 0 \tag{11}$$

Eq. (11) sets an outer radial limit on where emission at a given frequency can be originated, while the inner limit of the emitting region is set by the reabsorption. The optical depth decreases with $\mid N_{\parallel} \mid$; however for hot and large plasmas, the optical depth of the thermal resonance is sufficiently high to reabsorb the emission coming from $\omega > s\omega_{ce}$ and only a thin layer contributes to the observed intensity (see Fig. 1b) [10][11]. This emission is often called upshifted, because it comes from a region located to the right of the resonance location. In this scenario the emission from an optically thick harmonic is originated by electrons with energies near or slightly above the thermal energy, $E \geq T_e$ (the energy increasing with increasing observation angle). The main interest of measuring the emission at different angle to the magnetic field lies in the fact that by changing the observation angle, it is possible to change the range of energies at which the temperature of the distribution function can be probed. Moreover, the sign of the $N_{\parallel}$ selects the sign of $u_{\parallel}$, which makes the oblique view sensitive to asymmetries in the distribution function.

DIAGNOSTIC POTENTIAL OF ECE MEASUREMENTS

The spatial localization of the optically thick harmonics forms the basis of the ECE temperature measurements. Since the cyclotron frequency is determined by the magnetic field and the magnetic field in toroidal devices varies across the plasma in a known way, essentially inversely proportional to the major radius, the location of the origin of the emitted radiation at a given frequency is known and its intensity gives the local temperature at that location. Hence, for optically thick harmonics, the width and shape of the ECE spectrum corresponds to the width and shape of the T_e profile. The optical thickness depends on the plasma parameters (n_e, T_e), the magnetic field gradient, the harmonic number, polarization mode and the angle θ between the viewing direction and the magnetic field ($N_{\parallel} = \cos\theta$). For fixed magnetic field gradient and $N_{\parallel}$, the emitting region broadens with increasing T_e and decreasing n_e [12]. The overall radial resolution must be evaluated by taking into account the width of the emission layer, the finite bandwidth of the measurements and the antenna pattern. The first harmonic O-mode (designated by the notation 1O-mode) and the second harmonic X-mode (2X) have the larger optical thickness (with $\tau_{2X} > \tau_{1O}$) and are the most commonly used in present experiments. The condition of $\tau \gg 1$ is not necessary fulfilled at the plasma edge (optically thin plasma), which imposes a limit on how close to the last closed flux surface (R_{LCFS}) reliable T_e measurements can be obtained (see Fig. 2). In addition, edge measurements are strongly affected by the presence of small populations of suprathermal electrons. In this case, the unique correspondence between the emitting frequency and the spatial (and energy) location of the emission is lost and the measured intensity is not necessarily related to the local temperature.

Another important issue for the use of ECE diagnostics is the accessibility. The major radius range over which T_e measurements can be made using ECE diagnostics is limited by cut-off effects and harmonic overlap. The use of 2X-mode, which has twice the cut-off density of the 1O-mode, allows for a large range of accessible densities.

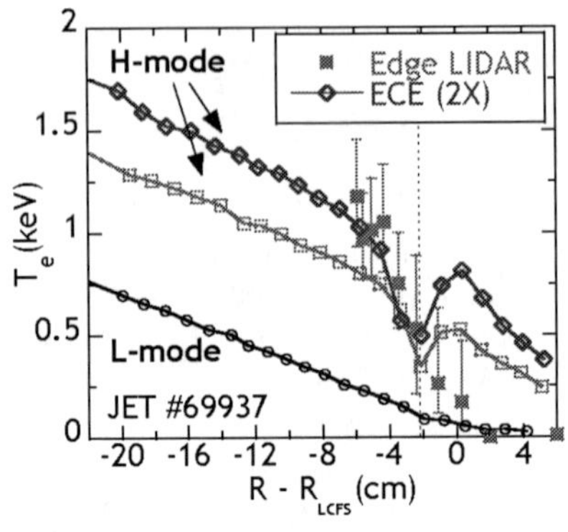

FIGURE 2. Edge T_e profiles, as measured by ECE (open symbols) and the edge LIDAR system (closed symbols) in JET, showing an enhancement of the T_{rad} (above the corresponding blackbody emission) measured by ECE close to the separatrix during H-mode (this effect is discussed in more detailed in [13])

Since, for 2X-mode measurements, $n_{e,cutoff} \propto B^2$, the cut-off represents no limitation in the operating density for high magnetic field tokamaks like ITER. Harmonic overlap occurs in small aspect ratio devices because of the large variation in toroidal magnetic field across the plasma, which allows two or more harmonics to resonate at the same frequency in different parts of the plasma. In JET, 3rd harmonic emission near the edge on the low-field side occurs at the same frequency as 2nd harmonic emission from radii inboard of the plasma center. However this limitation does not exist for the 1O-mode. The possibility of using the 1O-mode emission to access the high-field side (HFS) of the plasma has been recently explored in JET and simultaneous temperature measurements of the inboard and outboard pedestal region have been obtained [14]. These measurements are particularly relevant for investigating ELM dynamics, specially considering how poorly diagnosed the plasma in the inboard region is in large fusion devices due to the difficult access for diagnostics.

ECE MEASUREMENTS IN HIGH T_e PLASMAS

In the last years, ECE measurements in high T_e plasmas show features that appear to fall outside the standard description. In JET, an apparent disagreement between the central temperature measured by ECE (2X-mode) and Thomson Scattering (TS) have been observed in strong auxiliary heated (NBI+ICRH) plasmas, with ECE showing higher temperature values (up to 15% in cases where $T_{e0} > 7$ keV) in a narrow central region ($r/a < 0.35$) [15]. Similar discrepancy was reported in TFTR [16] and has remained unexplained so far.

From a formal point of view, agreement between the TS and ECE temperatures occurs only if the electron distribution function is Maxwellian. The peculiarity of the discrepancy observed in JET is that it appears in a region where the plasma is optically thick, which implies that the deviation of the electron distribution function from the Maxwellian must be localized at low energies (if this is the cause of the discrepancy).

Since 2nd harmonic ECE and TS measurements carry information on bulk electrons of different energy, discrepancies between ECE and TS temperature profiles can be considered as a proof of the existence of non-Maxwellian features in the bulk electron distribution function (if calibration errors are ruled out). Moreover, ECE spectra measured in those plasma conditions are also inconsistent with the electron bulk distribution being Maxwellian in the plasma core [17]: the central temperature deduced from the 3rd harmonic is lower than the one deduced from the 2X-mode. The 3rd harmonic emission is optically thick in these high temperature plasmas and therefore provides an independent measurement of the bulk electron temperature but related to electrons with higher energy [18]. Similar features in the ECE spectra has been reported in FTU [19] and DIII-D [20] during high power on-axis ECRH experiments. All of these experimental results appear to be consistent with the existence of a distortion in the bulk of the electron distribution function [11]. The detailed understanding of the origin of this intriguing experimental result is crucial when considering the application of the ECE techniques in ITER, where the temperature range over which the temperature measurements are expected will be significantly higher than those of existing tokamaks. Further work to identify the experimental conditions leading to the appearance of the described inconsistency between different T_e measurements in JET is ongoing.

As discussed in the previous section, an attractive and independent method to assess the existence of non-Maxwellian bulk electron distribution functions is oblique ECE. A diagnostic of this kind has been recently installed in JET [3]. This instrument allows the simultaneous measurement of the ECE spectrum at different angles ($\approx 10^o$, $\approx 20^o$) and with different polarizations (O- and X-mode). From the emitted spectra at different observation angles it will be possible to explore the distribution function in a wider energy range. For optically thick harmonics, there exist several frequencies for which on average, the emitting electrons are in the same position but have different energy, which, in principle, could permit an spatially resolved reconstruction of the distribution function.

ECE IN ITER

In the high-temperature (> 20 keV) plasmas predicted for ITER, the optical depth for the higher harmonics will increase considerably and the level and shape of the ECE spectra will be substantially different from that of the existing tokamaks. In the discussion that follows only thermal plasmas will be discussed. For the effect of suprathermal electron tails on the ECE spectra for ITER see [21]. Previous studies [6][7] have shown that access over a significant portion of the plasma cross-section can be obtained with the 1O-mode (with spatial resolution of 10-15 cm), while the use of the 2X-mode, more strongly affected by harmonic overlap, is restricted to the outboard plasma region. On the other hand, the best spatial resolution at the edge (4-6 cm) is achieved with the 2X-mode, although the stringent spatial resolution requirements for the edge measurements (5 mm) set by ITER can not be met [22].

Let us now examine the case of oblique emission. Shown in Fig. 3 are the ECE spectra for $\theta = 0^o$ (perpendicular observation) and $\theta = 10^o$ for parameters typical of the Scenario 2 in ITER along with the corresponding $T_e(f)$ profiles as obtained from

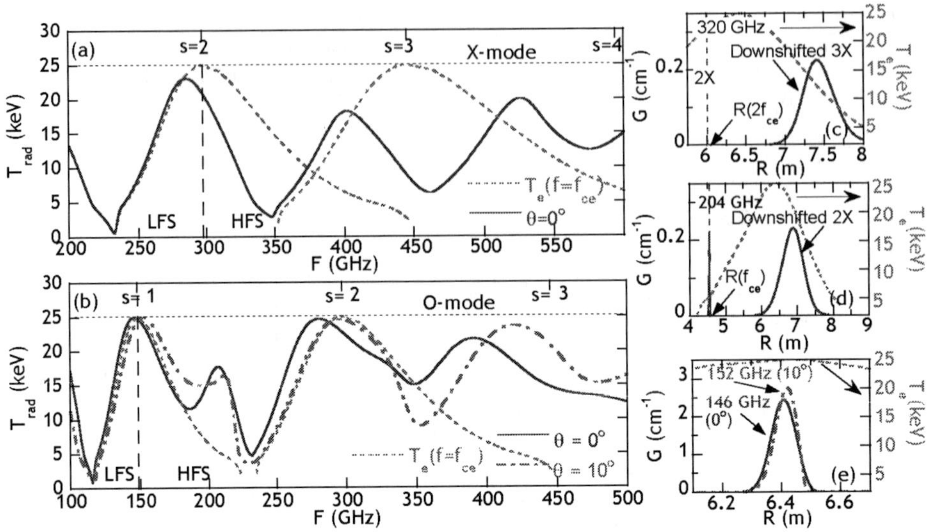

FIGURE 3. T_{rad} vs $f = \omega/2\pi$ for (a) X-mode and (b) O-mode in ITER (T_{e0}=25 keV, $n_{e0} = 10^{19}$ m^{-3}, B_0=5.4 T) for $\theta = 0^o$ and $\theta = 10^o$ in the case of O-mode and $G(R,\omega)$ factor for specific frequencies: (c) X-mode, f=320 GHz, (d) O-mode, f=204 GHz, and (e) two frequencies corresponding to emission from the plasma center for two different observation angles. The thin dashed lines are $T_e(R_{ce})$ vs $f = f_{ce}$.

$T_e(f = f_{ce})$. Those spectra have been calculated using the emission code SPECE [22]. The main observations are the following:

- $T_{rad}(f) < T_e(f)$ on the high frequency side of the 2X-mode. This can be understood by considering that the 2X-mode emission when propagating toward the antenna in the low field side is reabsorbed by the relativistically downshifted 3X-mode emission (even though the resonance radius for the 3rd harmonic is outside the plasma), being the observed radiation the 3X-mode contribution from the edge outboard plasma region (see Fig. 3c).
- $T_{rad}(f) > T_e(f)$ on the high frequency side of the 1O-mode. At the plasma edge, the observed intensity is dominated by the downshifted 2O-mode emission (from the high energy tail of the Maxwellian distribution) (see Fig. 3d).
- $T_{rad}(f) > T_e(f)$ in the lower frequency region for both modes. The optical thickness at the plasma edge is not high enough to reabsorb the downshifted emission from the Maxwellian tail of the hot core plasma.
- core T_e measurements for which $T_{rad}(f)$ matches up the corresponding T_e profile can be obtained from the 1O- and 2O-mode, although the spatial resolution of the 2O-mode is poorer (by more than a factor of 2) than that of the 1O-mode. The same applies to the oblique emission, being the size of the emitting region for the oblique view comparable to that of normal emission [22] (see Fig. 3e). Note also that the shift in the effective position of the emitted radiation generated by the relativistic broadening, and the Doppler effect in the case of the oblique spectrum, is higher

for the 2O-mode.

These results are of course profile dependent. The central region affected by harmonic overlap becomes wider for higher T_e or flatter profiles [6]. As for the impact of the antenna pattern on the overall spatial resolution, the effect of the beam size is negligible for perpendicular observation (for antennas located on the plasma midplane), whereas it tends to be significant for the oblique observation due to the Doppler effect [22].

CONCLUSIONS

The properties of the ECE radiation in hot plasmas and its potential for measuring the T_e profiles in ITER have been reviewed. We have shown that, for T_e<30 keV, core temperature measurements can be obtained from the 1O- and 2O-mode ECE for both normal and oblique observation (for $\theta < 20^o$). Despite the different spatial resolution of the measurements (the spatial resolution worsening with increasing observation angle and harmonic number), having more than one measurements of the central temperature in ITER may be useful as a general consistency check. Furthermore, since those measurements carry information on different regions in velocity space, their comparison would enable to detect distortions on the bulk of the distribution function.

The observed anomaly on the ECE spectra and the apparent discrepancies between the central T_e measured by ECE and TS in high-T_e plasmas seem to indicate that the assumption of a Maxwellian bulk may not be necessarily true in the present of strong auxiliary heating, contrary to what is usually assumed. Based on these experimental findings, we may speculate that non-Maxwellian bulk distribution functions could appear in plasmas with high level of auxiliary power whenever a sufficient amount of the heating power is absorbed by the bulk electrons. However, the extrapolation of these results to ITER remains uncertain because of the lack of a valid physical model for the origin of such a distortion in the distribution function. More experimental and theoretical work is necessary to assess the potential implications of a non-Maxwellian bulk for temperature measurements in ITER.

One point that has been stressed throughout the paper is the fundamental role played by the coupling between the energy and location of the resonant electrons in the evaluation of the T_e from ECE measurements. One intrinsic limitation of the diagnostic is that the information obtained with the ECE measurements can only be related to a narrow region in velocity space, which is determined by both the resonance condition and the reabsorption along the propagation path. In general, it is assumed that the plasma is Maxwellian and, therefore, the distribution function is well described with a single temperature value. However, this will not be the case when the distribution function deviates from the Maxwellian. In such cases, electron populations with different energy will have different average temperature, the temperature ceases to be a useful parameter and a description of the velocity distribution is needed. While the appearance of distortions in the ECE spectra is often sufficient to identify the non-Maxwellian features in the distribution function, the problem of inferring quantitative information on the electron distribution function from the shape of the ECE spectra is by no means straightforward. This is an ill-posed problem and does no have a unique solution. The typical approach is to choose an arbitrary distribution function with variable parameters and then, in a quantitative comparison between experiments and theory, these parameters are adjusted by a best fit between calculation and measurement. The accuracy of this procedure can be significantly improved by using as many independent measurements of the ECE spectrum (at different polarization, harmonics and observation angles) as possible.

ACKNOWLEDGMENTS

The author would like to thank D. Farina, L. Figini and C. Sozzi for providing the ITER spectra presented here. The author also wishes to thank V. Krivenski for giving information, criticism and suggestions on the subject of the oblique ECE. It is not the objective of this paper to give a complete overview of the current activities in the field, therefore there is no attempt to provide exhaustive references and I apologize to those whose work I have unwittingly overlooked. The work leading to this article was funded by the European Atomic Energy Community and is subject to the provisions of the European Fusion Development Agreement.

REFERENCES

1. A. Costley et al., *Phys. Rev. Letters* **33** 758-761 (1974)
2. H. Gasparino, Proc. 8th Joint Workshop on ECE and ECRH, Gut-Ising, Germany (1993) 19-44
3. C. Sozzi, et al, in this Proceedings
4. H. K. Park et al., *Phys. Rev. Letters* **96** 195004 (2006)
5. H. J. Hartfuss *Plasma Phys. Control. Fusion* **40** A231-A250 (1998)
6. D. V. Barlett, in *Diagnostics for Experimental Thermonuclear Reactors*, edited by P. Stott, G. Gorini and E. Sindoni, Plenum Press, New York (1996) 183-193
7. D. V. Barlett and H. Bindslev, in *Diagnostics for Experimental Thermonuclear Reactors*, edited by P. Stott, G. Gorini and E. Sindoni, Plenum Press, New York (1998) 171-180
8. M. Bornatici, R. Cano, O. de Barbieri, F. Engelmann, *Nuclear Fusion* **23** 1153 (1983)
9. G. Bekefi, *Radiation Processes in Plasmas*, Wiley, New York (1966)
10. V. Tribaldos, V. Krivenski, Proc. 8th Joint Workshop on ECE and ECRH, Gut-Ising (1993) 123-141
11. V. Krivenski, *Fusion Engineering and Design* **53** 23 (2001)
12. V. Tribaldos, *"Spatial Resolution for Electron Cyclotron Emission in JET"*, JET-Report, EFDA-JET-PR(01)44 (2001)
13. A. G. Peeters et al, Proc. 10th Joint Workshop on ECE and ECRH, Ameland (1997) 403
14. L. Barrera, in this Proceedings
15. E. de la Luna et al., *Review Sci. Instrum.* **74** 1414 (2003)
16. G. Taylor et al., Proc. 8th Joint Workshop on ECE and ECRH, Gut-Ising (1993) 277-288
17. V. Krivenski, E. de la Luna and G. Giruzzi, Proc. 19th Conference on Controlled Fusion and Plasma Physics, Montreaux (2002)
18. I. Fidone, G. Giruzzi and G. Taylor, *Phys. Plasmas* **3** 2331 (1996)
19. G. Bracco et al., Proc. 18th IAEA Fusion Energy Conference, Sorrento (2000)
20. R. W. Harvey et al., Proc. 13th Joint Workshop on ECE and ECRH, Nizhny Novgorod (2004) 119
21. M. Bornatici and U. Ruffina, *Plasma Phys. Control. Fusion* **38** 1561 (1996)
22. D. Farina, L. Figini, P. Platania and C. Sozzi. in this Proceedings

Measurements of Electron Velocity Distribution Function (invited paper)

C.Sozzi[1], E. De La Luna[2], D.Farina[1], J.Fessey[3], L.Figini[1], S.Garavaglia[1], G.Grossetti[1], S.Nowak[1], P.Platania[1], A.Simonetto[1], M.Zerbini[4], and JET EFDA contributors*

JET-EFDA, Culham Science Centre, OX14 3DB, Abingdon, UK

[1]*IFP-CNR, Associazione EURATOM-ENEA-CNR sulla Fusione, Via R.Cozzi 53, 20152 Milano ITALY*

[2]*Asociación EURATOM-CIEMAT, CIEMAT, Av.Complutense 22, 28040 Madrid, Spain*

[3]*EURATOM-UKAEA Fusion Association, Culham Science Centre, Abingdon, Oxon, OX14 3DB, UK*

[4]*Associazione EURATOM-ENEA sulla Fusione, C.R.E. Via Enrico Fermi 45 00044 Frascati (Roma), Italy*

Abstract. Deviations of the electron function velocity distribution from Maxwellian behavior in the high energy range have been extensively studied during the past two decades. A brief review of the experimental techniques and findings on the subject is given in this paper. There are indications that the electron distribution function in thermonuclear plasmas could be significantly different from the Maxwellian one even near the thermal velocity range when particular circumstances occur. These reasons motivate a renewed effort in the measurements that could detect low energies distortions, among which is the new Oblique Electron Cyclotron Emission diagnostics that has entered operations during 2006 experimental campaign at JET and aims to resolve relatively small differences in the ECE spectra taken at three toroidal angles. Such measurements allow insight into the characteristics of the electron distribution function. First results of such experimental system are discussed in this paper, along with emission simulations.

Keywords: Electron velocity distribution, Oblique Electron Cyclotron Emission.
PACS: 52.25.Tx, 52.70.Gw, 52.70.La

INTRODUCTION

There are several physics situation in which the features of the electron distribution function (edf) can play a role in modern devices for plasma magnetic confinement, and there have been in fact several experiments in which this characterization has been performed. The edf can depart from the canonical Maxwellian when intense electron heating and/or current drive techniques are applied [1,2], in presence of intense electric field at low plasma density [3], in disruptions and in magnetic reconnections related with magnetohydrodynamic activity [4,5]. It is well known that in such situations high energy tails in the function distribution can develop, producing characteristic signatures in the Electron Cyclotron Emission and in the X-rays Bremmstralung radiation.

---

* See the App. of M.L.Watkins et al., Proc. 21st IAEA Int. Fusion Energy Conference, Chengdu, China (2006)

CP988, *Burning Plasma Diagnostics*, edited by F. P. Orsitto, G. Gorini, E. Sindoni, and M. Tardocchi
2008 American Institute of Physics 978-0-7354-0507-3/08/$23.00

A more elusive phenomenon is the deformation of the electron distribution in the low energy (thermal) range. A few experimental evidences and extensive simulation work [6-8] indicate that the bulk of the electron population itself could be diverted, under particular situations, from the standard maxwellian behavior. If confirmed these observations imply a major rethinking of what concerns the electron kinetic measurements and their interpretation, such the impossibility in principle to give the definition of the temperature as a scalar plasma parameter.

In this paper a brief overview of the experimental techniques used to characterize the electron velocity distribution is given, along with the major findings of such experiments, followed by a description of the work that is being carried out at JET on the subject of the characterization of edf in the thermal energy range.

EDF STUDIES IN THE SUPERTHERMAL ENERGY RANGE

Most of the techniques for the characterization of the edf in the superthermal energy range were originally developed in connection with studies related with Lower Hybrid Current Drive. Such experiments demonstrated the formation of a pronounced, non isotropic fast electrons tail in the distribution function when LHCD is applied, which effectively sustains the plasma current with reduced or even absent inductive electric field [9-14]. More recently, with the development of high power sources in the range of frequency of the Electron Cyclotron resonances in modern toroidal devices, the perspective of a fine control of the current profile has been disclosed, based on the high localization of the EC absorption when compared with the LH one. Effects related with the absorption by superthermal electrons have been observed in ECRH/ECCD experiments, both when the superthermal fraction is generated by EC Resonance Heating and/or Current Drive itself, and when the ECRH/ECCD waves interact with the fast electron tail generated by other means [2]. The experimental techniques used to study the formation and relaxation dynamics of the tails and the absorption properties of fast electrons include soft and hard X emission [14, 15], EC Emission and EC Transmission in various geometrical arrangements [9-13,16]. A general problem with the measurements related to the edf is that a unique reconstruction of the edf itself is not possible without a priori assumption, and the measurements only supply constraints for the actual distribution [17]. Important pieces of information can be inferred by the comparison of different lines of sight of the ECE spectrum, particularly in the cases in which the propagation of the superthermal emission is not hided behind the thermal emission layer. The spatial resolution of the ECE diagnostics is rooted in the physical process of emission-reabsorption of the resonant electrons in a given spatial location and is the result of the interplay among the plasma kinetic parameters (density and temperature), the angle of propagation with respect to the magnetic field and the existence of multi-harmonics resonances and cut-off layers. X-rays based diagnostics instead can be arranged with high number of chords to produce tomography reconstruction with accurate spatial resolution. Table 1 summarizes typical diagnostic figures for the measurement techniques used in electron heating experiments or edf-related studies in toroidal devices. The figures here reported reflect the specific experimental conditions and do not represent absolute limits. Use of oblique ECE and Thomson Scattering in the thermal energy range will be discussed in the next section.

Table 1. Diagnostics for electron distribution measurements

Diagnostics	Energy range	Energy resolution	Spatial localization	Time resolution	Typical application
Soft X [15]	1-20 KeV	10%-20%	r/a~0.1 or less with tomography	~0.01ms	MHD events, Runaway el.
Hard X [14]	20KeV-1MeV	10%-20%	r/a~0.1 or less with tomography	1-10ms	Energetic runaway, LHCD tails,
Vertical ECE [10, 11]	30-250 KeV	15%	chord integrated	~10ms	Runaway, LHCD tails
Vertical microwave transm. [12]	50-200 KeV	10%	chord integrated	5ms	Relativistic EC absorption, LHCD tails
Oblique microwave transm. [9]	1-100 KeV depending on angle	10%-20%	chord integrated	1-5ms	LHCD tails
Oblique ECE [13]	0.1-200 KeV depending on angle	5%	depending on optical thickness and angle	~1ms	LHCD tails, bulk electrons, ECRH
HFS ECE [16]	10-50 KeV	~10%	r/a~0.04-0.25 depending on optical thickness	~0.01ms	ECCD tail, ECRH
Thomson Scatt. [18]	0.1-30 KeV	20%	r/a~0.1 or less	~0.05s	Bulk electrons

EDF STUDIES IN THE THERMAL ENERGY RANGE

Much less effort has been dedicated so far to the detailed study of the edf in the thermal energy range. Besides the question of principle, it has a direct impact on the value ascribed to the electron temperature, being the last essentially the perpendicular slope of the edf. Among the diagnostics listed in table 1, in principle both Oblique ECE (ObECE) and Thomson Scattering (TS) appear to be suitable to investigate this subject. In the TS diagnostics the spectral density function $S(\omega_s)$ of the light scattered by the electrons in the plasma is related with the velocity distribution of the electron population:

$$S(\omega_s, \vartheta, \vec{x}) = F\left[f(\vec{v}, \vec{x})\right] \quad (1)$$

where ω_s is the scattered angular frequency, θ is the scattering angle, $\vec{v}$ and $\vec{x}$ are the velocity and space vectors, F represents an integration operator in the normalized velocity space v/c, and f is the edf. The relation from S and f is then rather direct. In most of present diagnostic systems the scattered light is collected using a finite (sometime small) number of spectral channels and S interpreted assuming f as a Maxwellian. Provided that an adequate number of spectral channels are available and that a sufficient photon statistics is gathered significant deviations from Maxwellian should be detected [17].

A peculiar characteristic of the EC physics is that only resonant electron in a given position in real space and in momentum space contribute to the emission or absorption of EC waves. Being the resonance condition

$$\omega - k_{//}v_{//} = \frac{m\Omega_{ce}}{\gamma} \qquad (2)$$

where $n_{//}=ck_{//}/\omega$ is the parallel (to the magnetic field) index of refraction, m the harmonics number, $\Omega_{ce}=eB/mc$ is the resonance frequency and $\gamma=1/(1-v^2/c^2)^{1/2}$ is the relativistic factor, when ECE radiation is observed at an angle significantly different from perpendicular ($n_{//}\neq 0$), a Doppler shift is included in the resonance condition and a range of electron energies can contribute to the emission at a given frequency [18, 19]. The situation is described in figure 1, where a Maxwellian distribution (contour plots electron density in velocity space, in grayscale) is superimposed to a family of resonance curves (dashed lines with increasing $m\Omega_{ce}/\omega$). For perpendicular propagation (left) each resonant curve is close to a single electron energy. For oblique propagation (right) each resonance curve cross a wider range of electron energies, asymmetric with respect to the parallel velocity. Then, diagnosing different harmonics and lines of sight different regions of the electron population momentum space are sampled, and from the merging of those data details of the distribution function can be inferred.

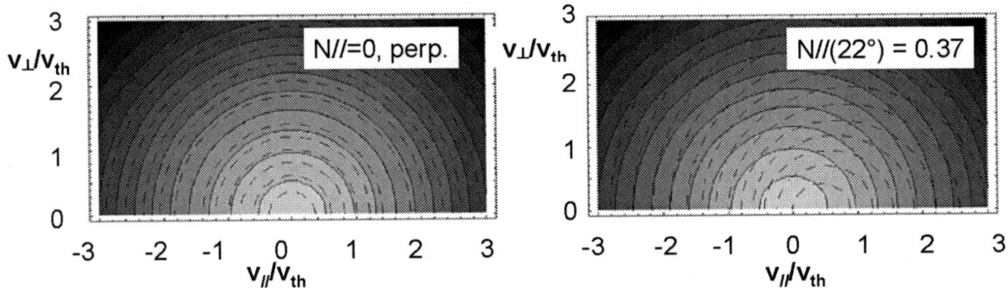

Figure 1. Contours of $f(v_e)$ for a Maxwellian distribution, on which the resonance curves for perpendicular (left) and oblique (right) are superimposed. Grey levels: electron density in velocity space. Dashed lines: resonance curves with increasing $m\Omega_{ce}/\omega$.

Experimental observations having reference to thermal energy range distortions of edf have been observed in TFTR [6], FTU [7] and JET [8] through discrepancies between the electron temperature measured with ECE and Thomson scattering in the plasma core, and through discrepancies between 2nd and 3rd harmonics ECE temperature peaks in optically thick plasmas or other anomalies noticed in the ECE spectrum. Moreover, the first Oblique ECE measurements performed on FTU under intense EC heating showed a discrepancy between perpendicular and oblique ECE measurements indicating again low energy non Maxwellian edf. Common features of the experimental situations in which such observations were performed are high electron temperature, high power heating and low plasma density. However, no extensive experimental studies pointing out detailed cause-effect relations have been performed yet.

OBLIQUE ECE AT JET

The Oblique ECE system of JET has been proposed mainly to perform systematic studies of edf in the thermal energy range, namely to resolve small deviations from Maxwellian-related ECE spectra taken at different toroidal angles. The practical opportunity came with the installation of the so called Microwave Access which includes six circular waveguides, 4 of which corrugated dedicated to Reflectometry diagnostics and 2 smooth dedicated to the oblique lines of sight for ECE. The Oblique ECE system (see figure 2) has three lines of sight respectively at 0° (X polarization only), ~10°, ~22° with respect to the perpendicular to the toroidal magnetic field. The antenna for the oblique lines are built in a single piece, with minimum overall dimensions. The perpendicular sight is positioned in a different octant of the vessel, also equipped with two other ECE diagnostics, a 96 channels radiometer and a perpendicular view Michelson interferometer. Each one of the two quasi optical oblique antennas collect ECE radiation along two linear polarizations (vertical and horizontal) corresponding respectively to mainly X and mainly O polarizations.

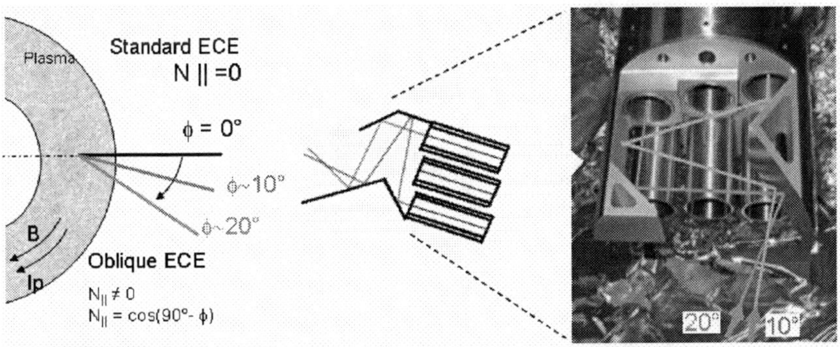

Figure 2. Left: lines of sight of Oblique ECE diagnostics with respect to plasma current and toroidal magnetic field. Centre: sketch of the quasi optical antennae (top view). Right: antenna block before installation in the vessel.

The spectral analysis is performed with a six channel Martin Puplett interferometer (one channel is presently unused), 75-800 GHz band, 7.5-15 GHz single line equivalent spectral resolution [20], being ~350 GHz the maximum frequency of interest for JET. A peculiar feature of the system is that the moving arm of all the channels is realized with a single helix-shaped mirror composed by four sectors. Rapid spinning on the helix (up to 2500 rpm in present configuration) allows time resolution up to 6 ms per radial profile. Due to oblique propagation, the polarization seen by the oblique antennas is elliptical, that is a combination of the pure O and X modes. The amount of "mode contamination", i.e. the fraction of X-mode coupled to the "vertical" polarization of the antenna, increases with the frequency varying from 96% (75 GHz) to 75% (350 GHz) at 10° and from 85% (75 GHz) to 60% (350 GHz) at 22°.

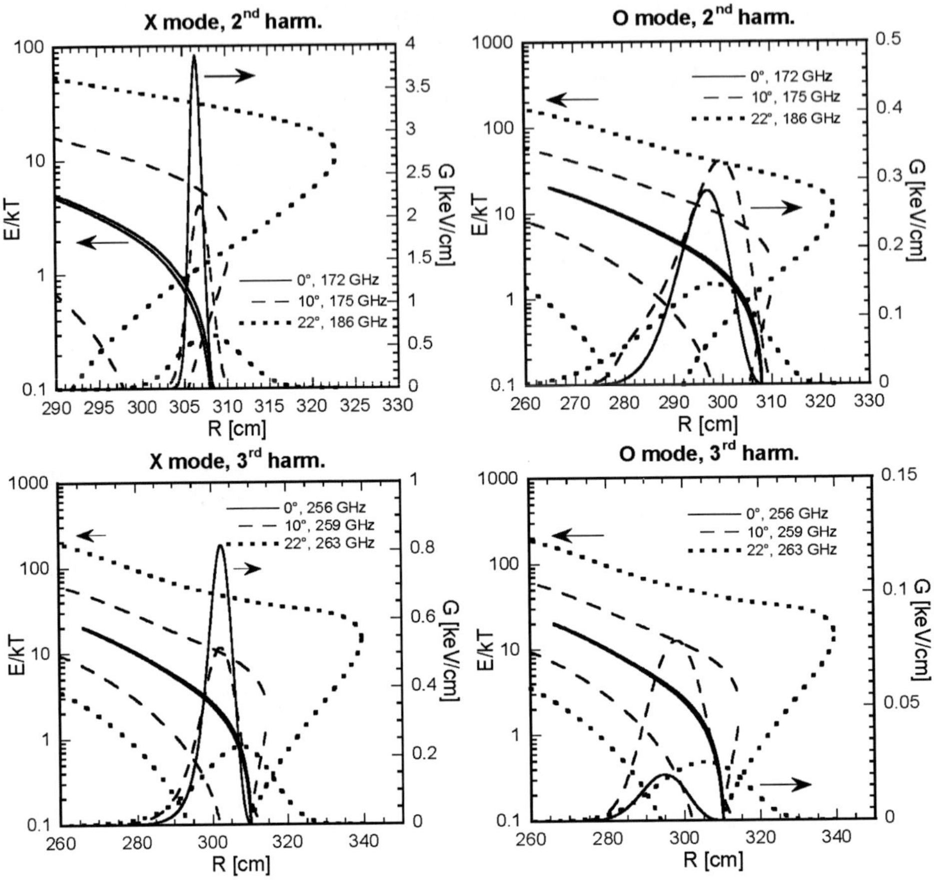

Figure 3. Energy of the emitting electrons (left axis of each plot) normalized to the electron temperature and line width *G* of the emission (right axis).

Plots in figure 3 show calculations performed with the code SPECE [21] for the shot 68394 (B_T=3.2T, I_p=2.3MA, $n_e dl$=9.1 $10^{19}m^{-2}$, 16MW NBI + 8MW ICRF). The line width $G= \alpha T_e e^{-\tau}$ of the ECE emission, where α is the absorption coefficient and τ the optical depth is plotted against the left hand *y* scale. Minimum and maximum branches of the resonance energy of the electron population contributing to a given spectral frequency are instead plotted against the right hand scale. Curves in figure 3 are plotted for the three frequencies respectively corresponding to the peak emission of 2nd and 3rd harmonic, X and O polarizations, at the three angle of sight of the Oblique ECE diagnostics. Calculations are done assuming purely Maxwellian energy distribution for the electrons. Examination of the energy curves for the different cases shows that for increasing angles of emission the energy of contributing electrons increases both in value and in range extension. In particular, perpendicular propagation corresponds to nearly monoenergetic electron population (in a given spatial location) in the thermal range for X polarization, 2nd harmonic or slightly higher in the other cases. The line width of emission tends to increase with the angle as well.

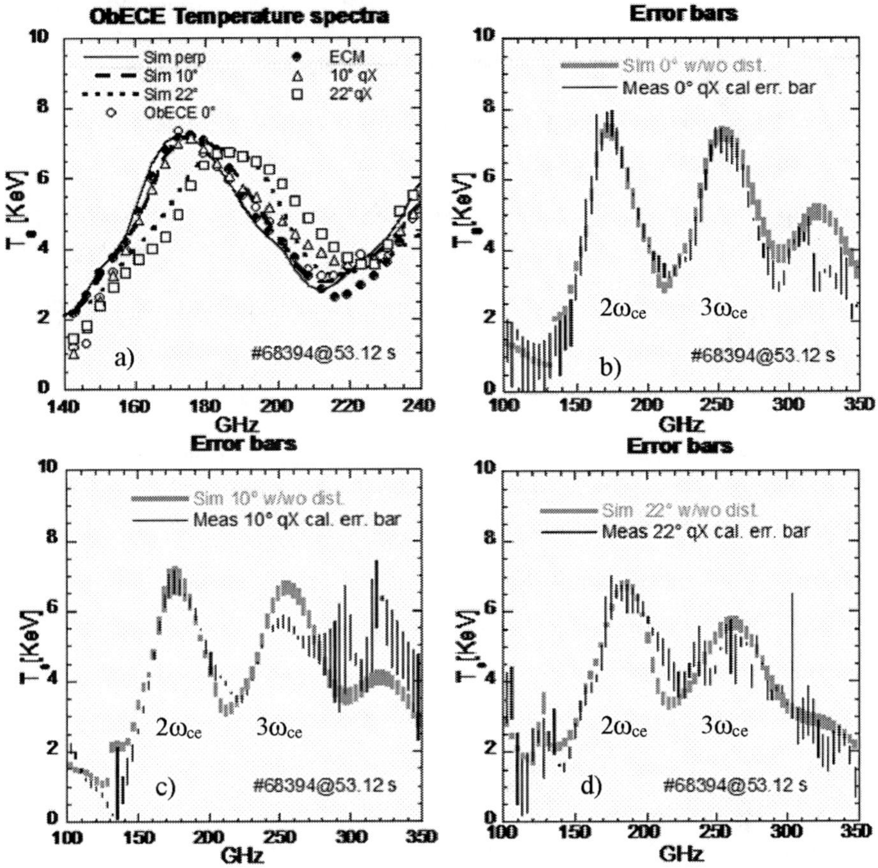

Figure 4. Top left: quasi X-mode, 2nd harmonic perpendicular and oblique measurements of the temperature spectra compared with Maxwellian simulation for the second harmonic peak. Other plots: temperature spectra with error bars due to the calibration procedure compared with simulated spectra for which a small fraction of non maxwellian population is assumed (see text).

In figure 4 the measured quasi X-mode temperature spectra for JET shot 68394 are shown. Spectra in a) show the expected shift in frequency of the peak of emission with angle. Decreasing height of the peak with angle is mainly related with the increasing O fraction, characterized by minor optical depth, reaching the antenna with vertical polarization. The other plots in figure 4 show the measured spectra with the uncertainty introduced by typical calibration errors compared with simulations in which different assumptions on the electron edf are taken. The error bars in the simulated spectra represent the expected variation in the emission if a small fraction of non-maxwellian electrons in the thermal energy range is switched on/off. The distortion here considered has parameters $u_d=\pm1.2$, $T_d=0.23$, $n_d=0.015$, where u_d is the location of the distortion in velocity space, T_d is the temperature and n_d is the fraction of electrons involved. This particular distortion is not necessarily the most probable, however it sets a reference value with respect to the diagnostics sensitivity. Work to reduce the error bar of the measurements and to extend the frequency range with reliable calibration is in progress.

CONCLUSIONS

A brief review of the motivations, techniques and results of the experimental studies on the electron function distribution has been given. The field is being revisited on sight of burning plasmas scenario with special attention to possible effects in the thermal energy range. Compact antenna design and well established theoretical basis enable Oblique ECE as an effective techniques for these studies. A dedicated Oblique ECE system has been installed at JET and first results have been discussed. Feasibility studies of oblique lines of sight in ITER based on JET design are being considered and reported elsewhere [21].

ACKNOWLEDGMENTS

A complete overview of the current and past activities in the field is beyond the scope of this paper. The authors apologize for unintended omissions. The work leading to this article was funded by the European Atomic Energy Community and is subject to the provisions of the European Fusion Development Agreement (EFDA Task Agreement JW2-TA-EP-MWA-036).

REFERENCES

1. S. Bernabei et al., Phys. Rev. Lett. **49**, (1982) 1255
2. S. Coda et al., PPCF **48**, (2006) B359-B369 and references therein
3. V.S. Vlasenkov et al., Nucl. Fusion **13**, (1973), 509
4. V. V. Plyusnin et al., Nucl. Fusion **46**, (2006) 277
5. P. V.Savrukhin, Phys. Rev. Lett. **86**, (2001) 3036
6. G Taylor et al., PPCF **36**, (1994), 523
7. O. Tudisco et al., 28th EPS conf. on Contr. Fusion Plasma Phys., Funchal, Pt,. ECA **25A** (2001) 1221
8. E. de la Luna et al., Rev. Sci. Instr. **74**, (2003) 1414 and V. Krivenski, Fus. Eng. Des. **53**, (2001) 23
9. F. Skiff et al., PPCF **36**, (1994) 1371
10. T.C. Luce at al., Rev. Sci. Instr. **59**, (1988) 1593
11. K. Kato and I.H. Hutchinson, Phys. Fluids **30**, (1987), 3809
12. J. L. Ségui and G. Giruzzi, PPCF **36** (1994) 897-910 and references therein.
13. S. Preische et al., Phys. Plasmas **3** (1996) 4065
14. Y. Peysson et al., Nucl. Instr. Methods **A 458** (2001) 269
15. L.C. Ingesson et al., Nucl. Fusion **38**, (1988) 1675
16. P.Blanchard at al, PPCF **44** (2002), 2231 and JFM van Gelder PPCF **40**, (1998) 1185
17. G.Zhuang et al., PPCF **47**, (2005) 1539
18. U. Gasparino, **EC-8** Proc. of 8th Joint workshop on ECE and ECRH, Gut Ising, Germany, (1992) 19
19. E. de la Luna, this Proceedings
20. S. Garavaglia, A. Simonetto, C. Sozzi et al., Fus. Eng. Des. **82**, (2007) 5
21. D. Farina, L. Figini, P. Platania and C. Sozzi, this Proceedings

Thomson Scattering and Burning Plasmas

M. J. Walsh

EURATOM/UKAEA Fusion Association, Culham Science Centre, Abingdon, Oxon OX14 3DB, UK

Abstract. In low temperature plasmas, tokamaks and other experimental fusion devices, the electron temperature and densities in the plasma can be measured with high accuracy by detecting the effect of Thomson scattering of light from a high-intensity laser fired through the device. It has been established as a robust technique to measure these parameters and in particular it can be used to provide detailed profiles right through the plasma.

Keywords: Thomson Scattering, Incoherent, ITER, Burning Plasma, LIDAR, detectors.
PACS: 42.55.-f, 42.55.Rz, 52.25.Os, 52.55.Fa, 52.70.-m, 85.60.Ha

INTRODUCTION

Thomson scattering is the scattering of electromagnetic radiation by a charged particle. The electric and magnetic components of the incident wave accelerate the particle, which in turn emits radiation and thus, the wave is scattered. This is an important phenomenon in plasma physics and was first explained by J.J.Thomson [1]. It can be split into coherent and incoherent scattering. In this paper, the emphasis will be on incoherent scattering.

The main cause of the acceleration of the particle is the electric field component of the incident laser light. The particle will move in the direction of the oscillating electric field, resulting in a dipole radiation pattern. The moving particle radiates most strongly in a direction perpendicular to its motion and that radiation is polarised along the direction of its motion. Hence, depending on where an observer is located, the light scattered from a small volume element will vary in magnitude.

Thomson scattering is routinely used in plasma physics to measure the electron temperature and density of the plasma. Effectively the velocity distribution is observed directly at a specific spatial location in the target plasma. The technique is quite robust and does not depend on other measurements to validate the data.

Making measurements using Thomson scattering has however several challenges. The small scattering cross-section and the existence of background plasma light requires the use of high-energy, pulsed laser sources. In turn, a high efficiency of light collection followed by a sensitive detection system is required. Finally, to achieve the measurement, calibration plays a critical role.

There exist many options for the realisation of the system. The chosen direction will be driven by experimental requirements, technical capability, engineering constraints, physical access and budget. Systems are usually based on either conventional imaging or the 'point and shoot' LIDAR systems. Specifications for the

CP988, *Burning Plasma Diagnostics*, edited by F. P. Orsitto, G. Gorini, E. Sindoni, and M. Tardocchi
2008 American Institute of Physics 978-0-7354-0507-3/08/$23.00

system will usually include spatial and temporal resolution, expected error and measurement parameter ranges. Technical risk and compromise will usually play a significant role as conflicting requirements may exist.

The current step to ITER is adding new challenges, namely those of very high plasma temperature (strong relativistic effects), difficult access both for design and maintenance and high reliability expectations. In recent years, many measurement systems have moved to the 'turn-key' concept, and Thomson scattering has to follow this trend. The development of more reliable lasers and a modular system approach with inherent redundancy is allowing the provision of very high performance yet highly reliable systems.

BASICS OF THOMSON SCATTERING

The temperature T features in the electron velocity distribution function F(v)=A $\exp(-\frac{1}{2}m_e[v-v_d]^2/kT)$ where v is the velocity in one direction and m_e is the electron mass and k is the Boltzmann constant. The quantity, v_d, is the drift velocity and represents a bias in the distribution. This could be an approximation of the effect of a plasma current for example. Generally v_d is small and can be neglected or can be eliminated by suitable choice of scattering geometry.

To understand the principle of Thomson scattering, a simple classical non-relativistic small-Doppler shift approach is outlined here. Consider a linearly polarised wave with incident wave-vector, $\mathbf{k_I}$ (laser direction) impacting on an electron (see Figure 1) and an observation vector $\mathbf{k_s}$ at an angle θ. This wave is re-emitted in a dipole radiation pattern as shown in Figure 1. The Doppler shift is given by $\Delta\omega=\mathbf{k}.\mathbf{v}$, where $\mathbf{k}$ is the differential scattering vector and $\mathbf{v}$ is the electron velocity vector (assumed unperturbed by the laser). Note that $\mathbf{v}$ is along $\mathbf{k}$. Noting that $\mathbf{k}=\mathbf{k_I}-\mathbf{k_s}$, and assuming that $|\mathbf{k_I}|\sim|\mathbf{k_s}|$ (for small Doppler shifts), it follows that $|\mathbf{k}|=2|\mathbf{k_i}|\sin(\theta/2)$.

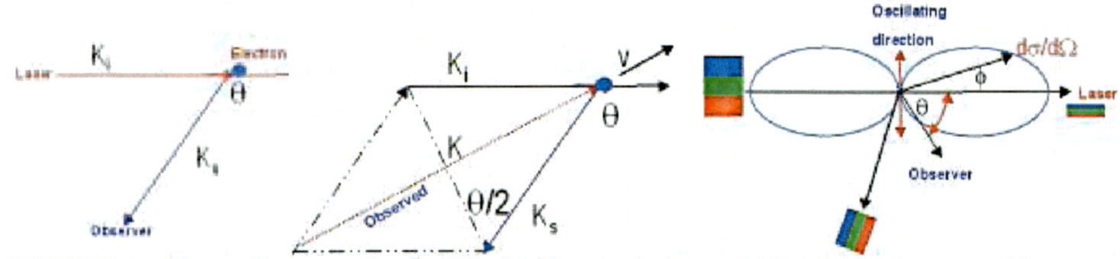

FIGURE 1. Vector diagrams showing the incident and observed electric field vectors. The second figure shows the resultant differential vector, **k** in the direction of the observer. Note the scattering angle, θ, definition. Third figure: from a polarised laser pulse, a dipole radiation pattern is observed. ϕ is the angle between the perpendicular to the laser polarisation vector and the viewing direction. For maximum collection efficiency, ϕ should be 0.

Using the fact that $\Delta\omega/\omega=-\Delta\lambda/\lambda$, the mapping from frequency to wavelength is accomplished. This results in a relationship between wavelength and the velocity observed along **k**, v_k, to be $\Delta\lambda/\lambda_0=2(v_k/c)\sin(\theta/2)$. So the wavelength shift is proportional to λ_0 and a function of the angle θ. This can be seen diagrammatically in the third diagram of Figure 1 above. Conventional imaging scattering is usually based around a scattering angle of 90°, while for LIDAR the scattering geometry is set at 180°.

Using the equation for the velocity distribution, it can now be seen that the velocity distribution and scattered wavelength distributions can be mapped from one to another where the wavelength distribution is now given by $F(\lambda)=B\exp[-\frac{1}{2}\Delta\lambda^2/\sigma^2]$ where σ, the width of the spectrum is given by $\sigma^2=[2\lambda_0\sin(\theta/2)]^2\ kT/m_ec^2$. Hence the temperature is proportional to the width of the spectrum squared and strongly dependent on the scattering angle chosen.

So far only individual electrons have been considered. In a plasma, there are electrons and ions and the Debye length λ_d, is the distance over which significant charge separation can occur. In some cases, the differential scattering vector is arranged so that those individual independent electrons are observed and multiple scattering events are not related. However, in some cases the individual events are well correlated so that the resultant scattering is representative of the ions and not the electrons. These regimes are differentiated by the terms coherent (sometimes co-operative) and incoherent scattering. The boundary between coherent and incoherent scattering can be identified by use of the α parameter defined as $\alpha=1/|\mathbf{k}|\lambda_d$, where $\alpha<<1$ represents the incoherent and $\alpha>>1$ represents coherent scattering. From above, this can be expanded as $\alpha = \lambda_0/[4\pi\sin(\theta/2)\ \lambda_d]$. For a typical JET plasma [2], it can be shown that the Debye length is of the order of $\lambda_d\sim70\mu m$. Using a scattering angle of $\theta=180°$ and a laser wavelength of $\lambda_0\sim0.7\mu m$, it is clear that the scattering is easily in the incoherent regime. Going to small angle or long wavelength is usually required to achieve the coherent scattering regime [3,4,5].

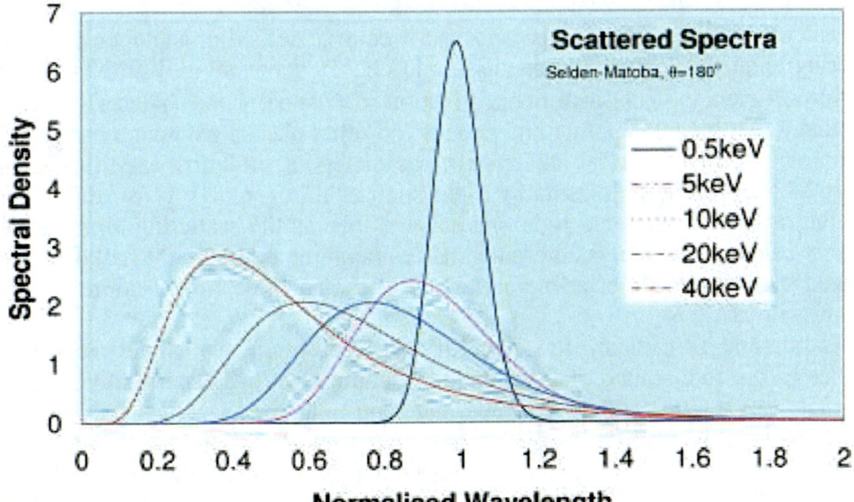

FIGURE 2. Spectral intensity versus normalised wavelength for a scattering angle of 180° showing different plasma temperatures.

While a classical model of the scattering system is good for illustration, this is not appropriate for hot plasmas and a more accurate analysis is required. The details of this are outside the scope of this paper [4,6,7,8]. In this case, the scattered electric fields can be related to the velocity and acceleration of the particle using the Lienard-Wiechert potentials. This is then used as the basis for the mapping of the data, and relativistic, transit time, and other relevant effects are inherently included. Figure 2 shows the theoretical spectral intensity curves for a range of temperatures at a scattering angle of 180° (back-scattering, as in LIDAR). The off axis peaking on the 'blue-side' or short wavelength side is clearly observed. This is a consequence of the more intense emission from electrons travelling fast towards the observer (sometimes known as the headlight effect). It can be seen here, that significant deviation from the non-relativistic treatment exists even at the lowest temperatures. This description also includes depolarisation which is essentially a change of orientation of the incident E-vector as seen by the electron [8]. Evaluation of these curves is straightforward with the use of a computer. The asymmetry of the spectrum can be an advantage to distinguish higher temperatures in the fitting process.

To measure the electron temperature and density in the plasma, the scattered distributions have to be detected and analysed. The scattered power spectrum is divided up in to a sufficient number of channels (typically 4-7) and this data can be analysed to produce a measurement of both quantities.

The number of scattered photoelectrons in any channel, i, can be calculated by:

$$NP^{*}_{TS,i}=N_{laser}n_e\ [d\sigma/d\Omega]_{TS}\ \Delta\Omega\ \Delta L\ T\ [QE/NF^2]\int\Phi_i(\lambda)\ F(\lambda)\ \lambda/\ \lambda_0\ d\lambda$$

where N_{laser}-number of laser photons incident, n_e-electron density, $[d\sigma/d\Omega]_{TS}$-differential scattering cross-section (or classical electron radius, r_e^2 for viewing perpendicular to the incident polarisation vector), $\Delta\Omega$-solid angle collected, ΔL-scattering length collected, T-transmission at chosen wavelength, $[QE/NF^2]$-effective quantum efficiency, $\Phi_i(\lambda)$-instrument function of channel I and λ_0-laser wavelength. Note that to get the temperature and density, no other plasma parameters are required. To eliminate the geometrical and system parameters, a similarity experiment can be performed. Replacing the plasma by a gas such as nitrogen and measuring either the Rayleigh or Raman scattered light can achieve this, as the scattering cross-section is precisely calculable if the gas pressure and temperature are known [9,10]. Usually the scattered spectrum is either a line or several lines and these can be approximated by Dirac functions.

It is possible to estimate the error in each measurement channel from Poisson statistics as the root square of the number of photo-electrons. The relative error then follows as $\sqrt{NP^*}/ NP^*= 1/\sqrt{NP^*}$. Plasmas also emit light due to other processes and this acts to increase the error in each channel. This background light is typically of the same order as the scattered light – if it were equal, it would increase the error in each channel by $\sqrt{2}$ (ignoring the detector noise). In typical laboratory plasmas, light from lines or continuum can play a significant role in the quality of the measurements produced. For this reason, most systems utilise lasers that have very short pulses e.g. less than 20ns, to minimise the number of background photons collected for the same number of laser photons, as it is the laser pulse energy rather than the power that is prescribed for any given laser.

THOMSON SCATTERING SYSTEMS

The realisation of a Thomson scattering system in the laboratory plasma brings with it many challenges. Over the years [11], developments in detectors and lasers have allowed significant advances in the field. Nowadays, it is common place to have high spatial resolution (from 30 to 60 points across the minor radius) systems [12,13,14,15,16,17]. This list is not exhaustive. Thomson scattering is also used in low temperature plasma physics. The typical challenges in the current tokamaks are often budget, lack of sensitivity, spatial resolution and time repetition. Typical implementations carve a careful path to provide a system that matches the physics requirements of the specific device.

Imaging Systems

In this case, the scattered light is imaged from a point on the laser beam as it goes through the plasma through a spectrometer and ultimately on to a set of detectors. Figure 3 shows a representation of this. This example is from the MAST Tokamak [13,18]. In this case, there are two different types of lasers. One set is outside the neutron shield while the other is inside. A collection lens directs the scattered light from the tokamak plasma onto fibre bundles and away to an adjoining area which houses the detectors and data acquisition. It is usual to keep a significant part of the system outside the tokamak area to allow easier maintenance of critical components. It can also be seen that the laser system while presently using 4 lasers has been designed to allow 8 lasers. This allows flexibility for future enhancements.

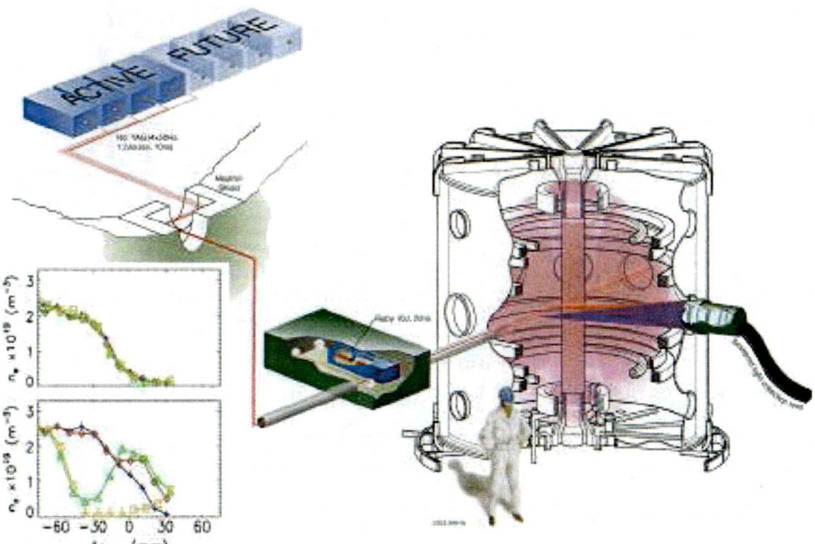

FIGURE 3. Example of a conventional scattering setup [13]. The data [18] inserted represents the electron density measured close to the last closed flux surface between ELMs (edge localised modes) and during an ELM (bottom). A set of 4 profiles is shown and the timing between the individual lasers was set to 5μs. The time sequence is blue, red, green, orange.

Non-Imaging or LIDAR Systems

LIDAR is an acronym for LIght Detection And Ranging. It is a 'point and shoot' method (input and collected light approximately collinear) and hence requires minimal access to the plasma. In this case, a very short burst of laser light is transmitted to the plasma and the back-scattered light is observed and analysed in both time and wavelength. The time of flight resolution gives the spatial information. This is illustrated in Figure 4 where 3 snapshots in time are chosen. The first shows the laser entering the plasma. Until this point no scattering takes place. Once the laser enters the plasma, the scattering begins and the scattered light that is going to be observed travels back along the direction of the laser beam at the same rate as the laser is traversing the plasma (middle graph). In the last part, the laser beam has traversed the full plasma and it can now be seen that while the laser traversed a length of L through the plasma, the scattered light is spread over a distance of 2L. Hence, the width in time of the light is given by dt=2L/c or 15cm/ns. It can be seen that to achieve a spatial resolution of 7cm, (ITER requirement, see Table 2) a time resolution of better than 460ps is required. This will be the combination of the detector and the laser response width times. By assuming that the laser response time is approximately matched to the detector response time, and the responses are Gaussian, a detector and laser pulse width of about 325ps is required. This value is lowered somewhat due to the fact that neither the laser nor the detector will be exactly Gaussian.

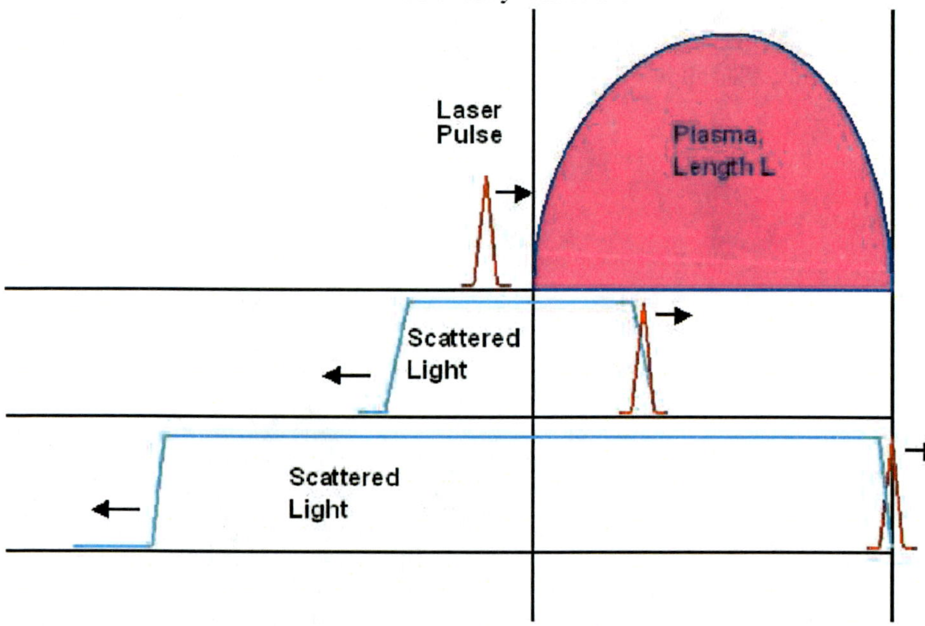

FIGURE 4. The LIDAR concept is illustrated above. Each of the diagrams shows a different position of the laser pulse as it traverses through the plasma.

An example of a LIDAR system (JET) is shown in Figure 5 [19]. In this figure, it can be seen that the lasers and detectors are outside the radiation shield. The light is transmitted to the tokamak approximately 30 metres from the laser room and the scattered light is collected back along the same path.

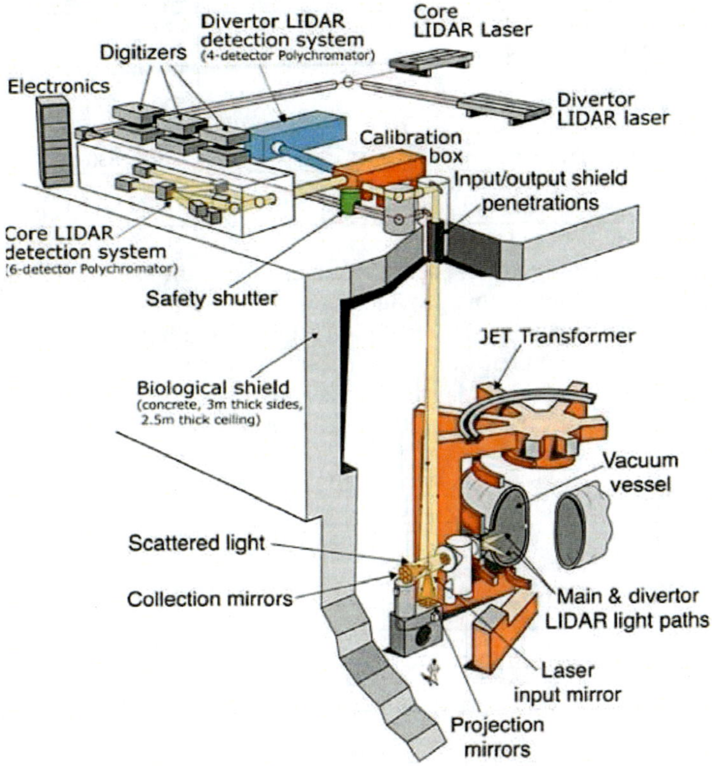

FIGURE 5. Example of the JET LIDAR system as outlined in [2].

ITER CORE THOMSON SCATTERING SYSTEM

For a burning plasma experiment like ITER, several new challenges arise, in particular, neutrons, material activation and long plasma discharges are present as can be seen from the typical ITER parameters shown in Table 1. Close to the tokamak, the access to the device will be infrequent and hence the design of the system must be for long term, low maintenance. The point and shoot principle of the LIDAR has several advantages here in particular as it allows the whole plasma cross-section to be observed from a single port. The high temperatures expected in ITER mean that the spectral broadening using back scattering is significant and due consideration has to be given to the issue of first mirror longevity since it is the short wavelength reflectivity that is affected first. Both passive and active cleaning and protection methods may be needed [21,22]. Imaging methods can also be used although they are better suited to non-core observation due to the difficulty of combining wide-angle viewing with high throughput.

TABLE 1. Typical ITER Parameters [20].

Parameter	Quantity
Plasma Current	15(17)MA
R (major radius), a(minor radius)	6.2m, 2m
Maximum shot length	4200 seconds
Peak burn duty cycle	25%
Toroidal field	5.3 Tesla
Heating power	73MW
Max power output	500 (700)MW
Q	10
Plasma surface area, Volume	$680m^2$, $830m^3$
Av neutron wall load to first wall	$0.56MW/m^2$
Integrated Full power lifetime	4700 (7800) hours

The specifications for the ITER core system are shown in Table 2. Compared to the present day LIDAR systems, this provides several new challenges. These include the higher temperatures, higher spatial resolution required and high time repetition rate of the observations. An example of an ITER operating scenario is shown in Figure 6. In this case, the expected temperature and density profiles can be seen. The short line inside the circle illustrates the scattering length of a/30.

TABLE 2. Requirements for ITER Core LIDAR (C.01) Thomson Scattering Systems.

Target requirements	Quantity
Electron Temperature range	0.5-40keV
Electron Temperature accuracy	10%
Electron Density range	$3x10^{19}$-$3x10^{20}m^{-3}$
Electron Density accuracy	5%
Spatial Extent	-0.9r/a to 0.9r/a
Spatial extent on midplane	a/30~7cm
Time repetition	100Hz(10ms)

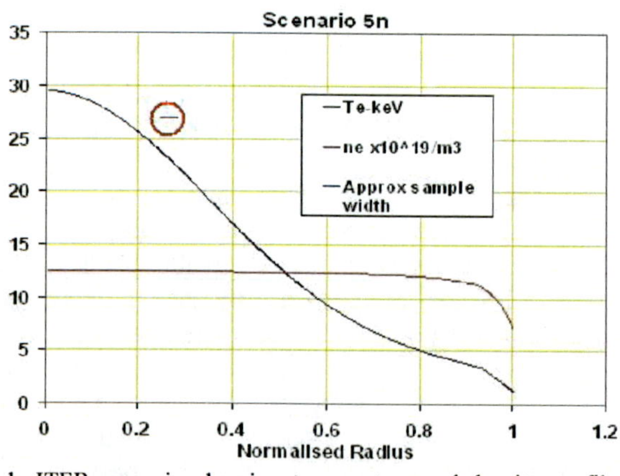

FIGURE 6. Example ITER scenario showing temperature and density profiles. The line inside the circle is indicative of the scattering length.

To realise the requirements above, an iterative design cycle including several disciplines is required. Once the physics requirements are defined, optical design, neutron shielding design, activation design, mechanical design, reliability, accessibility and performance need to be continually iterated. Figure 7 shows an example of what the core LIDAR system might look like in the vicinity of the vacuum vessel at ITER. This design is still undergoing development. The collection mirror labyrinth and laser flight path is clearly seen. Minimising activation in the area behind the port plug is important for future maintenance to allow relatively easy access to the replaceable components. Vacuum interfaces are a particular area of vulnerability and so the entrance of the laser beam has to be made sufficiently large to avoid excessive power densities.

There is a very small choice of suitable lasers for Thomson scattering in general and the specifications in Table 2 narrow this down further. Presently a 5J, 15Hz, 250ps Nd:YAG laser is the default specification. This has a first harmonic wavelength at 1.06μm. To achieve the 100Hz repetition rate, 7 lasers are proposed. The advantage of the multi-laser approach is that it provides redundancy in the case of a laser malfunctioning during the plasma discharges, as well as the option of operating the system in "burst mode" (cf fig 3).

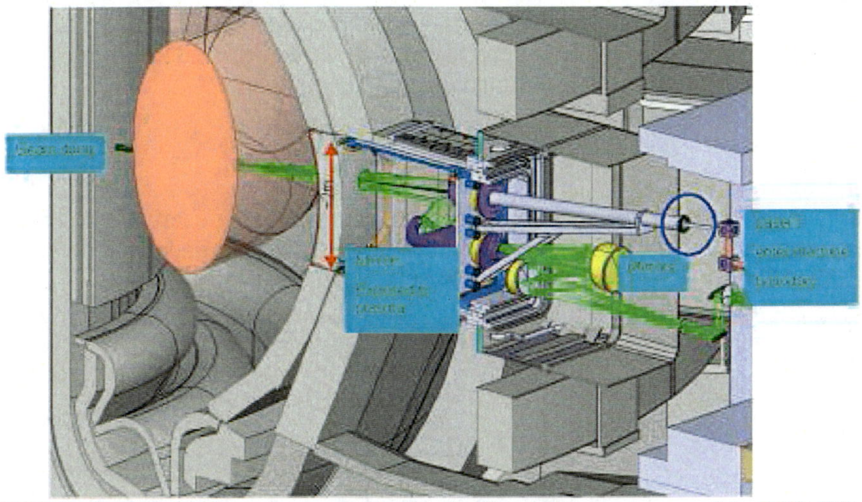

FIGURE 7. Design showing components in the vicinity of the vacuum vessel in ITER [21]. The collection mirrors and the laser launch concepts are clearly visible. This figure only shows one diagnostic in the port plug, however others will also exist.

The spatial resolution required is such that it is beyond current technology in some wavelength regions due to the need for very fast detectors with high sensitivity. Typically wavelengths below 750nm can be detected with suitable sensitivity and response time e.g. GaAsP and S20 photo-cathodes with a micro-channel plate electron multipliers. An example of a typical response of one of these detectors is shown in Figure 8 below. Between 750nm and 1100nm, detector variants exist but further development is required before they are usable on a LIDAR system. These photo-cathodes are based on GaAs and InGaAs [23,24,25].

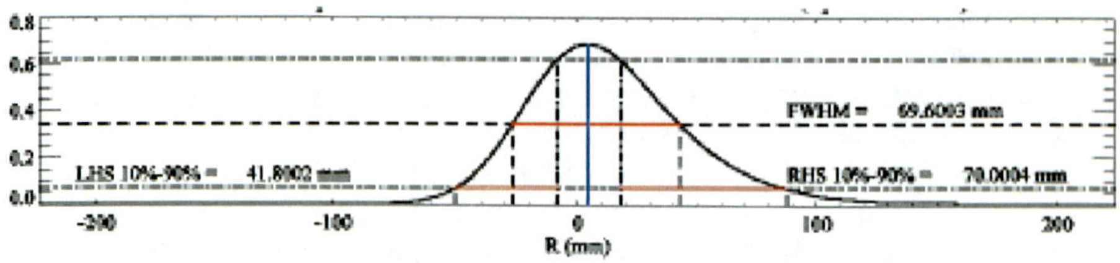

FIGURE 8. Response of an 11mm diameter, MCP-PMT GaAsP detector convolved with a laser pulse response of 300ps. The response is mapped to the space domain and has a FWHM of just below 7cm.

SUMMARY

Thomson scattering is a technique that has been used extensively in plasma physics as a reliable and robust diagnostic technique for measuring plasma temperature and density absolutely. The arrival of the burning plasma tokamaks provides new challenges that will need to be overcome. Some of these challenges are already being tackled for ITER and some will require further development. Looking ahead to DEMO, new ways to acquire measurements such as time integrated sampling, that are both diagnostic and plasma device compatible, will be required.

ACKNOWLEDGMENTS

This work was funded jointly by the United Kingdom Engineering and Physical Sciences Research Council and by the European Communities under the contract of Association between EURATOM and UKAEA. The views and opinions expressed herein do not necessarily reflect those of the European Commission. The ITER design work was carried out within the framework of the European Fusion Development Agreement.

REFERENCES

1. J. J. Thomson, Conduction of electricity through gases, Cambridge University Press, 1906
2. C. W. Gowers, B. W. Brown, H. Fajemirokun, P. Neilsen et al, RSI, Vol.66,1,Jan. 1995, pp.471.
3. A. W. DeSilva*, D. E. Evans & M. J. Forrest, *Nature* **203**, 1321 - 1322 (26 September 1964).
4. I. H. Hutchinson, Principles of Plasma Diagnostics, Cambridge University Press; 2 edition (2005).
5. H. Bindslev, J. A. Hoekzema *et al*, Phys. Rev. Lett. 83, 3206 - 3209 (1999).
6. O. Naito et al, Phys. Fluids B 5(11), November.1993, P.4256,[erratum, Phys. Plasmas 1(3), March, 1994, P.806].
7. A. C. Selden , Physics Letters A, Volume 79, Issue 5-6, p. 405-406, 1980.
8. A. C. Selden, (UKAEA reports, CLM-R220), 1982, ISBN: 0-85311-102-2:AEA Technology.
9. J. Howard, B. W. James and W. I. B. Smith, J. Phys. D: Appl. Phys., Vol. 12, 1979, pp1435-1440.
10. S. C. McCool, I. L. McCool *et al*, "Calibration of Thomson Scattering Density Measurements", The University of Texas, Fusion Research Center, FRCR #234
11. N. J. Peacock, D. C. Robinson, M. J. Forrest et al, *Nature* **224**, 488-490 (November 1969).
12. T. N. Carlstrom *et al*, Rev. Sci. Instrum, No. 10, Vol.63, October 1992, p.4901-4906.
13. M. J. Walsh et al, Rev. Sci. Instrum, No. 10, Vol.75, Part II, October 2004, p.3909-3911.
14. B. Kurzan et al, Plasma Physics and Controlled Fusion, Vol. 46, No 1, January 2004, p.299-317.

15. T.Hatae *et al*, Rev. Sci. Instrum. **70** 772-775 (1999).
16. R. Pasqualotto *et al*, RSI, October 2004, Volume 75, Issue 10, pp. 3891-3893.
17. C. J. Barth *et al*, Rev. of Sci. Instrum, January 2001, Volume 72, Issue 1, pp. 1138-1142
18. R Scannell *et al* 2007 Plasma Phys. Control. Fusion 49 1431-1446, July 2007.
19. H. Salzmann *et al*, Nuclear Fusion vol.27 no.11 November 1987 pp.1925-1928.
20. J.How, Editor:Project Integration Document, January 2007, ITER Organisation, Cadarache, France.
21. M. J. Walsh *et al*, RSI, No. 10, Vol.77, Part II, October 2006, p.10E524-1.
22. G. Razdobarin, M. Walsh *et al*, IAEA 2006, Fusion Energy 2006. Proceedings of the 21st IAEA Conference, Chengdu, China, 16-21 October 2006, IT/P1-25 (IAEA, 2007).
23. M. Kempenaars et al, Rev. Sci. Instrum, No. 10, Vol.75, Part II, October 2004, p.3891-3894.
24. M. Walsh, 'ITER Core LIDAR Thomson Scattering', Proc. 13th Int. Symp. On Laser-Aided Plasma Diagnostics, Takayama (2007).
25. L. Giudicotti, R. Pasqualotto *et al*, "Study of Fast, Near-Infrared Photodetectors for the ITER Core LIDAR Thomson Scattering", Burning Plasmas Workshop, Varenna 2007.

Advanced Interferometry Techniques for Burning Plasmas

D.L. Brower, W.X. Ding

Department of Physics
University of California at Los Angeles
Los Angeles, California 90095-1547 USA

V.V. Mirnov

Department of Physics
University of Wisconsin-Madison
Madison, Wisconsin 53706 USA

M.A. Van Zeeland, T.N. Carlstrom

General Atomics
San Diego, California 92121-1122 USA

Abstract. For future burning plasma experiments, all diagnostics must be re-evaluated in terms of their measurement capabilities and robustness to the new environment. This is certainly true for interferometry measurements where conventional approaches may not be ideal and interpretation may require modification due to high plasma temperatures. Optimizing these systems to provide maximum information will be crucial to understanding burning plasma dynamics. This paper will explore a variety of phase measurement techniques for the main body and divertor regions that can be utilized on burning plasma experiments like ITER and beyond.

Keywords: interferometry, polarimetry, Faraday rotation, diagnostics
PACS: 52.55.Fa, 52.70.Gw, 52.70.Nc

INTRODUCTION

For future burning plasma experiments, all diagnostic systems need to be re-evaluated in terms of their measurement capabilities and robustness to the new environment. This is particularly true for interferometry measurements where conventional approaches may not be ideal and interpretation may require modification. On ITER, it is presently planned to have three interferometer systems playing multiple measurement roles. First, a tangential interferometer-polarimeter is required primarily to measure the line-integrated electron density for fueling control through realtime density feedback. Second, a poloidally-viewing interferometry-polarimetry system is desired to determine the internal magnetic filed structure and current density for evaluation of plasma safety factor profile, q, and its temporal evolution. Both of the above mentioned systems also provide internal density and magnetic constraints for

CP988, *Burning Plasma Diagnostics*, edited by F. P. Orsitto, G. Gorini, E. Sindoni, and M. Tardocchi

equilibrium reconstruction. Third, a divertor interferometer is planned to measure the electron density distribution along each divertor leg and to monitor the strike point. Each of these systems will need high time response in order to track fast dynamic events associated with ELMs, pellets, and disruptions. In addition, at least for the core plasma systems, it is hoped that these instruments can also measure fluctuations associated with magnetohydrodynamic (MHD) instabilities like tearing modes (or NTMs), Alfvenic modes, and even broadband turbulence. Optimizing each of these systems to provide maximum information will be crucial to understanding burning plasma dynamics. This is particularly critical in view that many standard diagnostic techniques presently employed in high-performance plasmas may no longer be feasible in the burning plasma environment.

A burning plasma represents an unknown and certainly difficult environment in which diagnostic systems need to reliably operate. The environment is extremely harsh as high energy neutrons and high levels of radiation impact various diagnostic components. Erosion of and deposition on optical components, plasma dust accumulation, and radiation-induced EMFs are a few of the issues that must be addressed. In addition, technology issues are critical. For example, realtime feedback alignment of optical systems is required for the up to 1000 sec discharge pulse lengths in ITER where thermal expansion can be of order several centimeters. Diagnostic integration into the overall machine design and port structure also represent serious issues that must be dealt with. Finally, temperatures in burning plasmas will be in the range 10-25 keV where measurements have not previously been made and assumptions regarding the plasma dielectric response (e.g., cold plasma approximations) may no longer be valid. This will affect interpretation of experimental results and the ability of instruments to meet the specified measurement requirements.

The scope of this work does not allow us to address all of these issues. Herein, we will focus the discussion on a variety of phase measurement techniques for the main body and divertor regions that can be utilized on burning plasma experiments like ITER and beyond. These techniques will go beyond those typically considered for interferometry diagnostics. In addition, the consequences of finite electron temperature on the plasma dielectric response will be considered by evaluating the contributions of both dispersive and relativistic effects. Finally, we will address expansion of the diagnostic capabilities to measure new equilibrium and fluctuating quantities, including fluctuations and the fluctuation-induced fluxes.

INTERFEROMETRY TECHNIQUES

In this section we will briefly examine various interferometry techniques and their applicability to burning plasma experiments like ITER. In the context of this paper, a broad definition of interferometry is utilized whereby we refer to any phase measurement as an interferometry technique. This class of measurement then includes conventional interferometry (and variations like dispersion and differential interferometers), Faraday rotation, Cotton-Mouton effect, and Fizeau interferometry.

Conventional Interferometry

In the conventional interferometer setup (see Fig. 1), one uses probe (Laser1) and local oscillator (LO) beams (Laser 3) that mix in both the signal and reference detectors. Since the linearly-polarized laser beams are at difference frequency $\Delta\omega = \omega_1 - \omega_3$ (where $\omega >> \omega_{pe}$ *and* ω_{ce}), the output of each detector is a sinusoid waveform at $\Delta\omega/2\pi \sim 1$ MHz. By measuring the phase difference between the waveforms from the two mixers, one can determine the plasma induced phase shift due to the fact that the beam path for the signal mixer traverses the plasma. The resulting phase measurement ϕ can then be related to the plasma density according to the relation $\phi(rad) = 2.8\times10^{-15}\lambda\int n_e dL$, where λ is the laser wavelength, n_e the electron density and dL the path length through the plasma (mks units). In the above relation, it is assumed that (1) magnetic field can be ignored, (2) the plasma is stationary, (3) and temperature effects are unimportant.[1]

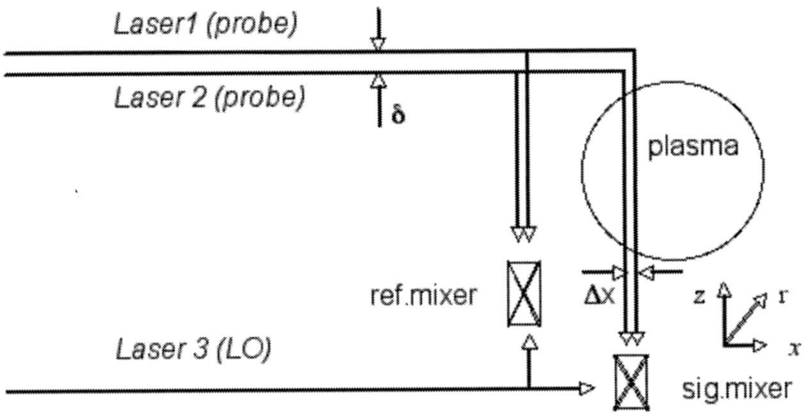

FIGURE 1. Experimental layout for various interferometer systems.

However, if path length changes (such as those due to thermal expansion or vibrations) do occur, an extra phase term that goes as $\Delta L/\lambda$ must be included. If this term is large, additional information is required to resolve the plasma density. This can by accomplished by adding an additional probe beam (Laser2 in Fig. 1) with spatial offset δ=0. If one now rewrites the measured phase for each probe beam, we end up with a system of 2 equations and 2 unknowns,

$$\phi_{\lambda_1} = c_I\lambda_1\int n_e dL + \frac{2\pi\Delta L}{\lambda_1} \quad and \quad \phi_{\lambda_2} = c_I\lambda_2\int n_e dL + \frac{2\pi\Delta L}{\lambda_2},$$

and can solve for the line-averaged plasma density.[2] This approach is commonly referred to as 2-color interferometry. The two laser wavelengths are typically chosen to be far apart (one more sensitive to plasma and the other to path length) as this optimizes the density resolution which can be written as

$$\delta n_e L = \frac{\delta\phi}{r_e(\lambda_1 - \lambda_2)},$$

where r_e is the classical electron radius, $\delta\phi$ is the diagnostic phase resolution and δn_e is the resulting density resolution. This approach is commonly used on high performance tokamak devices. A drawback of this approach is that often both plasma and the path length change terms are much greater than a fringe (>>2π phase change) making the system vulnerable to fringe skip errors and hence potentially unreliable for realtime density feedback control of burning plasmas. Various post-signal processing approaches[3] can be employed to minimize this error but it is still desirable to develop a more robust technique.

Dispersion Interferometer: One way the phase ambiguity associated with fringe counting errors can be remedied is by employing the dispersion interferometer technique.[4] This approach is a variation of the 2-color interferometer approach where Laser1 (Fig. 1) passes through a frequency doubler and converts part (depending on efficiency) of ω_1 into $2\omega_1(=\omega_2)$. These two beams are collinear (δ=0) and both traverse the plasma. After leaving the plasma, there is a second frequency doubler which again converts part of the remaining ω_1 into $2\omega_1(=\omega_2)$. The frequency doubler also doubles the phase making the phase difference between the 2 probing beams equivalent to $\Delta\phi = 2\phi_{\omega_1} - \phi_{2\omega_1} = \frac{3\lambda}{2} c_I \int n_e dL$. When phase is doubled, the path length change term is identical for each frequency and cancels out before the phase measurement leaving only the plasma term. This technique was successfully tested on the TEXTOR tokamak providing a line-integrated resolution of $\delta n_e dL \approx 2\times10^{17}\ m^{-2}$.[5] Since the path length change term is often the source of fringe counting errors, being many hundreds of fringes, this approach significantly improves system reliability. The plasma-induced phase change can still be >>1 fringe and a source of fringe counting error. However, in the ITER divertor, for most plasma scenarios envisioned the plasma-induced phase is <1 fringe (for 10.6 μm interferometer) making the dispersion interferometer immune to fringe counting errors even if the signal is temporarily lost.[6]

Differential Interferometer: Another approach to removing fringe counting errors is the differential interferometer.[7] From conventional interferometry, the density profile is found by performing a standard Abel inversion using

$$n_e(r) = -\frac{1}{\pi r_e \lambda} \int_r^a \frac{\partial\phi(x)}{\partial x} \frac{dx}{\sqrt{x^2 - r^2}},$$

where x is the impact parameter (cylindrical geometry). From this relation it is clear that the first spatial derivative of the phase, $\partial\phi/\partial x$, is required to obtain $n_e(r)$. Conventional interferometers measure $\phi(x)$ for multiple discrete chords, make a numerical fit to the available points, take the spatial derivative to infer $\partial\phi/\partial x$, and then perform an inversion. A difficulty in directly measuring $\partial\phi/\partial x$ is that the phase difference becomes smaller as two adjacent chords are brought closer together. However, phase noise due to vibrations and other effects can be minimized if both probe beams use the same optics. By taking advantage of the reduced phase noise, measurement of small phase difference becomes feasible. This approach was used to to directly measure $\partial\phi/\partial x$ at multiple spatial points and then invert to obtain the local density profile. Two separate laser beams (Laser1 and Laser2 in Fig. 1) with slight spatial offset δ and frequency difference Δω are coupled into a single mixer making a heterodyne measurement of the phase difference which is <1% of the total phase change experienced by each beam separately. This measure of the differential phase is made at multiple spatial points and can be inverted directly to provide the local density distribution.

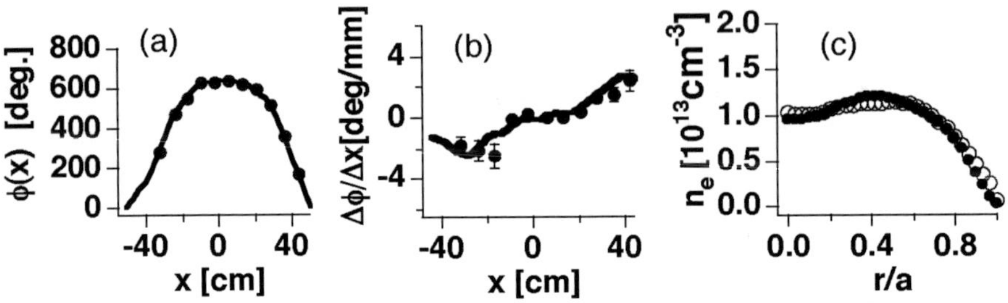

FIGURE 2. (a) Measured phase profile φ(x) for conventional interferometer (solid circles); Solid line is spline fit to measured points; (b) First derivative of phase, dφ/dx, is calculated and shown as solid line. Direct measurements by differential interferometry at the same chord locations as (a) are shown by solid circles for δ=1.2 mm; (c) Inverted density profile comparison using the conventional (open symbols) and differential (solid symbols) interferometer techniques.

The phase gradient $\Delta\phi/\Delta x$ from a differential interferometer measurement is shown in Fig. 2(b). The width of each Gaussian laser beam is ~10 mm and the spatial offset Δx(=δ)~1.2 mm. This parameter can also be obtained numerically by fitting φ(x) (in Fig. 2(a)) and taking the derivative as plotted in Fig. 2(b). Both give consistent results within experimental errors. For purposes of comparison, the density profiles obtained using the conventional interferometer and differential interferometer measurements are shown in Fig. 2(c). The agreement is not surprising since the two interferometric techniques produced the same $\Delta\phi/\Delta x$ profile. The advantage of the differential technique is that while the phase $\phi > 1000^{\circ}$, the differential phase $<4^{\circ}$ for all chords (even the steep gradient regions). Consequently, while the conventional interferometer is subject to fringe skips, the differential is not.

Faraday Rotation

Faraday rotation results from birefringence along the field in a magnetized plasma and can be expressed as the phase difference between R- (right hand circularly polarized) and L-wave (left hand circularly polarized) refractive indices according to

$$\psi_{Faraday} = \frac{1}{2}\int (k_L - k_R)dL = \int \frac{\omega}{2c}(N_L - N_R)dL \approx \int \frac{\omega_{pe}^2 \omega_{ce}\cos\theta}{2c\omega^2}dL \sim 2.62\times10^{-13}\lambda^2 \int n_e B_{//} dL,$$

where $B_{//}=B\cos\theta$ is the component of the magnetic field along the laser beam. (Referred to as Faraday rotation since an initially linear wave would have its polarization axis rotated according to the above relation.) Measurements can be realized schematically as shown in Fig. 1, where Laser1 and Laser2 have offset $\delta=0$ with R- and L- wave polarizations, respectively. Since the two beams are collinear, path length changes and vibration effects automatically cancel since the two beams have wavelength difference $\Delta\lambda/\lambda \approx 10^{-6}$ (i.e., frequency offset only for heterodyne detection). Plasma phase shifts for wavelengths of interest are $<<2\pi$, thereby making such measurements immune to fringe counting errors. For a poloidally-viewing system, Faraday rotation measurements provide information on the poloidal magnetic field and toroidal current density which is critical to evaluating the q profile in burning plasmas.[8] Simultaneous measurement of electron density can be made along the same chords.[9]

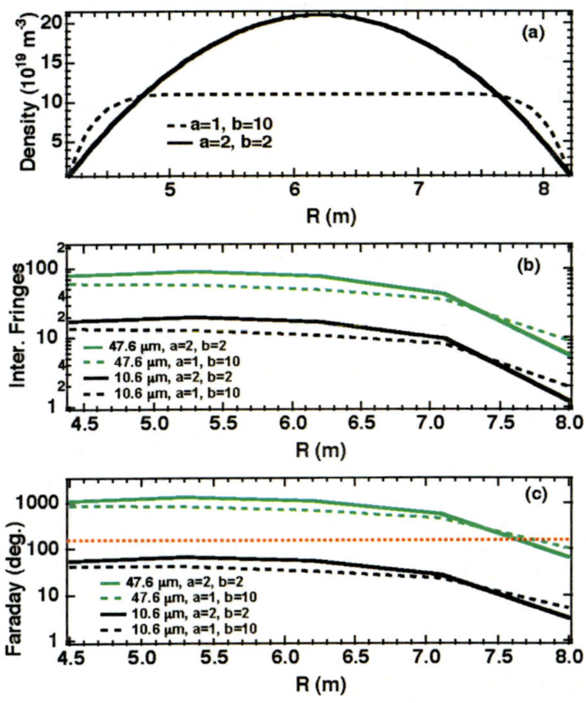

FIGURE 3. (a) Density profiles as a function of major radius for peaked profile (solid line) at ~2xn_{GW} (*a=2,b=2*) and flat profile (dashed line) at ~n_{GW} (*a=1, b=10*). Double-pass tangential (b) interferometer and (c) Faraday rotation phase [x2] as a function of tangency radius for 47.6 (green) and 10.6 μm (black) laser radiation. Horizontal red dashed line corresponds to a single fringe in (c).

For a tangentially-viewing system, Faraday rotation can be used to determine the electron density as the toroidal magnetic field is largely known for the tokamak configuration.[10] This approach has been selected for ITER in combination with a 2-color interferometer. To provide an estimate of the anticipated phase shifts, we take a density profile described by $n_e = a \times 10^{20}[1-(r/a_o)^b]\ m^{-3}$, where for simplicity we employ a circular plasma cross section with minor radius a_o = 2 m and major radius R_o = 6.2 m. The resulting density for cases of high-density peaked [a=2,b=2] and low-density flat [a=1,b=10] profiles are shown in Fig. 3(a). For ITER, the Greenwald density limit is n_{GW}~1.2 x 10^{20} m^{-3}, making steady-state operation at densities above 2-3xn_{GW} unlikely. However, during disruption mitigation phases of operation the density may reach as high as 10^{22} m^{-3}. The calculated interferometer and Faraday rotation phase shifts for laser radiation at 10.6 and 47.6 μm are shown in Fig. 3(b) and (c). Here we immediately see that the plasma-induced interferometer phase shifts (not including path length changes) are many fringes at each wavelength. For Faraday rotation, only the 10.6 μm choice gives phase change <2π for both density scenarios. This makes a 10.6 μm laser source the likely choice for ITER as it will be better suited to providing error-free operation.

Cotton-Mouton Effect

The Cotton-Mouton effect results from birefringence perpendicular to the field in a magnetized plasma and can be expressed as the phase difference in the O- (ordinary) and X-wave (extraordinary) refractive indices according to the relation

$$\psi_{CM} = \int (k_O - k_X)dL = \int \frac{\omega}{c}(N_O - N_X)dL \approx \int \frac{\omega_{pe}^2 \omega_{ce}^2}{2c\omega^3} dL \sim 2.45\times10^{-11} \lambda^3 \int n_e B_\perp^2 dL \quad ,$$

where $B_\perp$ is the component of the magnetic field perpendicular to the laser beam propagation direction. (The Cotton-Mouton effect changes an initially linear wave to elliptical polarization.) Measurements can be realized according to Fig. 1 where Laser1 and Laser2 have offset δ=0 with linear but perpendicular wave polarizations. Since the two beams are collinear, path length changes and vibration effects automatically cancel the same as for Faraday rotation. Typically plasma phase shifts, for wavelengths of interest, are <<2π, thereby making such measurements immune to fringe counting errors. For ITER, a Cotton-Mouton effect polarimeter is under consideration for the divertor legs as $B_\perp$ is the known toroidal field.[6] Estimates of anticipated phase shifts are shown in Fig. 4. Longer wavelength is necessary to obtain a sufficiently large phase shift. This measurement has been successfully demonstrated on the WVII-AS stellarator.[11]

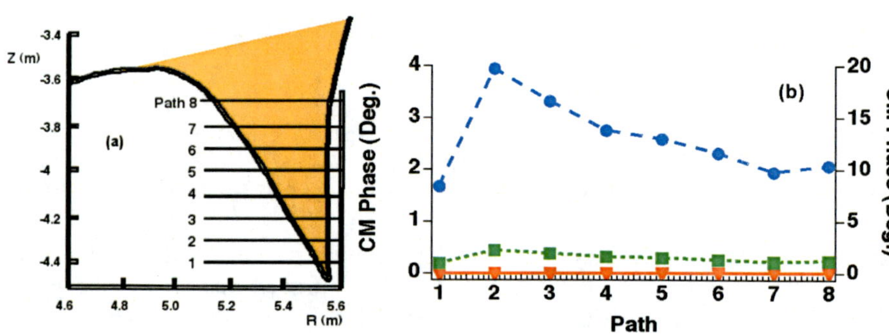

FIGURE 4. (a) Lines-of-sight for ITER outer divertor leg, (b) Estimate of Cotton-Mouton phase for λ= 118 (circles), 57 (squares) and 10.6 (triangles) μm radiation for baseline case -1.8 x 10^{21} m^{-3} (left vertical axis) and high density case - 1 x 10^{22} m^{-3}(right vertical axis).

Fizeau Interferometer

The Fizeau effect is the relativistic phase shift of an electromagnetic wave associated with movement of a dielectric medium. In a moving electron fluid, the phase shift experienced by an electromagnetic wave including Fizeau effect is

$$\phi' = \phi + \alpha = \phi + \frac{1}{\omega c^2}\int v_e \omega_{pe}^2 dL = \phi + 1.875\times10^{-23}\lambda\int n_e v_e dL \ ,$$

where v_e is the electron velocity along the laser beam and $\phi(rad) = 2.8\times10^{-15}\lambda\int n_e dL$, the conventional interferometer phase shift. The additional term α is the Fizeau phase shift and is dependent of the direction and magnitude of electron velocity. By devising a counter-propagating interferometer,[12] the Fizeau effect phase can be isolated. In such a configuration, Laser1 and Laser2 in Fig. 5 would be linearly-polarized, counter-propagating, colinear (δ=0) beams with phase difference proportional to 2α. A single-pass tangential Fizeau interferometer phase shift for ITER parameters using a 10.6 μm laser is ~0.5-2°, depending on plasma conditions. Multiplying the Fizeau phase shift by the electron charge provides a line-integrated measure of the toroidal electron current density. The Fizeau measurement is insensitive to path length changes.

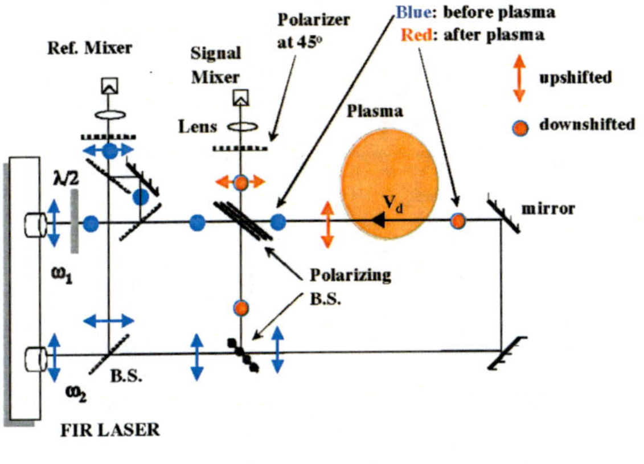

FIGURE 5. Experimental arrangement for measuring the Fizeau effect.

Finite Temperature Effects

All interferometer and polarimeter phase shifts discussed to this point are based on the cold plasma dispersion relation (i.e., T_e=0). ITER will operate at electron temperatures in the range T_e~10-25 keV motivating us to reexamine the cold plasma assumption. Finite thermal (or nonrelativistic dispersive) effects essentially increase the electron plasma frequency and, therefore, the measured phase is larger than for a cold plasma. However, relativistic effects dilate the electron mass which decreases the plasma frequency and reduces the measured phase. The combined effect has recently been approximated analytically by Mirnov[13] for temperatures and frequencies appropriate for ITER, where the phase correction factors become approximately $(1-\frac{3}{2}\frac{T_e}{m_e c^2})$ for interferometry and $(1-2\frac{T_e}{m_e c^2})$ for Faraday rotation. Measured phase shifts for interferometry and polarimetry including temperature effects can be written as

$$\phi = c_I \int n_e(1-\frac{3}{2}\frac{T_e}{m_e c^2})dl = c_I \int n_e dl - \frac{3}{2}\frac{c_I}{m_e c^2}\int n_e T_e dl \quad (1)$$

$$\psi_{Faraday} = c_F \int n_e B_{//}(1-2\frac{T_e}{m_e c^2})dl = c_F \int n_e B_{//} dl - \frac{2c_F}{m_e c^2}\int n_e T_e B_{//} dl \quad (2)$$

where c_I and c_F are constants. For T_e=15 keV, the correction is 4.4% for interferometery and 6% for polarimetry, respectively. This means that an interferometer interpreted using the cold plasma dispersion relation would underestimate the central density by 4.4%.

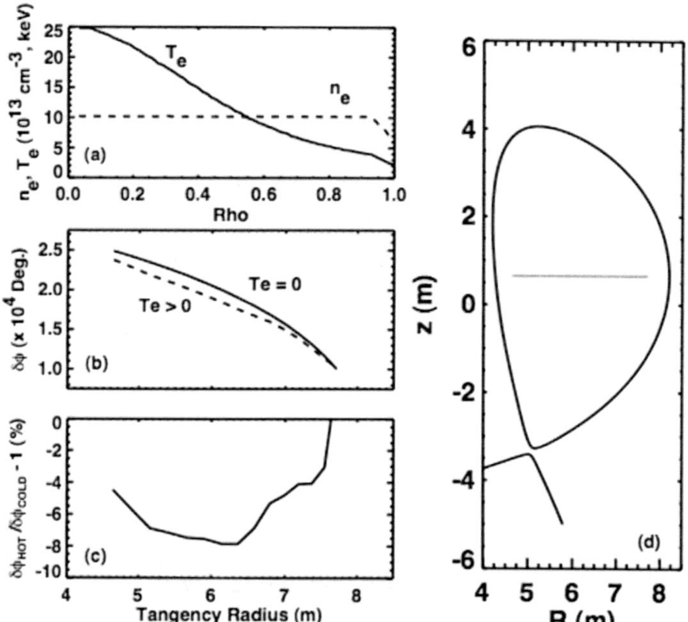

FIGURE 6. GENRAY calculations for the ITER. (a) Input electron temperature and density profiles. (b) Interferometric phase shift for cold plasma (solid) and fully relativistic dispersion relation (dashed). (c) Relative difference between finite temperature and cold plasma interferometric phase shift shown in (b). (d) Equilibrium shape of ITER scenario 2 plasma.

Equation (1) above has been verified numerically using the GENRAY code and the full hot plasma relativistic dispersion relation[14] for a variety of plasma conditions and wavelengths. An example GENRAY calculation of the interferometer phase shift for a 25 keV ITER plasma and a 57 micron probing beam is shown in Figure 6(b). For reference, the electron density and temperature profiles, as well as the equilibrium profile used are shown in Figs. 6(a) and (d), respectively. Both the finite temperature (dashed) and cold plasma (solid) results are given in Fig. 6(b), while the relative difference is shown in Fig. 6(c). The interferometry phase error varies from 0 to 8% across the profile. The net effect of the finite temperature effect is a decrease in the interferometric phase shift – as expected from Eq.(1).

Since temperature is known from Thomson scattering, finite T_e effects can be corrected. Alternatively, for a tangential Faraday system where B_ϕ is largely known, the above two equations have only two unknowns and hence provide information on both the plasma electron density and temperature. One can then directly determine the plasma electron pressure profile. Finite temperature effects must be considered for all interferometry techniques used in the high-temperature burning plasma environment.

Fluctuation Measurements

For all the interferometry techniques discussed, it is possible to make measurements with both high phase resolution and high time response. This combination permits measurement of both fast changes to the equilibrium profiles as well as fluctuations. As an example, a fast, laser-based Faraday rotation diagnostic has been employed to measure both the line-integrated wavenumber and frequency spectrum of radial magnetic field fluctuations in the high-temperature plasma core, as shown in Fig. 7. Such measurements can be used on ITER or future burning plasmas to monitor core MHD fluctuations associated tearing modes, fast particle driven Alfvenic modes and even broadband turbulence. Combined Faraday rotation and interferometry measurements can be used to get local information on δn, $\delta b_{r,\theta,\phi}$, $\delta j(\omega,k)$, their coherence, phase relation and even the fluctuation-induced flux.[15,16,17]

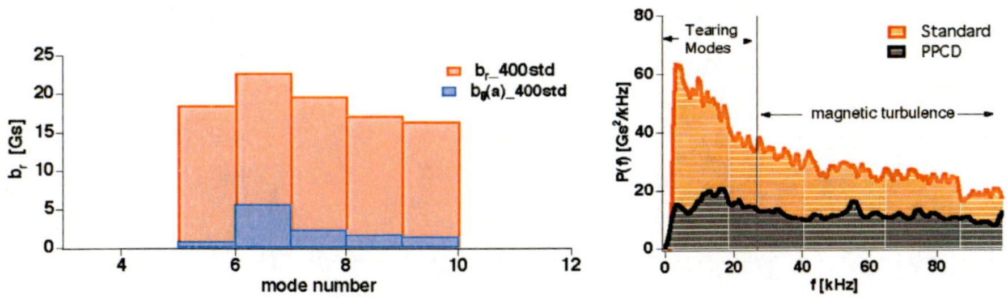

FIGURE 7. Radial magnetic field fluctuation line-integrated wavenumber and frequency spectra for standard (std) and high confinement (PPCD) plasmas.

REFERENCES

[1] I.H. Hutchinson, Principles of Plasma Diagnostics, Cambridge University Press (Cambridge, UK) 2002.

[2] Yasunori Kawano, et al., Rev. Sci. Instrum. **63**, 4971(1992).

[3] A. Murari et al., Rev. Sci. Instrum. **77** 073505 (2006).

[4] V.P. Drachev, et al., Rev. Sci. Instrum. **64**,1010(1993).

[5] P.A. Bagryansky, et al., Rev. Sci. Instrum. 77,053501(2006).

[6] D.L. Brower, W.X. Ding, B.H. Deng, Rev. Sci. Instrum. 77,10E911(2006).

[7] W.X. Ding, D.L. Brower, B.H. Deng, T. Yates, Rev. Sci. Instrum. 77, 10F105(2006).

[8] A.J.H. Donne, et al., Rev. Sci. Instrum. **75**, 4694(2004).

[9] B.H. Deng, D.L. Brower, et al., Rev. Sci. Instrum. 77,10F108(2006).

[10] R.T. Snider, et al., Rev. Sci. Instrum. **68**,728(1997).

[11] Ch. Fuchs and H.J. Hartfuss, Phys. Rev. Lett. **81**,1626(1998).

[12] D.L. Brower, et al., Rev. Sci. Instrum. **75**, 3399-3401(2004).

[13] V.V. Mirnov, W.X. Ding, D.L. Brower, M.A. Van Zeeland, T.N. Carlstrom, accepted for publication in Phys. of Plasmas (September, 2007).

[14] A.P. Smirnov, R.W. Harvey, Bull Amer. Phys. Soc. **39** 1626 (1994).

[15] N.E. Lanier, et al., Phys. Plasmas **8**,3402-3410(2001).

[16] W.X. Ding, D.L. Brower, et al., Phys. Rev. Lett. **99**,055004(2007).

[17] D.L. Brower, et al., Rev. Sci. Instrum. **74**, 1534-1540(2003).

Measurements of Alfvén Eigenmodes in Fusion Plasmas

M.A. Van Zeeland* and G.J. Kramer†

*General Atomics, P.O. Box 85608, San Diego, California 92186-5608, USA
†Princeton Plasma Physics Laboratory, Princeton, New Jersey 08543, USA

Abstract. Ensuring appropriate measurements of Alfvén eigenmodes in burning plasma experiments will allow the validation of predictions of alpha driven instabilities in plasmas dominated by self heating from fusion produced alpha particles and provide the control capability for optimizing performance, reducing the alpha particle losses due to these modes, and minimizing potential first wall damage. This paper gives a brief overview of several of the relevant Alfvén eigenmodes in burning plasma experiments as well as the theoretical tools currently being used to model them. A procedure is then outlined in which predicted Alfvén eigenmodes from the ideal MHD code NOVA can be used to simulate the performance of diagnostics on future devices with respect to measurements of these modes. A preliminary application of the technique to ITER's tangential CO_2 interferometer is discussed and it is found that the system should be more than adequate to resolve Alfvénic fluctuations with amplitudes in the range of those commonly encountered on current devices.

Keywords: Alfvén eigenmodes, interferometer.
PACS: 52.55.Fa, 52.70.Gw, 52.70.Nc

INTRODUCTION

In fusion plasmas and tokamaks in particular there exists a large variety of Alfvénic instabilities. These instabilities may impact the success of ITER and other burning plasma experiments since they can resonate with fast ions and be driven unstable, causing enhanced transport of the energetic particles necessary for heating [1,2]. Present tokamak experiments with Alfvénic instabilities show flattening of neutral beam fast ion profiles and loss of injected beam ions during periods of strong Alfvénic activity [2,3]. This type of redistribution or loss of fusion born alpha particles in a burning plasma experiment could reduce the performance of these devices and potentially damage the first wall.

Recent advances in core fluctuation diagnostics are enabling great progress in understanding fast ion driven Alfvén eigenmode (AE) instabilities and the resulting transport of the fast ions. Unprecedented measurements are being made by observation of the perturbed density [4,5], electron temperature [6], and magnetic field as well as indirectly through their effects on fusion performance and fast ion confinement [7]. Detailed measurements and modeling of these instabilities on existing devices show that for several classes of AEs, the mode structure is well predicted by ideal MHD codes such as NOVA [8]. By extending this finding to ITER and other future fusion experiments one can simulate the expected diagnostic performance needed to

CP988, *Burning Plasma Diagnostics*, edited by F. P. Orsitto, G. Gorini, E. Sindoni, and M. Tardocchi

adequately resolve Alfvénic activity. Ensuring appropriate measurements of AEs in burning plasma experiments will allow the validation of predictions of alpha driven instabilities in plasmas dominated by self heating from fusion produced alpha particles and provide the control capability for optimizing performance, reducing the alpha particle losses, and minimizing potential first wall damage. In addition, such measurements may provide key information on the bulk plasma through so-called Alfvén spectroscopy [9].

This paper is organized as follows: In section, "Alfvén eigenmodes and diagnostic observations," a brief review of several of the AEs relevant to burning plasma experiments is given and some of the recent advances in AE fluctuation diagnostics are discussed. In the "Modeling technique" section, a technique for extrapolating the performance of these diagnostics using NOVA calculated eigenmodes is given. The "Initial application to ITER's tangential interferometer" section, initial results from application of this technique to the interferometer component of ITER's tangential interferometer and polarimeter (TIP) [10] system are presented.

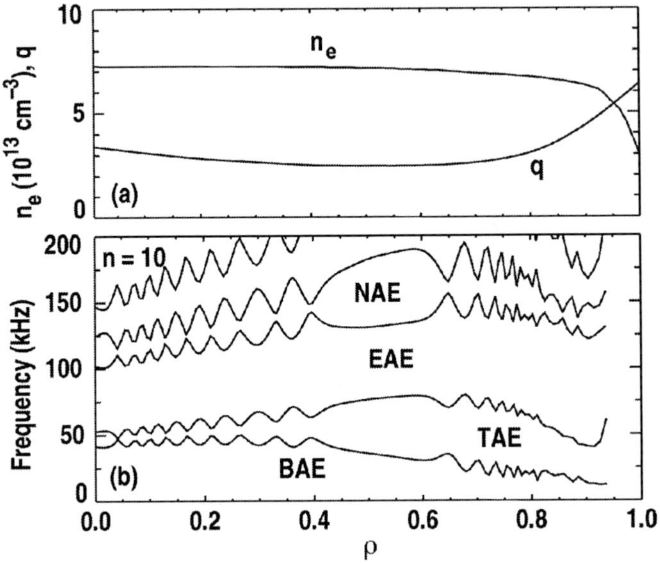

FIGURE 1. (a) Input profiles of electron density and safety factor for NOVA calculations used in this paper. (b) NOVA calculated $n = 10$ Alfvén continuum for equilibrium similar to ITER Scenario 4 plasma. Shape is shown in figure 4, elongation = 1.85, triangularity = 0.43, $B_T = 5.3$ T, $R_m = 6.34$ m, $a = 1.86$ m.

ALFVÉN EIGENMODES AND DIAGNOSTIC OBSERVATIONS

In tokamaks, the stable Alfvén continuum contains gaps associated with the coupling of various poloidal harmonics induced by, among other things, such geometric factors as finite aspect ratio, ellipticity, and triangularity. These gaps result in spectral regions in which spatially extended eigenmodes can exist and avoid interaction with the continuum. Shown in figure 1(b) is the NOVA calculated $n = 10$ Alfvén continuum for a plasma similar to that expected in ITER scenario 4. The assumed input profiles of electron density (n_e) and safety factor (q) are shown in figure 1(a). The lowest four gaps labeled are the beta-induced Alfvén eigenmode (BAE) gap [11], toroidcity-induced Alfvén eigenmode (TAE) gap [12], ellipticity-induced Alfvén eigenmode (EAE) gap [13,14] and triangularity-induced Alfvén eigenmode (NAE) gap [13,14].

As mentioned, discrete eigenmodes can exist in each gap and to a rough approximation have frequencies given by:

$$f_{AE} = \left| \frac{n\Delta m}{2m + \Delta m} \right| \frac{V_A}{2\pi R} \tag{1}$$

where V_A is the local Alfvén speed, R is the major radius, n, m are toroidal and poloidal mode number respectively, and Δm is an integer that specifies which poloidal harmonics are coupling, i.e. m and $m + \Delta m$. For the TAE, EAE, and NAE, $\Delta m = 1, 2,$ and 3 respectively. The eigenmode is peaked near the q surface $q_{AE} = (2m + \Delta m)/2n$. The radial displacement of the largest five harmonics for an example $n = 10$ TAE (that will be the subject of the next section) is shown in figure 2, where several poloidal harmonics have been labeled. As expected for the standard TAE, there are two dominant harmonics in this particular eigenmode ($m = 26, 25$) and it is peaked near the $q = (25 + 1/2)/10 = 2.55$ surface. It is important to point out that NOVA is a linear code and the predicted AEs are the linear eigenmodes; as a result no prediction for the amplitudes are possible using these tools alone. This is the reason for the normalized, unitless displacement amplitude.

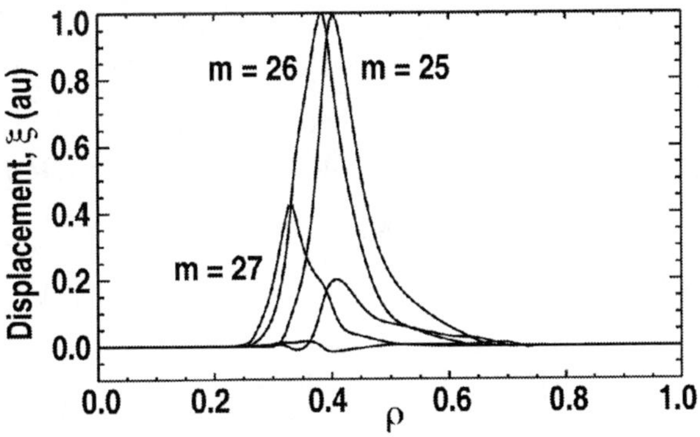

FIGURE 2. Displacement for $n = 10$ TAE. Major poloidal mode numbers $m = 25$, 26, and 27 are labeled.

With regard to experimental observation of Alfvén eigenmodes, many recent diagnostic upgrades on devices around the world are allowing unprecedented measurements of these instabilities. Among them, the standard CO_2 interferometer similar to the TIP system proposed for ITER, is proving to be extremely valuable [5].

These systems are producing high quality fluctuation data for both coherent and incoherent fluctuations with wavenumbers $k_\perp < 2/a_o$, where a_o is the Gaussian beam-waist, and with frequencies up to several MHz. They can operate in almost all tokamak plasma conditions, rarely suffer from refraction, are based on well developed CO_2 laser and detector technology and are able to take advantage of modern digital phase demodulation techniques. An example fluctuation spectrum from the DIII-D CO_2 interferometer system is shown in figure 3 where several different types of Alfvén eigenmodes as well as other coherent fluctuations are present. In the figure, both TAEs and ELMs have been pointed out. The TAEs are driven by 80 keV neutral beam ions and have an appreciable Doppler shift due to plasma rotation — which varies significantly in the time window shown. It is likely that the lower frequency fluctuations around 50 kHz between t = 550-1000 ms are what has been called the BAE in DIII-D [11]. These data serve to show the level of detail obtainable with modern CO_2 interferometry.

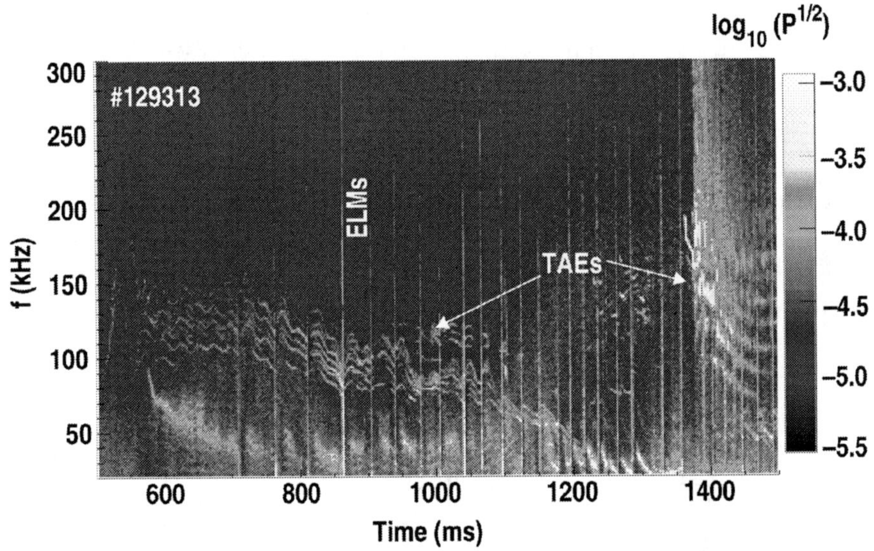

FIGURE 3. CO_2 interferometer data from DIII-D discharge 129313.

MODELING TECHNIQUE

As stated in the introduction, NOVA and other ideal MHD codes have proven to be very successful at predicting many aspects of Alfvén eigenmodes in existing tokamaks, in particular, the eigenmode structure [6]. Making use of this fact, one can extrapolate to ITER and other future fusion experiments and simulate the expected diagnostic performance needed to adequately resolve Alfvénic activity as well as the expected performance given current state-of-the-art hardware. This process is currently beginning for the TIP diagnostic and will be described here.

Shown in figure 4 is the calculated density perturbation structure for the $n = 10$ TAE shown in figure 2, where the vector displacement has been used to infer the density perturbation profile. An $n = 10$ TAE was chosen since toroidal mode numbers in this range are expected to be unstable in ITER [15]. The top view represents a cross section through the plasma midplane and over-plotted are the approximate locations of the five planned ITER TIP chords.

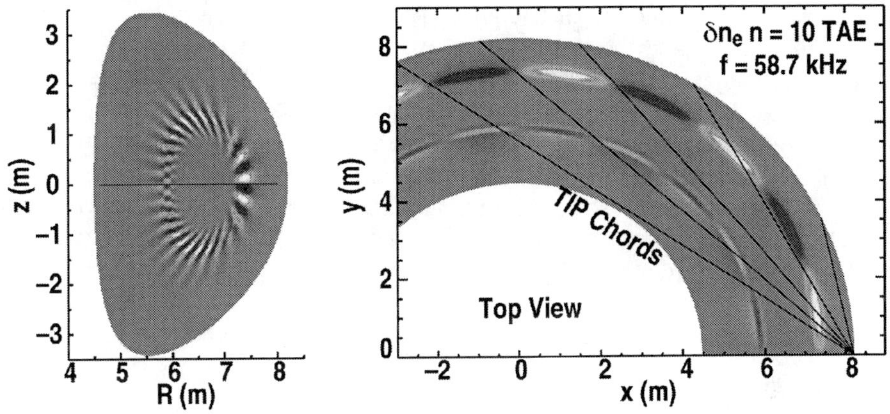

FIGURE 4. NOVA calculated density perturbation, δn_e. (left) R-Z cross-section of density perturbation due to $n = 10$ TAE shown in figure 2. (right) Top view of cross-section through midplane of TAE density perturbation. Blue/yellow = positive/negative density perturbation. Solid lines are approximate locations of ITER TIP chords.

In order to predict the TIP response to this eigenmode, the equilibrium electron density (n_e) and parallel magnetic field ($B_{\|}$) as well as fluctuating electron density (δn_e) and fluctuating parallel magnetic field ($\delta B_{\|}$) can be mapped onto the ray trajectories. The interferometer phase change (ϕ), polarimeter measured Faraday rotation (ψ) and their fluctuating quantities ($\delta\phi$ and $\delta\psi$ respectively) are given by:

$$\phi = k_{\phi}\lambda \int n_e dL \tag{2}$$

$$\delta\phi = k_{\phi}\lambda \int \delta n_e dL \tag{3}$$

and

$$\psi = k_{\psi}\lambda^2 \int n_e B_{\|} dL \tag{4}$$

$$\delta\psi = k_{\psi}\lambda^2 \left(\int \delta n_e B_{\|} dL + \int n_e \delta B_{\|} dL \right) \tag{5}$$

where λ is the wavelength used, $k_{\phi} = 2.82\times10^{-15}$ and $k_{\psi} = 2.62\times10^{-13}$, in MKS units [16]. The amplitude of the fluctuating signal depends on the phase of the wave and is found by evolving the mode through $2\pi/n$ in phase. The end result of this process is the creation of a so-called "synthetic diagnostic" for simulating the TIP system.

INITIAL APPLICATION TO ITER'S TANGENTIAL INTERFEROMETER

With respect to the interferometric phase shift, this process has been carried out for the TAE of figure 4 and the results are shown in figure 5 where both ϕ and $\delta\phi$ are plotted versus tangency radius. For these calculations $\lambda = 10.59$ μm was used as this is expected to be the case for the longer wavelength in ITER's TIP system. The peak density perturbation amplitude of the eigenmode itself was set to be $\delta n_e/n_e = 1\times10^{-3}$ on the midplane. This is consistent with the typical mode amplitudes observed in present devices [5]. By comparing the equilibrium and fluctuating quantities, one can see that $\delta\phi/\phi \approx 10^{-4}$, which is an order of magnitude smaller than the peak normalized density perturbation. Also, from this figure, it is clear that the line integration creates a significant $\delta\phi$ at almost all tangency radii, irrespective of the fact that the eigenmode itself has almost no density perturbation on the midplane between R = 6.2-7 m.

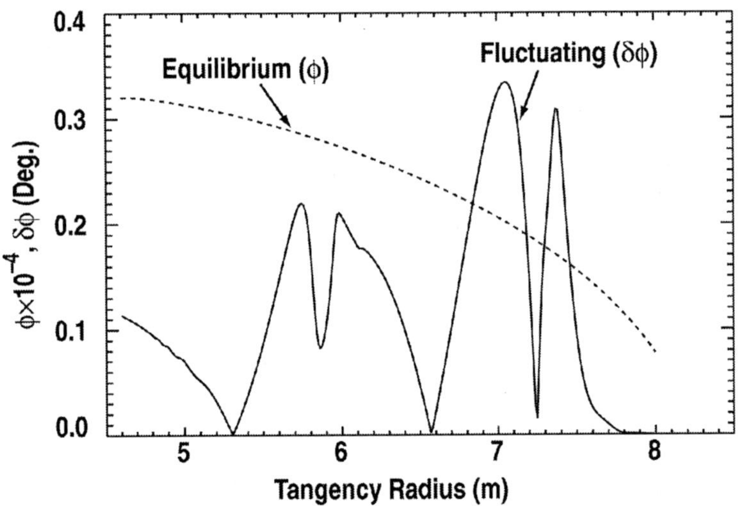

FIGURE 5. Fluctuating ($\delta\phi$) and equilibrium (ϕ) interferometric phase shift as a function of tangency radius for the n = 10 of figure 4.

In terms of actual expected diagnostic performance, phase resolution noise floors from existing devices can provide an excellent starting point. For this purpose, the digital phase demodulation system developed at DIII-D for the standard two-color CO_2 interferometer will serve as a reference. Typical noise floors in the frequency range of the TAE investigated (f = 58.7 kHz) are approximately O(10^{-5}-10^{-6}) Deg^2/ kHz, implying for a 1 kHz wide mode a resolvable fluctuation-induced phase shift of

$\delta\phi_{min} \approx O(10^{-3})$ Degrees. This noise floor is approximately two orders of magnitude below the predicted fluctuating interferometric phase signals shown in figure 5, for which a peak $\delta n_e/n_e = 1\times10^{-3}$ was assumed.

CONCLUSIONS AND FUTURE WORK

In terms of the ability of the interferometer in ITER's TIP system to measure density fluctuations resulting from AEs and other similar MHD, these results are extremely encouraging. In fact, they should serve to increase the emphasis of the TIP system for fluctuation studies. A potential problem with this analysis is the role that broadband density fluctuations will play in the measured signal. It is expected that, as with existing systems, a significant component of the measured fluctuation spectra will be due to broadband density fluctuations at the plasma edge. Depending on the level of these fluctuations they may set the resolvable noise floor rather than the phase demodulation system itself.

This paper is intended to provide a brief outline for predicting the ability of diagnostic systems to measure Alfvénic activity on ITER and other devices. The method itself can easily be adapted, however, to other types of fluctuations. A much more detailed publication will follow in which this analysis is extended to include both interferometry and polarimetry measurements using the TIP system and the poloidally viewing 118 μm polarimeter system [10], as well as temperature fluctuation measurements with the ECE system. Also, the sensitivity of this analysis to equilibrium parameters such as safety factor and density profile will be considered.

ACKNOWLEDGMENTS

This work was supported by the U.S. Department of Energy under DE-FC02-04ER54698 and DE-AC02-76CH03073.

REFERENCES

1. G. Y. Fu and J. W. Van Dam, *Phys. Fluids B* **1**, 1949 (1989).
2. K. L. Wong, *Plasma Phys. Control. Fusion* **41**, R1 (1999).
3. H. H. Duong, W. W. Heidbrink, T. W. Petrie, R. Lee, R. A. Moyer, J. G. Watkins, *Nucl. Fusion* **33**, 749 (1993).
4. R. N. Nazikian, et al., *Phys. Rev. Lett.* **91**, 125003 (2003).
5. M. A. Van Zeeland, et al., *Nucl. Fusion* **46**, S880 (2006).
6. M. A. Van Zeeland, et al., *Phys. Rev. Lett.* **97**, 135001-1 (2006).
7. W. W. Heidbrink, et al., *Phys. Rev. Lett.*, In submission (2007).
8. C. Z. Cheng and M. S. Chance, *J. Comput. Phys.* **71**, 1124 (1987).
9. A. Fasoli, et al., *Plasma Phys. Control. Fusion* **44**, B159 (2002).
10. A. J. H Donne, et al., *Nucl. Fusion* **47**, S337 (2007).
11. A. D. Turnbull, et al., *Phys. Fluids B* **5**, 2548 (1993).
12. C. Z. Cheng and M. S. Chance, *Phys. Fluids* **29**, 3695 (1986).
13. R. Betti and J. P. Freidberg, *Phys. Fluids B* **3**, 1865 (1991).
14. G. J. Kramer, et al., *Phys. Rev. Lett.* **80**, 2594 (1998).
15. N. N. Gorelenkov, et al., *Nucl. Fusion* **43**, 594 (2003).
16. W. X. Ding, et al., *Rev. Sci. Instrum.* **75**, 3387 (2004).

Development of laser-aided plasma diagnostics and related technology

K. Kawahata, T. Akiyama, R. Pavlichenko, K. Tanaka, K. Nakayama[1] and S. Okajima[1]

National Institute for Fusion Science, Toki 509-5292, Japan
[1]College of Engineering, Chubu University, Kasugai 487-8501, Japan

Abstract. Laser-aided plasma diagnostics, aiming for establishment of reliable density measurement in next step magnetically confined fusion devices, are under development at the National Institute for Fusion Science. A new type of two color laser (57.2/47.6-μm CH_3OD) interferometer has been developed and its original function, vibration subtraction, was confirmed in a test stand. The line integrated density measurements by the polarimeter were demonstrated at Compact Helical System by the Cotton-Mouton polarimeter and at the LHD by the Faraday rotation polarimater.

Keywords: plasma diagnostics, laser interferometer/polarimeter, far infrared laser, two color.
PACS: 52.70.kz

INTRODUCTION

Measurement of the refractive index of plasmas by using electromagnetic waves is a well-established technique for measuring electron density profiles in high temperature plasmas. In the large fusion devices such as JET[1], JT-60U[2], TFTR[3] and LHD[4], far infrared laser diagnostics have widely been used for density profile measurement. The wavelength of the FIR lasers used in these devices is around 100 to 200 μm, which is optimized from the viewpoints of the plasma refraction and mechanical vibration effects for target plasmas. Recent progress of particle fueling using a high repetitive-rate pellet injection and a supersonic pulsed gas injection makes it possible to produce high density plasmas of more than 10^{20} m^{-3}. In these high density plasmas, a steep density gradient is formed leading to the fringe jumps on the density traces measured by fringe counters. In order to overcome this difficulty there are several innovative approaches, which are classified into two categories, the interferometry and the polarimetry. In this paper, development of laser-aided diagnostics including hardware development necessary for future fusion devices and some typical experimental results will be described.

DEVELOPMENT OF LASER SOURCES

For high-density operation of the Large Helical Device (LHD) and for future large fusion devices such as ITER, we have been developing short wavelength far infrared

CP988, *Burning Plasma Diagnostics*, edited by F. P. Orsitto, G. Gorini, E. Sindoni, and M. Tardocchi

laser oscillation lines by using a CO_2-laser-pumped FIR laser. For ITER application, a poloidal polarimeter system[5] is being designed for measuring the current density profile. A 118.8-μm CH_3OH laser has been proposed as a probing light source for the polarimeter, since the 118.8-μm CH_3OH laser line is the shortest far infrared laser oscillation line with the highest power at the present. If we consider shorter wavelength lines, the next candidate is the 10.6-μm CO_2 laser linc. Apart from this limitation, let us now examine the optimum wavelength for the ITER polarimeter.

When a linearly polarized electromagnetic wave, whose electric field vector is parallel to the toroidal magnetic field, transverses a poloidal plasma cross section perpendicular to the plasma current, the phase of the wave is changed by φ and its plane of polarization rotates by angle Ω due to the plasma[6].

$$\varphi = 2.82\times10^{-15}\lambda[\mathrm{m}]\int n_e(z)\left[\mathrm{m}^{-3}\right]dz \tag{1}$$

$$\Omega = 2.62\times10^{-13}\lambda[\mathrm{m}]^2\int n_e(z)\left[\mathrm{m}^{-3}\right]B_{\parallel}[\mathrm{T}]dz \tag{2}$$

where λ is the wavelength of probing laser light, n_e is the electron density, $B_{\parallel}$ is the component of the poloidal magnetic field parallel to the laser light. In order to obtain the current density profile from the values of Ω, the density profile should be measured by the interferometry or by using the Cotton-Mouton effect. The ellipticity ε due to the Cotton-Mouton effect is given by

$$\varepsilon = 2.45\times10^{-11}\lambda[\mathrm{m}]^3\int n_e(z)\left[\mathrm{m}^{-3}\right]B_t\,[\mathrm{T}]^2\,dz \tag{3}$$

here B_t the perpendicular component of the magnetic field (toroidal magnetic field). Another important factor to optimize the wavelength is the beam refraction effect caused by plasma density gradient, whose magnitude is proportional to $\lambda^2\nabla n$. Let us now estimate the magnitude of the each wave parameter in the ITER polarimeter along the chord No.5 in Ref. 5 (for the double pass). Figure 1 shows the calculated phase shifts, Faraday rotation angles, Cotton-Mouton phase shifts and the beam displacement at the position of the No. 5 c.c. mirror as a function of the laser wavelength. From the figure, we can see that laser sources around 50 μm would be another candidate for the ITER polarimeter if powerful oscillation lines were available, since the Faraday rotation angle of 14 degrees at 57 μm can be measured with high accuracy, and laser beam bending effect is small.

For the development of short wavelength far-infrared laser sources with high power, we have constructed an optical pumped laser system[7] which consists of two twin type FIR laser cavities with different parameters to cover wide range of oscillation frequency, a pump CO_2 laser and coupling optics of the CO_2 laser light into the FIR laser cavities. The CO_2 laser is of the intra-cavity type composed of a water-cooled laser tube of 10.8 mm inner diameter having four electrodes, a grating of 150 lines/mm and a ZnSe output coupler(20 m radius curvature, 55 % reflectivity at 10.6 μm) mounted on a coaxial PZT. The cavity length is fixed at about 3.0 m by using

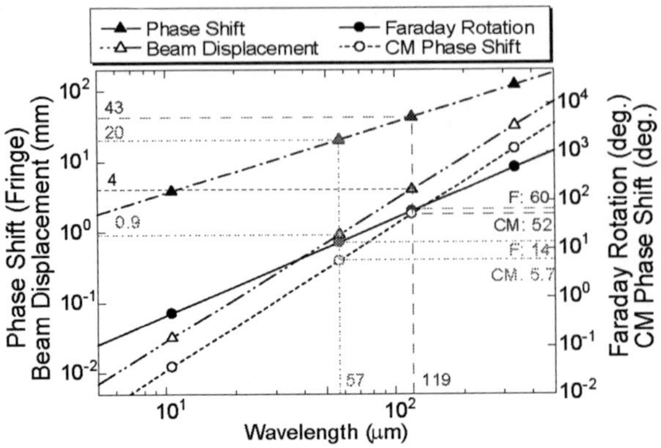

FIGURE 1. Calculated phase shifts, Faraday rotation angles, Cotton-Mouton phase shifts and the beam displacement as a function of the laser wavelength for the chord No.5 in the figure 1 in Ref. 5.

four Neoceram glass rods which have a low thermal expansion coefficient(- 1 x 10^{-7}/deg.). A gas mixture of 33%/67% in CO_2/N_2 and He are used for the CO_2 laser operation. The achieved output power of 10P(20) oscillation line is about 250 W so far. We have achieved a lot of high power oscillation lines from 40 µm to 100 µm in wavelength by changing FIR laser molecules and pump CO_2 laser lines as is listed in Table 1. Among these oscillation lines, the most powerful line was found to be a 57.2-µm CH_3OD laser line and is able to oscillate simultaneously at a 47.6 µm[8]. This two color laser enable us to construct a new two color laser interferometer[9]. This two color interferometer system is unique compared with a conventional two color interferometer system, where two independent lasers are used to be combined.

TABLE 1. CO_2 laser pumped FIR laser oscillation lines from 40 to 100 µm.

CO_2 laser			FIR laser								
Line	Length (m)	Power (W)	Molecule	Wavelength (µm)	Pol.	Press. (Pa)	He (Pa)	Power (mW)	Length (m)	Tube dia. (mm)	Output mirror* Si hybrid coupler
9R(18)	3.0	129	CH_3OH	61	⊥	36		9.5	2.9	25	14
9R(10)	2.4	83	CH_3OH	96.5	∥	93		68	2.2	22	8
9R(8)	2.4	65	CH_3OH	77.3	∥	101		1.5	2.2	22	8
9P(34)	2.4	75	CH_3OH	70.5	⊥	27		65	2.2	34	8
10R(16	3.0	123	CH_3OH	63.0	∥	60		54	2.9	25	14
9R(34)	2.4	63	CD_3OH	52.9	⊥	47		10	2.2	34	8
9R(28)	2.4	81	CD_3OH	55.6	⊥	40		4	2.2	34	8
	2.4	78	CD_3OH	42.6	∥	33		4	2.2	34	8
10R(18	2.4	89	CD_3OH	41.4	⊥	24		13	2.2	34	8
	2.4	107	CD_3OH	43.7	∥	40		6	2.2	34	8
10R(16	3.0	113	CD_3OH	81.6	∥	47		25	2.9	25	14
9R(8)	3.0	153	CH_3OD	57.2	⊥	59	16	1.6 W	2.9	25	14
	3.0	153	CH_3OD	47.6	∥	59	16	0.8 W	2.9	25	14
9P(30)	2.4	95	CH_2DOH	44	∥	17		6	2.2	25	8
10P(46)	2.4	43	CH_2DOH	49.1	⊥	27		17	2.2	25	8
9R(30)	2.4	65	NH_3	67	∥	55		97	2.2	22	8

* Diameter of coupling hole (mm)

One of the key issues to realize this diagnostic instrument is the development of high quality detectors with fast and sensitive heterodyne detection. There are three candidates used in FIR wavelength regime, a GaAs Schottky barrier diode mixer, a InSb He-cooled detector and a gallium-doped germanium detector. The GaAs Schottky barrier diode mixer has widely been used in many FIR laser diagnostics since the detector can be operated at room temperature with high sensitivity. The sensitivity of the detector, however, is decreased with operating frequency due to the cutoff frequency bellow several THz. The InSb He-cooled detector with magnetic tuning has been applied[1] on JET tokamak to simultaneously detect a 195-μm DCN laser and a 118.8-μm CH_3OH laser. The spectral responsivity of the detector, which is normally limited in the frequency region of 60 to 600 GHz, can be tuned up to about 3 THz by means of a magnetic field, but not over than 3 THz. We finally chose a gallium-doped germanium detector operating at liq.He temperature. Figure 2 shows the frequency spectrum of two color beat signals at 0.5 MHz for 57.2 μm and 1.25 MHz for 47.6 μm. It can be seen in this figure that the signal-to-noise ratio is excellent - about 40 dB. The beat frequency of each laser oscillation line can be set at the optimum value by changing the pressure of the FIR laser cavity and by tuning the cavity length. So far a single mode beat frequency of up to ~ 2MHz has been achieved without large reduction of laser oscillation power.

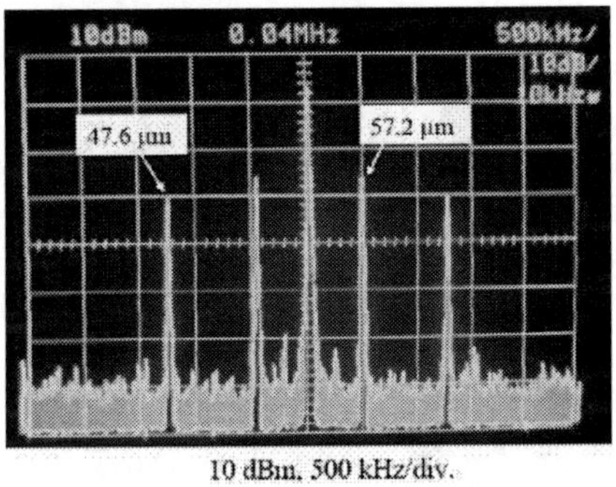

FIGURE 2. Two color beat signals detected by the gallium-doped germanium photoconductor. The spectrum analyzer trace shows two color beat signals corresponding to a 57.2 μm beat of 0.5 MHz and a 47.6 μm beat of 1.25 MHz.

TWO COLOR FAR INFRARED LASER INTERFEROMETER

The conventional two color laser interferometer systems use two independent laser interferometers for vibration subtraction as shown in Fig.3. Probing laser beams (ω_1 and ω_2) are combined and made coaxial at the DM_1 dichromatic mirror, and divided into two beams at BS_1, a probe beam and a local beam. The probe beam is combined with the local beam which also contains two laser beams ($\omega_1 + \Delta\omega_1$ and $\omega_2 + \Delta\omega_2$) at BC_1 after passing through the plasma, and then separated into two laser components by using the DM_4 dichromatic mirror, and finally detected by the proper detectors (D_3 and D_4). By using 4 output signals we can get information about the fringe shifts

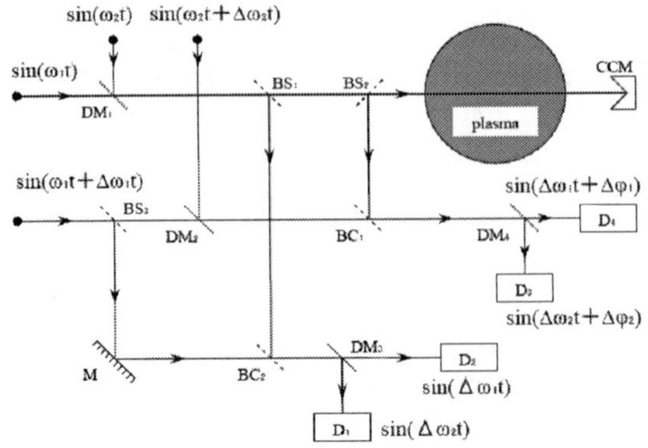

FIGURE 3. Schematic drawing of the conventional two color laser interferometer. Two laser oscillation lines are combined and separated with the dichromatic mirrors (DM). BS: Beam Splitter, BC: Beam Combiner, M: Beam Reflection Mirror, D_2, D_4: Detector for ω_1, D_1, D_3: Detector for ω_2.

caused by the plasma and by the mechanical vibrations. In this type of the two color system, the fringe shift caused by the mechanical vibrations couldn't be canceled out completely since the optical path difference between two independent interferometers is remained and effects due to optically dispersive components are large when the wavelength of two laser sources differ largely.

In order to overcome these difficulties, new types of two color interferometers[10-12] have been developed to establish more reliable vibration compensations. Figure 4 shows the schematic drawing of a new type of two color laser interferometer system under development at NIFS, which is a simpler system compared with Fig.3. The optical arrangement is similar to a single heterodyne interferometer system. The laser source is a twin optically-pumped CH_3OD laser, which simultaneously oscillates at 57.2 and 47.6 μm. Since each laser line has different polarization, a Martin –Puplett

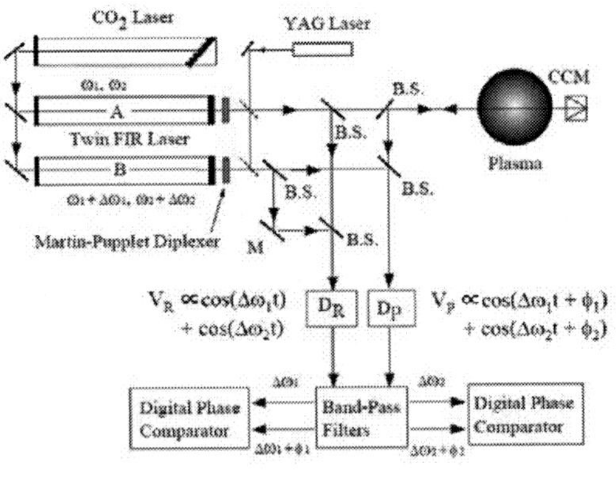

FIGURE 4. Schematic drawing of a new type of the two color laser interferometer of the Michelson type. The optical components are the same as a conventional heterodyne interferometer with single laser oscillation line. The detector output includes two heterodyne beat signals, $\Delta\omega_1$ and $\Delta\omega_2$.

diplexer is used to make a co-polarized beam. By tuning the cavity length the beat frequency of each heterodyne interferometer can be set at an optimum value. The main components to construct the system are beam splitters and beam combiners made of a silicon etalon with high resistivity, which is a low absorption material (absorption coefficient α = 0.56 cm^{-1} at 47.6 μm) in short wavelength FIR regime compared with quartz (α_o, α_e = 6.6 cm^{-1}, 4.9 cm^{-1}at 47.6 μm). The interference signals are detected by Ge:Ga photoconductive detectors, D_1 and D_2. The detected signals are separated electronically into two wavelength components. Each frequency component is fed into a phase comparator for phase measurement.

In order to develop a new type of two color interferometer, a single channel two color interferometer system is constructed and operated in the laboratory, where the main key issues to establish this innovative diagnostics are as follows: 1) development of a laser control system to achieve stable simultaneous oscillation at 57 and 48 μm, 2) development of optical components to construct short wavelength FIR laser diagnostics. An example of vibration subtraction test is shown in Fig.5. In this case the reflecting mirror is placed on a piezo-electric transducer to confirm the idea of the two color laser interferometer and to find out what kinds of problem need to be solved. The position of the mirror is modulated in triangular form (Fig.5(a)) at 10 Hz with the amplitude of ~118 μm. The displacements of the mirror measured by 57 and 48 μm laser interferometers are shown in Fig.5(b). The difference of these displacements measured is also plotted in Fig.5(b), which shows the effectiveness of the vibration compensation using two color laser oscillations.

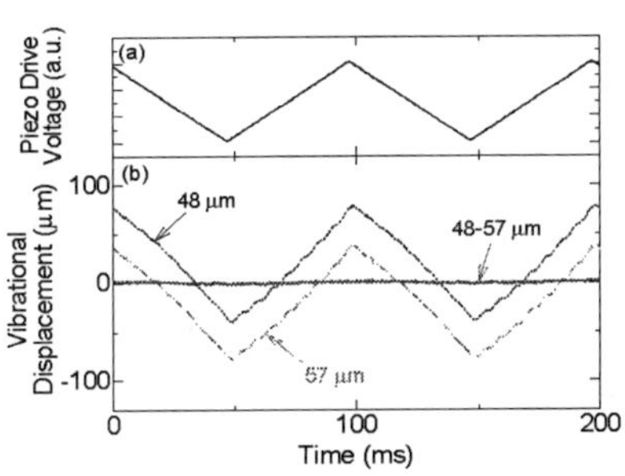

FIGURE 5. (a) Waveform of the applied voltage on the piezo-electric transducer. (b) The measured displacements of the vibration mirror by a 57 μm and 48 μm CH_3OD laser interferometers, and their subtraction. The mechanical vibration is simulated by using a piezo-electric transducer.

POLARIMETER FOR DENSITY MEASUREMENTS

The second category is polarimetry, where the Faraday rotation and Cotton-Mouton effects influence on the polarization of the laser beam which propagates through a magnetically confined plasma. The magnitude of the both effects depends on the magnetic field strength and electron density. When the magnetic field strength is known as in low beta helical device plasmas, electron density information is available from the measurements of rotation angle or the ellipticity of the laser beam. In comparison with the conventional interferometry, this technique can be considered to deliver more robust measurements of the line integrated electron density, since it does not depend on the past time history of the discharge when the rotation angle is set to be less than 2π radians. We have already demonstrated the usefulness of Faraday rotation polarimetry for density profile measurement on LHD[13]. Another type of polarimeter, Cotton-Mouton polarimater, has been proposed for ITER for density profile measurements aiming to achieve a reliable diagnostic for long pulse discharges. On the Compact Helical System (CHS) a Cotton-Mouton polarimeter with a 337-μm HCN laser was developed[14] in order to see the accuracy and limitations of polarimetry since there are very few applications[15,16] for electron density measurement on the magnetically confined plasmas. Figure 6 shows the optical system of the Cotton-Mouton polarimeter with a HCN laser installed on CHS. For both probe and reference beams, combined beams of O- and X-mode beam polarization, each with a slightly frequency offset is used. The frequency of one beam is shifted by 1 MHz with a super rotating grating (SRG) [17]. The Cotton-Mouton effect can be measured as a phase difference between the probe and reference beat signal. By selecting the orthogonal polarization component from probe and reference beams and mixing them, usual interferometry can be done simultaneously with the same chord as that of the polarimeter. The phase difference due to the Cotton-Mouton effect along the plasma center chord is 8.2° when a magnetic field strength at the magnetic axis is 1.96 T and the electron density is 1.2×10^{20} $(1 - \rho^8)$ m^{-3} in CHS. Hence the Cotton-Mouton polarimeter is free from fringe jump errors owning to having a smaller phase shift than 2π.

Figure 7 shows an example of measurement results of the Cotton-Mouton polarimeter. The response time determined from the band width of the band-pass filter is 1 ms. Except for the beginning of the discharge, the line averaged electron density evaluated from the polarimeter is well consistent with that from the interferometer. While the interferometer suffered from fringe jump errors twice (t = 135 and 144 ms) at the end of the discharge, when the amplitude of the beat signal decreased significantly, the Cotton-Mouton polarimeter showed just spike noises there and the correct value after the amplitude of the beat signal recovered. The beat signal contains side-band frequency noises (1 MHz ± (rotation frequency of SRG 500 Hz) × integer, less than -6 dB), which are attributed to frequency shift with SRG, which causes phase noises in the polarimeter. While the phase noises are negligible compared with the phase shift of the interferometer (more than several fringes), they are significant in the Cotton-Mouton

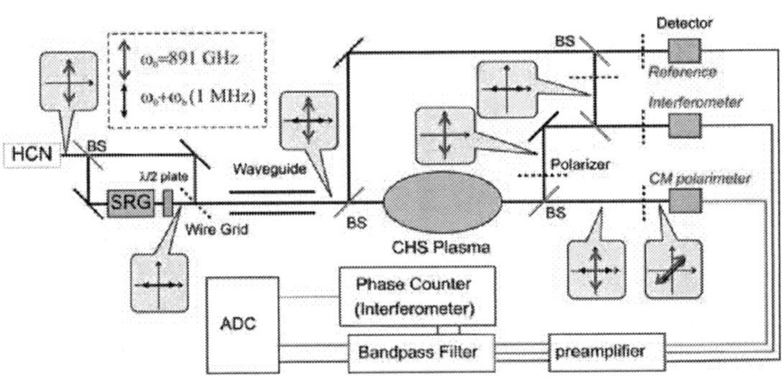

FIGURE 6. Optical system of the Cotton-Mouton Polarimeter installed on CHS. Since the polarization component parallel to a receiving antenna in the SBD is detected, SBDs are tilted by $\pi/4$, and they remove polarizers.

polarimeter because of the small phase shift. Hence, high phase resolution in the polarimeter requires higher purity of the beat signal than that in the interferometer. Cross talking between signals of the Cotton-Mouton polarimeter and the interferometer and back talking (multi-reflection between a detector and optical components) of the polarimeter are also causes of modulations of phase difference similar to polarimeters based on the Faraday effect. The inverse time evolution from t=0 to 20 ms might be related to that. Electrical signal separation in a data acquisition system and suppression of multi-reflection are indispensable.

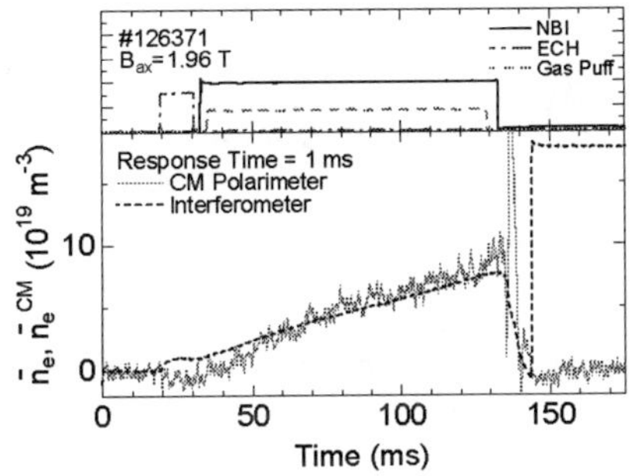

FIGURE 7. Comparison of the line averaged electron density evaluated from the Cotton-Mouton Polarimeter with that from the interferometer.

ACKNOWLEDGMENTS

This work was supported in part by a Grant-in-Aid for Science Research from the Japanese Ministry of Education, Culture, Sports, Science and Technology, "Priority area of Advanced Burning Plasma Diagnostics" (16082208) and also by the NIFS budget code of NIFS06ULHH501.

REFERENCES

1. G. Braithwaite, N. Gotardi, G. Magyar, *et al.*, Rev. Sci. Instrum. **60**, 2825-2834 (1989).
2. T. Fukuda and A. Nagashima, Rev. Sci. Instrum. **60**, 1080-1085 (1989).
3. D.K. Mansfield, H.K. Park, L.C. Johnson, *et al.*, Applied Optics **26**, 4469-4474 (1987).
4. K. Kawahata, K. Tanaka, Y. Ito, *et al.*, Rev. Sci. Instrum.**70**, 707-709 (1999).
5. A.J.H. Donne, M.F. Graswinckel, *et al.*, Rev. Sci. Instrum.**75**, 4694 -4701 (2004).
6. F. De Marko and S.E. Segre, Plasma Phys. 14, 245-252 (1972).
7. Okajima, K. Nakayama, H. Tazawa, K. Kawahata, *et al.*, Rev. Sci. Instrum. **72**, 1094-1096 (2001).
8. K. Nakayama, H. Tazawa, S. Okajima, K. Kawahata, *et al*, Rev. Sci. Instrum. **75**, 329-332 (2004).
9. K. Kawahata, T. Akiyama, *et al.*, Rev. Sci. Instrum.**77**, 10F132 (2006)
10. J. Irby, R. Murray P. Acedo and H. Lamela, Rev. Sci. Instrum. **70**, 699-702 (1999).
11. V. P. Drachev, Yu. I. Krasnikov and P. A. Bagryansky, Rev. Sci. Instrum. **64**, 1010-1013 (1993).
12. P. A. Bagryansky, A.D. Khilichenko, A.N. Kvashinin, *et al.* , Rev. Sci. Instrum. **77**, 053501 (2006).
13. S T. Akiyama, S. Tsuji-Iio, R. Shimada, *et al*, Rev. Sci. Instrum. **74**, 2695-2703 (2003).
14. T. Akiyama, K. Kawahata, Y. Ito, *et al.*, Rev. Sci. Instrum. **77**, 10F118 (2006).
15. Ch. Fuchs and H. J. Hartfuss, Phys. Rev. Let. **81**, 1626-1629 (1998).
16. A. Boboc, L. Zabeo , *et al.* , Rev. Sci. Instrum. **77**, 10F324 (2006).
17. T. Maekawa, T. Minami, K. Makino, *et al.*, Rev. Sci. Instrum. **62**, 304-307 (1991).

Fast Ion Collective Thomson Scattering Diagnostic for ITER - Progress of Design

S.B. Korsholm, H. Bindslev, V. Furtula, F. Leipold, F. Meo, P.K. Michelsen, S. Michelsen, M. Salewski, and E.L. Tsakadze

Association EURATOM-Risø National Laboratory, Technical University of Denmark DK-4000 Roskilde, Denmark

Abstract. In the era of high power and burning plasma fusion experiments with significant populations of fast particles, the diagnosis of fast ion dynamics becomes an important topic. In ITER, populations of fast ions due to ICRH and NBI, as well as fusion born alphas will carry a significant fraction of the free energy of the plasma. This may affect instabilities and transport in the plasma. A key candidate for diagnosing confined fast ions in ITER is the technique of collective Thomson scattering (CTS). A fast ion CTS system with a probing frequency of 60 GHz has been proposed for ITER. Based on diagnostic experience from particularly TEXTOR and ASDEX Upgrade, work is now progressing towards a final design of a fast ion CTS diagnostic for ITER. The biggest challenge of the diagnostic design is the HFS receiver located in the restricted space behind the blanket modules. Calculations and a series of mock-up measurements have brought the design towards a four mirror quasi-optical solution. The development as well as the present design will be presented.

Keywords: ITER, plasma diagnostics, collective Thomson scattering, mm-wave
PACS: 52.40.Db, 52.55.Pi, 52.70.Gw

INTRODUCTION – THE ITER CTS DIAGNOSTIC

ITER provides the unique opportunity to study the dynamics of alpha particles in a fusion grade deuterium/tritium plasma. Dynamics of fast ions and fusion alpha particles can be diagnosed by collective Thomson scattering (CTS), and in 2003, an extensive physics feasibility study concluded that the optimal configuration of a fast ion CTS diagnostic for ITER would be a sub-harmonic system ($\omega_{probe} < \omega_{ce}$) having a probe frequency of 60 GHz [1, 2]. The diagnostic fulfils the ITER measurement criteria for fast ions (40 ms temporal resolution, a/10 spatial resolution, and a minimum of 16 resolved nodes in velocity space). The diagnostic is technologically robust with no moveable components near the plasma, and can be built with present day technology. It consists of two 60 GHz 1 MW gyrotron probes and two times ten receivers located on the low field side (LFS) and high field side (HFS), enabling diagnosis of the perpendicular and parallel fast ion dynamics, respectively. The two probe beams will be injected from the LFS equatorial port #12, next to the LFS receiver system, while the HFS receiver will be located behind blanket modules – see sketches of the systems and the scattering geometries in Figure 1.

CP988, *Burning Plasma Diagnostics*, edited by F. P. Orsitto, G. Gorini, E. Sindoni, and M. Tardocchi

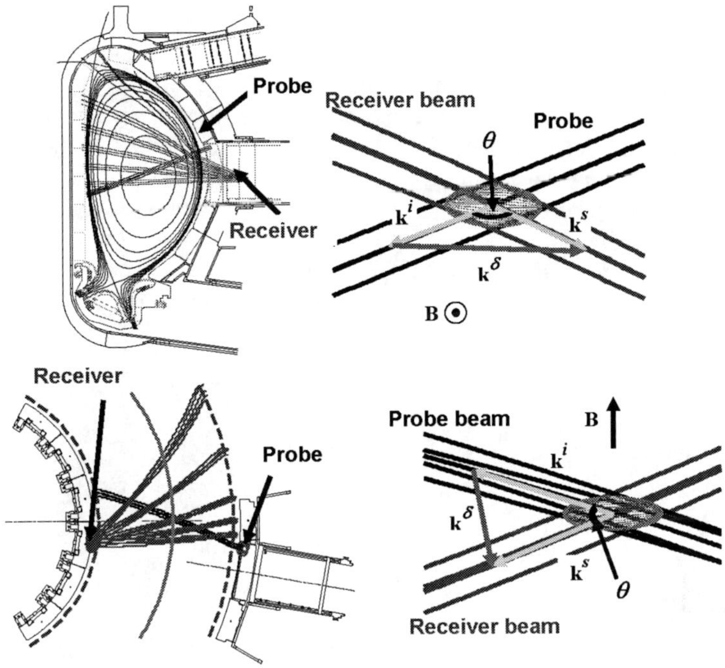

FIGURE 1. Sketch of the LFS system of the fast ion CTS diagnostic for ITER (top left), and the corresponding scattering geometry (top right) showing that dynamics perpendicular to B is resolved. And sketch of the HFS system of the fast ion CTS diagnostic for ITER (bottom left), and the corresponding scattering geometry (bottom right) showing that dynamics parallel to B is resolved. The shown scattering volumes are just a few examples and not all the actual volumes.

In addition to measurements of the fast ion dynamics, the technique of collective Thomson scattering offers a number of additional interesting measurements on ITER, like those of the fuel ion ratio and the poloidal and toroidal bulk ion drift velocities. More information on the fast ion CTS diagnostic and the additional possibilities can be found in [3]. However, the most challenging part of the diagnostic is the HFS receiving antenna due to its location in the equatorial plane between blanket modules. The remainder of this article concerns the progress towards a final design of this antenna.

DEVELOPMENT OF THE HFS CTS ANTENNA

The HFS antenna viewing the plasma consists of a quasi-optical transmission line with one or more mirrors, and a set of approximately ten horns viewing these quasi-optics. From the horn array, waveguides transmit the received signal along the vacuum vessel wall behind the blanket modules and exit the tokamak via an upper port. The properties of the receiving beams, and hereby the locations and localization of the measurement volumes, are determined by the quasi-optical mirror antenna.

The initial design task was to calculate the vertical and horizontal extent of the gap between the blanket modules, in order to fulfill the requirements to the Gaussian beams of the receiver. In the vertical direction a gap being too narrow would result in an increased opening angle of the beam, leading to a poorer signal level, while the horizontal extent of the gap should be sufficient to accommodate the 10 antenna beams without affecting the beam properties. To support calculations of the required gap, the first mock-up of the HFS quasi-optical transmission line was manufactured. At that time, 110 GHz microwave equipment was available; hence, the mock-up was downscaled by the ratio of 60GHz/110GHz. The setup of the mock-up is shown in Figure 2, while a deeper description of the mock-up and the results is found in [4, 5].

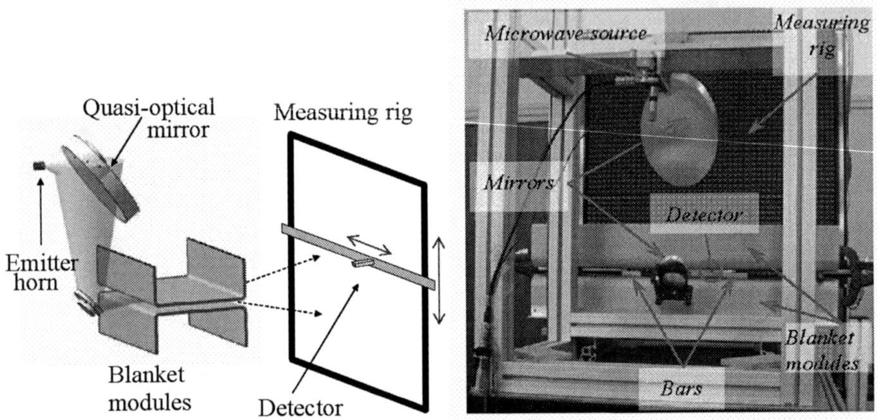

FIGURE 2. Schematic (left) and photo (right) of the experimental setup of mock-up I viewed from the emitter/horn.

The beams of the HFS antenna have to be astigmatic. The mirrors of the first mock-up did not include this astigmatism, and the vertical and horizontal problems were decoupled by two sets of mirrors [4, 5]. The second mock-up consisted of two astigmatic mirrors, which produced a Gaussian beam with the needed astigmatism at the gap entrance between the blanket modules (see Figure 3). The actual geometry was not realistic with respect to ITER, but the purpose was to study the propagation of astigmatic beams and to study the effect on the off-axis beams (see Figure 4). Furthermore, sensitivity to alignment of the horns was investigated.

FIGURE 3. Photo of mock-up II with the line representing the path of the beam from the horn to the slit via the two mirrors.

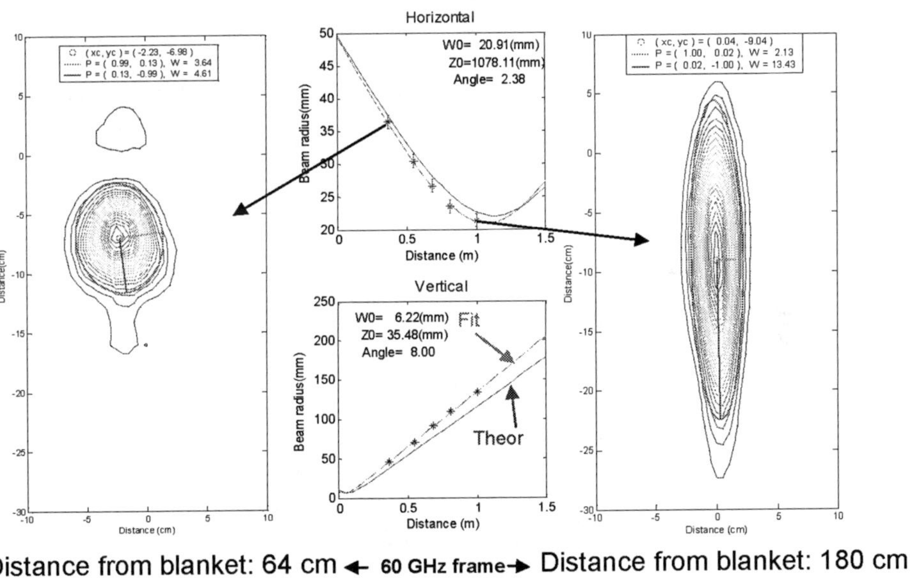

FIGURE 4. Results of measurements of the astigmatic Gaussian beams from mock-up II. Note that the numbers in the graphs are dimensions in 110 GHz frame (which is the frequency of source).

Extended numerical studies of the properties of Gaussian beams in the quasi-optical transmission line and the measurements on mock-up II showed that the two-mirror systems have very astigmatic beams causing too much distortion of the off-axis beams. Furthermore, the two-mirror antenna has a large degree of focusing, causing the system to be very sensitive to misalignment. Hence, it was found that the best option for the HFS CTS receiver is an antenna consisting of four quasi-optical astigmatic mirrors. Based on the findings of the preceding mock-ups and simulations, a third mock-up was designed and produced in 1:1 scale, along with a mock-up of the surrounding blanket modules in a realistic geometry. The goals of the third mock-up are: 1) to demonstrate an engineering solution to the HFS CTS antenna, 2) to study effects of misalignment, 3) to investigate properties of off-axis beams and investigate different horn configurations, and 4) to measure the signal throughput. The design drawing of the mirror system and a photograph of the actual mock-up III is shown in Figure 5. Measurements are currently being performed and will be reported elsewhere.

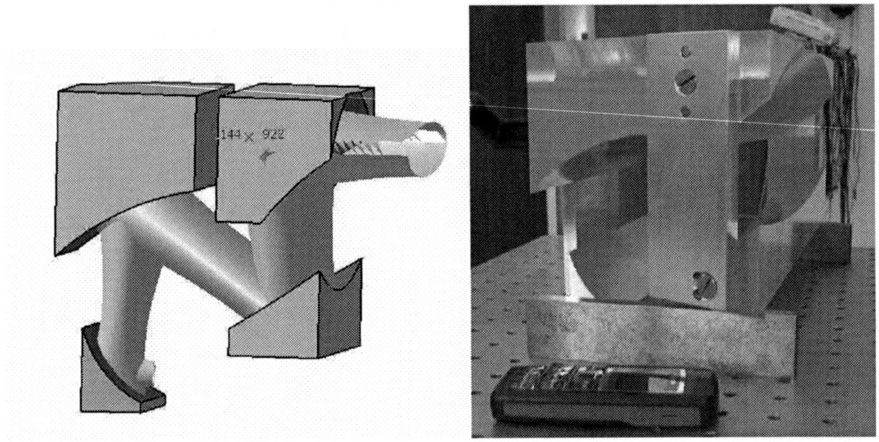

FIGURE 5. Design drawing of mock-up III (left) and a photo (right) of the 4-mirror antenna. The 4-mirror assembly will focus the elliptical beam at the slit in the blanket to a circular beam at the horn. The horn is not shown in the figure.

ACKNOWLEDGMENTS

This work was supported by the European Communities under the contract of Association between EURATOM and Risø, and was carried out within the framework of the European Fusion Development Agreement (EFDA). The views and opinions expressed herein do not necessarily reflect those of the European Commission.

REFERENCES

1. H. Bindslev, F. Meo, and S. B. Korsholm, ITER Fast Ion Collective Thomson Scattering, Feasibility study and conceptual design (URL: http://sitecoremedia.risoe.dk/research/Fusion_CTS/Documents/CTS_Feasibility.pdf)
2. H. Bindslev, F. Meo, E.L. Tsakadze, S.B. Korsholm, P. Woskov, Rev. Sci. Instrum. **75**, 3598-3600 (2004)
3. URL: http://cts.risoe.dk
4. S. B. Korsholm, H. Bindslev, F. Meo, F. Leipold, P.K. Michelsen, S. Michelsen, S.K. Nielsen, E.L. Tsakadze, P. Woskov, E. Westerhof, J.W. Oosterbeek, J. Hoekzema, F. Leuterer, D. Wagner, Rev. Sci. Instrum., **77**, 10E514 (2006)
5. E. Tsakadze, H. Bindslev, S.B. Korsholm, A.W. Larsen, F. Meo, P.K. Michelsen, S. Michelsen, A.H. Nielsen, S. Nimb, B. Lauritzen, E. Nonbøl, and N. Dubois, *Fusion Sci Technol*, **53**, 69-76 (2008)

Comparison of Inboard-Outboard Pedestal Temperature Measurements in JET Using ECE Diagnostics

L. Barrera[1], E. de la Luna[1], L. Figini[2] and JET EFDA Contributors[3]

JET-EFDA, Culham Science Centre, OX14 3DB, Abingdon, UK

1) Laboratorio Nacional de Fusión, Asociación EURATOM-CIEMAT, Madrid, Spain

2) Istituto di Fisica del Plasma, CNR, EURATOM-ENEA-CNR Association, Milano, Italy

3) See the Appendix of M.L.Watkins et al., Fusion Energy 2006 (Proc. 21st Int. Conf. Chengdu, 2006) IAEA, (2006)

Abstract. Despite considerable effort, both theoretically and experimentally, a complete physical model to describe the particle and energy losses during ELMs is far from complete. On the experimental front, improved description of the spatial structure (poloidal asymmetry, radial distribution) and the dynamics of the ELM crash is a key requirement to answer some of the basic outstanding questions concerning the physics of ELMs. A significant number of diagnostics is now capable of fast measurements of the pedestal profile during an ELM, however, there is a lack of data from the inboard midplane, so assumptions of poloidal symmetry on the flux surfaces have often to be made. The aim of this work is to explore the capabilities of the electron cyclotron emission (ECE) diagnostics to provide simultaneous measurements of the edge temperature for both inboard and outboard plasma midplane. Access to the inboard region of the plasma is achieved in JET by using 1st harmonic/O-mode polarization, as it is not affected by harmonic overlap with the 2nd harmonic. This paper focuses on the validation of the inboard ECE data and the identification of the limitations of the measurements and the data analysis.

Keywords: ECE, HFS, ELMs, Temperature profile.

PACS: 52.25.Tx, 52.70.Gw

ELECTRON TEMPERATURE MEASUREMENTS IN JET

Several diagnostics provide the electron temperature profile in JET: the electron cyclotron emission (ECE) heterodyne radiometer [1], a core and edge LIDAR Thomson Scattering system [2] and a recently installed High Resolution Thomson Scattering (HRTS) system [3]. Both the HRTS and the edge LIDAR systems have a nominal spatial resolution of 1.5 cm (at the pedestal region), for the core LIDAR the resolution is ~12 cm. The ECE radiometer consists of 96 closely spaced channels (~1 cm for the magnetic field gradient at JET) with an spatial resolution of ~3-6 cm depending on the plasma conditions and the harmonic number (including spectral resolution of the instruments, relativistic broadening of the cyclotron emission and antenna pattern effects). For the experiments reported here a sampling rate of 5 kHz was used. The radiometer is cross-calibrated against an absolutely calibrated scanning Michelson Interferometer (2nd harmonic, X-mode) during the ohmic phase of the discharge. All of these diagnostics provides local values of the electron temperature at various poloidal locations. In order to compare the radial temperature profiles from the different diagnostics, measurements are mapped onto the mid-plane using EFIT reconstruction. The objective of this paper is to report on a new set of recent measurements obtained at JET using high spatial and temporal resolution ECE data from both high-field side (HFS) and low-field-side (LFS) midplane. Given the relative

CP988, *Burning Plasma Diagnostics*, edited by F. P. Orsitto, G. Gorini, E. Sindoni, and M. Tardocchi
2008 American Institute of Physics 978-0-7354-0507-3/08/$23.00

novelty of the technique, this paper focuses on the validation of the new ECE data measured at the HFS and the identification of the limitations of the measurements and the data analysis.

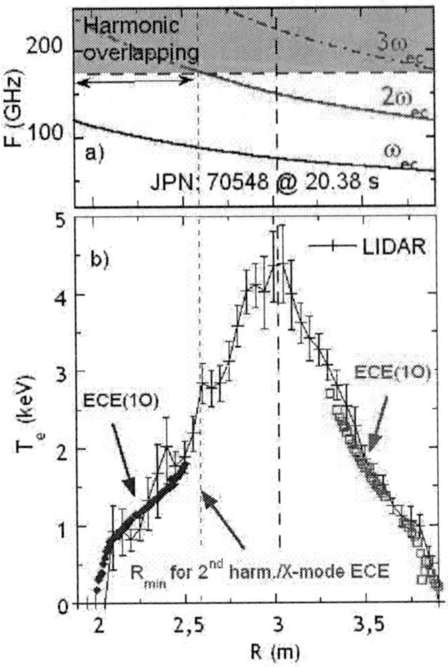

FIGURE 1. a) Harmonics of the electron cyclotron frequency showing the region where 2nd and 3rd harmonic overlap, b) T_e profile measured by LIDAR and ECE: 2nd harmonic/X-mode is used for access to the LFS region and 1st harmonic/O-mode for access to the HFS region.

SIMULTANEOUS HFS/LFS T_e MEASUREMENTS BY ECE

The radiometer in JET collects the emission using an antenna located on the low field side of the torus. It consists on 6 independent heterodyne receivers in the frequency range 69-139 GHz. Each of the receivers can be independently set to measure O-mode or X-mode polarized radiation (for first and second harmonic respectively) which allows good radial coverage of the outer half of the plasma for a large range of toroidal fields (up to 4 T). ECE measurements using the 2nd harmonic have an intrinsic limitation due to the harmonic overlap with the 3rd harmonic, so the maximum radius at which reliable temperature measurements can be obtained in JET is limited to R>2.6 m. However, for 1st harmonic O-mode polarization this limitation does not exist and access to the inboard plasma region can be achieved (see Fig. 1). By using a combined O- and X-mode operation and for a sub-set of toroidal fields, simultaneous access to the plasma edge at both the high-field-side (HFS) and the low-field-side (LFS) region is possible.

Analysis of the ECE Measurements from the Inboard Region

One difficulty in the analysis of the ECE data from the HFS is that the standard scheme of cross calibrating the radiometer data (1st harmonic/O-mode) against the Michelson data (2nd harmonic/X-mode) is not valid anymore, and a different technique must be implemented. Two independent calibration methods have been compared: cross-calibration with the Michelson interferometer using specific toroidal field ramp pulses and comparison to the data measured at the plasma LFS during L-mode. With the first method, the radiometer data is calibrated using pulses in which the O-mode channels are located in the LFS region (R<3 m) and therefore they can be cross-calibrated against the Michelson interferometer. This requires to carry out the calibration pulses (ohmic pulses with a magnetic field ramp) at a higher magnetic field than that used for the pedestal measurements and to select the appropriate radiometer configuration (the calibration factors depend on all the components along the heterodyne detection stage, i.e. the mixer, the IF amplifier, the IF filter and the video detector). However, due to the variability of the measured calibration factors (mainly due to lack of stability on the receiver's sensivity), it was necessary to develop an alternative calibration technique. With this second technique the radiometer channels located at the plasma HFS are calibrated by comparison with the channels at the plasma LFS during L-mode (assuming that the T_e is constant on a given flux surface). In order to obtain a reliable calibration the following conditions must be fulfilled: the method is restricted to the L-mode phase (reliable equilibrium reconstruction) where the gradients are nearly lineal (in this case the possible differences in calibration due to the different spatial resolution for O-mode and X-mode are minimized) and in the region where T_e >150 eV (to guarantee high enough optical thickness). The errors for the calibration factors obtained using the second method (when it is compared with the absolute calibration technique) give typical uncertainties of less than 10%.

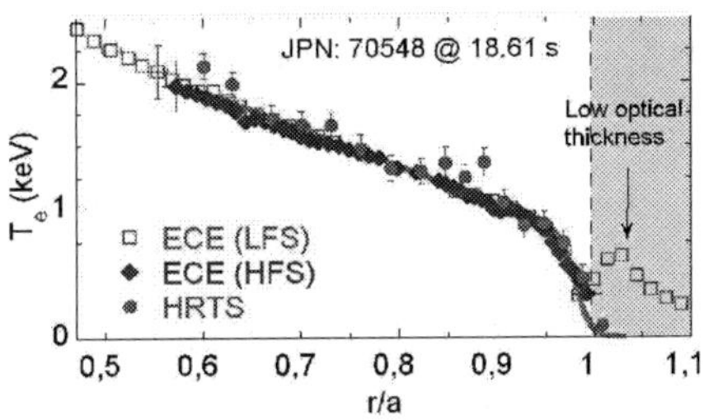

FIGURE 2. T_e profile measured by ECE (for both HFS and LFS) and HRTS for a low density ELMy H-mode discharge. The position of the ECE profiles has been calculated increasing the total magnetic field by 0.7%.

Figure 2 shows an example of the comparison between the T_e profiles measured by ECE (with the HFS calibrated using the second method described above), both in the LFS and the HFS, across the pedestal region and the data obtained with the HRTS system in JET for an ELMy H-mode plasma (B_0=2.7 T, I_p=2 MA, P_{NBI}=8.7 MW, P_{ICRH}=1.75 MW, n_{e0}=6·10^{19} m^{-3}). In general, good agreement is obtained between ECE and TS systems for relatively low density plasmas (where refraction effects are negligible). It is a general observation that the ECE profiles on the LFS are systematically shifted approximately 5 cm with respect to the TS profiles. This shift can be corrected by including an error of <1% in the value of the toroidal field, which is well within the uncertainty of its calibration. We have found that this correction also brings into alignment the inboard and outboard ECE profiles as it can be seen in Fig. 2.

One of the main limitations for the use of the 1st harmonic O-mode ECE data is that it is closer to its cutoff frequency than the 2nd harmonic X-mode, which makes it more susceptible to cutoff and refraction effects. This limiting factor is stronger at lower magnetic fields (B=2.7 T was found the best option in JET).

Refraction effects become important as the cutoff density is approached. In those conditions, the trajectory of the cyclotron radiation is bent causing the emission at a given frequency to come from a different region than expected from its nominal location along the antenna axis. This effect can be calculated using a beam tracing code [6]. In figure 3 an example of such kind analysis is shown for a high density H-mode plasma (B_0=2.7 T, I_p=2.5 MA, P_{NBI}=13.9 MW, P_{ICRH}=2 MW, n_{e0}=$9\cdot10^{19}$ m^{-3}). Figure 3a shows a clear asymmetry between the LFS and HFS temperature profiles mapped onto flux coordinates assuming that the emission comes from the cold resonance position along the antenna axis. When the location of the emission is corrected taking into account the refraction experienced by the emitted radiation, the resultant profiles are approximately equal in shape but they are shifted. This shift can be finally removed by including the error in the magnetic field mentioned previously.

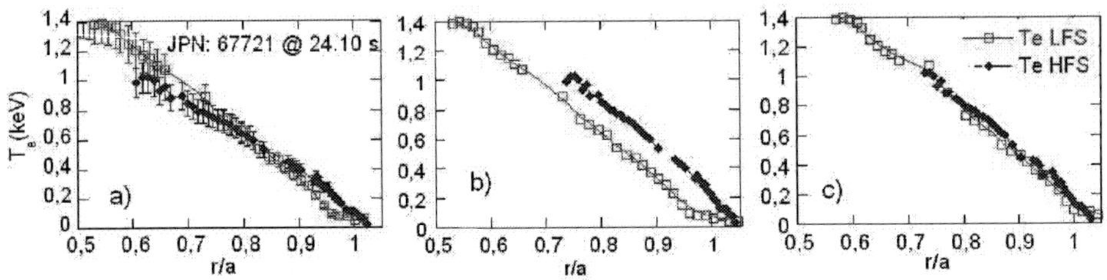

FIGURE 3. Temperature profiles measured by ECE and mapped onto flux coordinates for the LFS (red, open symbols) and HFS (blue, closed symbols): a) assuming that the emission location is given by the cold resonance along the antenna line of sight, b) correcting the emission location taking into account the refraction effects and c) the same than b) including magnetic field error of ~1%.

CONCLUSIONS

In JET, a new set of ECE measurements of electron temperature profiles across the pedestal region at both HFS and LFS has been recently obtained. The access to the inboard side of the plasma is achieved by using 1st harmonic O mode emission which it is not affected by harmonic overlap. While it may seem obvious to use the O-mode electron cyclotron radiation to have access to the HFS, this task has hardly been taken on in the literature. Part of the problem is that the use of X-mode has been favored due to the better spatial resolution. In the case of JET the achievable spatial resolution on the HFS, calculated using the SPECE code [6], is poorer than for LFS data (<3 cm in the LFS and <6m in the HFS pedestal region), but the measured data have shown that it is good enough for a comparative analysis of the inboard/outboard pedestal region in ELMy H-mode plasmas [4]. The difficulty of access to the inboard plasma region in large fusion devices makes this new ECE data measured at JET especially relevant, in particular for improving our understanding of ELM dynamics.This information can also provide a very valuable input for the equilibrium reconstruction before and after the ELM. Analysis of the temperature drop caused by the ELMs crash in both the inboard and outboard plasma region is in progress [4] and will be published elsewhere. For ITER, the access to the inboard plasma midplane is restricted due to harmonic overlapping, effect that becomes more important as T_e increases [5,6]. Further analysis is in progress to asses the possible use of O-mode ECE from the inboard region during the first operation phase in ITER (with reduced plasma parameters).

ACKNOWLEDGMENTS

The work leading to this article was funded by the European Atomic Energy Community and is subject to the provisions of the European Fusion Development Agreement.

REFERENCES

[1] E. de la Luna *et al*, Rev. Sci. Instrum. 75 (2004) 3831
[2] C. Gowers *et al*, J. Plasma Fusion Res. 76 (2000) 874
[3] R. Pasqualotto *et al*, Rev. Sci. Instrum. 75 (2004) 3891
[4] E. de la Luna *et al*, 34th EPS Conf. on Plasma Physics Controlled Fusion (Warsaw, Poland) 2007
[5] L. Figini *et al*, This conference
[6] D. V. Bartlett *et al*, Workshop on Diagnostic for Experimental Fusion Reactors (1997)

SPECE: a code for Electron Cyclotron Emission in tokamaks

D. Farina, L. Figini, P. Platania, C. Sozzi

Istituto di Fisica del Plasma, CNR, Euratom-ENEA-CNR Association, Milano, Italy

Abstract. The code SPECE has been developed for the analysis of electron cyclotron emission (ECE) in a general tokamak equilibrium. The code solves the radiation transport equation along the ray trajectories in a tokamak plasma, in which magnetic equilibrium and plasma profiles are given either analytically or numerically, for a Maxwellian plasma or a non thermal plasma characterized by a distribution function that is the sum of drifting Maxwellian distributions. Ray trajectories are computed making use of the cold dispersion relation, while the absorption and emission coefficients are obtained solving the relevant fully relativistic dispersion relation valid at high electron temperature. The actual antenna pattern is simulated by means of a multi-rays calculation, and the spatial resolution of the ECE measurements is computed by means of an algorithm that takes properly into account the emission along each ray of the beam. Wall effects are introduced in the code by means of a heuristic model. Results of ECE simulations in a standard ITER scenario are presented.

Keywords: ECE (Electron Cyclotron Emission), ITER
PACS: 52.25.Os, 52.70.Gw

THE SPECE CODE

The SPECE code computes the solution of the equation of radiation transfer along the ray trajectories in a general tokamak equilibrium, using the relativistic formulation for emission and absorption. Within the WKB approach, wave propagation is computed solving the ray equations

$$\frac{d\mathbf{x}}{ds} = -\frac{\partial D/\partial \mathbf{N}}{|\partial D/\partial \mathbf{N}|}\bigg|_{D=0}, \qquad \frac{d\mathbf{N}}{ds} = \frac{\partial D/\partial \mathbf{x}}{|\partial D/\partial \mathbf{N}|}\bigg|_{D=0}, \tag{1}$$

being $D(\omega, \mathbf{N}, \mathbf{x})$ the cold dispersion relation for EC waves with frequency ω and refractive index vector $\mathbf{N}$. The transfer equation reads [1]

$$n_r^2 \frac{d}{ds}\left[\frac{I_\omega}{n_r^2}\right] = \beta - \alpha I_\omega\,, \tag{2}$$

where I_ω is the radiation intensity, α and β the absorption and emission coefficients and n_r the ray refractive index. Introducing the radiative temperature $T_{rad} = 8\pi^3 c^2/\omega^2 (1/n_r^2)(\beta/\alpha)$, and the optical depth $\tau(s) = \int_0^s \alpha(s')ds'$, the solution of Eq. (2) for the intensity I_ω emerging from the plasma is

$$I_\omega = I_{\omega 0} e^{-\tau_0} + \frac{\omega^2}{8\pi^3 c^2}\int_0^L T_{rad}(s)\alpha(s) e^{-\tau(s)} ds\,, \tag{3}$$

CP988, *Burning Plasma Diagnostics*, edited by F. P. Orsitto, G. Gorini, E. Sindoni, and M. Tardocchi

where $I_{\omega 0}$ is the intensity entering the plasma, and $\tau_0 = \int_0^L \alpha(s)ds$ the optical depth relevant to the one-pass optical path L. In terms of the single pass integration, the antenna temperature $T_A = (8\pi^3 c^2/\omega^2)I_\omega$ is given by $T_A = \int_0^L G(s)ds$, with $G(s) = T_{rad}(s)\alpha(s)e^{-\tau}$. Within the relativistic kinetic theory, the radiative temperature can be expressed as the ratio between the sums over the cyclotron harmonics of suitable resonant integrals in momentum space

$$T_{rad} = mc^2 \frac{\sum_n \int d\mathbf{u}[f(\mathbf{u})/\gamma]\, |\Theta_n|^2\, \delta(\gamma - n\Omega/\omega - N_\parallel u_\parallel)}{\sum_n \int d\mathbf{u}[-\hat{L}[f(\mathbf{u})]\, |\Theta_n|^2\, \delta(\gamma - n\Omega/\omega - N_\parallel u_\parallel)}, \qquad (4)$$

where Ω is the cyclotron frequency, $\mathbf{u} = \mathbf{p}/mc$, $\gamma = \sqrt{1+u^2}$, $\Theta_n(u_\perp, u_\parallel, \Omega, k_\perp, \mathbf{E})$ is the polarization term and the differential operator reads $\hat{L}f = 1/u_\perp \partial f/\partial u_\perp - N_\parallel/(\gamma u_\perp)(u_\parallel \partial f/\partial u_\perp - u_\perp \partial f/\partial u_\parallel)$. As it is well known, from Eq. (4) $T_{rad} = T_e$ is obtained in the case of a Maxwellian distribution function.

The code is specialized to the case in which the electron distribution function is either a Maxwellian distribution f_m with density n_e and temperature T_e or a superposition of dritfing Maxwellian functions, $f = (1 - \sum_i \eta_i) f_{m,b} + \sum_i \eta_i f_{m,i}$, with $f_{m,i}$ the Lorentz transformed Maxwellian distributions with drift velocity $v_{0,i}$, temperature $T_{e,i}$ and density $n_{e,i} = \eta_i n_e$. The absorption coefficient $\alpha = 2(\mathrm{Im}\mathbf{k}) \cdot \mathbf{v}_g/v_g$ is computed by solving the relativistic dispersion relation for the EC waves at the required Larmor radius expansion [2], with dielectric tensor $\boldsymbol{\varepsilon} = \mathbf{I} + 4\pi i/\omega(\boldsymbol{\sigma}_{m,b} + \sum_i \boldsymbol{\sigma}_{m,i})$. The conductivity tensors for the drifting Maxwellians $\boldsymbol{\sigma}_{m,i}$ are obtained from the conductivity tensor for a Maxwellian distribution by means of Lorentz transformations [3, 4].

The radiative temperature is obtained by Lorentz transforming the Kirchoff law

$$T_{rad} = \frac{\tilde{\alpha}_b T_{e,b} + \sum T_{e,i}\tilde{\alpha}_i/[\gamma_{0i}(1 - N_\parallel v_{0,i}/c)]}{\tilde{\alpha}_b + \sum \tilde{\alpha}_i}, \qquad (5)$$

where $\gamma_{0i} = [1 - v_{0,i}^2/c^2]^{-1/2}$ is the relativistic factor, and $\tilde{\alpha}_j = \mathbf{E}^* \cdot \boldsymbol{\varepsilon}^{AH}(f_j) \cdot \mathbf{E}$ is computed in terms of the antihermitian part of the relativistic dielectric tensor relevant to f_j and of the electric field obtained from the solution of the relativistic dispersion relation with the full distribution function.

The antenna pattern is simulated using a set of M rays, uniformly distributed in radial and angular direction with respect to the antenna axis. The weight $q_i = \exp[-2(\theta_i/\theta_0)^2]$ is assigned to the ith ray in order to describe a Gaussian beam in the far field approximation, being θ_i the angle between the ray and the antenna axis, and $\theta_0 = \tan^{-1}[2c/(\omega w_0)]$ the beam divergence, with w_0 the beam waist. To determine the effective radial resolution of the emission, the function $G_i(s)$ is computed on the ith ray versus the spatial coordinate along the ray and mapped (moving along the magnetic surface) onto the radial coordinate in the plasma midplane R, so that $G_i(R) = G_i(s)|ds/dR|$. Finally, the weighted contributions of the M rays are summed up to obtain $G(R) = \sum_i q_i G_i(R)/\sum_i q_i$. The radial resolution is estimated in terms either of the function $G(R)$ or of its integral function $\Gamma(R) = \int_0^R G(R')dR'$, being $\Gamma(R_A) = T_A$. In the following, we shall make use of the function $\Gamma(R)$ to estimate the mean emission radius R_{50}, and the radial resolution $\Delta R_{95} = |R_{2.5} - R_{97.5}|$, with R_P such that $\Gamma(R_P) = (P/100)T_A$ [5].

To simulate the emitted intensity measured by a polarized antenna, the radiation is decomposed into the two components corresponding to the radiation emerging from the plasma as pure O and X mode.

The effect of wall reflection and polarization scrambling (relevant in case of thin harmonics) is described by means of the empirical reflection and polarization transfer fraction coefficients r and p, extending the formula in [6] to the case of finite τ_O. The total emission $\bar{I}_\omega^{O,X}$ is calculated from the single-pass intensity (I_ω^O, I_ω^X) as

$$\bar{I}_\omega^{O,X} = \frac{[1-r(1-p)e^{-\tau_{X,O}}]I_\omega^{O,X} + rpe^{-\tau_{O,X}}I_\omega^{X,O}}{1-r(1-p)(e^{-\tau_X}+e^{-\tau_O})+r^2(1-2p)e^{-\tau_X}e^{-\tau_O}}. \tag{6}$$

Expression (6) has been derived assuming that the same radiative temperature is seen on both sides of the plasma (i.e., on inner and outer walls). Corrections may be required in the cases of high temperature plasmas and/or harmonic overlap. Within the same framework, the model has been generalized taking into account both inner and outer radiation (we skip here the explicit analytical formula for sake of simplicity). We note that the above models are valid in the absence of cutoffs inside the plasma volume.

This code, originally developed for the Oblique Electron Cyclotron Emission diagnostic recently implemented at JET [7], can be conveniently used to perform simulations relevant to ITER scenarios.

ECE SIMULATIONS IN ITER

The capabilities of ECE diagnostics in different ITER scenarios have been investigated with the SPECE code. Here we present a few results relevant to Scenario 2 for perpendicular (0°) and oblique views (10° and 20°) locating the antenna at $R = 9.72$ m and $z = 0.56$ m with waist 0.7 m in front of it. The "pure" XM and OM spectra are shown in Fig. 1. As already investigated in the literature, good spatial resolution is achieved only in the case of 1st harmonic O-mode (OM1) and 2nd harmonic X-mode (XM2). Plasma centre is accessible only with OM1, while plasma edge is well resolved both with OM1 and XM2 (XM2 has higher resolution). For OM1 and XM2, the emission is localized

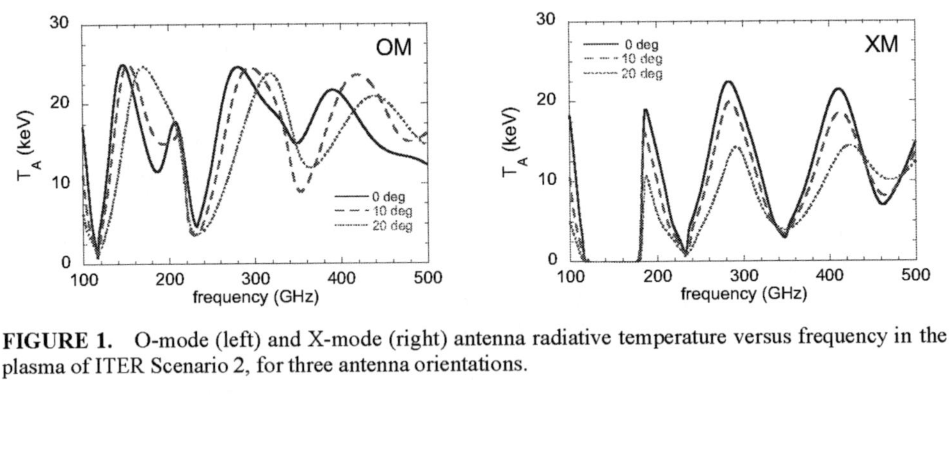

FIGURE 1. O-mode (left) and X-mode (right) antenna radiative temperature versus frequency in the plasma of ITER Scenario 2, for three antenna orientations.

at $R \lesssim R_c(\omega)$, with $n\Omega(R_c) = \omega(1-N_\parallel^2)^{1/2}$. This is responsible for the frequency shift in the oblique spectra (Fig. 1). Simultaneous ECE measurements performed at various angles with respect to the magnetic field may give some information on possible distortions of the electron distribution function, since the resonant electrons contributing to the emission have different energies [8]. In particular, for the case of OM1 (i.e., in the frequency range $\sim 120-170$ GHz) it is found that the energy of the resonant electrons is lower than thermal, thermal, and mildly suprathermal at 0°, 10°, and 20° respectively.

The effective spatial resolution has been studied for OM1 and XM2 in the frequency ranges where the emission is well localized in space (see Fig. 2). The simulations have been performed both with a single ray and with different antenna beamwidths. It is found that the effect of the antenna pattern is almost negligible in perpendicular direction. At oblique propagation, ΔR_{95} increases with the beamwidth due to the Doppler effect, and this effect is more evident at OM1 since the beamwidth is larger at lower frequencies.

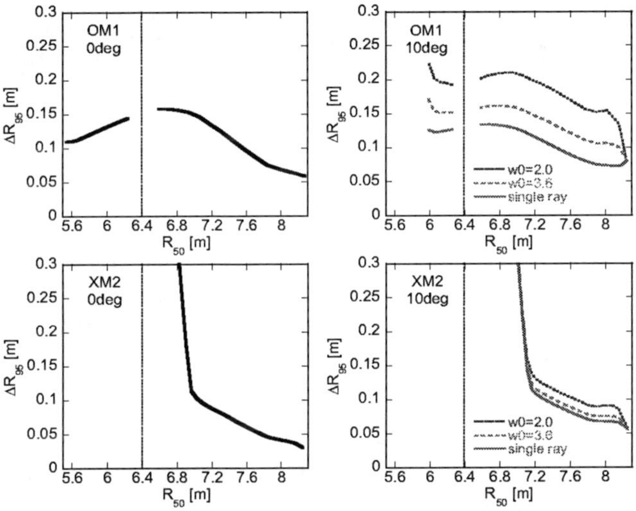

FIGURE 2. Radial resolution ΔR_{95} versus mean emission radius R_{50} for OM1 (top) and XM2 (bottom) at 0°, 10°, and 20° in the case of a single ray (solid line) and for two waists $w_0 = 3.6$ cm (dashed line) and 2.0 cm (dotted line). At 0° all the lines overlap. The vertical dotted line represents the magnetic axis.

REFERENCES

1. G. Bekefi, *Radiation Processes in Plasmas*, John Wiley & Sons, New York 1966.
2. D. Farina, *Fusion Science & Technology* **53** January Issue (2008), in press.
3. D. B. Melrose, *Plasma Phys.* **15**, 99 (1973).
4. S. Tamor, *Nucl. Fusion* **19**, 455 (1979).
5. D. V. Bartlett, and H. Bindslev, in *Diagnostics for Experimental Thermonuclear Fusion Reactors 2*, edited by P. E. Stott et al, Plenum Press, New York, 1998, pp. 171–180.
6. A. E. Costley et al., *Phys. Rev. Lett.* **33**, 758 (1974).
7. C. Sozzi, et al., in *Proc. of 34th EPS Conf. on Plasma Physics*, Warsaw, 2007, Paper P-5.079; C. Sozzi, et al., this workshop.
8. E. de la Luna, et al., *Review of Scientific Instruments* **74**, 1414 (2003).

Compact high-speed high-sensitivity second-harmonic interferometer for electron density measurement

F. Brandi, P. Marsili and F. Giammanco

Physics Department, University of Pisa, Largo Bruno Pontecorvo 3, 56127 Pisa, ITALY

Abstract.
A second-harmonic interferometer based on a CW Nd:YAG laser is presented. The device is extremely compact and versatile and is developed for line-integrated electron density measurements in large plasma machines. A sensitivity in measuring line-integrated electron density of 4×10^{13} cm^{-2} with time resolution of 8 μs, and 10^{13} cm^{-2} with time resolution of 50 μs, is demonstrated.

Keywords: electron density, second-harmonic interferometer
PACS: 52.70.-m,07.60.Ly

INTRODUCTION

The second-harmonic interferometer (SHI) [1, 2] is a common-path, two-color interferometer that measures the line-integrated dispersion at the fundamental and second-harmonic wavelength of a laser beam. When used with plasma the SHI measures the line-integrated electron density (LIED), i.e., $\int_L N_e(z)dz$, where L is the overall plasma length, and N_e is the free electron density. Being a common-path interferometer the SHI is insensitive to vibration when compared with conventional two-harm interferometers. This property makes the SHI attractive as diagnostic tool in noisy environments, for example large plasma machines or combustion plants. When compared with typical two-color interferometers the SHI has the advantage of using a single laser source. The principles of operation of the SHI are the following: a second harmonic beam (SHB) is generated in a single pass frequency doubler; the fundamental and SHB co-propagate through the dispersive sample; a second SHB is generated in a second frequency doubler; finally, interference takes place between the two colinear SHBs. A detailed explanation of the SHI optical design and principles is given in Ref. [3].

The schematic of the presented compact device is shown in Figure 1. Two compact Optical Units (OUs) host two second-harmonic generation temperature stabilized crystals (BIBO) and several optical components. Due to their reduce volume and flexible design the OUs, can be conveniently adapted to different experimental set-up, as for example large plasma chambers with strong space constrains. The first OU is connected to the laser with a 10 meter long single-mode polarization maintaining fiber delivering about 0.5 W of narrow-band linearly polarized light at $\lambda = 1064$ nm. The second OU is connected to the optical detectors with multimode optical fibers. The optical fiber connections allow to place the sensitive electrical devices (laser and detectors) far away from the experiment. This is very important when the interferometer has to operate in

CP988, *Burning Plasma Diagnostics*, edited by F. P. Orsitto, G. Gorini, E. Sindoni, and M. Tardocchi

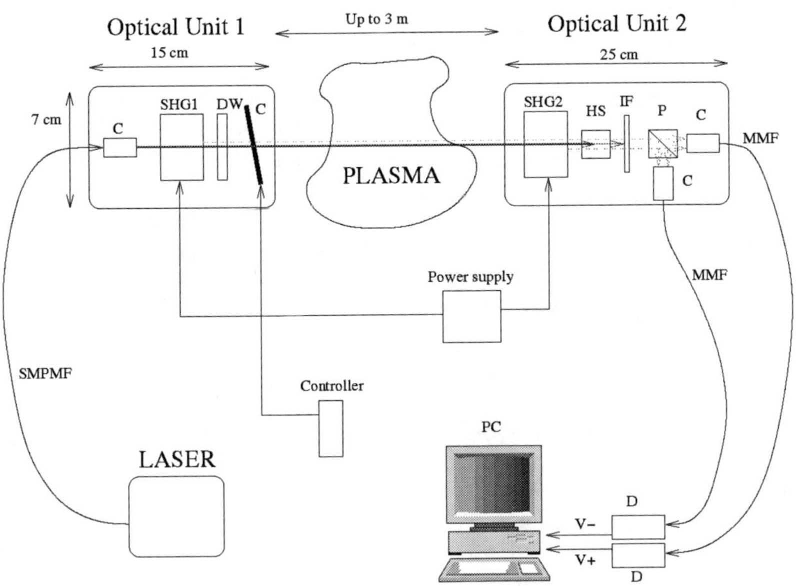

FIGURE 1. Schematic of the SHI: SMPMF, single mode polarization maintaining fiber; C, collimator; SHG, second harmonic generation; DW, dual-wavelength waveplate; C, compensator; HS, harmonic separator; IF, interferential filter; P, polarizer; MMF: multi-mode fiber; D: detector.

environments with high electromagnetic noise. Full remote operation of the interferometer is obtained with a motorized phase compensator, a tilted BK7 optical window, that enables to control the spurious phase shift of the interferometer.

The detector signals are

$$V_{\pm} = \alpha_{\pm}\frac{P_0^2}{2}[\beta_1+\beta_2 \pm 2\sqrt{\beta_1\beta_2}\cos(\phi+\phi_0)], \tag{1}$$

where $\phi = -4.5\times10^{-17}$ LIED[cm^{-2}] is the phase to be measured, $\alpha_{\pm}$ are the detector responsivities, ϕ_0 is the off-set phase due to spurious dispersions plus the compensator phase, P_0 is the laser power, and $\beta_i P_0^2$ with $i = 1,2$ are the intensities of the SHBs. The recorded signal is

$$R \equiv \frac{V_+ - V_-}{V_+ + V_-} = \frac{\alpha \pm V\sin(\Delta\phi+\Delta\phi_0)}{1\pm\alpha V\sin(\Delta\phi+\Delta\phi_0)}, \tag{2}$$

which is insensitive to fluctuations in the laser power, and where $\alpha \equiv \frac{\alpha_+-\alpha_-}{\alpha_++\alpha_-}$, $\Delta\phi_0$ is the un-compensated off-set phase, i.e., $\phi_0 = \frac{\pi}{2}(2p+1)+\Delta\phi_0$ with $p = 0,\pm1,\pm2$, the sign depends upon the value of p, and $V = \frac{2\sqrt{\beta_1\beta_2}}{(\beta_1+\beta_2)} \leq 1$ is the fringe visibility. The value of α can be adjusted to be less than 0.1, while $\Delta\phi_0$ can be set to the mrad level by tilting the compensator window. For $\alpha, (\phi+\Delta\phi_0) \ll 1$, Eq.(2) can be approximated by $R = \pm V(\phi+\Delta\phi_0)+\alpha$.

PERFORMANCE

The sensitivity and time resolution of the interferometer depend on the laser power and on the detectors configuration. The performance of the presented SHI are tested with $P_0 = 0.5$ W, and using a low-noise photodiode in combination with a low-noise variable gain transimpedance. The signals are digitalized with a 16 bit DAQ card.

The device can be self-calibrated, in order to determine the actual values of α and V, by recording R while scanning the compensator angle, and performing a least-square fit using Eq.(2), see Figure 2. The well known dispersion of BK7 glass and a costant step size of the piezo motor of the compensator are assumed for the fit. Actually, such step size is not constant over the entire scan range of two fringes, but it is sufficiently constant over half-fringe, where the fit is performed. The fit gives $\alpha = -0.028$, $V = 0.954$, with an accuracy of few %.

The SHI is tested with a 28 cm long Xe plasma excited in a 10 ms pulsed microwave cavity running at 50 Hz. The total length of the Xe cell is 71 cm. The cell is connected to vacuum pumps and the Xe pressure is 0.6 mbar. The OUs are set 1.5 m apart on two separate tables, and hold by gimbal tilting optical mounts for optical alignment.

In Figure 3(a) a 2 second measurement of the interferometer phase at 20 KS/s is reported in the upper panel, while the simultaneous measurement of emitted light is reported in the lower panel. The actual sensitivity and time resolutions of the device are demonstrated in Figure 3(b): the upper (lower) panel shows the interferometer phase measured at 125 KS/s (20 KS/s), corresponding to 50 μs (8μs) time resolution, over a single plasma pulse, and a phase noise of about 2 mrad (0.5 mrad) is obtained resulting in a LIED sensitivity of 4×10^{13} cm^{-2} (10^{13} cm^{-2}). The actual evolution of the measured phase can be understood as follow. With the rising of the micro-wave intensity, about 0.4 % of the xenon in the oven ionizes in less then 1 ms, thus inducing the fast rise of ϕ.

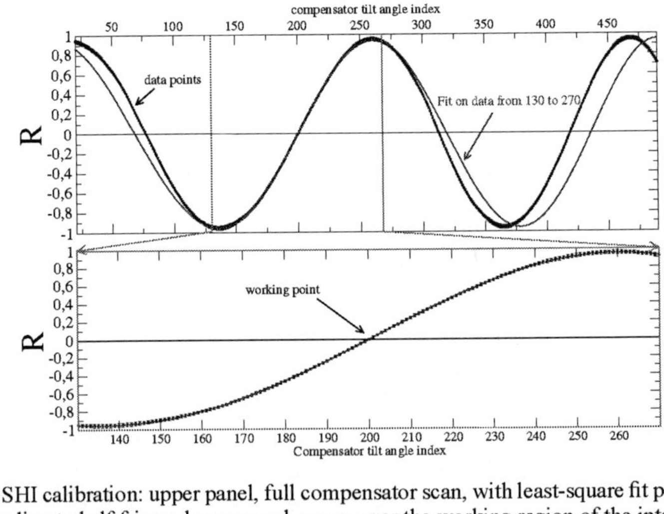

FIGURE 2. SHI calibration: upper panel, full compensator scan, with least-square fit performed on data point corresponding to half fringe; lower panel, zoom over the working region of the interferometer.

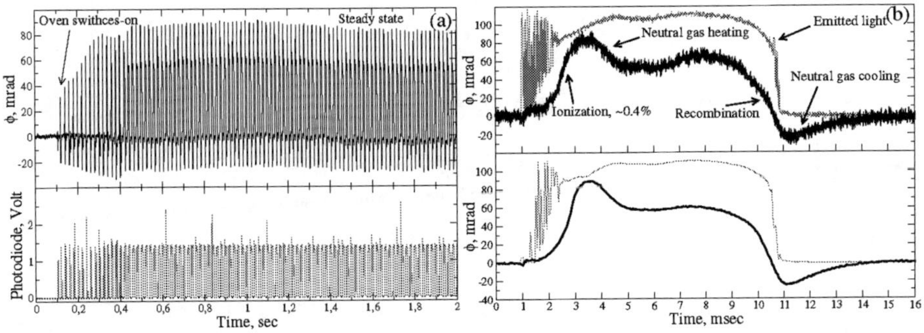

FIGURE 3. Interferometer performance: (a) upper panel, measured phase over 2 sec acquisition time at 20KS/s; lower panel, simultaneous measurement of emitted light using a fast photodiode. (b) upper panel plasma pulse measured at 125 KS/s; lower panel, plasma pulse measured at 20 KS/s; the red/thinner curves are the emitted light signals for timing reference.

In the next few ms, the residual neutral gas in the oven heats up and moves towards the coolest part of the vacuum tubing outside the oven, thus the neutral gas density along the laser path decreases, inducing a decrease of ϕ because of the normal dispersion of neutral Xe ($\Delta n \sim 1.9 \times 10^{-5}$ [4]). At the end of the micro-wave pulse the electrons recombines with the ions, thus inducing a fast decrease of ϕ. However, when the micro-wave pulse is over the value of ϕ falls below the initial value, indicating that the neural density is less then 0.6 mbar. The initial pressure is restored when the neural gas cools down with time scale of few ms.

CONCLUSIONS

A novel SHI based on a CW Nd:YAG laser is presented. The device, developed for free electron measurements in large plasma machines, has the following characteristics: insensitive to vibrations; compact and modular; insensitive to laser intensity noise; remotely operated and controlled; self-calibrating. A sensitivity in line integrated electron density of 4×10^{13} cm^{-2} with time resolution of 8 μs, and 10^{13} cm^{-2} with time resolution of 50 μs, is reported. Such performance is maintained over a measuring period of seconds, and it is stressed that the time resolution and sensitivity of the SHI can be further increased by increasing the fundamental laser power. Finally, the modular design of the interferometer allows for an easy upgrade to multi-chord configuration.

REFERENCES

1. F. Hopf, A. Tomita, and G. Al-Jumaily, *Opt. Lett.* **5**, 386–388 (1980).
2. K. Alum, Y. Koval'chuk, and G. Ostrovskaya, *Sov. Tech. Phys. Lett.* 7, 581–582 (1981).
3. F. Brandi and F. Giammanco, *Opt. Lett.* **32**, 2327–2329 (2007).
4. S. Velasko and D. Eimerl, *Appl. Opt.* **25**, 1344–1349 (1986).

The Interferometer As An Optical Mixer: A Novel Approach To Far Infrared Spectroscopy

M. Zerbini and E. Franconi[(*)]

Associazione Euratom-Enea sulla Fusione, CRE Frascati, FUS-MAG
Via E. Fermi, 45 - 00044 Frascati – Italy
[(*)]Dip. FIM

Abstract. In Tokamaks, the Electron Cyclotron Emission (ECE) radiation frequency spans from 80 GHz up to 1 THz for high magnetic field (Bt> 5 T) and higher harmonics (n>2). This frequency ranges between Microwaves and Far Infrared (FIR), moulding an intrinsically hybrid nature for ECE diagnostics[(1)]. Signal detection is carried out in two surprisingly different ways: electronics and optics. The two detection methods, beyond the chief external differences, are conceptually the same. The dualism often reflects the mindset of the laboratory performing the measurement. In this paper we discuss a different approach to an integrated instrument. A traditional quasi-optical interferometer is the first down-conversion stage with intrinsic broad band capability. Its output signal can then be analysed with a radio-style detection, in the amenable kHz frequency range.

Keywords: Electron Cyclotron Emission, Plasma Diagnostics, Interferometer, Mixer, Heterodyne, Homodyne, Far Infrared.
PACS: 52.70.Gw

DETECTION & MIXING

A receiver is a device, such as a part of a radio, television set, telephone or measurement instrument, that collects radio frequency (RF) signals and converts them to perceptible forms, such as sound, light or a voltage. There are many different ways of detecting the intelligence from the electromagnetic carrier[(2)]. In principle they all consist of a change in frequency, or conversion, which translates a signal at one frequency to another (usually lower) where processing it is more amenable, effective and inexpensive. In order to change the carrier frequency to a lower value more suitable for detection, it can be mixed with a reference high frequency voltage (LO) in a non linear device: the Mixer. The result is a lower frequency output similarly modulated, called the intermediate frequency (IF). Depending on the origin of the LO, the method is called either Homodyne or Heterodyne. On Homodyne the LO signal is self-generated by dividing the input in two components, then changing one of the parameters (usually the phase) and mixing them to analyse the resulting beating frequency, usually lower. The heterodyning is achieved by mixing the modulated carrier voltage with a locally generated high frequency LO. Superheterodyne reception is a form of multistage heterodyne reception in which frequency conversion takes

CP988, *Burning Plasma Diagnostics*, edited by F. P. Orsitto, G. Gorini, E. Sindoni, and M. Tardocchi

place two or more times before the modulated carrier voltage is fed to the final detector.

ECE INSTRUMENTATION DUALISM

A broad-band optical ECE diagnostic usually consists of a polarizing interferometer, capable to detect frequency in the 80-1000 GHz range(3). As shown in Fig. 1, by scanning the optical path difference X with a moving mirror following the law: $X(t)= V_{scan}*t_{scan}$, every single input frequency ν_0 produces in output a purely sinusoidal interferogram of electric frequency $f= \nu_0 * V_{scan}/c$. For broad-band radiation, so called "white", the interferogram is the linear superposition of the individual sinusoidal components:

$$I(x) = 2\int_{\nu} I(\nu)\cdot(1+\cos(2\pi\frac{\nu}{c}V_{scan}\cdot t))d\nu \qquad (1)$$

The Intensity spectrum $I(\nu)$ is obtained via Fourier Transform of the interferogram.

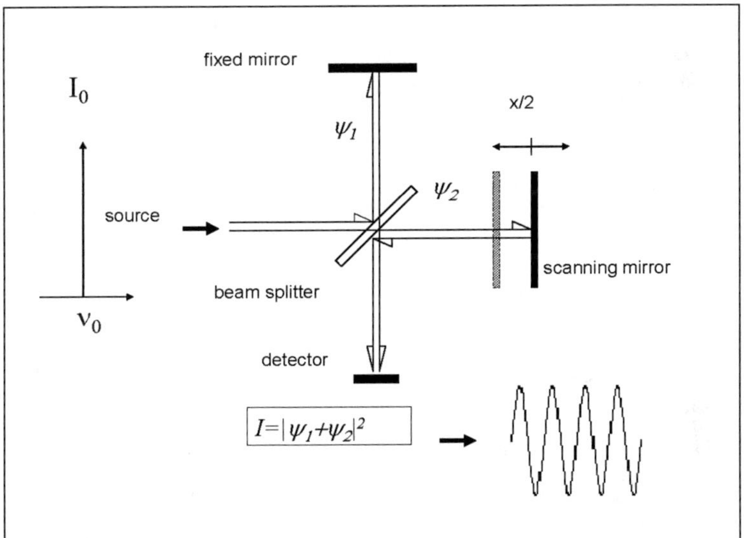

FIGURE 1. Michelson interferometer. The physical displacement of the mirror is X/2.

In the electronic approach(4), the ECE radiation is downconverted, usually by heterodyning, and frequency-resolved with a number of mixing stages. The time depended intensity for each spectral frequency is available at the hardware output of the receiver.

It can be easily shown, with the help of Fig. 2 and a few trigonometric relations, that the Michelson interferometer (eq. 1) and the Homodyne detection, lead to a varying part of the signal, or IF, amplitude modulated (AM) with the phase difference introduced by the modulating device. A Michelson interferometer is the optical equivalent of a Homodyne detector.

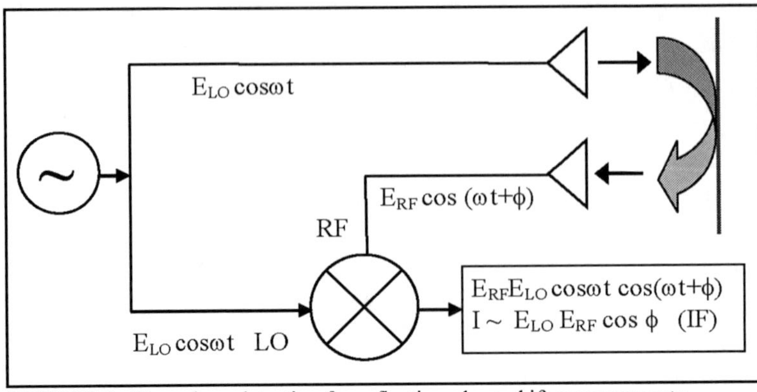

FIGURE 2. Homodyne detection for reflection phase shift measurements

A SUPERHETERODYNE INTERFEROMETER

If V_{scan} is constant, the low (electric) frequency components of the Michelson AM output (interferogram) can be Heterodyne analysed. Or, given the low frequency involved (kHz), a high-Q (quality factor) filter can be used to extract the amplitude of the desired components (Fig. 3). Such multistage, hybrid receiver can be called Superheterodyne Interferometer. This method offers a better time resolution of the traditional Fourier Transform analysis, which requires the full interferogram for spectral analysis. A few wave cycles are sufficient to "clip" the ECE emission amplitude.

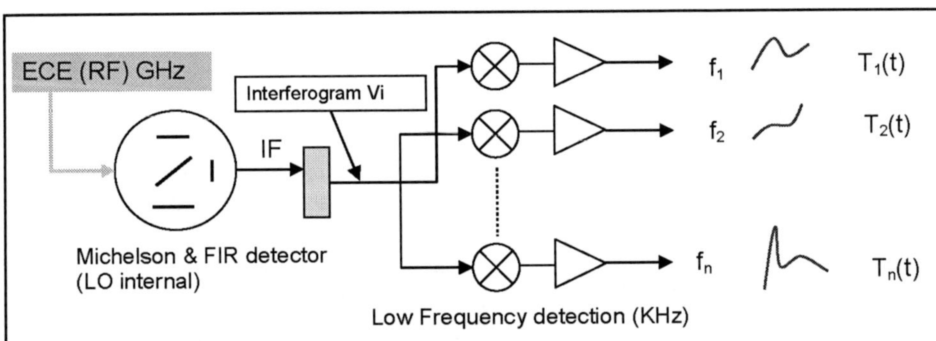

FIGURE 3. Superheterodyne Interferometer concept. T(t) is the temperature vs. time output.

DIAGNOSTIC PARAMETERS

A typical value for V_{scan} is 8 m/s, corresponding to a 5 ms time resolution Michelson, with 4 cm path difference (ref 3, FTU). For a generic 3 T plasma , the ECE frequency in second harmonic will be in the range 100-250 GHz. The resulting IF or electric frequency range is $f_{el} = \nu_0 * V_{scan}/c = 2.5 - 7$ kHz. The time resolution is determined by the number of wave cycles necessary to clip a reasonable average value for each

component. With 3 cycles the estimated time resolution for the above range, with $T=1/f$ is $\Delta T = 3*T = 1.2 - 0.5$ ms.
A simple, heuristic solution for the final detection stage is the High-Q Operational Amplifier filter(5) in Fig. 4. Better performances and higher Q can be obtained using a heterodyne stage. Note that Michelson Interferometers with non-constant scanning speed (shakers) introduce a FM component which would require a more sophisticated detection system (modern Phasitron or Foster-Seeley Discriminator).

f = 4 KHz (150 GHz, ~ 2.5T magnetic field)
B = 60 Hz (2 GHz spectral bandwith)
R1 = R2 = 12 Ω, R3 = 100 kΩ, C = 0.05 μF

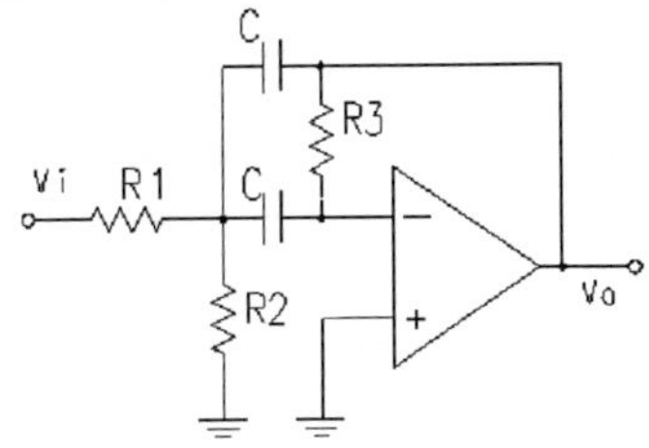

$$\omega_o = \frac{1}{C\sqrt{(R_1 \| R_2)R_3}} \qquad \beta = \frac{2}{CR_3} \qquad f_o = \frac{\omega_o}{2\pi} \qquad B = \frac{\beta}{2\pi}$$

FIGURE 4. High-Q filter for final detection stage. $Q=f_0/B=64$ with the parameters in the example.

FINAL REMARKS

The ECE ultimate problem is to detect the absolute intensity of electro-magnetic radiation, with frequency around 300 GHz (1 mm wavelength), as a function of time. This work is an attempt to show that a fresh approach can be found to a age-old problem. The aim is to stimulate the interest for developing new solutions, not only for ECE diagnostic. Very often plasma diagnostics are taken for granted when major fusion projects are planned. New design and development are not adequately taken in to consideration, assuming that one day, when necessary, the required solution will magically appear. But this is unlikely to happen, since nothing is achieved without hard work, skilful organization and proper funding. The hidden cost of the apparent saving on diagnostics development will be transferred to the future experiments, in terms of inefficiency, delay and loss of data quality.
The instrumentation discussed in this work has great relevance for Fusion, but also for Astrophysics, Nanostructures spectroscopy and Solid State physics.

ACKNOWLEDGMENTS

The authors wish to thank J. Fessey for the useful comments about the detection system.

REFERENCES

1. A. Costley, *Diagnostics for Fusion Reactor Conditions* (Varenna 1982), EUR 8351-I, p. 129 & 149.
2. R. V. Pond, *Microwave Mixers*, McGraw-Hill, 1948.
3. P. Buratti and M. Zerbini, *Rev.Sci.Instrum* **66**, 4208 (1995).
4 H.J. Hartfuss et al. *Plasma Phys. & control. Fusion* **39**, 1693 (1997).
5 J. Millman and C. C. Halkias, *Integrated Electronics*, McGraw-Hill, 1982, p. 556.

Models comparison for JET polarimeter data

C Mazzotta[1], F P Orsitto[1], A Boboc[2], E Giovannozzi[1], M Brombin[3], A Murari[3], O Tudisco, L Zabeo[2], and EFDA-JET Contributors (*)

JET-EFDA, Culham Science Centre OX14 3DB, Abingdon, UK.
[1]Centro Ricerche Energia Frascati, Euratom-ENEA Association, Frascati, Italy
[2] Association EURATOM-UKAEA Culham Science Centre Abingdon OX14 3DB (UK)
[3]Consorzio RFX, Euratom-ENEA Association, Padova, Italy
(*)See the Appendix of M.L.Watkins et al., Fusion Energy 2006 (Proc. 21st Int. Conf. Chengdu, 2006) IAEA, (2006)

A complete comparison between the theory and the measurements in polarimetry was done by using the Far Infrared Polarimeter at JET. More than 300 shots were analyzed, including a wide spectrum of JET scenarios in all critical conditions for polarimetry: high density, high and very low fields, high temperatures.
This work is aimed at the demonstration of the robustness of the theoretical models for the JET polarimeter measurements in the perspective of using these models for ITER like plasma scenarios .
In this context, an assessment was performed on how the line-integrated plasma density along the central vertical chord of FIR polarimeter could be evaluated using the Cotton-Mouton effect and its possible concrete use to correct fringe jumps of the interferometer.
The models considered are: i) the rigorous numerical solution of the Stokes propagation equations, using dielectric tensor evaluated from JET equilibrium and Thomson scattering [1,2]; ii) two types of approximated solutions [2,3] and iii) the Guenther empirical model [4] that considers the mutual effect between Cotton-Mouton and Faraday rotation angle. The model calculations have been compared with polarimeter measurements for the Cotton-Mouton phase shift.
The agreement with theory is satisfactory within the limits of experimental errors [3].

Keywords: Polarimetry, line-integrated plasma density, Cotton Mouton effect, plasma physics.
PACS: 52.70 Kz

Introduction: The JET Polarimeter System

The JET polarimeter, even if sharing the FIR interferometer optical path, takes independent measures by employing different processing electronics and using polarimetric optical principia [1]. In fact the polarization of an electromagnetic wave passing through the plasma changes producing Cotton Mouton and Faraday rotation effects. In particular the Cotton Mouton phase shift angle is related to the plasma behavior similar to a birifrengent medium. This paper presents calculations of Cotton-Mouton angle using different models and compare them with the entire set of available JET polarimetric measures, which includes a wide range of plasma conditions.

Solution of Stokes Equations (Rigorous Solution)

We can briefly consider the three components [1,2] of the Stokes vector $\vec{S}$ when the laser beam is propagating along a vertical chord (z axis) in a poloidal plane. The components of $\vec{S}$ can be expressed in terms of ellipticity (χ), related to the Cotton Mouton phase shift angle (Φ), and Faraday rotation angle (Ψ), through the following equations: $S_1 = \cos 2\chi \cos 2\psi$, $S_2 = \cos 2\chi \sin 2\psi$, $S_3 = \sin 2\chi$ while the propagation of

CP988, *Burning Plasma Diagnostics*, edited by F. P. Orsitto, G. Gorini, E. Sindoni, and M. Tardocchi
2008 American Institute of Physics 978-0-7354-0507-3/08/$23.00

polarization is defined by the Stokes equation: $\frac{d\vec{S}}{dZ} = \vec{\Omega} \times \vec{S}$, $\vec{\Omega} = \vec{\Omega}(n(z,t), \vec{B})$.

The $\vec{\Omega}$ vector depends upon plasma parameters: the components in the Stokes space are: $\Omega_1 \propto n_e B_t^2$, $\Omega_2 \propto n_e B_r B_t$, $\Omega_3 \propto n_e B_z$; where $\vec{B}$ is the magnetic field whose components for Tokamak configuration are *r* for radial, *t* for toroidal and *z* for vertical, and n_e is the electron density.

If it is supposed, as in the present paper that the two effects are not mutually interacting, and using the definition of the Stokes vector it is possible to obtain simple relations for Faraday rotation (Ψ) and Cotton Mouton phase shift (Φ) angle: $S_2 / S_1 = \tan 2\psi_{meas}$, $S_3 / S_2 = \tan \phi_{meas}$.

In particular case of JET, for each vertical chord, the two measurements of Faraday rotation and Cotton Mouton angle can be expressed respectively as follows: $\Delta\Psi \approx \lambda^2 \int n_e B_{p\|} dz$, $\Phi \approx \lambda^3 B_t^2 \int n_e dz$.

Approximated Solutions: Type 1, Type 2 and Empirical Model

Defining the quantities $W_i = \int_{z_0}^{z_1} \Omega_i(z) dz$, in the hypothesis $W_i << 1$ a simple approximation to solve [2] the equation of propagation is found to be $W_3 = C_3 \int n_e B_{\|} dz = -1/\tan 2\psi$ and $W_1 = C_1 \int n_e B_T^2 dz = \tan\phi$. In this approximation, called Type I, the Cotton-Mouton phase shift angle results proportional to the line integral of plasma density. To check JET polarimeter data, a first task was to extract the line-integral density from the Cotton-Mouton measurement and to compare it with the one delivered by the other two diagnostics operating at JET: LIDAR Thomson Scattering (TS) and FIR interferometer. The figures 1.a, and 1.b show that the agreement is very good. Moreover comparing the two graphics we can roughly estimate the error bar associated to the polarimeter density measures, it results of the order of $\approx 2\ 10^{19} m^{-2}$, 0.5 10^{19} larger with respect to the LIDAR TS resolution. The proportionality coefficient C_1 resulting is lightly overestimated (0.0019) respect of the expected one (0.0018).

Introducing the relation [2,3] $\Omega_3 > \Omega_1 >> \Omega_2$, that is almost always verified in tokamak configuration plasmas, the solution to the Stokes equations can be found: $s_1 = -\sin W_3$, $s_2 = \cos W_3$ and $s_3 = \int_{z_1}^{z_2} \Omega_1(z') \cos W_3(z') dz$. It means that Faraday rotation effect is much larger than the Cotton-Mouton, this model is called Type 2 approximation.

Finally, if α is the measured Faraday rotation angle, ϕ_{CM} the measured Cotton-Mouton phase shift angle and ϕ the pure Cotton-Mouton angle, the differential equation $d\phi = d\phi_{CM} - (\sin 2\phi / \tan 2\psi)\, d\alpha$ gives the correction applied to the measured Cotton-Mouton to obtain the pure Cotton-Mouton angle. This is the so called empirical model [4],

Models Comparison

The fig.1.c shows the comparison between Cotton-Mouton experimental data from polarimeter vs rigorous numerical solution, while the fig.1.d shows the experimental data vs Type 1 solution. The data for all the models are calculated from EFIT equilibrium together with density taken from LIDAR TS. In fig.1.e, the impressive accordance between Type 1 vs Type 2 model indicates that both are good approximations and that the assumptions done for Type 2 model are correct. For the Empirical model we have to distinguish two cases: the general case is shown in fig.1.f, i) it is observed a worse approximation for the most part of measures respect of other models, ii) nevertheless if it is considered a selected sub-set where the difference in terms of density is less then one fringe jump for high density measures, the model applied gives better results [5].

Figure 1: Models comparison: Density vs Cotton Mouton Effect

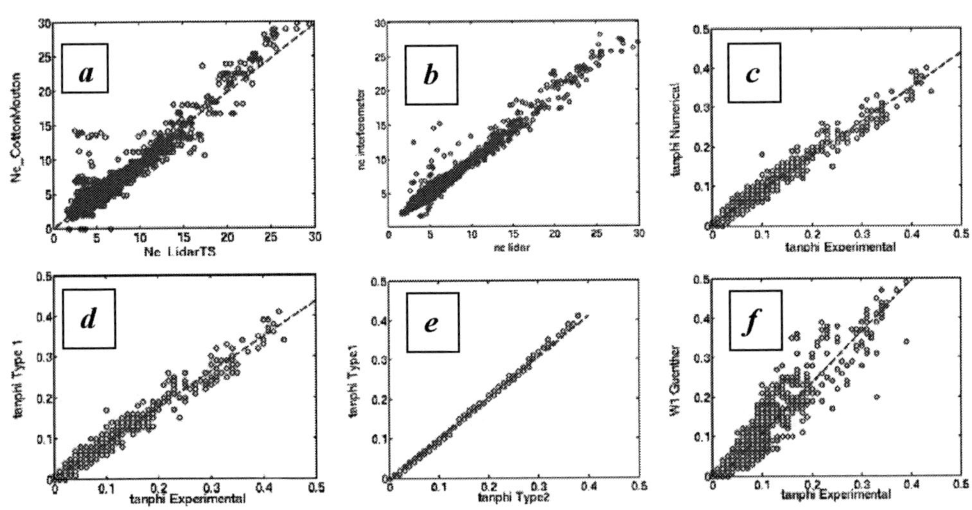

FIGURE 1. From left at high, *a*) electron density comparison between LIDAR TS and polarimeter measurements, the density is expressed in $10^{19}m^{-2}$; the resolution is scarcely lower respect of the interferometer ones (*b*). At right *c*) the tangent of Φ measured by polarimeter vs numerical solution: it is almost identical to the fig.*d* of the Type 1 approximation. In fig.*e* Type 1 vs Type 2 solution has an impressive accordance. Fig.*f* the Empirical model is in general worse than the others ones.

Fringe Jumps Correction

In fig.2.a an example of attempt to correct JET interferometer fringe jumps by using polarimeter data is shown for the central chord #3 shot #61049. At JET a post processing software operates from many years in order to counter the problems of the fringe jumps but with limitations, in fact several fringe jumps are slipped. This work is under development at present, in fact using polarimetry not all the signals result recoverable due to different reasons, for example the timing resolution of polarimeter is lower than interferometric signal, so the presence of more fringe jumps in a restricted interval of time can cause elaboration problems.

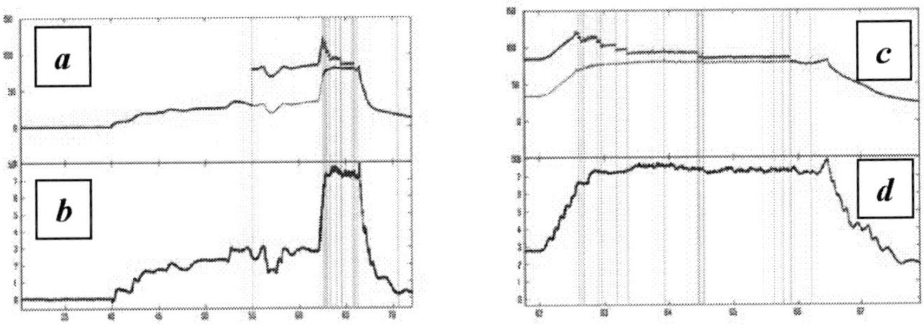

FIGURE 2. At left figure *a*) electron density signal from interferometer: dark green line from JET ppf signal, blue is the same signal but with the correction of the polarimetry. At low, figure *b*) signal from polarimeter is the blue line, all the verticals light green lines indicate possible fringe jumps resulted comparing with the interferometric signal (fig.*a* at high). Subsequent controls return only ten real fringe jumps, the most evident is at 55 s, enlarging the imagine (figures *c,d*) less evident fringe jumps are corrected in the interval 62-68 secs.

Conclusion

A wide statistical analysis, more than 300 JET shots covering a wide range of plasma conditions, leads to the comparison of all the existing models for polarimetry. Polarimeter measures result reliable and in accordance with models.

Cotton Mouton phase shift can provide information on the line integral density also using approximated solutions and it can be used to correct interferometer fringe jumps.

Acknowledgements

The work leading to this article was funded by the European Atomic Energy Community and is subject to the provisions of the European Fusion Development Agreement.

REFERENCES

1. Max Born and Emil Wolf -Principles of Optics- Pergamon Press (1980) pp23-31
2. S.E.Segre and V Zanza– PPCF 48(2006)339
3. F P Orsitto et Al, 33rd EPS-ECA Vol.30I, P-1.073(2006)
4. K Guenther et Al.- PPCF 46(2004)1423
5. M Brombin et Al, 34rd EPS, Warsaw, Poland , july 2-6, 2007

Study of Fast, Near-Infrared Photodetectors for the ITER Core LIDAR Thomson Scattering

L. Giudicotti*†, R. Pasqualotto*, A. Alfier*,
M. Beurskens¶, M. Kempenaars¶, and M. J. Walsh¶

JET-EFDA, Culham Science Centre, OX14 3DB, Abingdon, UK
**Consorzio RFX, Associazione EURATOM-ENEA sulla Fusione, Corso Stati Uniti 4, 35127 Padova, Italy*
†Department of Electrical Engineering, Padova University, Via Gradenigo 6/A 35131 Padova, Italy
¶EURATOM/UKAEA Fusion Association, Culham Science Centre, OX14 3DB, Abingdon, UK

Abstract. A key component for the ITER core LIDAR Thomson Scattering (TS) diagnostic would be a detector with good sensitivity in the 850-1060 nm near infrared (NIR) spectral region. Covering this spectral region becomes necessary if a Nd:YAG laser system operating at $\lambda = 1.06$ μm is used as the laser source, which is a very attractive choice in terms of available energy, repetition rate, reliability and cost. In this paper we review the state of the art of two types of detectors available for the above spectral range: the transferred electron (TE) InGaAs/InP hybrid photodiode and the $In_xGa_{1-x}As$ microchannel plate (MCP) image intensifier and we describe the advancements necessary for a possible application in the ITER LIDAR TS. In addition we describe the preliminary characterization of new GaAsP fast MCP photomultipliers (PMTs) suitable for the detection of the visible part of the LIDAR TS spectrum in JET and ITER.

Keywords: LIDAR, Thomson scattering, photoemissive detectors.
PACS: 52.70.-m, 52.25.Os, 85.60.Ha, 95.85.Jq

INTRODUCTION

In the study of a new LIDAR Thomson Scattering (TS) system for measurements of electron temperature and density in the core of the ITER plasma, the need for a detector with adequate quantum efficiency (QE) in the 850-1060 nm wavelength range has been soon recognized. Such a detector would permit to use a Nd:YAG laser system operating at λ=1.06 μm as the laser source, instead of less efficient Ti:sapphire or 2nd harmonic Nd:YAG systems, with clear benefits in terms of available energy, repetition rate and reliability. The key specifications for the ITER LIDAR TS detectors are an active area diameter $D \geq 11$ mm, an equivalent quantum efficiency EQE $\geq 6\%$, and a pulse response time $\tau \leq 330$ ps FWHM.[1] Here EQE is defined as QE/k_F and k_F is the excess noise factor that accounts for any additional noise introduced after the primary detection. At present these specifications can be aimed at only by photoemissive detectors and the two more promising devices so far available are the transferred electron (TE) InGaAs/InP hybrid photodiode and the $In_xGa_{1-x}As$ image intensifier. In

CP988, *Burning Plasma Diagnostics*, edited by F. P. Orsitto, G. Gorini, E. Sindoni, and M. Tardocchi
2008 American Institute of Physics 978-0-7354-0507-3/08/$23.00

this paper we briefly review these two near infrared (NIR) detector technologies and discuss the advancements necessary for their use in the ITER core LIDAR TS system.

THE TE InGaAs/InP HYBRID PHOTODIODE

The transferred electron effect is the electric field induced transfer of electrons from a lower to higher energy valley of the conduction band. In a TE photocathode the extra energy acquired from the electric field is used by photogenerated electrons to overcome the energy vacuum barrier, thus increasing the probability of photoemission.[2] A typical transmission-mode TE photocathode (Fig. 1) is constituted by an absorbing layer and a separate electron emitter. The absorbing layer is made of InGaAs or InGaAsP, lattice matched to InP, and has a long wavelenght cut-off up to λ=1.7 μm, dependent on the material composition. At zero bias the diffusion of photogenerated electrons to the emitter is prevented by a internal ~0.6 eV energy barrier. When a reverse voltage is applied, the electric field of the Ag-InP Schottky barrier depletion region extends back into the absorbing layer and the photogenerated electrons can then drift-diffuse into the emitter InP layer, where they can undergo transfer to the upper valley and are eventually emitted into the vacuum.[3] This detection mechanism gives values of QE in the NIR comparable to those of the best visible photocathodes (Fig. 2). This technology has been pioneered by the US company Intevac[4] that at present provides TE hybrid photodiodes suitable for photon counting, two active laser imagers for (military) night vision applications, and an imaging camera for NIR spectroscopy. In a TE hybrid photodiode (Fig. 3) the photoelectrons emitted from the photocathode are accelerated by an external electric field onto a GaAs photodiode where they produce secondary electrons proportionally to the acquired energy.[5-9] This electron bombardment (EB) gain mechanism has a typical excess noise factor k_F = 1.1, thus providing the cleanest amplification so far available for photoemissive detectors. A standard TE hybrid photodiode for photon counting at λ=1.06 μm has a photocathode of 1 mm diameter, and a 0.5 mm diameter, GaAs Schottky avalanche photodiode (SAPD) at the anode. The TE InGaAsP/InP photocathode has a QE in excess of 25% at λ = 1.06 μm, a long wavelength cut-off at λ = 1.33 μm and a dark current i_D = 0.3 pA at –20 °C.[10] Pulse linearity up to 30000 photons/pulse (for 50 ps

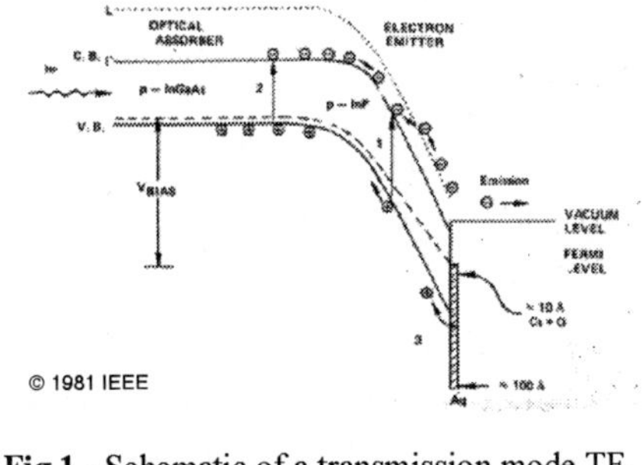

Fig.1 - Schematic of a transmission mode TE photocathode with bias applied (from ref. 2)

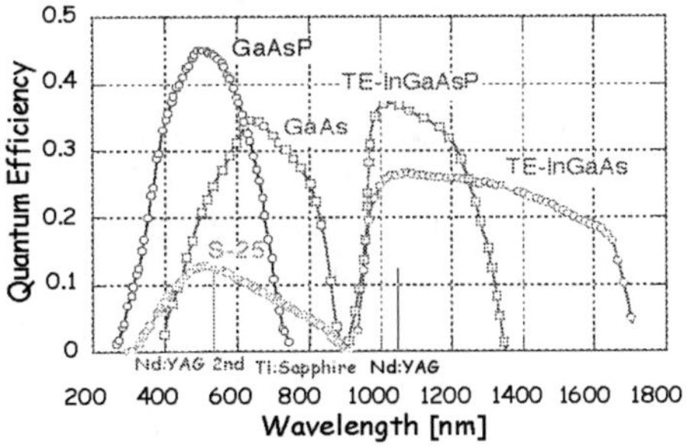

Fig. 2 - Quantum efficiency of visible and NIR photocathodes of interest for ITER LIDAR TS.

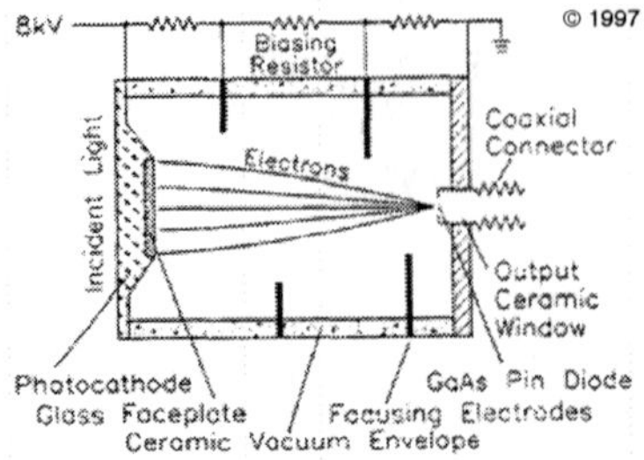

Fig. 3 - Schematic of a TE hybrid photodiode (from ref. 5)

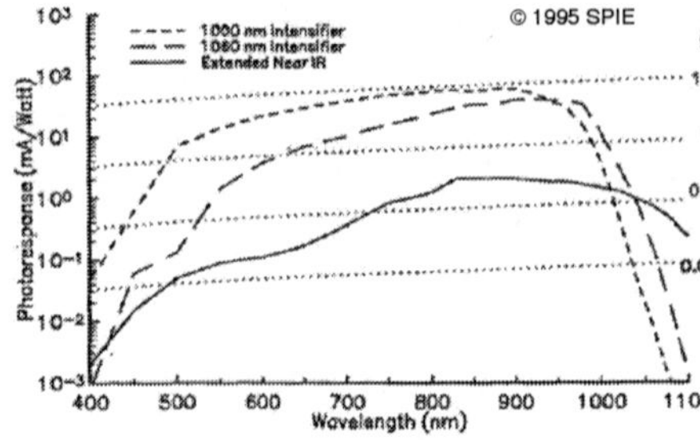

Fig. 4 - Responsivity of $In_xGa_{1-x}As$ photocathodes with different composition (from ref. 10)

pulses) and a gating on-off time of 20 ns have been demonstrated. The pulse response of this detector to a 50 ps source is typically 450 ps FWHM and it is determined by the response time of the SAPD and the time jitter of the photocathode.[3,10] For a larger active area the pulse response becomes dominated by the transit time spread (TTS) of the photoelectrons emitted from different regions of the photocathode. The TTS is ~ 300 ps for a 11 mm photocathode and ~ 800 ps for 18 mm.[3] Therefore the development of a TE hybrid photodiode with 11 mm diameter suitable for the ITER LIDAR TS requires an improved design of the electrostatic focusing and possibly an improvement in the speed of the SAPD and of the photocathode. In spite of these challenging difficulties, the excellent EQE and the large wavelength range make this detector the most interesting NIR device for a future application in ITER LIDAR TS.

THE $In_xGa_{1-x}As$ IMAGE INTENSIFIER

The GaAs photocathodes used in GEN III night vision image intensifiers have a long wavelength cut-off at λ = 900 nm, determined by the 1.1 eV bandgap. But GaAs can be considered as the end point of the ternary system $In_xGa_{1-x}As$ in which the bandgap can be lowered and the sensitivity extended into the IR, by increasing the Indium fraction x. Unfortunately the peak QE also decreases almost exponentially due to an internal interfacial barrier whose height remains approximately constant as the bandgap decreases. These photocathodes are available in commercial image intensifiers and are usually optimized for maximum sensitivity at λ=1.06 μm. This is obtained at x ~ 15% – 20%, resulting in a peak QE<1%. But for LIDAR TS with a ND:YAG source, detection up to the laser wavelength is not necessary and a value x~10% with a cutoff just above 1000 nm and higher QE is more suitable. Fig. 4 shows the spectral sensitivity of two image intensifiers with different composition, developed by Northrop Grumman.[10] Similar photocathodes were also available in image intensifiers by ITT Night Vision. It appear possible that by careful optimization of the composition, $In_xGa_{1-x}As$ can provide a QE ~5% in the spectral region 920 nm – 1000 nm. Due to the low value of x these photocathodes are very similar to GaAs, that has already demonstrated good linearity, speed of response and dark current suitable for LIDAR TS [11,12] and can be easily included in fast MCP PMTs.

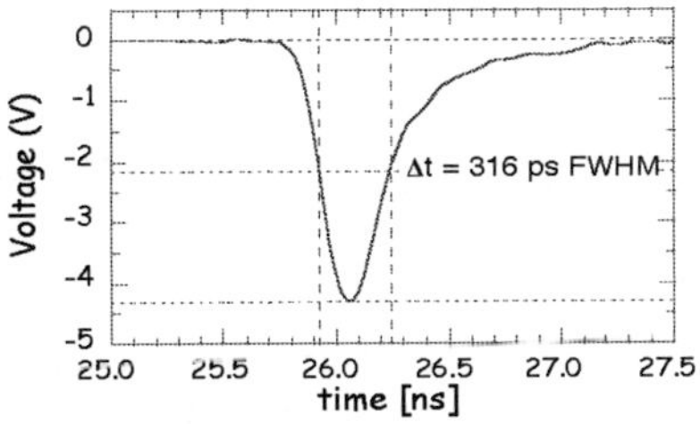

Fig. 5 - Measured pulse response of a fast GaAsP MCP photomultilplier

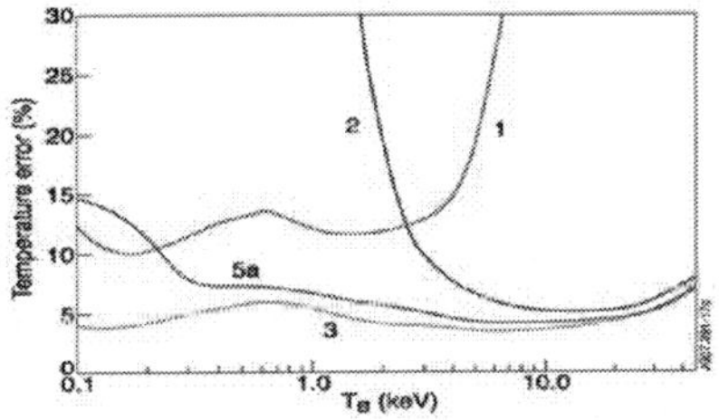

Fig. 6 - Te measurement error calculated for different detector/laser sets: 1) λ=532 nm, GaAsP+GaAs; 2) λ=1064 nm, GaAsP+GaAs; 3) λ=1064 nm, GaAsP+GaAs+TE–InGaAsP/InP; 5a) λ=1064 nm, GaAsP+GaAs+InGaAs.

TEST OF GaAsP FAST MCP PHOTOMULTIPLIERS

Besides the NIR range, in the ITER LIDAR TS a set of detectors is also necessary for covering the visible spectral range down to 300 nm. The JET LIDAR TS systems at present use fast MCP PMTs with either a GaAs photocathode, suitable for operation with 500 nm < λ < 850 nm or a S25 photocathode whose sensitivity extends down to 400 nm, but with lower QE.[11] A relatively new photocathode material suitable for the short wavelength region of the TS spectrum is GaAsP, which shows peak QE~40% (λ = 500 nm) and a QE~10% at λ = 300 nm. This material has never been used in LIDAR TS so far, therefore a set of fast MCP PMTs with a 10 mm diameter GaAsP photocathode has been developed for operation in the JET edge LIDAR TS system, serving also as a test bed for a possible future operation in ITER. The characterization of these detectors is presently under way and preliminary results indicate an EQE~10%, and a pulse response $\tau < 330$ ps (Fig. 5). Fig. 6 shows the expected error on the measured temperature calculated by a simulation of the detected signals and of the associated noise, for different detector sets. These results confirm that with the new NIR detectors and the existing visible MCP PMTs an almost uniform sensitivity over the entire 0.5 – 40 keV ITER design temperature range can be obtained.

The work leading to this article was funded by the European Atomic Energy Community and is subject to the provisions of the European Fusion Development Agreement.

REFERENCES

1. M. J. Walsh, M. Beurskens et al., *Rev. Sci. Instrum.* 10E523 (2006).
2. J. S. Escher, P. E. Gregory et al., *IEEE Electr. Dev. Lett.* **2**, 123 (1981).
3. V. W. Aebi, K. A. Costello et al., *Proc. SPIE* Vol. 2022, 34 (1993).
4. Intevac Inc., 3560 Bassett Street Santa Clara, CA 95054, http://www.intevac.com
5. S. M. Bradbury, R. Mirzoyan et al., *Nucl. Instrum. Meth. A*, **387**,45, (1997).
6. R. A. La Rue, J. P. Edgecumbe et al., *Proc. SPIE* Vol. 2022, 64 (1993).
7. R. A. La Rue, K. A. Costello et al., *IEEE Trans. Electr. Dev.* **44**, 672 (1997).
8. R. A. La Rue, G. A. Davis et al., *IEEE Electr. Dev. Lett.* ***20***, 126 (1999).
9. X. Sun, M. A. Krainak et al., *Proc. SPIE* Vol. 6583, 65830D-1 (2006).
10. J. P. Estrera, T. W. Sinor et al., *Proc. SPIE* Vol. 2551,135 (1995).
11. R. Pasqualotto and P. Nielsen, *Rev. Sci. Instrum.* **74**, 1671 (2003).
12. M. Kempenaars, P. Nielsen et al., *Rev. Sci. Instrum.* **75**, 3894 (2004).

Dust Measurement With Thomson Scattering In FTU

E. Giovannozzi[1], C. Castaldo[1], G. Maddaluno[1], S. Ratynskaia[2], A. Rydzy[1]

[1)] *Associazione EURATOM-ENEA sulla fusione, CR Frascati, c.p. 65, 00044 Frascati, RM, Italy*
[2)] *Royal Institute of Technology (KTH), SE-100 44 Stockholm*

Abstract. The size distribution of dust particles present in FTU after disruptions has been evaluated in the range 0.05 μm - 0.1 μm using the Thomson scattering system. Rayleigh approximation has been used to estimate the dust size. The distribution, although the radii considered do not differ more than an order of magnitude, behave like a power law. The laser energy density far exceed that necessary to vaporize, at least partially, the particles. This can affect the previous estimate of the particle sizes, and suggests that the effective radii might be larger than the estimated values. A preliminary analysis on the broad band signal is presented.

Keywords: Tokamak, Dust, plasma, Thomson scattering.
PACS: 52.70.Kz , 52.27.Lw

INTRODUCTION

The Thomson Scattering System (TS) installed in FTU (major radius 0.93 m, minor radius 0.3 m, B_t up to 8 T, I_p up to 1.6 MA) consists of two Nd:YAG lasers (1.5 J pulse energy, 10 ns pulse width, 1064 nm wavelength, repetition rate 30 Hz each) as laser sources and 19 spectrometers looking vertically from the edge of the plasma (z = -0.3 m) up to and beyond the plasma center (z = 0.1 m). Each spectrometer is equipped with 5 spectral channels (made with interferometric filters and avalanche photodiodes), one is used for Rayleigh scattering (tuned on the laser wavelength), and the other four are used for TS measurement during plasma shots. Dust has been observed only after disruptions, as no spectrometer is looking at the scrape off layer (SOL) where dust can be present even during discharges. The temperature and density of the plasma inside the separatrix are so high that only few dust particles can survive. The edge spectrometers are also affected by a large amount of stray light, so Rayleigh measurement are not available for position beyond mid plasma radius. A database of 1322 discharges that terminates with a disruption has been analyzed. Dust particles were detected by the Thomson scattering, in the majority of the discharges (70 %) [1]. The FTU limiter is made of TZM (Tungsten, Zirconium and Molybdenum) alloy, and the wall is of stainless steel, so the dust is expected to be metallic. Optical data for metals are taken from [2]. The Thomson scattering system has been calibrated with Raman and Rayleigh scattering in hydrogen [3].

CP988, *Burning Plasma Diagnostics*, edited by F. P. Orsitto, G. Gorini, E. Sindoni, and M. Tardocchi

RAYLEIGH SCATTERING

Dust particles, hit by the laser, scatter the light elastically. It can be assumed that the grains have diameters order of magnitude smaller than the laser wavelength, as will be confirmed a posteriori. Thus the Rayleigh approximation can be used, the relevant perpendicular cross section being:

$$\sigma = \left(\frac{2\pi}{\lambda}\right)^4 a^6 \frac{n^2-1}{n^2+2} \cong \left(\frac{2\pi}{\lambda}\right)^4 a^6 \qquad (1)$$

where a is the particle radius λ is the wavelength, and n is the complex index of refraction. The scattered energy depends not only on the particle size, but also on the local energy density, at least as long as we can assume that the particle radius is smaller than the width of the laser beam. Assuming that only one particle is present in the scattering volume (this is reasonable in our case if the particle density is of the order of or lower than $1\times10^7\ \mathrm{m}^{-3}$, since the scattering volume is of the order of $10\ \mathrm{mm}^3$) the measured signal is:

$$S = KE\sigma \qquad (2)$$

Where E is the local energy density, K is a proportionality factor, and σ is the particle cross radius. The value of K for each spectrometer has been obtained with Rayleigh scattering calibration in hydrogen.

Assuming as in [1] a power law for the distribution of the particle cross sections:

$$P_\sigma(\sigma) = \frac{\alpha-1}{\sigma_0^{1-\alpha}}\sigma^{-\alpha} \qquad \sigma > \sigma_0 \qquad (3)$$

we get:

$$P_s(S) = S^{-\alpha}\, K^{\alpha-1}\frac{\alpha-1}{\sigma_0^{1-\alpha}}\int_{E_{\min}}^{E_0} P_e(E)E^{\alpha-1}dE \qquad S > KE_0\sigma_0 \qquad (4)$$

Where $P_e(E)$ is probability distribution of the local laser energy. For a gaussian beam we have:

$$P_e(E) = \frac{1}{\log R}\frac{1}{E} \qquad E_{\min} < E < E_0 \qquad (5)$$

In the formula $R = E_0/E_{\min}$ and $\log R = D^2/w^2$ where w is the laser beam diameter and D is the maximum diameter for detectable scattering. The E_{min} has been introduced to keep the distribution of the laser energy finite at low energy. This is not really a problem as can be seen by splitting the integral in (4) in two part, and calculating the energy limiting E_β where a given fraction of the particles is hit:

$$\beta = \frac{\int_{E_\beta}^{E_0} P_e(E)E^{\alpha-1}dE}{\int_{E_{\min}}^{E_0} P_e(E)E^{\alpha-1}dE} \qquad E_\beta \cong \sqrt[\alpha-1]{1-\beta}\,E_0 \qquad \alpha > 1 \qquad (6)$$

The condition on α keeps the integral limited toward zero, and we have replaced E_{min} with zero in the formula. This give us a statistical cut-off, as it is more probable to have a small particle hit in a part of the laser beam with a high energy density than a larger particle in a part of the laser beam with lower energy density even if the latter area is bigger. Assuming $\alpha = 1.8$ we get $E_{50\%} = 0.42\, E_0$ and $E_{90\%} = 0.06\, E_0$ corresponding to a distance from the center of the laser beam of $d_{50\%} = 0.7\, w$ and $d_{90\%} = 1.2\, w$ where w is the beam radius.

There is also a physical cut-off, as at the very edge of the laser beam the energy density is so low that dust particles should have a diameter comparable to the laser beam to give an appreciable signal. This considerations lead us to estimate $\log R \approx 6$.

A subset of hundreds discharges where the percentage of dust after a disruption is less than 10 % has been extracted from the original database. This was done in order to limit the effects related to a double hit inside a single laser pulses.

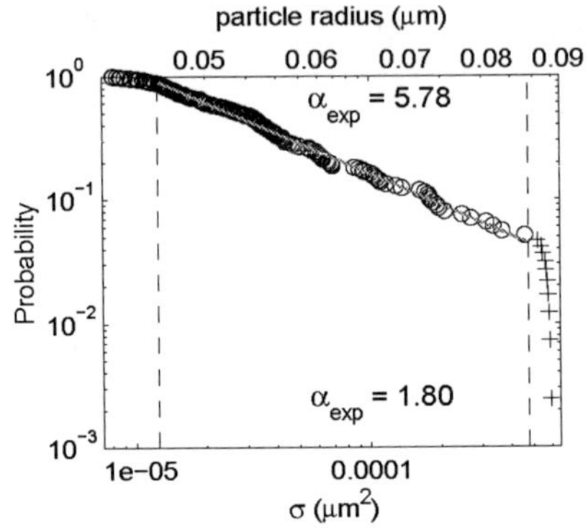

FIGURE 1. Experimental cumulative distribution function of particle cross sections and radii using the Rayleigh approximation in the central spectrometer. Data to the right of the last dashed vertical line (+) are saturated, but are necessary to evaluate the probability of size larger than the saturation value, their radii is meaningless.

The experimental distribution function of the signals has been compared with a power law (Fig. 1) distribution, an average laser energy density has been assumed given by the energy of the laser pulse divided by its diameter. The tails of the distribution seems to follow a power law, even though more than a order of magnitude in the signal would be required to asses the true distribution function [4,5].

BROAD BAND SPECTRUM

In some cases signals were detected in the channels devoted to the Thomson scattering measurement. They are spectrally far from the laser wavelength and have enough rejection not to be contaminated with stray light. The energy in the laser beam exceed of almost an order of magnitude that necessary to vaporize a small molybdenum dust particle (less than a micron). This would also affect the previous Rayleigh measurement as the particle would quickly vaporize and only a fraction of the laser beam energy would be scattered by the solid particle.

Particles greater than few microns would likely survive laser ablation, due to their large radius the scattering at the laser wavelength would saturate the corresponding channel. The particle would be heated up to the vaporization temperature and would emit thermal radiation with a spectrum close to that of a black body. However black body emission by solid or liquid particles cannot explain the signals in all analyzed cases. For smaller particles the light likely comes from the cloud of ablated material. Fig. 2 shows the region in particle radius and temperature that are compatible with the measured signals in shot #28952 at time 1.18 s.

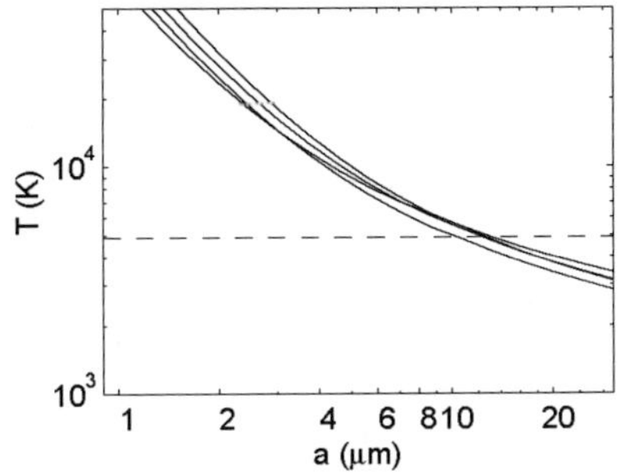

FIGURE 2. The region along the continuous lines is the allowed region assuming a black body emission lasting as the laser pulse for each spectral channel (shot #28952, t = 1.18, central spectrometer). The region above those lines is the allowed region if the emissivity is less than one (gray body). The horizontal dashed line is the vaporization temperature of Mo.

CONCLUSIONS

Dust scattering has been observed after disruptions in FTU with the TS system, mostly in the Rayleigh channels at the laser wavelength in more than 90 % of the discharges. The particle sizes seem to follow a power law distribution, and their radius can be estimated. The unequal distribution of the laser intensity across the beam is not changing the power law of the signal distribution. The particle radii distribution seems to have a larger exponent than that from [5], even though a smaller radii range has been analyzed.

In some cases signals are also present in the other channels, the signals resembling that of a black body. A preliminary analysis shows that the signals could be due to black-body emission by very large ($a > 10$ μm) molybdenum particles heated up to the vaporization temperature, or due to emission by the ablation cloud of smaller particles. A more detailed analysis of the particle vaporization is under investigation.

REFERENCES

1. E. Giovannozzi et al 2006 33rd EPS Conf. on Control. Fusion and Plasma Phys. Rome
2. M.A. Ordal, et al. Appl. Opt. **24**, 24, 4493 (1985).
3. F. Orsitto, et al. Appl. Opt. **34**, 15, 2712 (1995)
4. M. Rytgaard, *Estimation in the Pareto distribution* Astin Bull. **20**, 2 (1990) 201
5. W.P.West Plasma Phys. Control. Fusion **48** (2006) 1661–1672

SPECTROSCOPY, OPTICAL/IR DIAGNOSTICS AND NEUTRAL PARTICLES

Passive Spectroscopy Bolometers, Grating- And X-ray Imaging Crystal Spectrometers

M. Bitter[a], K. W. Hill[a], S. Scott[a], S. Paul[a], A. Ince-Cushman[b], M. Reinke[b], J. E. Rice[b], P. Beiersdorfer[c], M. F. Gu[c], S. G. Lee[d], Ch. Broennimann[e], and E. F. Eikenberry[e]

[a]*Princeton Plasma Physics Laboratory, Princeton, NJ, 08543*
[b]*Plasma Science and Fusion Center, Massachusetts Institute of Technology, Cambridge, MA 02139*
[c]*Lawrence Livermore national Laboratory, Livermore, CA 94550-9234*
[d]*Korea Basic Science Institute, Taejeon 305-333, Korea*
[e]*Paul Scherrer Institut, CH5232 Villigen-PSI, Switzerland*

Abstract: This tutorial gives a brief introduction into passive spectroscopy and describes the working principles of bolometers, a high-resolution grating spectrometer, and a novel X-ray imaging crystal spectrometer, which is also of particular interest for profile measurements of the ion temperature and plasma rotation velocity on ITER and future burning plasma experiments.

Keywords: Plasma diagnostic techniques and Instrumentation
PACS: 52.70.Kz, 52.70.La

INTRODUCTION*

Magnetically-confined fusion plasmas contain measurable amounts of impurity ions in addition to the hydrogen isotopes which constitute the bulk of the plasma ions. The impurity ions lead to enhanced energy loss and to reduction in the density of the bulk plasma ions, and diminish the fusion reactivity. Spectroscopic measurements of the continuum and line radiation emitted by impurity ions are therefore important to understand and minimize the effects of impurities on plasma performance, but they are also important for the diagnosis of plasma parameters, as e.g. the electron temperature and density, ion temperature, particle transport, and particle influx rates. 'Passive' spectroscopic diagnostics exploit radiation from atomic processes in the plasma, while 'active' spectroscopic diagnostics involve injected particles and laser beams.

The spectral regions of the continuum and line radiation relevant to spectroscopy of tokamak plasmas are defined in Table 1. It is convenient to group the spectroscopic diagnostics by wavelength region and measuring techniques, as they apply to the core and edge of the plasma – see Table 2. In this tutorial, we describe only three passive diagnostic systems: bolometers, a high-resolution grating spectrometer, and a novel X-ray imaging crystal spectrometer. For further information on passive spectroscopic diagnostics is given in ref. [1] and the literature cited therein.

CP988, *Burning Plasma Diagnostics*, edited by F. P. Orsitto, G. Gorini, E. Sindoni, and M. Tardocchi

TABLE 1: Spectral Regions relevant to Spectroscopy of Tokamak Plasmas

Spectral Region	Wavelength/Energy Region
Near Infrared	700-1200 nm / 1-2 eV
Visible	400-700 nm / 2-3 eV
Ultraviolet (UV)	200-400 nm / 3-6 eV
Vacuum Ultraviolet (VUV)	30-200 nm / 6-40 eV
Extreme Ultraviolet (EUV)	10-30 nm / 40-120 eV
Soft X-Ray	0.1-10 nm / 120-12000 eV

TABLE 2: Parameter/Technique for Passive Spectroscopy of Plasma diagnostics

	Impurity species identification	Impurity densities	Z_{eff}	Electron temperature	Ion temperature	Flow velocity	Electron density	Impurity transport	Particle influxes	H/D/T ratio	MHD modes	Runaway electrons	Described in Section
VUV/EUV grating spectroscopy	✸	✸						✸					II.A
VUV/EUV filtered detectors (inc. multilayer mirrors)		✸						✸					II.B&C
Visible bremsstrahlung			✸				❖						II.D
Pulse height analysis	✪	✸	✪	✪									III.B
Filtered SXR diodes		✸						✸			✸		III.C
SXR crystal spectroscopy	❖	✸		❖	✸	✸		❖					IV.B
Edge/divertor visible/UV/near-IR spectroscopy	✪	✸		❖	✸	✸	❖	✸	✸	✸			V.B&C
Visible/UV filtered detectors & filtered cameras		✸		❖			❖	✸	✸				V.D
Fabry-Perot interferometry and visible spectroscopy										✸			V.E
IR synchrotron radiation imaging												✸	VI

✸ Primary (technique is well suited to measurement)

✪ Backup (technique provides useful data but has some limitations compared to primary)

❖ Supplementary (technique provides data that can be used to check or constrain other measurements but is not a reliable measurement itself)

BOLOMETERS

Bolometers monitor the total radiated power from the plasma over a wide spectral range. Compared to VUV spectrometers, they have poor spectral resolution, but since they are relatively inexpensive they are usually equipped with an array of detectors to obtain good spatial coverage. A tomographic reconstruction of the radial profile of the total radiated power can thus be obtained by an Abel-like inversion of the data from multiple sightlines. Figure 1 shows a picture of the bolometer on NSTX. It consists of an array of 16 diode detectors, which record the radiation from the plasma through a pinhole along sixteen, evenly spaced, sightlines, over the plasma diameter in the horizontal mid-plane. The diodes of the AXUV-16ELO/G photo-diode array [2], which is used on NSTX, have an extremely thin dead layer < 1 μ and, therefore, a high quantum efficiency, near the theoretical limit, for XUV photons from 1 to 8000 eV. The detector response is proportional to the photon energy flux.

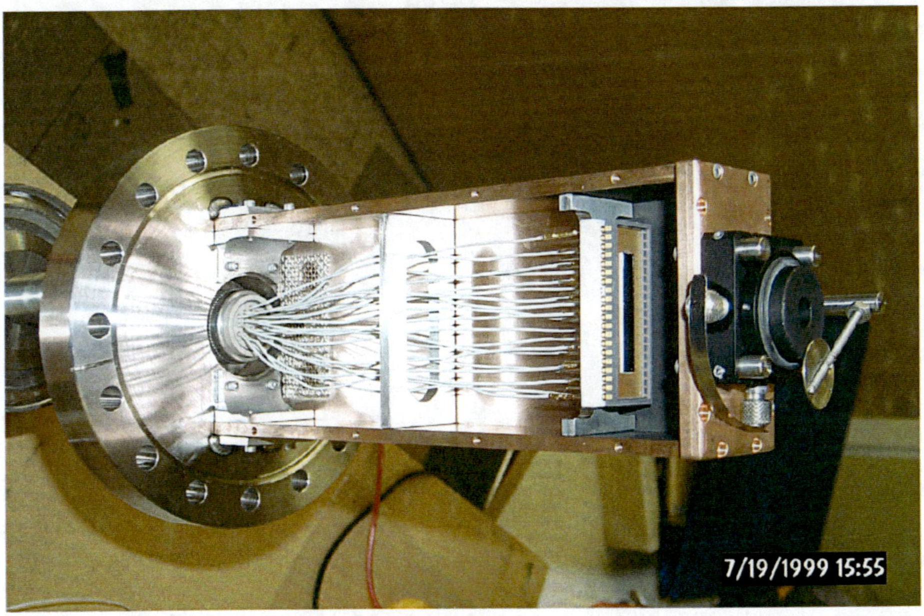

FIGURE 1. Sixteen-channel, horizontally viewing, bolometer on NSTX.

Figure 2 shows an Abel-inverted power density profile from an NSTX discharge with neutral-beam heating. The main metallic contributor to the core radiation is iron as seen from XUV spectrometers. An inward impurity pinch leads to a central power density of 0.35 Wcm^{-3} and a metallic contribution to Z_{eff} of 0.6. Except for two large events, this discharge had an 'Edge-Localized-Mode (ELM)'-free period for 0.35 s. The bursts of radiation at the inner major radius result from interactions of plasma ions with the center stack.

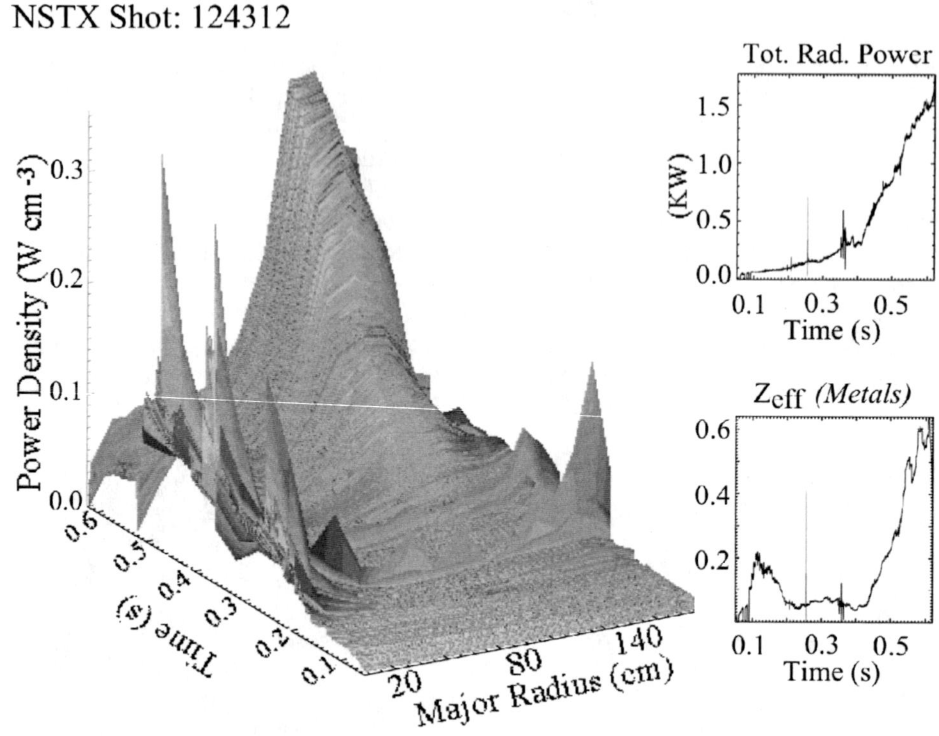

FIGURE 2. Abel-inverted profile of the total radiated power from NSTX

FLAT-FIELD GRATING SPECTROMETERS

A compact grazing-incidence spectrometer has been implemented on NSTX for monitoring impurity line emission in the wavelength range from 6 to 135 Å [3]. This wavelength range is longer than that accessible by a crystal spectrometer [e.g., 4, 5], and shorter than the 100 to 1100 Å range [6, 7], which is covered by the 'Survey, Poor Resolution, Extended Domain (SPRED)' instrument on NSTX. Figure 4 shows the instrument on NSTX, where it was installed at the end of the pump duct. Its line of sight follows a radial line to the center stack of NSTX within the horizontal mid-plane. The instrument is an adaptation of a spectrometer originally used on the Livermore electron beam ion trap [8 - 10]. It employs a variable line spacing grating mounted on a rotary table (average line spacing of 2400 *l*/mm [11]), which affords flat-field focusing with a focal distance of about 23 cm. A 30 μm entrance slit provides a line width of 0.1 Å, i.e., a resolving power of 500 at 50 Å. A 25 mm by 25 mm, liquid nitrogen cooled charge couple device (CCD) camera is used for recording and covers in excess of 50 Å in a single setting. In fast readout mode, the instrument provides a spectrum every 90 ms.

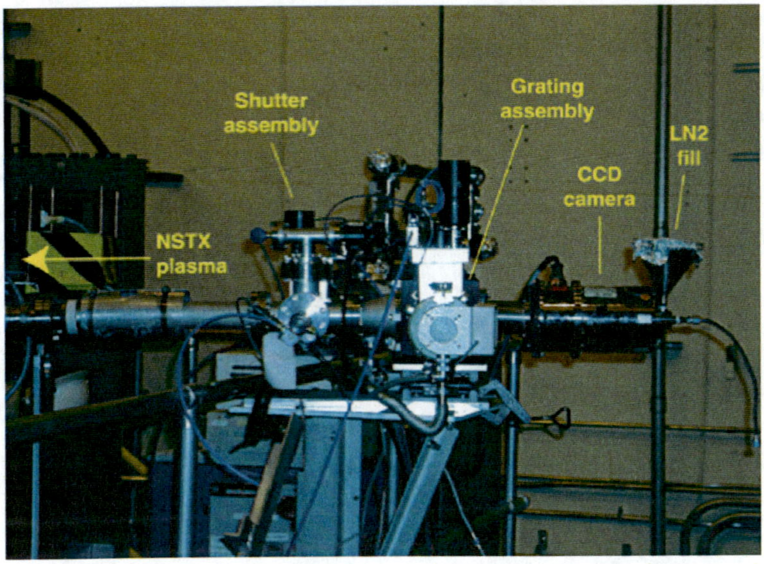

FIGURE 3. Compact grazing-incidence spectrometer on NSTX

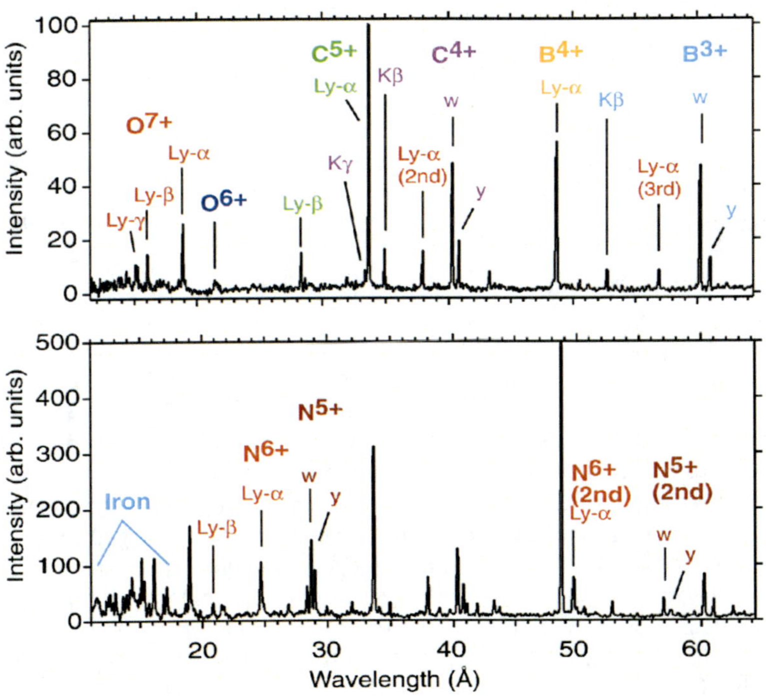

FIGURE 4. Typical K-shell lines and iron L-shell lines (bottom panel)

The instrument typically monitors the emission from hydrogen-like and helium-like boron, carbon, and oxygen, as illustrated in Fig. 4 (top). These three elements are standard impurities in NSTX plasma discharges. Observation of lines from other elements indicates unwanted inflow of impurities from a variety of possible sources. Figure 4 (bottom) shows lines from L-shell iron ions as well as from hydrogen-like and helium-like nitrogen entering the plasma during high harmonic fast wave current drive experiments. Other unwanted species detected so far with the grazing-incidence spectrometer include molybdenum, nickel, and titanium.

In some experiments, trace elements have been deliberately introduced into the plasma. For example, argon was injected for ion temperature measurements with the crystal spectrometer; neon was injected for transport measurements using the soft x-ray array; and lithium was evaporated to coat the inner walls. All these elements have been observed with the grazing-incidence spectrometer.

X-RAY IMAGING CRYSTAL SPECTROMETERS

High-resolution X-ray crystal spectrometers have been used for Doppler measurements of the central ion temperature and central toroidal rotation velocity of tokamak plasmas since the 1970's, by analyzing the line spectra from highly charged ions of medium-Z elements from argon to krypton. These elements are present in tokamak plasmas as indigenous impurities, or they are injected for diagnostic purposes. Most widely used are Johann crystal spectrometers, which consist of a cylindrically bent crystal and a one-dimensional position sensitive detector – see, e.g., Fig. 28 in ref. [1]. The Johann configuration has optimal focusing properties for the so-called meridional rays in the main dispersion plane, perpendicular to the crystal, and it therefore allows the simultaneous observation of an entire spectral range; but it does not provide any focusing for the 'sagittal' rays, which are oblique to the main dispersion plane, so that it is not possible to obtain spatially resolved spectra from a tokamak plasma with the standard Johann configuration. This longstanding problem for the X-ray spectroscopy of tokamak plasmas has now been solved by the use of spherically bent crystals, which also provide focusing for the sagittal rays. The concept of these new 'X-ray imaging crystal spectrometer' was described in ref. [12], and it was also experimentally verified on Alcator C-Mod and Textor [13, 14]. The proof-of-principle experiments were, however, conducted with two-dimensional, position-sensitive, multi-wire proportional counters, which had a count rate limitation of 400 kHz that was far too low to satisfy the requirements for a good photon statistics and time resolution of the spectral data. This situation has now changed thanks to the advent of the 'PILATUS II' detector, a pixilated semiconductor diode array of 35 mm x 85 mm, which consists of a photo-sensitive ('sensor') chip and a similar readout chip that includes on each pixel an amplifier, comparator, counter, and a memory, so that it is possible to do single photon counting at a rate of 1 MHz per pixel [15]. Here we present results from a prototype X-ray imaging crystal spectrometer, which has recently been implemented on Alcator C-Mod. The spectrometer consists of two

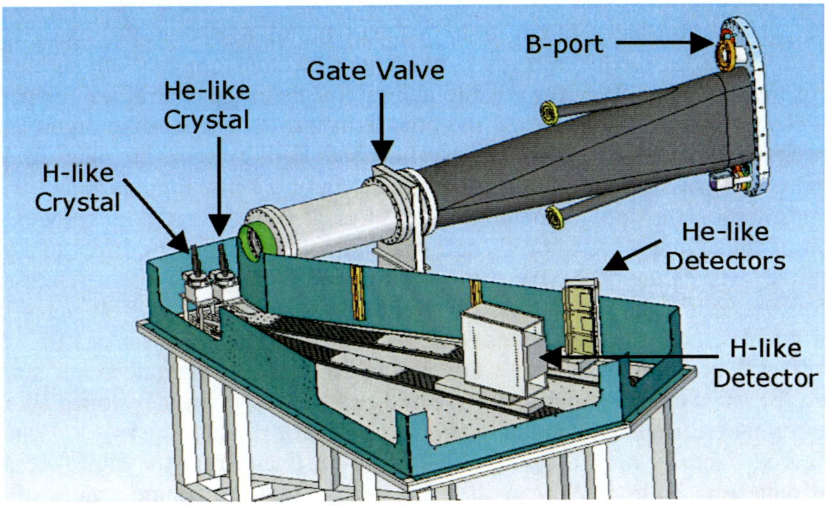

FIGURE 5. Layout of X-ray imaging spectrometer on Alcator C-Mod

FIGURE 6. PILATUS II detector module

spherically bent (102) quartz crystals with a 2d-spacing of 4.56216 Å and radii of curvature of 1444 mm and 1365 mm, for the observation of He-like and H-like argon at 3.9494 Å and 3.7311 Å, and four PILATUS II detector modules – see Figs. 5 and 6. Three PILATUS II detector modules record spatially resolved spectra of He-like argon from the whole, elongated, 72 cm high, plasma column, as shown in Fig. 7. The 4th PILATUS II detector module records H-like argon spectra from a 20 cm high, central region of the plasma to allow for reliable measurements of the central ion temperature at high electron temperatures, when the emissivity profile for He-like argon lines is hollow. Spatially resolved spectra of He-like and H-like argon are now routinely observed on Alcator C-Mod with a time resolution of 20 ms. Figures 8 and 9 show radial profiles of the ion temperature and toroidal plasma rotation velocity as a function of time from an Alcator C-Mod discharge with a plasma current of 800 kA and auxiliary rf heating of 2 MW during the period from 0.7 to 1.2 s. The profiles were constructed from 48 sightlines by dividing the detector in 48 elements in the vertical direction, which corresponds to a spatial resolution in the plasma of 1.5 cm. The profiles shown in Figs. 8 and 9 were derived from chord-integrated data, since the software for an Abel inversion of the spectral data is still being developed. We note that the central electron temperature during the period of rf heating was 4 keV and that the observed intensity profile for the He-like argon lines, from the chord-integrated spectral data, was hollow. We can therefore expect that the central values of these profiles will be changed by a final analysis, using Abel-inverted data. It is noteworthy that the concept of our X-ray imaging crystal spectrometer has already been adopted for the design of the crystal spectrometers for ITER [16] and that the dimensions of the present prototype are similar to those planned for the instruments on ITER.

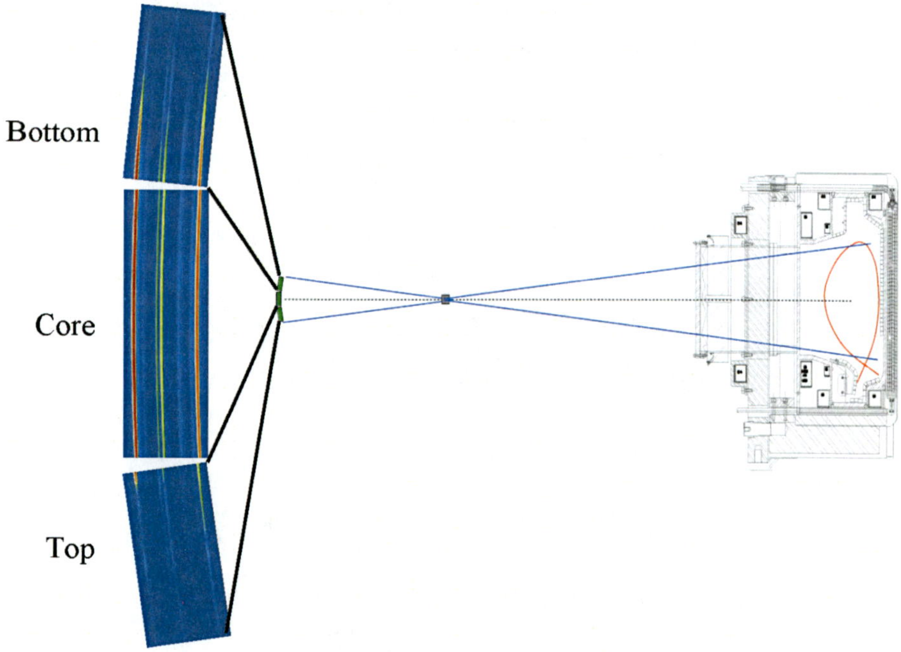

FIGURE 7. Spatially resolved He-like argon spectra from Alcator C-Mod

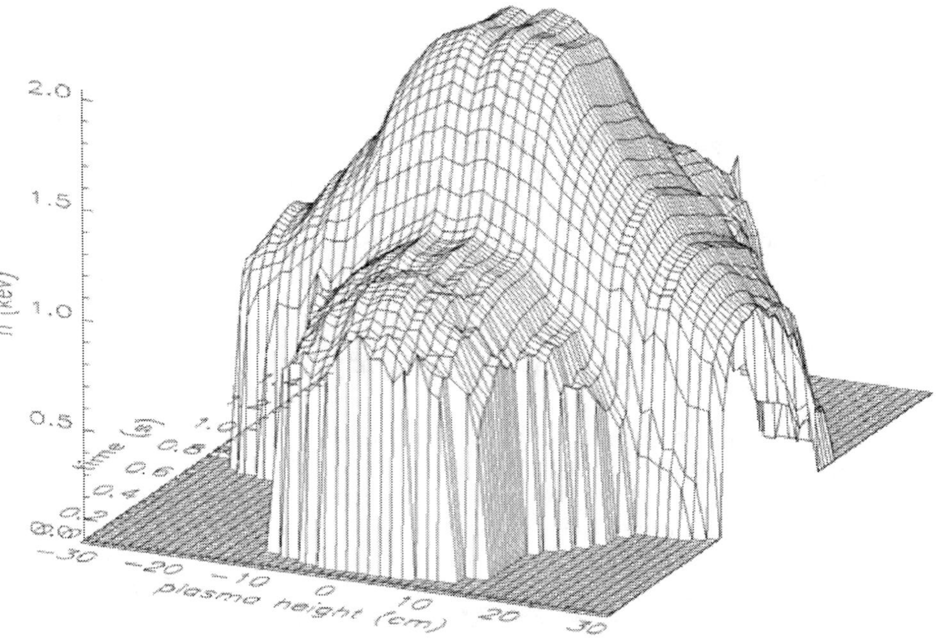

FIGURE 8. Ion-temperature profile as a function of time

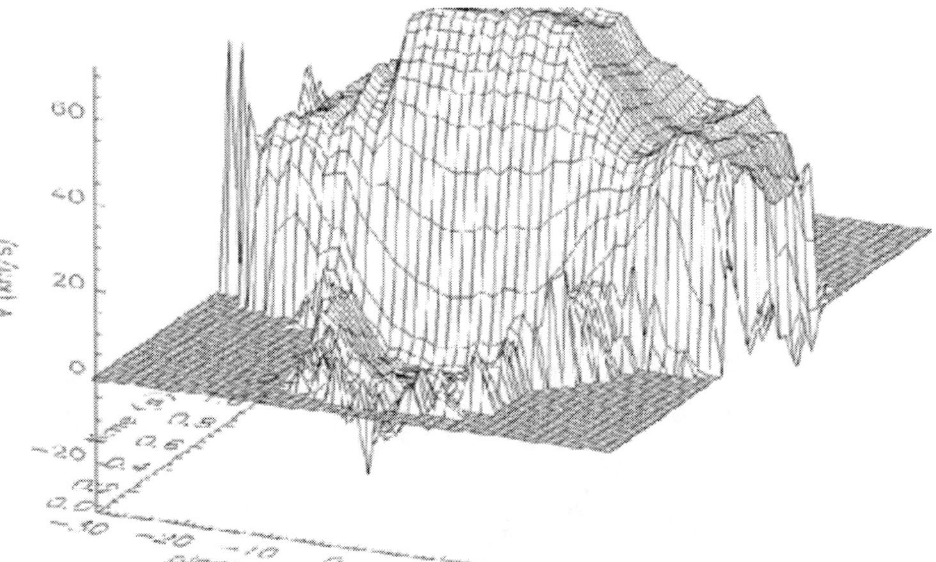

FIGURE 9. Profile of the toroidal rotation velocity as a function of time

CONCLUSIONS

'Passive' spectroscopic diagnostics play an important role in determining effects of impurities and in measurements of plasma parameters. The diagnostic techniques are being continuously improved to obtain better spatial and time resolution and to meet the requirements for burning plasma experiments. New developments in X-ray spectroscopy may lead to an attractive alternative to 'active' spectroscopic diagnostics, such as the challenging beam-based Charge-Exchange Recombination Spectroscopy, for the measurements of ion-temperature and plasma rotation velocity profiles in those experiments.

ACKNOWLEDGEMENTS

Work by the Princeton University Plasma Physics Laboratory was performed under the auspices of the Department of Energy under Contract No. DE-AC02-76CHO3073; work by the Lawrence Livermore National Laboratory was performed under the auspices of the Department of Energy under Contract No. DE-AC52-07NA27344; the work at Massachusetts Institute of Technology was supported by the Department of Energy contract No. DE-FC02-99ER54572 and the Department of Energy Initiative for the Development of Diagnostics for Magnetic Fusion Energy Research Contract-1083.

REFERENCES

*This introduction are drawn from ref. [1].

1. B. C. Stratton, M. Bitter, K. W. Hill, D. L. Hillis, and J. T. Hogan, *'Passive Spectroscopic Diagnostics for Magnetically-Confined Fusion Plasmas'*, to be published in the February 2008 issue of Fusion Science and Technology,
2. http://www.ird-inc.com/axuvarr.html
3. P. Beiersdorfer, et al.., *Rev. Sci. Instrum.* **77**, 10F306 (2006).
4. P. Beiersdorfer, et al., *Rev. Sci. Instrum.* **77**, 1974 (2003).
5. P. Beiersdorfer, et al. *Rev. Sci. Instrum.* **60**, 895 (1989).
6. R. J. Fonck, et al., Appl. Optics **21**, 2115 (1982).
7. B. C. Stratton, et al., *Rev. Sci. Instrum.* **57**, 2043 (1986).
8. S. B. Utter, et al., *Rev. Sci. Instrum.* **70**, 284 (1999).
9. P. Beiersdorfer, et al., *Rev. Sci. Instrum.* **70**, 276 (1999).
10. A. T. Graf, *et al., Can. J. Phys.* (in press).
11. N. Nakano, et al., *Appl. Optics* **23**, 2386 (1984).
12. M. Bitter, *et al., Rev. Sci. Instrum.* **70**, 292 (1999).
13. M. Bitter, *et al., Rev. Sci. Instrum.* **75**, 3660 (2004).
14. G. Bertschinger, *et al., 33rd EPS Conference Rome, 2006*, vol. 30I, P-2.153.
15. http://pilatus.web.psi.ch/pilatus.htm
16. R. Barnsley, et al., *Rev. Sci. Instrum.* ***75**, 3743 (2004)*

Active Beam Spectroscopy

M.G. von Hellermann[1], E.Delabie[1], R.J.E.Jaspers[1], W.Biel[2], O.Marchuk[2], H.P.Summers[3], A.Whiteford[3], C.Giroud[4], N.C.Hawkes[4], K.D.Zastrow[4]

[1]FOM-Institute for Plasma Physics Rijnhuizen, Association EURATOM-FOM, member of Trilateral Euregio Cluster, PO Box 1207, 3430 BE Nieuwegein, The Netherlands*
[2]Institut für Plasmaphysik, Forschungszentrum Jülich GmbH, Association EURATOM-FZJ, member of Trilateral Euregio Cluster, 52425 Jülich, Germany
[3]Strathclyde University,Dept. Appl. Physics, Glasgow,UK
[4]UKAEA Culham Laboratory Euratom Association, Abingdon OX14 3EA, UK

Abstract. Charge eXchange Recombination Spectroscopy (CXRS) plays a pivotal role in the diagnostics of hot fusion plasmas and is implemented currently in most of the operating devices. In the present report the main features of CXRS are summarized and supporting software packages encompassing "Spectral Analysis Code CXSFIT", "Charge Exchange Analysis Package CHEAP", and finally "Forward Prediction of Spectral Features" are described.

Beam Emission Spectroscopy (BES) is proposed as indispensable cross-calibration tool for absolute local impurity density measurements and also for the continuous monitoring of the neutral beam power deposition profile. Finally, a full exploitation of the 'Motional Stark Effect' pattern is proposed to deduce local pitch angles, total magnetic fields and possibly radial electric fields.

For the proposed active beam spectroscopy diagnostic on ITER comprehensive performance studies have been carried out. Estimates of expected spectral signal-to-noise ratios are based on atomic modelling of neutral beam stopping and emissivities for CXRS, BES and background continuum radiation as well as extrapolations from present CXRS diagnostic systems on JET, Tore Supra, TEXTOR and ASDEX-UG.

Supplementary to thermal features a further promising application of CXRS has been proposed recently for ITER, that is a study of slowing-down alpha particles in the energy range up to 2 MeV making use of the 100 keV/amu DNB (Diagnostic Neutral Beam) and the 500 keV/amu HNB (Heating Neutral Beam). Synthetic Fast Ion Slowing-Down spectra are evaluated in terms of source rates and slowing-down parameters

Keywords: Active beam spectroscopy, diagnostics of hot fusion plasmas, CXRS, MSE, BES.
PACS: 52.25Vy, 52.55Fa, 52.55Pi, 52.70Kz

INTRODUCTION

Active Beam Diagnostics have developed over the last three decades [1-14] into a mature diagnostic tool for fusion plasmas. A comprehensive package of active beam based spectroscopy tools for ITER has been recently submitted and evaluated. The feasibility study encompasses CXRS (*Charge Exchange Recombination*

* www.fom.rijnh.nl

CP988, *Burning Plasma Diagnostics*, edited by F. P. Orsitto, G. Gorini, E. Sindoni, and M. Tardocchi

Spectroscopy) for the measurement of the main impurity ion densities (including helium ash), ion temperatures and toroidal as well as poloidal plasma rotation. *Beam Emission Spectroscopy* is proposed as indispensable cross-calibration tool for absolute local impurity density measurements [21] and also for the continuous monitoring of the neutral beam power deposition profile. Finally, a full exploitation of the '*Motional Stark Effect*' pattern is proposed to deduce local pitch angles, total magnetic fields and possibly radial electric fields.

More recently, a further promising application has been proposed, that is a study of slowing-down alpha particles in the energy range of 0.1 to 0.7 MeV and 1.6 to 2.4 MeV respectively making use of the *3.6MW 100 keV/amu* H^0 *DNB (Diagnostic Neutral Beam)* and the *17MW 500keV/amu* D^0 *HNB (Heating Neutral Beam)*. An important asset of the proposed slowing-down scheme is the potential investigation of anisotropic features in the alpha velocity distribution function making use of top and equatorial observation periscopes.

Performance studies and estimates of expected spectral signal-to-noise ratios are based on atomic modelling of neutral beam stopping and emissivities for CXRS, BES and background continuum radiation as well as extrapolations from present CXRS diagnostic systems. Single high-étendue and high-resolution spectrometers for each radial channel are proposed for thermal CXRS and BES/MSE and high-étendue broad-band spectrometers for CXRS on slowing-down features.

Basic Features of Active Beam Spectroscopy

The following section gives a very brief summary of the main features of Active Beam Aided Spectroscopy which is a synonym for the following two reactions between plasma ions, plasma electrons and impurity or bulk ions. We show here as an example the reaction for plama ions i.e. the bulk plasma deuterons:

CXRS: $D^{+1}(T) + D^0(E) \rightarrow D^*(T) + D^+(E)$

$$I_{CXRS} \cdot = \frac{1}{4\pi} n_d \cdot Q_{CXRS} \cdot \int n_b ds \qquad (1)$$

Here n_d is the local deuteron density, Q_{CXRS} the effective atomic emission rate [9-11] and the line of sight integration extends across the beam path (cf. Fig.3)

and **BES**: $D(E) + e \rightarrow D^*(E) + e$

$$I_{BES} \cdot = \frac{1}{4\pi} n_e \cdot Q_{BES} \cdot \int n_b ds \qquad (2)$$

The local neutral beam strength is determined by the beam stopping processes (c.f. [9-11]), that is ionization by electrons and ions and by charge exchange losses.

$$n_b \cdot = n_b(0) \cdot \exp\left[- \int_{beampath} n_e \sum_{z=1}^{z=18} c_z \cdot \sigma_{stop,z} \right] \qquad (3)$$

Combining BES and CXRS and making use of multiple instruments linked to the same viewing chords allows the replacement of the exponential beam stopping calculation by local beam densities derived from BES. This method avoids not only the exponential error propagation along the beam path but takes also care of uncertainties in absolute calibration of the optical observation system shared by BES and CXRS we obtain therefore:

$$\frac{n_z}{n_e} = \frac{I_{CXRS}}{I_{BES}} \cdot \frac{Q_{BES}}{Q_{CXRS}} \tag{4}$$

There are two caveats. The atomic emission rates Q_{CXRS} and Q_{BES} do depend on the collision energy, i.e. beam energy, electron density n_e, plasma temperature T_i and also impurity ion composition n_Z. So the apparent simplicity of eqn. 4 hides in truth an iterative process. However, both the beam-stopping calculation as well for the ratio of Q_{BES}/Q_{CXRS}, the iteration usually takes only a few steps and converges rapidly. The second caveat is the wavelength variation of the calibration factor for the case the CXRS spectrum not being at the same wavelength as the Balmer-Alpha Spectrum. In this case we use the underlying bremsstrahlung as cross-reference:

$$\frac{n_z}{n_e} = \frac{I_{CXRS}}{I_{BES}} \cdot \frac{Q_{BES}}{Q_{CXRS}} \cdot \frac{I_{cont}(\lambda_{BES})}{I_{cont}(\lambda_{CXRS})} \cdot \frac{\lambda_{CXRS}}{\lambda_{BES}} \cdot \frac{g_{ff}(\lambda_{CXRS})}{g_{ff}(\lambda_{BES})} \tag{5}$$

To illustrate the two examples of beam induced spectra we show an historical first spectrum at JET in 1988 which shows nicely the dramatic changeover from passive background spectrum to the beam induced CXRS D_α spectrum representing the bulk plasma ions (deuterons) and the Doppler shifted Beam Spectrum.

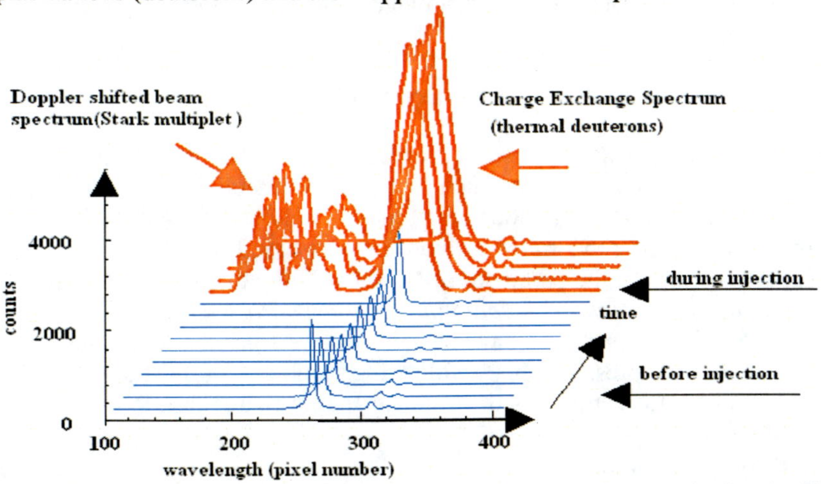

FIGURE 1. Beam induced Balmer Alpha Spectra as measured on JET in 1988. The Doppler shifted beam spectrum represents the MSE pattern of high energy beam neutrals with energies E, E/2 and E/3 respectively travelling across the magnetic field (cf. [6])

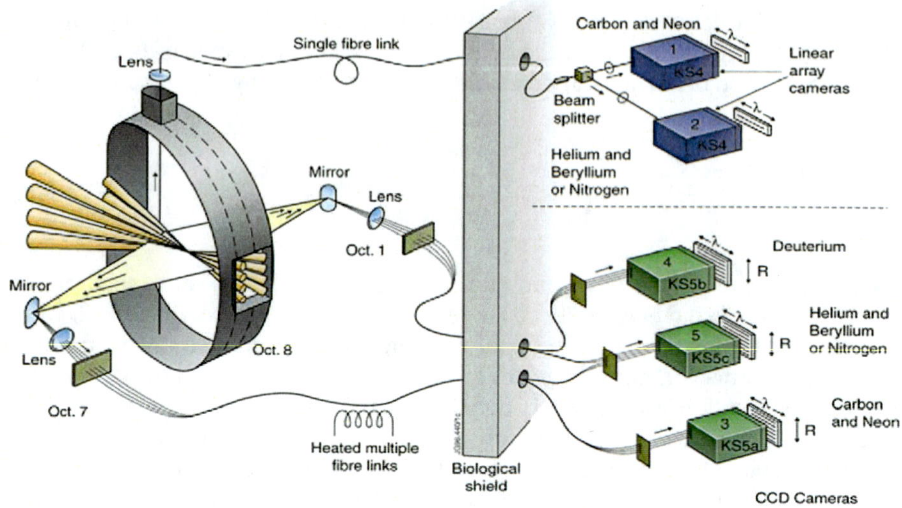

FIGURE 2. The CXRS diagnostic on JET. Several multi-chord periscopes aimed at the JET heating neutral beams are used for the radial profile reconstruction of the main low-Z ions as well as bulk ion CX spectra.

The JET diagnostic evolved in the course of time from an originally single chord diagnostic into a comprehensive instrumentation with several periscopes and accompanying spectroscopic instruments (Fig.2). This example illustrates the multiplicity of instruments required for a self-consistent evaluation of spectroscopic data based on atomic modelling of beam target interactions in a fusion plasma and its ionic species.

Supporting Software Packages

A successful quantitative exploitation of observed CX spectra on fusion devices depends on a number of key ingredients such as a powerful, well focused diagnostic neutral beam, perfect instrumentation, established calibration and alignment procedures, but more importantly, *suitable analysis tools* based on advanced atomic modelling of spectral features, the parameterization of its components, and finally *the efficient and reliable extraction of meaningful physics parameters*. An extensive collection of CXRS software packages have been developed since the mid-eighties, in particular at JET, which are presented in the following section as an illustration. In the current preparation phase for ITER, it is the ultimate goal to share those tools with the participating partners.

a) Spectral Analysis Codes

An efficient tool for the extraction of physics parameters such as local ion temperature, plasma rotation, intensity and continuum radiation level is a challenging task for the routine application of CXRS. The main basis for the development of this tool is an advanced atomic modelling [9-11] of the emission rates and modelling of the beam target interaction, which allows subsequently the parameterization of complex features [15], and finally the link to physical meaningful plasma data. Following the earlier developments at JET a recent update and review of the spectral analysis code "KS4FIT" has led to modernized and more efficient, platform independent software packages (e.g. the IDL based, ADAS supported version "CXSFIT" [16]).

A typical performance criterion for a suitable spectral code is the availability of physics results in between subsequent plasma pulses. Stretching this point even more, once analysis tools can be used for the creation of substantial data bases, the next step towards real-time neural-network based spectral analysis is very close [17]. Several advanced code features based on the JET experience allow a great flexibility in the treatment of standard and non-standard cases. Each plasma scenario is typically characterized by its combination of active and passive spectral features, as a result a comprehensive collection of representative examples have been developed at JET. Moreover, a pulse number linked documentation of all code settings allows the replay of already analysed spectra at any time.

b) Charge Exchange Analysis Package (CHEAP)

A key issue for the self-consistent evaluation of active CX spectra is a comprehensive modelling of the beam target interaction, this approach requires in the presence of several impurity ions the monitoring of all CX spectra representing the main low-Z impurity ions, such as He^{+2}, Be^{+4}, C^{+6}, N^{+7}, Ne^{+10} or Ar^{+18}. The latter is a typical representative for a medium-Z seeded plasma to induce radiation cooling at the plasma edge. The CHEAP concept is based on a number of key principles such as:

1) *Assumption of flux symmetry of main plasma data (ion and electron temperature and densities, plasma rotation)*
2) *Mapping of all plasma data on a common magnetic flux surface grid*
3) *Mapping of beam-path on magnetic flux surface coordinates*
5) *Monitoring of the main low-Z impurity ions*
6) *Absolute calibration of spectroscopic instruments*
7) *Use of BES as a complementary measurement for local beam densities*
8) *Analytical presentation of neutral beam profile*
9) *Analytical presentation of Beam-Target interaction and its anisotropic Slowing-down distribution function*
10) *Starting from initial estimates for the impurity composition, local atom emission rates are calculated. New ion densities are derived from absolute photon fluxes and the process is then iterated until convergence is achieved*

c) Simulation of Active and Passive Spectral Features

An import step in the preparation of the ITER CXRS diagnostic has been the development of suitable tools for the simulation of CX spectra, or the forward prediction of active and passive spectral features [15]. This tool package is based on the one hand on advanced atomic modelling and on the other hand on the extrapolation from established CXRS diagnostic systems on JET, TEXTOR, Tore Supra and ASDEX-UG. The universal simulation package which encompasses existing and future planned diagnostics (EAST, SST, W7-X and ITER) allows not only a minute prediction of spectral feature, but more importantly, the optimization of key ingredients for the CXRS and BES performance that is:

1) Specification of neutral beam

2) Optimization of periscope optics

3) Optimization of spectral instruments

A schematic presentation of the simulation is shown in the following figure:

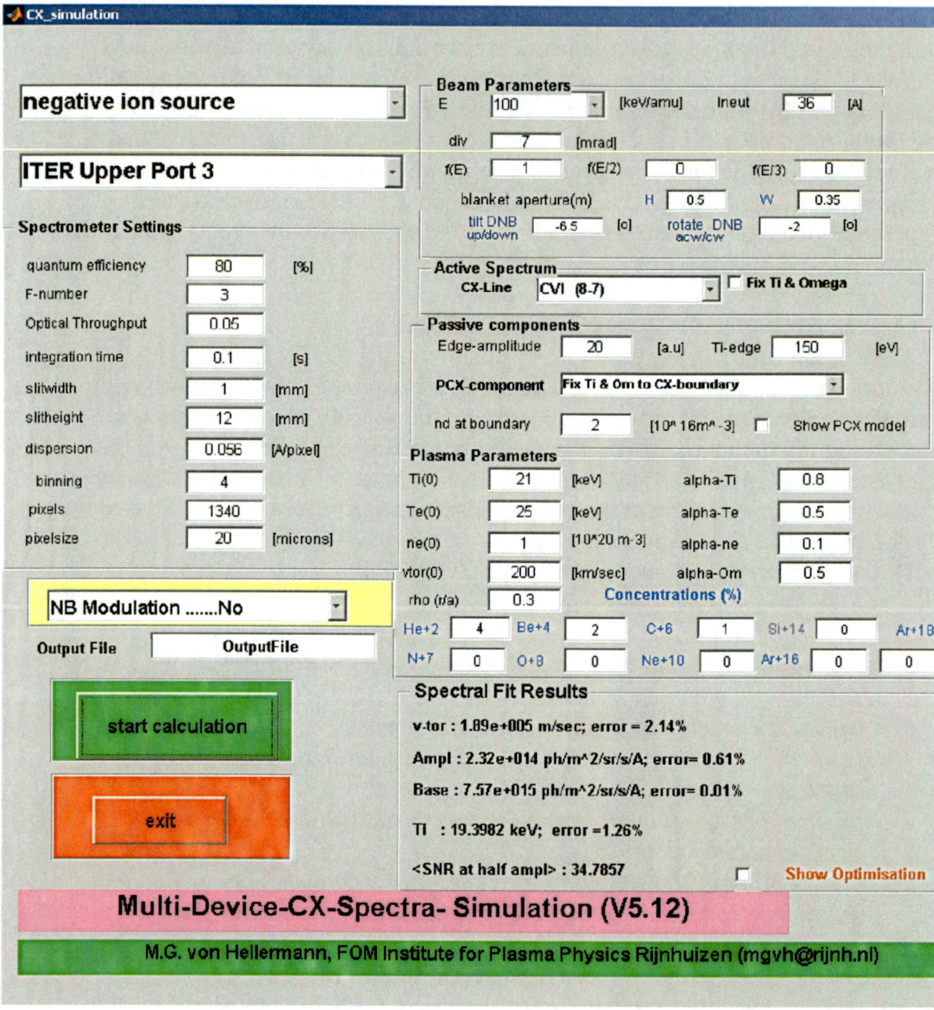

FIGURE 3. The Matlab Simulation Package "**CX_simulation**" input panels contains several input modules for a) neutral beam type, b) device selection, c) beam parameters, d) CX transition, e) plasma parameters and f) instrumentation. The presented example is ITER CXRS diagnostic, using a negative ion DNB, the ITER U-port-3 for the case of the CVI spectrum and its passive background features. Noise levels represent the photon noise from active and passive signals. A least-square routine is finally applied to retrieve from the noisy spectrum the original plasma parameters and associated fit-errors.

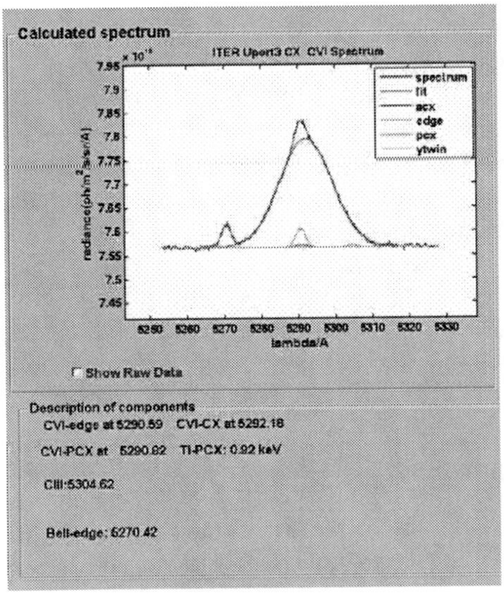

FIGURE 4. Simulated CVI CX spectrum and its passive background components comprising: CVI – PCX (passive CX), CVI-edge, BeII and CIII respectively.

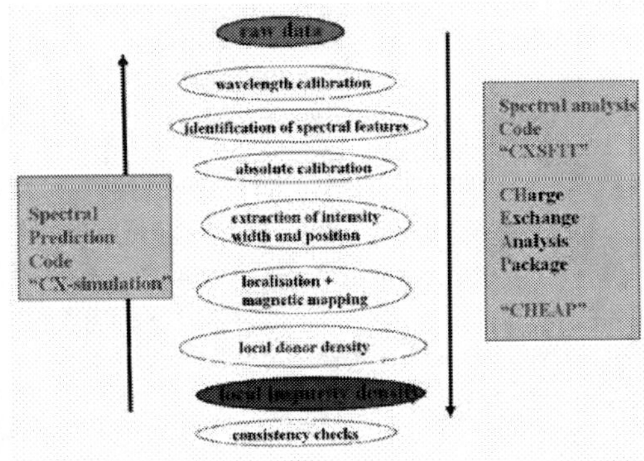

FIGURE 5. Schematic presentation of supporting software packages. Spectral prediction code : "CX_simulation", extraction of spectral features : "CXSFIT" , self-consistent reconstruction of ion densities: "CHEAP".

Active Beam Spectroscopy for ITER

The CX simulation package as describes in the previous section has enabled a detailed performance study for the future ITER CXRS system [18-25].The proposed system comprises CXRS (Charge Exchange Recombination Spectroscopy) and BES (Beam Emission Spectroscopy), which will exploit a Diagnostic Neutral Beam injected radially into the plasma:

A 'Core CXRS' system (EU) which covers dominantly the inner half of the confined plasma, and an 'Edge CXRS' system (RF) which ensures a high radial resolution in the outer half of the plasma. The Core system utilises an upper port periscope (U-port-3), and two periscopes located in an equatorial port (E-port-3) are envisaged for the Edge CX system. The equatorial port is shared with a dedicated MSE (Motional Stark Effect) system for the measurement of magnetic field pitch angles making use of the Heating Neutral Beams. A complementary MSE approach is proposed for the exploitation of intensity line ratios as measured by the BES data from upper and equatorial ports.

Obviously, the main players in the predicted ITER CXRS performance are the achievable values for the required DNB specifications, the optical transmission of the observation periscopes and the optical throughput of the spectroscopic

instrumentation. A serious challenge is the maintenance of the "First Mirror" and its durability during ITER operation (cf. [27]), a similar challenging task is the performance of the proposed DNB (100keV/amu, 36A, 7mrad) which stretches today's technology in terms of neutral beam current, beam divergence and duty cycle. The following performance studies are developed for a standard ITER plasma scenario-2 with a flat electron density profile ($10^{20}m^{-3}$) an electron temperature of 25keV, ion temperature of 21keV, v_{tor}=200km/s and a magnetic field of 5.3 Tesla.
Ongoing discussions suggest a beryllium and tungsten-only impurity composition. In this case, the main ion temperature monitor will be the BeIV CX spectrum at 465nm. Alternatively, a low-level of Neon (0.1%) may serve as a CXRS temperature indicator. The main goal for the ITER CXRS, the measurement of core helium ash densities, will even benefit from enhanced levels of beryllium providing twice the same ion temperature information in the same spectrum (cf.[31]).
For MSE, a complementary scheme to the proposed HNB based MSE polarization measurement, is the use of line intensity ratios in the DNB induced MSE spectrum. ITER relevant pilot experiments presently performed on TEXTOR are presented on this conference [26]. Figs.8 and 9 show the predicted MSE spectrum and the error analysis for the MSE evaluation, where the pitch angle is derived from the intensity ratio I_π/I_σ as measured by the Upper-port-periscope and the expected error levels across the radial profile for pitch angle, Lorentz field and MSE amplitudes. For a dt-plasma the CX feature comprises both d and t, and therefore the local fuel density ratio d/t may be derived from the combined CXRS/BES spectrum.

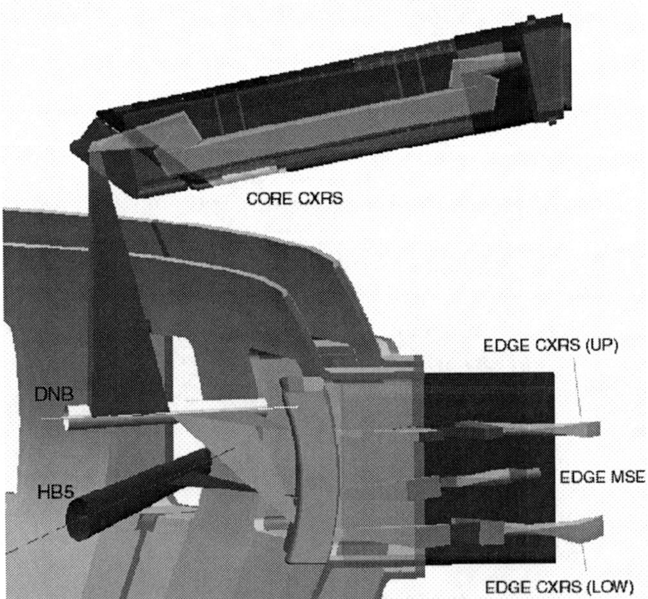

FIGURE 6 ITER Core and Edge CXRS periscopes. The DNB is injected approximately radially (angle of 7°) it reaches the magnetic axis at R=6.2m, z=0.6m. The periscopes consist of optical labyrinths in order to avoid a direct neutron streaming from blanket to the end of periscope. A crucial part is the "First Mirror" (c.f. [26,27]) close to the blanket aperture which will be mounted on a heat-exchange system at the end of an extractable tube to allow replacements in case of irreversible damages.

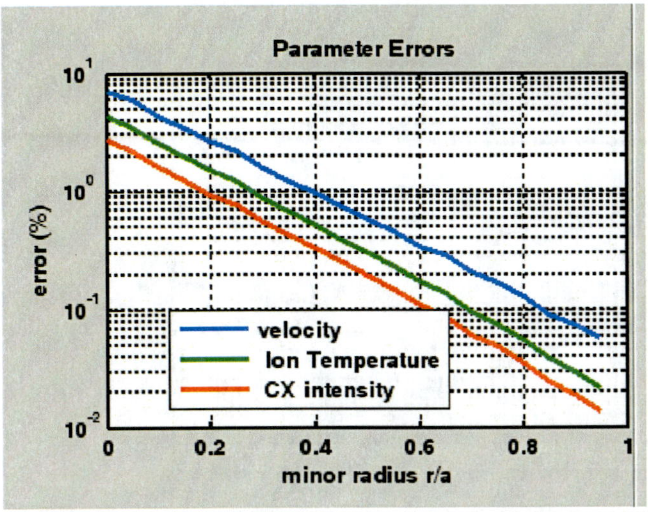

FIGURE 7. Performance analysis for ITER based on evaluation of predicted CVI spectra ($c(C^{+6})$=1%, $n_e=10^{20}m^{-3}$, $T_i(0)$=21keV), $T_e(0)$=25keV, v_{tor}=200km/s). Integration time 0.1sec. In the case of a tungsten and beryllium scenario for ITER, alternative CXRS ion temperature monitors such as low level Ne^{+10} will be used.

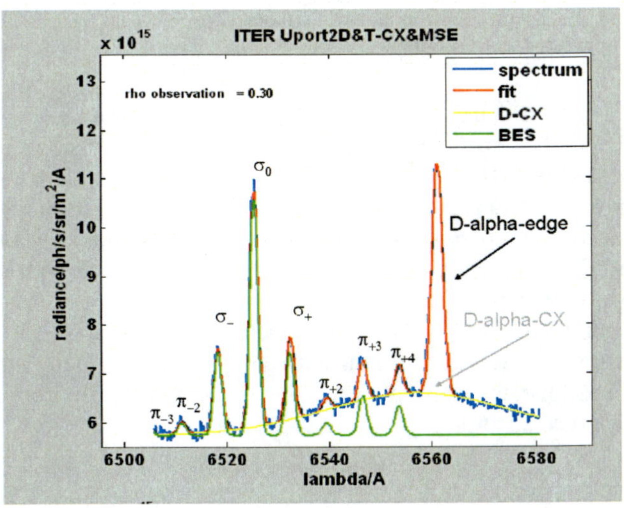

FIGURE 8. "**MSE_Simulation**" Predicted MSE (**E_L=v_{DNB} x B**) and D-alpha-CX spectrum for ITER U-port-3 for a magnetic field of 5.2Tesla. The local pitch angle and safety factor q is derived from MSE intensity ratio I_π/I_σ , the local field from the Lorentz split. The huge offset represents the underlying continuum. Both continuum and edge lines will be subtracted in the case of neutral beam modulation or use of a passive viewing lines not directed at the beam.

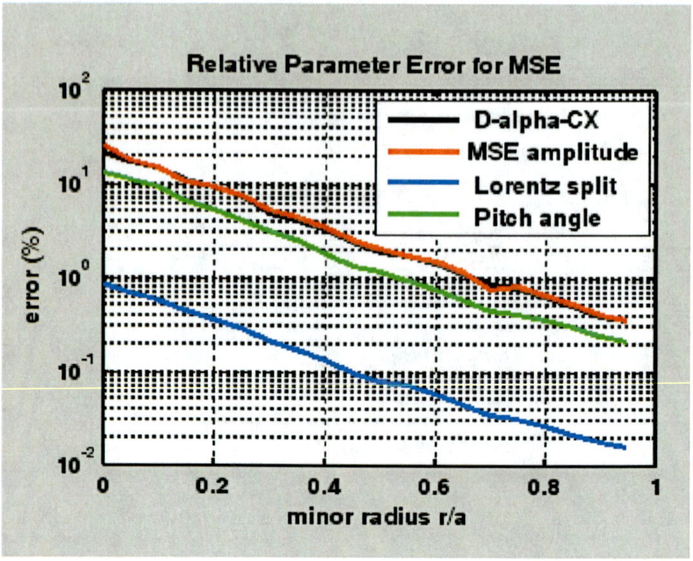

FIGURE 9. Error analysis of DNB induced Balmer Alpha spectrum (ITER U-port 3, τ=0.1s, slit-width =0.1mm). The most accurate parameter extracted from the spectrum is the Lorentz split representing the local total magnetic field. The pitch angle is used for the deduction of the q-profile, the MSE amplitude represents the local DNB density and finally the CX amplitude represents the bulk ion deuteron density. In the presence of d and t the local isotope ratio d/t may be deduced from the respective CX intensities.

Non Thermal Slowing-Down Features

The use of CXRS has been proposed in several papers as a promising option for the diagnosis of fast ion distributions (cf. [8,13,28,29,31]). Two cases are considered, alpha slowing down spectra and beam fast ion spectra [13,28]. For ITER the potential use of the proposed CXRS diagnostic in connection with the 100 keV/amu DNB and the 500 keV/amu HNB is briefly summarized here. In this paper we present as an illustration the isotropic case of slowing-down 3.5 MeV alpha particles as predicted for ITER. The source rate profile is based on nuclear reaction rates [30]. The alpha velocity distribution function depends on source rate, critical velocity and slowing-down time. Each parameter can, in principle, be extracted from the comparison of modelled and observed spectrum.

$$S_{alpha} = n_d \cdot n_t \cdot \sigma v_{dt}(T_i) \tag{6}$$

$$g_{alpha}(v) = \frac{S_{alpha} \cdot \tau_s}{v_{crit}^3 + v^3} \tag{7}$$

$$v_{crit}^3 = \frac{3}{4}\sqrt{\pi}\left(\frac{2T_e}{m_e}\right)^{3/2} \sum_j \frac{n_j}{n_e} Z_j^2 \frac{m_e}{m_j} \tag{8}$$

$$\tau_s = \frac{3v_e^3 m_e m}{16\sqrt{\pi} e^4 Z^2 n_e \ln\Lambda} \quad (9)$$

Finally, the observed spectrum which is the result of a 3-dim integration in velocity space representing the convolution between collision velocity dependent emission rate and the slowing-down function is given by [13]:

$$f_{obs}(v_z) = \int_0^\infty v'^2\, dv' \int_0^\pi d\theta' \sin\theta' \int_0^{2\pi} d\varphi' g_{slow}(v',\theta',\varphi') Q_{cx}(v_r)\delta(v_z - v'\cos\theta') \quad (10)$$

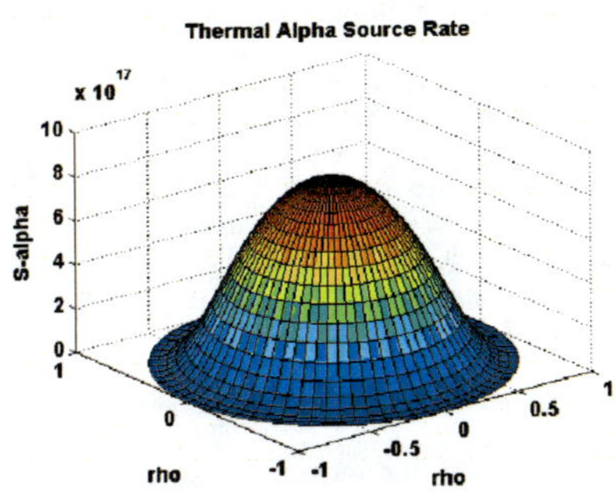

FIGURE 10 Alpha Source Rate profile for ITER based on d-t nuclear rates [30]

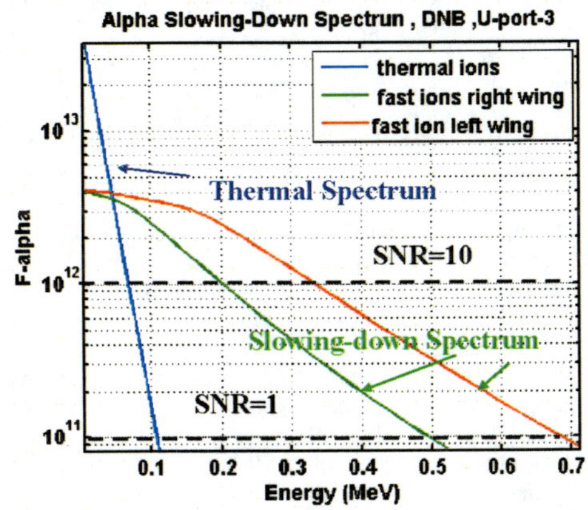

FIGURE 11 Predicted HeII CX spectrum in logarithmic scale for the Maxwellian thermal feature (blue) and the slowing-down alpha feature (red) for the case of an ITER standard plasma. The x-axis scales as $\Delta\lambda^2$, or energy. The SNR values refer to dedicated broad-band instrumentation (2nm/mm, and integration time of 0.1s). For the higher energy range up to 2MeV the use of the MSE periscope in E-port-3 aimed at the 500keV/amu HNB is proposed [31].

ACKNOWLEDGMENTS

The authors are indebted to the support by the CXRS teams of JET, Tore Supra, TEXTOR and ASDEX-UG as well as the international members of the ITPA expert group on Beam Aided Diagnostics

REFERENCES

1. R.C Isler, *Phys.Rev.Lett.* **38**, 1359(1977)
2. V.V Afrosimov., Y.S.Gordeev, A.N. Zinoviev, A.A Korotkov., *JETP Lett.***28**, 501(1979)
3. R.C.Isler, L.E Murray., *Appl. Phys. Lett.* **42**, 355(1983)

4. R.J Fonck., R.J.Goldston, R Kaita., D.Post, *Appl. Phys. Lett.* **42**,239(1983)
5 R.J Groebner., N.H Brooks., K.H Burrell., L Rottler., *Appl.Phys. Lett.* **43**,920(1983)
6. A. Boileau, M.G. von Hellermann, W. Mandl, H.P. Summers, H.Weisen, A. Zinoviev, *J.Phys.B:At.Mol.Opt.Phys.*,**22**, L145(1989)
7. A.Boileau, M.G. von Hellermann, L.D Horton., J. Spence, H.P Summers. *Plasma Phys. Contr. Fusion* **31**,779(1989)
8. M.G. von Hellermann, W.Mandl, H.P.Summers, A.Boileau, R. Hoekstra, F.J. de Heer, J.Frieling, *Plasma Physics and Controlled Fusion,* **33**, 1805 (1991
9. H.P. Summers, P. Thomas, R. Giannella, M.G. von Hellermann, W.Dickson, K. Lawson, W. Mandl, P. Briden *Z.f.Phys.D suppl.* **21**,17(1991) and http://adas.phys.strath.ac.uk
10 . H.P.Summers et al. *Physica Scripta* **T92**,80(2001)
11. H.P.Summers , M.G. von Hellermann, Atomic and Plasma Material Interaction Processes in Controlled Thermonuclear Fusion, Ed. Janev, *'Elsevier Science Publishers* ',1993, 87-117
12. M.G. von Hellermann, H.P.Summers , Atomic and Plasma Material Interaction Processes in Controlled Thermonuclear Fusion, Ed. Janev, *'Elsevier Science Publishers* ',1993, 135-164
13. M.G. von Hellermann, W.G.F.Core, J.Frieling, L.D.Horton, R.W.T. König, W.Mandl, H.P.Summers, *Plasma Phys. Contr. Fusion,* **35**,799, 1993
14. R C Isler *Plasma Phys. Control. Fusion* **36** No 2 (February 1994) 171-208
15. M.G. von Hellermann, W.Biel, G.Bertschinger, C.Jupen, M.O'Mullane, H.Summers, A.Whiteford, *Physica Scripta,* T120, 2005
16. A.Whiteford, M.G. von Hellermann, L.D.Horton, K-D Zastrow, "CSFIT User Manual", May 2007, refer to allan@phys.strath.ac.uk
17. J.Svensson, M.G. von Hellermann, R. König, *Plasma Physics and Controlled Fusion,* **41**, 315(1999)
18. M.G. von Hellermann, W.G.F. Core, A.Howman, C.Jupen, R.W.T.König, M.F. Stamp, H.P.Summers, P.R.Thomas, K-D. Zastrow, 1st International Workshop on ITER Diagnostics, Varenna, Sept. 1995, Diagnostics for Experimental Fusion Reactors I, *Plenum Press N.Y.* 1996, Ed. P.E.Stott et al., page 321
19. R.W.T. König, H. Anderson, P.Breger, W.G.F.Core, J.P.Feist, N.Hawkes, M.G. von Hellermann, A.C Howman, A.C.Maas, P.D.Morgan, M.F.Stamp, H.P.Summers, J.Svensson, M.Tunklev, P.R.Thomas, K-D. Zastrow, 2nd International Workshop on ITER Diagnostics, Varenna, Sept. 1997, Diagnostic for Experimental Fusion Reactors 2, Ed. P.Stott et al., *Plenum Press, N.Y.* 1998, page 371
20. D.M. Thomas, K.H.Burrell, M.R.Wade, R.T.Snider, 2nd International Workshop on ITER Diagnostics, Varenna, Sept. 1997, Diagnostic for Experimental Fusion Reactors 2, Ed. P.Stott et al., *Plenum Press, N.Y.* 1998, page 361
21. M.G. von Hellermann et al. Proc.Adv.Diag.for Magn. & Inert. Fus. Varenna 2001,205-208, *Plenum Publ. N.Y.*
22. M.G. von Hellermann et al. Proc. 4'th ITPA Topic Group Meeting on Diagnostics, Padova ,17-21 February, 2003
23. S.Tugarinov et al. *RF Meeting on ITER/BPX Relevant Diagnostic Developments,* St. Petersburg,15 July 2003
24. M.G. von Hellermann, EFDA contract 01.649 "Active Charge Exchange Recombination Spectroscopy (CXRS) and Beam Emission Spectroscopy (BES + MSE) with Diagnostic Neutral Beam (DNB)", December 2003, not publicly available (contact mgvh@rijnh.nl)
25. M.G. von Hellermann, R.Jaspers, W.Biel, O.Neubauer, N.Hawkes, Y. Kaschuck,V.Serov, S.Tugarinov, D.Thomas, W.Vliegenthart, *21st IAEA Conference on Fusion Energy,* Chengdu 2006, IT/PI-26
26. R.J.E. Jaspers, M.G. von Hellermann, E. Delabie, J.E. Boom, A.J.H. Donné et al, *this conference*
27. D.V.Orlinski, V.S.Voitsenya, K.Yu.Vukolov, *Plasma Devices and Operations,* **15**, 33-75, 127-146(2007)
28. W W Heidbrink, Y Luo, K H Burrell, R W Harvey, R I Pinsker and E Ruskov *Plasma Phys. Control. Fusion* **49** No 9 (September 2007) 1457-1475
29. H.Bindslev, S.K.Nielsen, L.Porte, J.Hoekzema, S.B. Korshholm et al. *Phys. Rev. Lett.* **97**, 205005 (2006)
30. H.S. Bosch, G.M. Hale, *Nuclear Fusion,* **32**, 611(1992)
31. M.G. von Hellermann, *12th ITPA Topical Meeting on ITER diagnostics,* Princeton, March 2007

Neutral Particle Analysis on ITER and requirements for DEMO

V.I.Afanasyev, M.I.Mironov, S.V.Konovalov[a], A.V.Khudoleev, M.P.Petrov, S.S.Kozlovsky[b], V.G.Nesenevich, B.V.Lyublin[c], S.Ya.Petrov, A.I.Kislyakov, F.V.Chernyshev, A.D.Melnik

A.F.Ioffe Physical-Technical Institute of Russian Academy of Sciences, Politechnicheskaya 26, St. Petersburg, 194021, Russia, e-mail: val@npd.ioffe.rssi.ru
[a]Russian Research Center "Kurchatov Institute", Kurchatov sq.1, 123182, Moscow, Russia
[b]St.Petersburg State Technical University, Politechnicheskaya 29, St.Petersburg, 195251, Russia
[c]Efremov Scientific Research Institute of Electrophysical Apparatus, St.Petersburg, 189632, Russia

Abstract. The use of the neutral particle analysis (NPA) to study the ion component of plasma on ITER has been reviewed. Both thermal (10-200keV) and supra-thermal (0.2-4 MeV) energy ranges of neutral fluxes are studied in respect to feasibility of the fusion fuel isotopic composition measurements. Influence of heating neutral beams and diagnostic neutral beam on the measurements is also shown. Possible application of the NPA to measure the energy distribution function of fusion alpha particles is discussed. Low- and high- energy NPA monitors have been proposed for DEMO machine to control the DT fuel isotope ratio.

Keywords: ITER, neutral particle analyzer, DT isotope ratio, knock-on ions, fusion alphas
PACS: 52.25.Ya, 52.55.Fa, 52.55.Pi, 52.70.Nc

INTRODUCTION

In this paper we revise the abilities of the NPA measurements for ITER steady-state (SS) operation scenario [1] in respect to the main objective of the NPA diagnostic on ITER, which is to measure the DT composition of fusion plasma [2]. D^0- and T^0-neutral flux measurements within thermal energy range (10 – 200 keV) are proposed to provide D/T density ratio control in ITER plasma in the absence of any non thermal ions [3]. The neutral fluxes at supra-thermal energies (0.4 – 4 MeV) caused by "knock-on" collisions of fusion alphas and thermal ions [4] are also considered to be promising to measure D/T ratio in plasma core. The presence of the diagnostic (DNB) and of the heating (HNB) neutral beams can disturb the measurements [5, 6]. We present the calculation of the contribution of the beam particles into NPA fluxes and derive the energy ranges suitable for the measurements of the D/T fuel ratio in plasma core and edge. To control the D/T fuel ratio in DEMO fusion reactor, specific NPA monitors are proposed which integrate neutral spectra over the determined energy ranges.

Possible application of the NPA diagnostic to measure the confined DT alphas distribution function is also discussed.

CP988, *Burning Plasma Diagnostics*, edited by F. P. Orsitto, G. Gorini, E. Sindoni, and M. Tardocchi

EXPERIMENTAL LAYOUT ON ITER AND NPA PARAMETERS

At present the design engineering of all first priority ITER diagnostics is in active phase, including NPA system. The system consists of low- (LENPA) and high- (HENPA) energy analyzers [7]. Prototypes of the analyzers were successfully tested and used on JET [8, 9]. On ITER tandem of the analyzers is supposed to be located at Port No.11 of the machine (see Fig.1, where the location of the heating beams and the diagnostic beam is also shown).

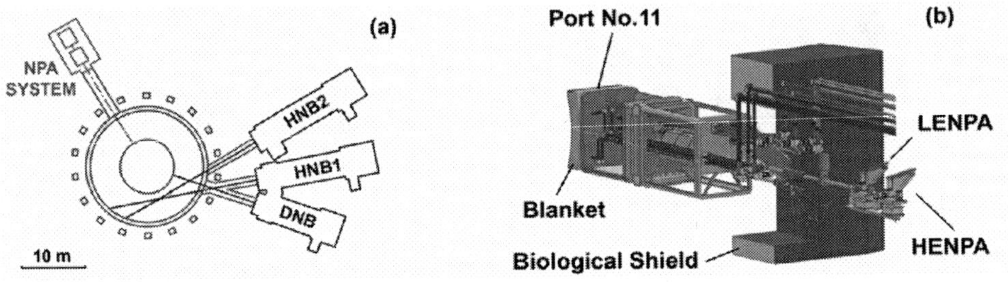

FIGURE 1. (a) Schematic layout of the NPA and NBI systems on ITER (top view); (b) Design project of NPA system integrated in ITER.

HENPA and LENPA both view plasma along the major radius close to equatorial plane of the torus. To reduce n,γ – background the analyzers share the same flight-line (20 cm in diameter at the blanket face) in which their line-of-sights are shifted by 10 cm from each other. Therefore both analyzers can operate independently. The basic characteristics of the analyzers are given in Table 1.

TABLE 1. LENPA and HENPA parameters.

Analyzer	LENPA	HENPA
Diameter of input diaphragm	1.5 cm	2 cm
Stripping in diamond-like carbon (DLC) foil [10], thickness	150Å	300Å
Acceleration of stripped secondary ions	+100 kV	-
Analyzing fields configuration	E‖B	E‖B
Detector type	Photomultiplier + scintillation layer CsI(Tl)	Photomultiplier + scintillation layer CsI(Tl) or semiconductor detectors
Detector sensitivity to n,γ – radiation	~ 10^{-7} counts/neutron	~ 10^{-7} counts/neutron
Energy range	10 – 200 keV (D^+ and T^+)	0.1 – 2 MeV (D^+ and T^+) 0.1 – 4 MeV (He^{2+})
Energy Resolution	~ 10 %	~ 10 %
Detection Efficiency	~ 3 – 25 %	~ 10 – 60 %
Mass rejection	> 100	> 1000

The energy parameters and the detection efficiency were evaluated with use of the especially designed computer code, which simulates the electromagnetic dispersion system of the analyzers and provides the ion trajectory calculations [8]. The energy dependences of the detection efficiency for different particle species are presented in Fig.2a and Fig.2b for LENPA and HENPA, respectively.

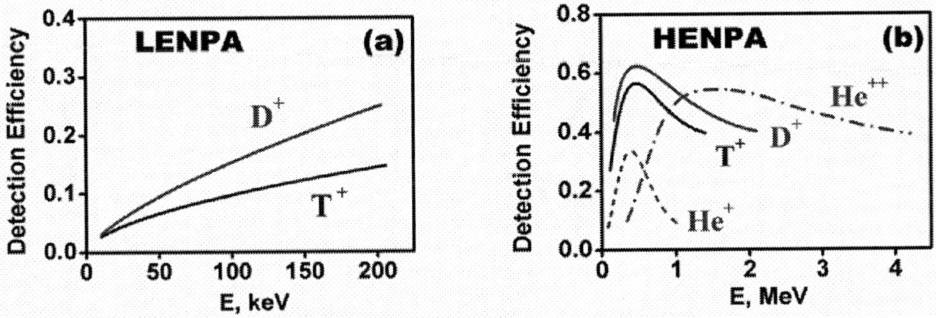

FIGURE 2. Energy dependence of the detection efficiency of LENPA (a) and HENPA (b) for different neutral particle species.

These expected parameters of the devices were taken into account below for the evaluation of the NPA counting rates produced by the different neutral fluxes emitted from plasma.

MODELING OF NEUTRAL PARTICLE FLUXES

DOUBLE code was used to model the neutral fluxes. The code is a Monte-Carlo version of the analytic method described in [11] for calculation of the radial neutral distributions in plasma and of the emitted neutral fluxes. DOUBLE provides simulation for multi-component plasma including the main light impurities. To calculate the HNB and DNB distribution functions and contribution of the beam particles into NPA measurements we used DRIFT part of the HYBRID code [12, 13], which is a pure Orbit Following Monte Carlo code. No acceleration technique, such as MAPPING part of HYBRID or acceleration of Coulomb scattering process etc. was used in the present calculations.

For modeling we used the equilibrium and the plasma profiles for SS operation scenario of ITER (S4) [1] with the main parameters as follows: $n_{eo} = 0.7 \times 10^{20}$ m^{-3}, $T_{io} = 33$ keV, $T_{eo} = 30$ keV, $Z_{eff} = 1.7$, $n_D/n_T = 1$, $n_{He^{2+}}/n_e = 7.5$ %, $n_{Be^{4+}}/n_e = n_{C^{6+}}/n_e = 1$ %, $P_{HNBtotal}$ (D^0, 1 MeV) = 33 MW. Perpendicular deuterium DNB (200 keV, 5 MW) was also assumed.

Thermal Energy Range

Typical radial dependence of the neutralization rate of D^+ and T^+ ions with the energy of 100 keV for S4 operation scenario is shown in Fig.3. It is seen that for the central region of ITER plasma the neutralization of the deuterium and tritium ions of thermal energies range occurs mainly due to charge-exchange reaction on He^+ ions. Near plasma periphery the charge-exchange reaction on D^0 and T^0 neutrals become also important.

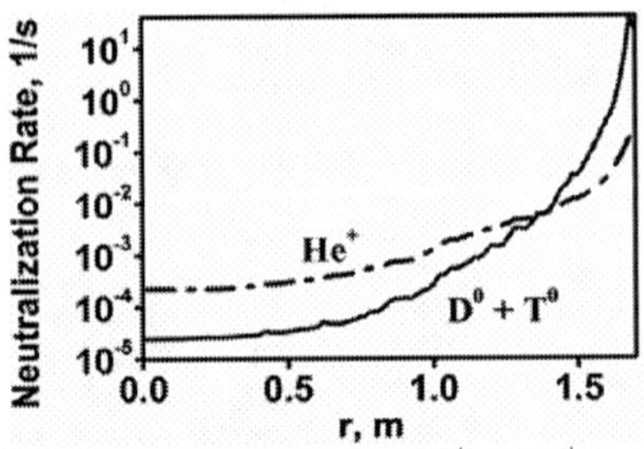

FIGURE 3. Radial distribution of neutralization rate of D^+ and T^+ ions with the energy 100 keV for different neutralization processes.

Expected counting rates of D^0 and T^0 atoms in LENPA are presented in Fig.4a in comparison with the contribution of the DNB and HNB particles and of the "knock-on" deuterons and tritons (formation of "knock-on" ions is described in the section below).

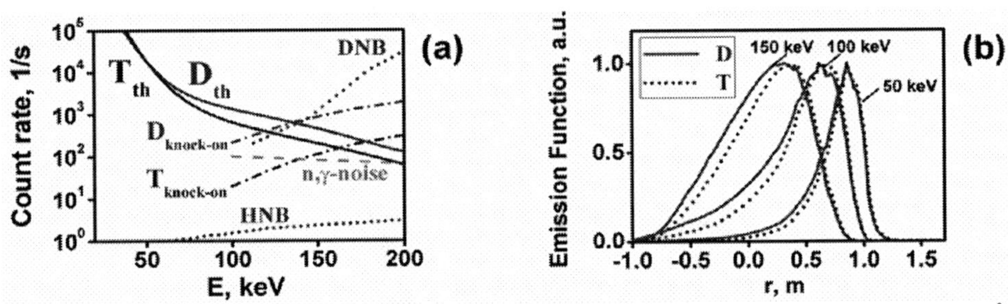

FIGURE 4. (a) Energy dependence of D^0 and T^0 counting rates in LENPA due to bulk ions, DNB (D^0, 200 keV), HNB (D^0, 1 MeV), and D,T knock-ons; (b) Emission functions for D^0 and T^0 atoms of different energies.

We consider that reliable D/T ratio measurements can be achieved in the energy range up to 100 keV only, where the thermal neutral flux is free from the non-thermal particles. This provides > 1kHz of the particle count rate which gives better than 10% of statistical accuracy for the time window 100 ms. From the calculated emission functions of the fluxes (Fig.4b) we see that NPA measurements in the energy range up to 100 keV can give D/T ratio near the plasma core at r/a ~ 0.4 (a – minor radius).

Supra Thermal Energy Range

Impurity induced neutralization of the high energy ions in plasma discovered on JET [9, 14] is considered to be the dominant neutralization mechanism for fast ions of hydrogen and helium isotopes in ITER [15]. However the absolute value of H- and He- like impurity ions densities in ITER plasma is not clear. To reveal the scale of uncertainty we give in the Table 2 the densities of carbon and beryllium ions calculated in corona equilibrium and zero-dimensional approaches [14]. For the latter particle confinement time was taken as $\tau_p \sim 5\tau_E$ = 15.5 s [1]. There is no large discrepancy between the models for H-like impurities while for He-like ions the difference is dramatic ~ $10^4 - 10^5$.

TABLE 2. Density of H-like and He-like impurity ions calculated in corona model and zero-dimensional model.

Ion type	Impurity ions	Corona model, cm^{-3}	Zero-dimensional model, cm^{-3}
H-like	Be^{3+}	$8.2x10^5$	$3.0x10^6$
	C^{5+}	$1.2x10^7$	$1.9x10^7$
He-like	Be^{2+}	$1.5x10^0$	$6.8x10^5$
	C^{4+}	$2.3x10^2$	$2.0x10^6$

We consider that careful experimental check has to be done to resolve this uncertainty, which will help us to avoid possible mistakes in the NPA development for ITER.

Fusion Alpha Particles

To estimate the maximal possible count rate of neutralized fusion alphas in HENPA we calculated the densities of C^{4+} and Be^{2+} ions in zero-dimensional approximation (see Table 2) [16].

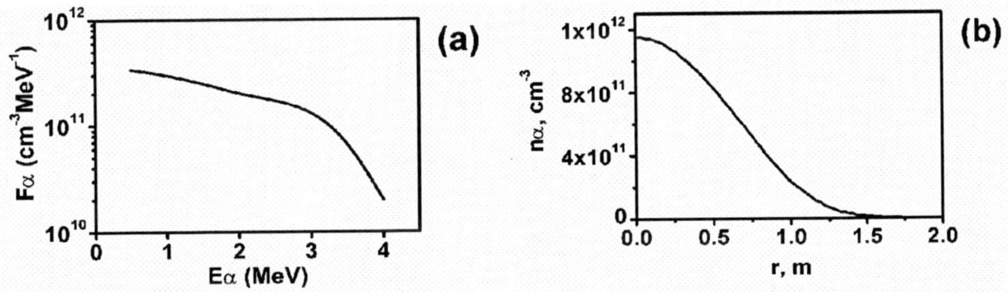

FIGURE 5. Energy (a) and radial (b) distribution functions of fusion alphas.

Energy and radial distribution of alphas (see Fig.5) were assumed to be classical (no spatial diffusion and losses) [18]. The resulting count rates of the secondary He^+ and He^{++} ions, which can be detected by HENPA are shown in Fig.6.

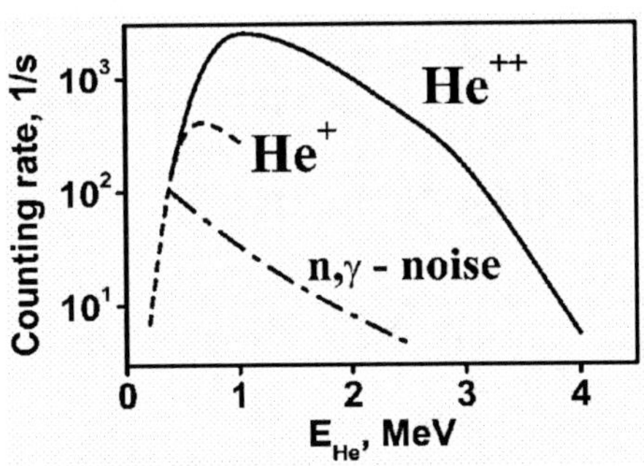

FIGURE 6. Counting rates of He^+ and He^{++} ions in HENPA produced by the neutral alpha-particle flux emitted from plasma. Density of He-like impurity ions is evaluated with the use of zero-dimensional model.

From these calculations we can conclude the following: if the level of the He-like impurity densities happens to be lower than the predicted zero-dimensional value by factor of 100 still being 10^3 times higher than basic corona value then the NPA measurement of fusion alphas will be impossible. Importance of this problem for NPA diagnostic development requires a clear experimental evidence of either presence or absence of such enhanced level of the densities of He-like impurities in ITER relevant plasma.

"Knock-on" Deuterons and Tritons

Another interesting application of the NPA in MeV energy range is to detect so-called neutralized "knock-on" D^+, T^+ ions [4]. These ions appear in fusion plasma as a result of close elastic collisions of fusion alphas with thermal D^+ and T^+ ions [17, 18]. To estimate the count rate of neutralized "knock-on" particles, which can be detected by HENPA we used the result of calculation of "knock-on" tails presented in [17] (see Fig.7a). The level of H-like impurity densities was assumed to be equal to the corona model values, i.e. minimal value. Even in this case we find that the count rates of neutralized "knock-on" ions exceed the noise level in the energy range of 0.2 – 2.0 MeV and reach the value ~ 0.1 – 10 kHz (see Fig.7b). This opens the possibility to measure fuel D/T ratio in the plasma core on the basis of "knock-on" D/T ratio.

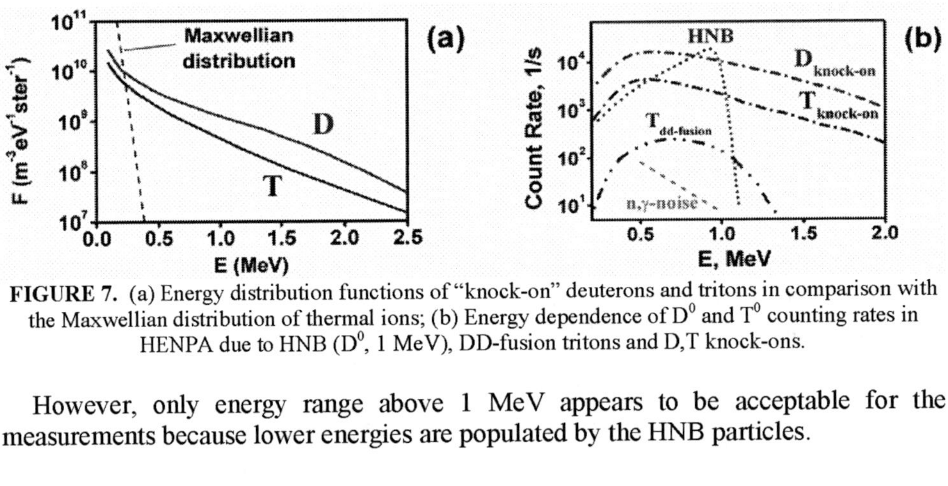

FIGURE 7. (a) Energy distribution functions of "knock-on" deuterons and tritons in comparison with the Maxwellian distribution of thermal ions; (b) Energy dependence of D^0 and T^0 counting rates in HENPA due to HNB (D^0, 1 MeV), DD-fusion tritons and D,T knock-ons.

However, only energy range above 1 MeV appears to be acceptable for the measurements because lower energies are populated by the HNB particles.

APPLICATION OF NPA FOR DEMO

Since there is no strong energy dependence of D/T flux ratio for the supra-thermal energy range it can greatly simplify the NPA method for such application as D/T ratio fuel control in the industrial reactor conditions. We consider prospects to design a specific high-energy monitor (HEM), which integrates the neutral fluxes over 0.5 – 1 MeV/amu energy range (1 – 2 MeV for D^0 and 1.5 – 3 MeV for T^0, respectively). In Fig.8a the layout of monitor is shown for the conventional E||B electromagnetic configuration with the stripping of atoms in DLC foil. The magnet focuses all ions of chosen energy range into one point. The condenser provides ion mass separation. Detector system is significantly simplified because it is enough to have only two detectors for D and T ion species. As an example we have studied the feasibility of the monitor for DEMO (variant A) [19]. HEM operation region which is defined as a range of ion temperature and D/T ratio values where the monitor provides the time resolution ~ 1 s and the counting rate ~ 100 Hz is presented in Fig.8b.

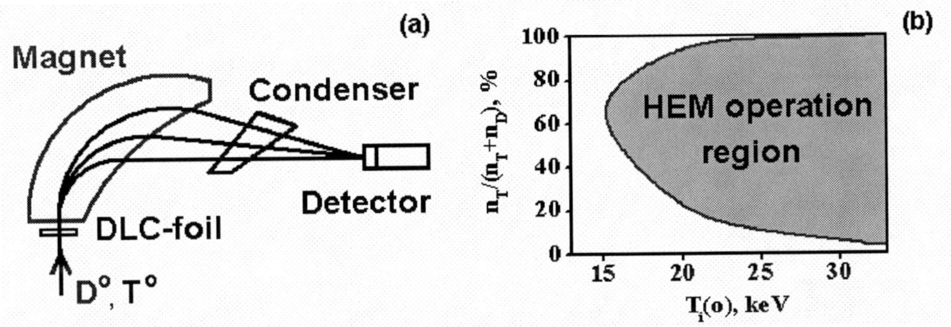

FIGURE 8. (a) Schematic layout of high energy monitor (HEM) for DEMO; (b) Range of ion temperature and D/T ratio where HEM provides reliable measurements with 1s time resolution. Monitor integrates the neutral fluxes over energy range 1 – 2 MeV for D^0 and 1.5 – 3 MeV for T^0 respectively.

Similar monitor can be designed for thermal energy range because there will be no diagnostic beam on DEMO which can disturb the measurements. In this case the most preferable energy range for the flux integration is 40 – 50 keV/amu (80 – 100 keV for D^0 and 120 – 150 keV for T^0, respectively).

CONCLUSIONS

For ITER steady state operation scenario (S4):

- Measurement of D^0 and T^0 neutral fluxes in the thermal energy range up to 100 keV can provide an information about D/T isotope ratio in the plasma region r/a ~ 0.4 – 0.9 with the count rate > 1kHz and 10% of statistical accuracy for 100 ms time window. Range above 100 keV is populated by the non-thermal particles.
- Energy distribution function of fusion alphas can be measured only if the density of He-like impurity ions (Be^{2+} and C^{4+}) in plasma center will exceed the coronal value by factor of ~ 10^3 - 10^4. Experimental evidence is required to prove the high He-like impurity densities in ITER relevant plasma.
- Measurement of D^0 and T^0 neutral fluxes in the supra-thermal energy range above 1 MeV ("knock-on" tails) can provide an information about D/T isotope ratio in burning plasma core with the count rate ~ 1kHz. Range below 1 MeV is populated by the HNB particles.

For DEMO Variant A:

- Low- and high- energy NPA monitors which integrate the fluxes over 40 – 50 keV/amu and 0.5 – 1 MeV/amu have been proposed for D/T fuel ratio control for industrial tokamak reactor DEMO.

ACKNOWLEDGMENTS

The authors are grateful to A.R.Polevoi for very useful discussions in respect to a heating beam impact on the NPA measurements. The work was supported by Russian Federation Contracts N.4z.41.19.07.511 and N.4a.41.19.07.534; by Russian Foundation for Basic Researcher project No. 06-02-16189; by International Science and Technology Center (ISTC) project No. 2482.

REFERENCES

1. ITER Technical Basis. Chapter 4, p. 11. http://www.iter.org/pdfs/PDD4.pdf.
2. M.P.Petrov, V.I.Afanasiev, A.I.Kislyakov, S.S.Kozlovski, B.V.Ljublin, S.Ya.Petrov, E.V.Suvorkin, Problems of Atomic Science and Technology **4** 80-83 (2002).
3. A.I.Kislyakov, M.P.Petrov and E.V.Suvorkin, *Plasma Phys. Contr. Fusion* **43**, 1775-1783 (2001).
4. A.V.Khudoleev, V.I.Afanasyev, A.V.Dech, A.I.Kislyakov, *27th EPS Conference on Contr. Fusion and Plasma. Phys.*, Budapest **24B**, 412-415 (2000).
5. M.P.Petrov, G.T.Razdobarin, I.N.Chugunov, *Intern. Conf. and School on Plasma Phys. and Cont. Fusion*, Alushta (Crimea), September 11-16, (2006)
6. A.V.Khudoleev, M.I.Mironov, *XII Russian Conference on Plasma Physics Diagnostics*, Zvenigorod, June 3-9 (2007), p.95.
7. V.I.Afanasiev, A.I.Kislyakov, S.S.Kozlovski, E.G.Kuzmin, B.V.Ljublin, M.P.Petrov, S.Ya.Petrov, *Plasma Devices and Operations*, **12** No.3, 209-215 (2004).
8. V.I.Afanasyev, A.Gondhalekar, P.Yu.Babenko, P.Beaumont, P. De Antonis, A.V.Detch, A.I.Kislyakov, S.S.Kozlovski, M.I.Mironov, M.P.Petrov, S.Ya.Petrov, F.V.Tschernyshev and C.H.Wilson, *Rev. Sci. Ins.* **74** No.4, 2338-2352 (2003).
9. M.P.Petrov, V.I.Afanasyev, S.Corti, A.Gondhalekar, A.V.Khudoleev, A.A.Korotkov, A.Maas, *19th EPS Conference on Contr. Fusion and Plasma Phys.*, Innsbruck **16C** 1031-1034 (1992).
10. V.K.Liechtenstein, V.I.Afanasyev, P.Y.Babenko, E.A.Gridneva, T.M.Ivkova, V.A.Kurnaev, E.D. Olshanski, M.P.Petrov, O.L.Vaisberg, K.Yu.Vukolov, *29th EPS Conference on Contr. Fusion and Plasma. Phys.*, Montreux, 17-21 June 2002, paper ID P-2.128.
11. Yu.N.Dnestrovskij, S.E.Lysenko, A.I.Kislyakov, *Nucl. Fusion* **19**,293 (1979).
12. S.V.Konovalov, S.V.Putvinski, V.S.Saplakhidi, A.P.Smirnov, *in Proc. of the IAEA Technical Committee Meeting on Alpha Particles in Fusion Research*, Kiev, USSR, 1989, V1, p.107.
13. S.V.Konovalov et al, *JAERI-Research* 94-033 (1994).
14. A.A.Korotkov, A.Gondhalekar and A.J.Stuart, *Nucl. Fusion* **37** 35-51 (1997).
15. M.P.Petrov, R.K.Fisher, *in "Diagnostics for Experimental Thermonuclear Reactor"*, Plenum Press, New York, 495-505 (1996).
16. A.A.Korotkov, A.M.Ermolaev, *22nd EPS Conference on Contr. Fusion and Plasma Phys.*, Bournemouth **19C**, 389-292 (1995).
17. D.Ryutov, *Phys. Scr.* **45** 153-158 (1992).
18. L.Ballabio, G.Gorini, J.Kallne, *Phys.Rev.E* **55** No.3, 3358-3368 (1997).
19. European Power Plant Conceptual Study, http://www.efda.org/eu_fusion_programme/downloads/scientific_and_technical_publications/PPCS_overall_report_final.pdf

Fast visible camera installation and operation in JET

J. A. Alonso[1], P. Andrew[2], A. Neto[3], J. L. de Pablos[1], E. de la Cal[1], H. Fernandes[3], J. Gafert[4], P. Heesterman[2], C. Hidalgo[1], G. Kocsis[5], A. Manzanares[1], A. Murari[6], G. Petravich[5], L. Rios[1], C. Silva[3] and P.D. Thomas[2] and JET EFDA contributors*

JET-EFDA, Culham Science Centre, OX14 3DB, Abingdon, UK
[1]*Asociación EURATOM-CIEMAT, Av. Complutense 22, 28020 Madrid – Spain*
[2]*EURATOM-UKAEA Association, Culham Science Center, Abingdom – UK*
[3]*Associação EURATOM-IST, Av. Rovisco Pais, 1049-001 Lisbon – Portugal*
[4]*Max-Planck IPP, Boltzmannstrasse 2D-85748 Garching – Germany*
[5]*EURATOM-HAS Association, KFKI-RMKI, P.O.Box 49, H-1525 Budapest-114 - Hungary*
[6]*Consorzio RFX-Associazione EURATOM ENEA per la Fusione, I-35127 Padova – Italy*

This article is a summary of the measurements of the recently installed wide-view fast visible camera in the Joint European Tokamak JET. Here we limit ourselves to a description of the different phenomena and leave for forthcoming articles a more extensive analysis of every phenomenon.

1. Camera system and technical issues

A fast visible Photron APX-RS camera has recently been installed on JET. The camera is located near the equatorial plane of octant 8 and is looking through an endoscope designed for infrared imaging [1]. It uses a CMOS detector with a maximum recording speed of 250 kHz and a minimum exposure time of 1 us. In the present set-up a commercial camera lens and no filters are used. The camera was tested against high radiation fluxes and rapidly varying magnetic fields. The integrated neutron fluence in the camera position after four months of operation is in the range of 4×10^{10} n/cm^2. So far there is no evidence of permanently damaged pixels. Several disruptions were recorded the camera withstanding field variations of up to 4 T/s.

2. Observed phenomena and first estimations

2.1 ELM physics and plasma-wall interaction - The fast camera has proven to provide useful information to study different fusion plasma relevant issues including plasma wall interaction, ELMs and disruption physics. Figure 1 shows the expulsion of macroscopic particles from the wall after the occurrence of an ELM.

* See Apendix of M.L. Watkins et al., Fusion Energy 2006 (Proc. 21st Int. Conf. Chengdu, 2006, IAEA)

CP988, *Burning Plasma Diagnostics*, edited by F. P. Orsitto, G. Gorini, E. Sindoni, and M. Tardocchi
2008 American Institute of Physics 978-0-7354-0507-3/08/$23.00

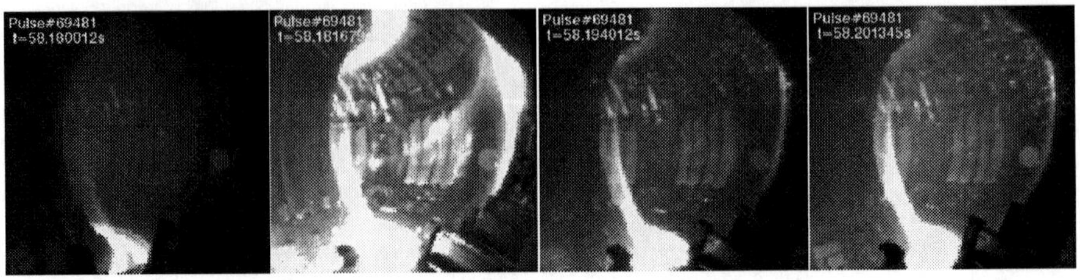

Figure 1. Particle expulsion from a poloidal limiter. Images are taken at 3 kfps with 166 us exposure time.

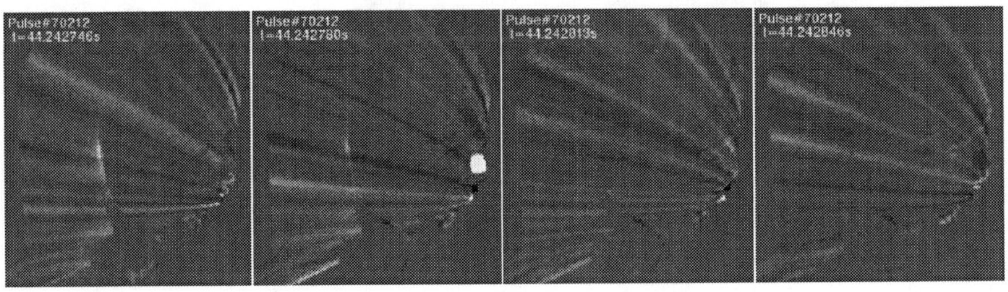

Figure 2. Filament development during an ELM (30 kfps, 25us exp. time). The background images (calculated as the pixel median over 500 us) has been subtracted from the original frames to highlight the filamentary structure.

Some 6 ms after the sudden radiation increase caused by an ELM, a shower of solid fragments is seen to come out of an outer wall poloidal limiter and seem to be accelerated in the toroidal co-current direction (that is also the direction of the parallel plasma flows for this configuration). These particles can be seen for 50 ms.

ELMs are commonly seen in the images as a sudden increase of radiation. For high power type-I ELMs like the one in figure 2, an associated filamentary structure can be observed as in other machines [2], [3]. In the high frequency recording of an ELM in figure 4, a helical interaction pattern can be seen on the wall within 40 us following the onset of the ELM (the associated D_{α} rise). This kind of interaction pattern has also been seen with infrared cameras in JET [4]. It should be noted that an ELM impact can produce localized particle desorption from the wall elements. These particles are transported parallel to the magnetic field and can also be seen as filamentary structures (figure 4, third and fourth frames) which are thus different from the ELM own structure.

2.2 ELM radial / parallel dynamics - Very fast framing can be achieved if the active part of the detector is reduced. Images in figure 3, show a 100 kfps frame sequence of the divertor during an ELM. The radiation increase starts very close to the

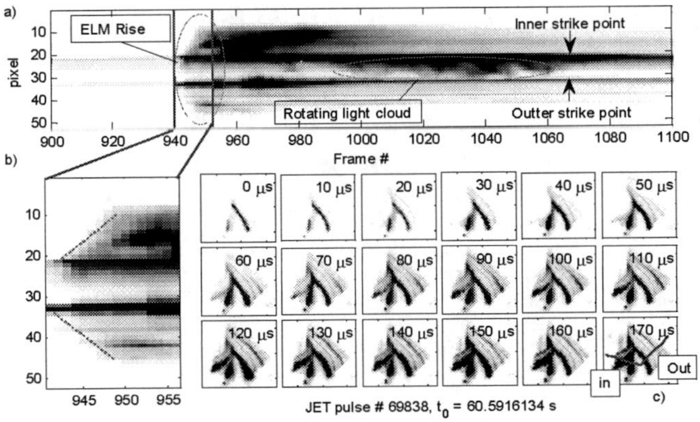

Figure 3. (a) Time trace of the pixels on the blue line in the last frame of (c) -approx. in a poloidal section. The radiation onset shows a delayed arrival to the ELM energy flux to the different divertor positions. (b) Close-up of the ELM rise. (c) Sequence of the divertor at the ELM onset (@ 100 kfps 9 us exposure).

strike points (figure 3). The D_α rise progressively reaches further divertor positions up to the limiters shadows. Two possible causes of this delayed arrival have been identified as the radial decay of the parallel flow because of the SOL temperature profiles [4] and the intrinsic ELM radial velocity. The relative importance of these two factors is under investigation.

From figure 3 it is apparent that the volume light emission from inner divertor is stronger that that of the outer, consistently with the measured higher neutral density. An intensively radiating cloud localized in between the strike points is seen after the impact of some ELMs and appears to rotate in the toroidal direction (figure 3.a).

2.3 Plasma Disruptions - During disruptions (figure 5) the plasma detaches from the divertor as the current is lost and high particle fluxes are directed to the wall. MHD activity can also be seen as fillamentation. In the course of a disruption a highly radiating MARFE-like ring could be seen which travels from the upper plate, where it formed, to the equatorial plane in the HFS in ~5ms, where it keeps radiating for ~32 ms. After this time the plasma terminates violently.

3. Conclusions and future plans

The recently installed fast visible camera in JET has produced the first results. Several physical phenomena like particle expulsion from the wall, ELM filamentation, MARFE formation and disruptions have been observed. The apparent toroidal rotation velocities of the structures during ELMs and disruptions were estimated from the images to be in the order of 10 km/s. ELMs can be seen to reach the first wall less than 40 us after the start of the light rise giving, for a typical wall clearance of 40 mm, radial propagation velocities in the range of 1 km/s. More extensive analysis of the individual phenomena, here only briefly described, will be carried out in future articles. Investigations of the pellet-plasma interaction will also be studied by making use of the camera system capabilities.

The study of lower intensity phenomena, like the continuous plasma wall interaction and the generation and dynamics of impurities, requires a higher signal level. Image intensifiers are being tested for its possible installation on JET fast imaging system, which would allow the use of interference filters so as to investigate these and other phenomena. Image analysis techniques for optical flow estimation, spatial-temporal structure detection and IR/visible data merging are under development [6].

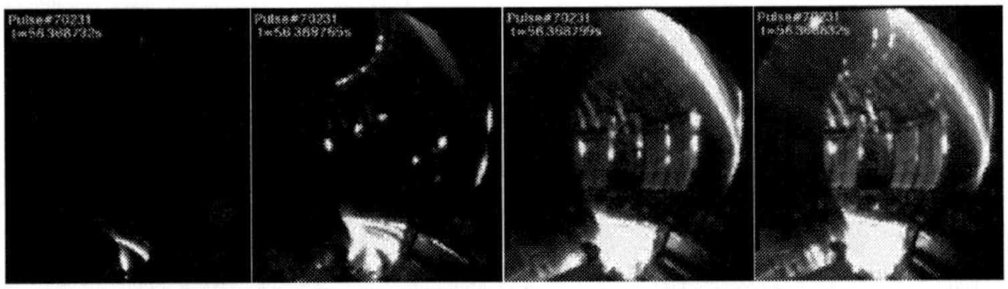

Figure 4. ELM onset and interaction with the wall (@ 30 kfps 25 us exposure). A helical wall-interaction pattern can be seen less than 40 us after the start of the D_α rise.

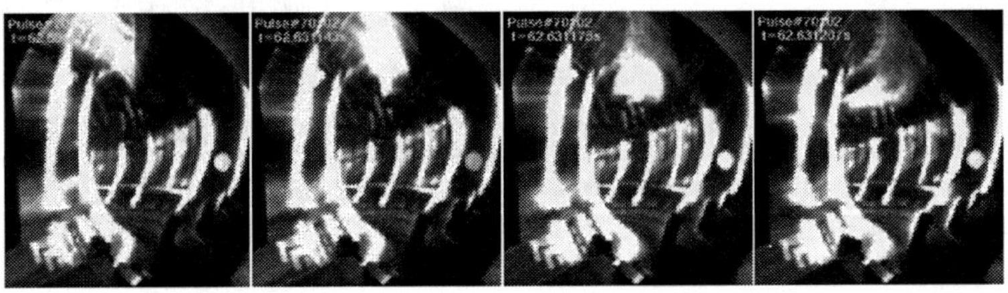

Figure 5. JET interior during a plasma disruption (@ 30 kfps, 25us exp. time). The magnetic configuration is largely lost and the plasma interacts strongly with the inner and outer walls. A light cloud is seen to move toroidally in the counter-current direction . It moves ~90 deg in the toroidal direction in the lapse of 3 frames. For R=3m this gives

Acknowledgments

The work leading to this article was funded by the European Atomic Energy Community and is subject to the provisions of the European Fusion Development Agreement.

Bibliography

[1] E. Gauthier et al, Proc. of the 17th PSI Conference, Heifei, China 22-26/05/2006

[2] A. Kirk, H. R. Wilson *et al.*, Phys. Rev. Lett. **92**, 245002 (2004)

[3] R. J. Maqueda, R. Maingi, *et al.*, J. Nucl. Mater. 363-365, (2007)

[4] P. Andrew *et al.*, 34th EPS Conference on Plasma Physics, Warsaw 2007.

[5] R. Pitts *et al.*, Nuclear Fusion **46** (2006) 82

[6] S. Hong, A. Murari *et al.*, 34th EPS Conference on Plasma Physics, Warsaw 2007.

Measurements of He-like Argon on TEXTOR, a benchmark Experiment for the ITER X-ray Spectrometer

G. Bertschinger, O. Marchuk and the TEXTOR team

Institut für Energieforschung-4 Plasmaphysik, Forschungszentrum Jülich GmbH, Association EURATOM-FZJ, member of the trilateral Euregio Cluster, 52425 Jülich, Germany

Abstract. On TEXTOR, measurements of He-like argon lines with an imaging Bragg spectrometer have been performed. The instrument had a layout similar to the x-ray spectrometers planned for ITER and was comparable in size and Bragg angle. The experiments showed that the required accuracy in the measurement of the ion temperature is obtained, for the plasma rotation, additional measurements of absolute line position and relative measurements with secondary standards such as x-ray fluorescence lines are required.

Keywords: **x-ray spectroscopy, ITER**
PACS: **52.25.Os, 52.55.Fa, 52.70.La**

INTRODUCTION

For ITER, x-ray spectrometers are foreseen to measure the ion temperature, as well as the toroidal and poloidal rotation of the plasma. Emission lines of highly ionised impurities, such as He-like krypton for the plasma core, and argon for the plasma edge are suitable. The Lyman series of intrinsic impurities of the iron group, such as iron or nickel are applicable too. The energy range of the spectrometers will be between 3 and 15 keV. In this range, crystal spectrometers operating in reflection mode (Bragg spectrometers) are the most efficient. On most of the fusion devices, spectrometers in Johann mount have been used. Bitter et. al [1] proposed to apply doubly bent crystals for fusion devices in order to image the plasma onto the detector and to improve the throughput of the device. Spherical, as well as toroidal crystals have been discussed. The imaging properties of spherical crystals have been used for high resolution spectroscopy and have been analysed in detail [2,3,4]. The imaging is highly astigmatic, a point on the on the focal plane is focused into two line foci, the meridional focus on the Rowland circle and the sagittal focus outside the Rowland circle. For extended sources such as tokamaks, excellent wavelength resolution and simultaneously good spatial resolution is achieved, if the plasma is positioned close to the sagittal focus. The spatial resolution does not suffer from the astigmatism, if the sagittal focal lines are parallel to the flux tubes.

Due to the rotational symmetry of the spherical crystal, the Bragg condition is fulfilled on a cone and the Rowland circle transforms to a "Rowland sphere". The

CP988, *Burning Plasma Diagnostics*, edited by F. P. Orsitto, G. Gorini, E. Sindoni, and M. Tardocchi

intersections of the "Bragg cone" with the detector plane are conic sections and for typical Bragg angles the conic sections are ellipses. The distance between the meridianal focus and the sagittal focus depends on the Bragg angle. For a Bragg angle of 45°, the sagittal focal length is infinite and for Bragg angles below 45°, no focus is obtained.

For toroidal crystals, the sagittal focus can be chosen independent of the meridional focus, but the symmetry of a spherical crystal is lost and the wavelength resolution degrades rapidly outside the midplane of the spectrometer. Depending on the chosen parameters, the toroidal instrument is a single chord devices with optimised throughput compared to the traditional Johann spectrometer.

In order to keep the number of spectrometers reasonably low, spherical crystals have been chosen for ITER, an array of about 6 spectrometers is required to measure the temperature profile and the poloidal and toroidal plasma rotation. The Bragg angles are around 50°, the diameter of the Rowland circle will be around 3 m. The imaging properties of an imaging Bragg spectrometer are shown in figure 1, including the focal lengths for meridional and sagittal focusing.

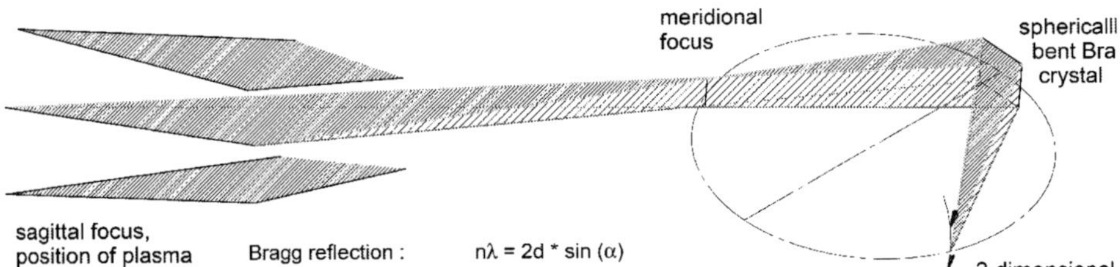

Bragg reflection : $n\lambda = 2d * \sin(\alpha)$

Meridional focus : $f_m = 2 r_r * \sin(\alpha)$

Sagittal focus : $f_s = - f_m / \cos(2\alpha)$

FIGURE 1. Principle of an imaging Bragg spectrometer with spherically bent crystal.

Measurement

On TEXTOR the horizontal branch of the existing Johann spectrometer / polarimeter has been upgraded to an imaging device by using a spherical quartz crystal, cut 11-20, crystal size 4*10 cm^2 and a bending radius of 3.85 m. For the Ar XVII resonance lines at 3.1 keV, this results in a Bragg angle of 54°. The spectral resolution $\lambda/\Delta\lambda$ of the crystal was about 5000, the 2-dimensional x-ray Multi-Wire-Proportional-Detector had a sensitive area of 10 × 30 cm^2 and a linear resolution of 0.4 mm. The overall resolution of the spectrometer was then 4300. The spatial resolution in the plasma was 3 cm, the size of the flange at the torus vessel limited the range in the plasma to about ± 9 cm, i. e. about 20% of the central plasma can be observed. Due to the symmetry of the spherical crystal, each part of the spectrometer is representative for the whole, and therefore the performance of the spectrometer can be tested, even though the observed part in the plasma is smaller than the angular range of the spectrometer planned for ITER.

In typical tokamak plasmas, the variation of the plasma parameters within the q=1 surface is rather small and only minor changes in the plasma parameters are expected.

In order to have a reasonable variation, we have chosen a discharge with low plasma current and hence high q at the plasma edge. Under these conditions, the plasma shrinks and the plasma properties vary measurably in the observed region. The spectra have been recorded during the flat top phase of the discharge, the gas feed was adjusted to keep the count rate on the detector below saturation. The overall intensity was about 10 000 photons per vertical spectrum slice.

As expected, the focal lines are curved. A scatter plot of the data is shown in fig. 2, the raw data as well as the data where the bending of the spectral lines has been corrected with the theoretical ellipses. The spectra at different vertical positions differ significantly, and the vertical spectral intensity is broader for the Li-like lines than for the He-like lines. The experimental spectra have been fitted by a theoretical model, which included all relevant atomic processes such as collisional excitation of He- and Li-like ions, dielectronic recombination of He-like ions as well as radiative and charge exchange recombination of the H-like ions and inner shell ionisation of Li-like ions. The radial variation of the plasma parameters has been taken into account by integration over the lines of sight.

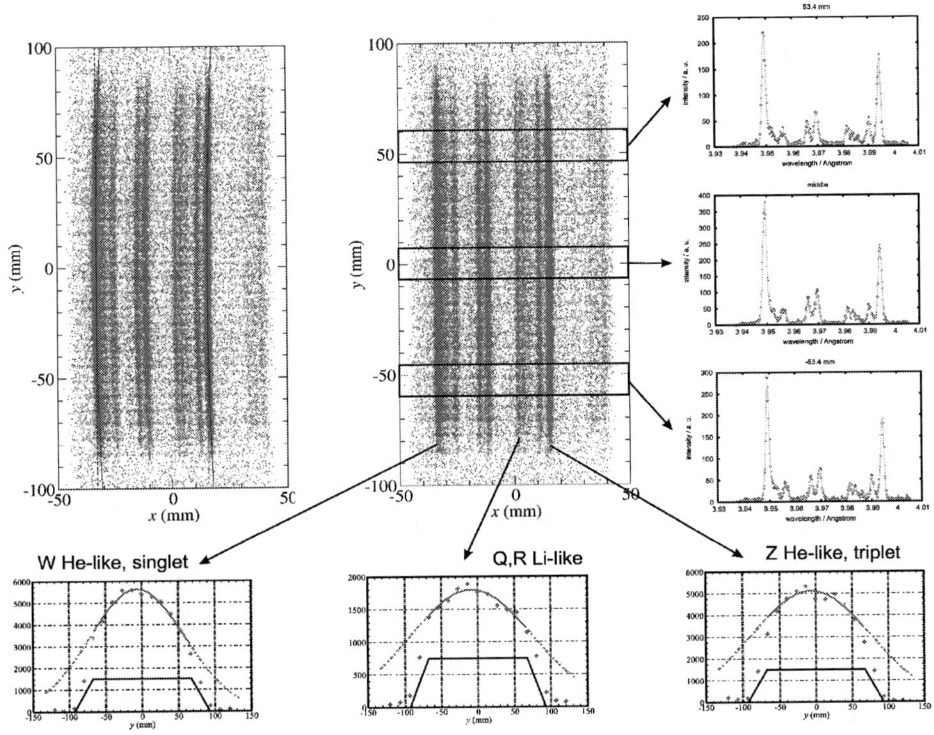

FIGURE 2. Scatter plot of the 2-D spectrum of He-like argon. x corresponds to wavelength and y to the vertical extend in the plasma. Left: uncorrected. Center: curvature corrected . Right: spectra at different vertical positions, data points and fits. Lower: variation of intensity along different spectral lines. Vignetting due to the input flange on TEXTOR.

Results and Discussion

The plasma parameters which can be derived from He-like spectra are shown in figure 3. The data with the highest precision are the electron temperature and the relative abundance Li / He-like ions. Due to the limited count rate, the statistical errors in the ion temperature and the toroidal plasma rotation are larger but still in the error margin of about 10%. The variation of the toroidal rotation is larger than expected and probably due to errors in the adaption of the ellipses. In the future, this will be improved by checking the resolution with x-ray reference lines or by reversing the toroidal magnetic field and the plasma current, which reverses the direction of rotation too. For ITER, reversal of magnetic field and plasma current will not be possible, therefore procedures to check for line positions with absolute accuracy $> 10^5$ must be developed. Recently, the wavelengths of the resonance lines of He-like argon have been measured with absolute accuracy of 5 ppm. Such accuracy is required for all spectral lines which will be used on ITER.

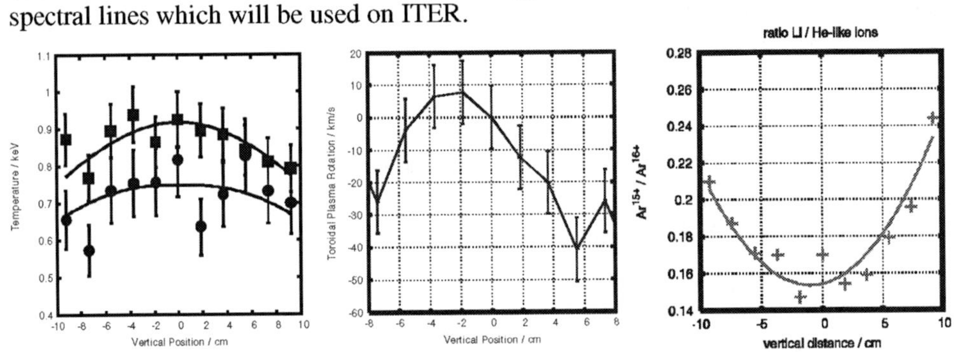

FIGURE 3. Electron temperature (boxes), ion temperature (dots), plasma rotation and relative abundance Li-/He-like Ar at different vertical positions. The variation of the electron temperature is clearly indicated. The ion temperature is lower and broader. The measured variation on the plasma rotation is probably due to errors in the correction of line curvature, indicating that additional absolute wavelength calibration is necessary.

ACKNOWLEDGMENTS

The authors recognize the loan of the x-ray detectors from Princeton Plasma Physics Laboratory, loan agreement PPPL 94-18-05

REFERENCES

1. M. Bitter et al, *Rev. Sci. Instrum.* **66,** pp. 530 (1995)
2. J. Eggs and K. Ulmer, *Z. Angew. Physik* **20(2),** pp 118, (1965)
3. D. Gotta, Hyperfine Interact. **118,** pp 35, (1999)
4. G. Bertschinger, M. Bitter, D. Rusbüldt, *Advanced Diagnostics FOR MAGNETIC AND INERTIAL FUSION*, pp 269,.Kluwer Academic / Plenum Publishers ISBN 0-306-47297-X (2002)

Advanced X-ray Imaging Crystal Spectrometer for Magnetic Fusion Tokamak Devices

S. G. Lee[1)], J. G. Bak[1)], M. G. Bog[1)], U. W. Nam[2)], M. K. Moon[3)], and J. K. Cheon[3)]

1) National Fusion Research Institute, Taejeon 305-333, Korea
2) Korea Astronomy and space Science Institute, Taejeon 305-348, Korea
3) Korea Atomic Energy Research Institute, Taejeon 305-353, Korea

Abstract. An advanced X-ray imaging crystal spectrometer is currently under development using a segmented position sensitive detector and time-to-digital converter (TDC) based delay-line readout electronics for burning plasma diagnostics. The proposed advanced XICS utilizes an eight-segmented position sensitive multi-wire proportional counter and supporting electronics to increase the spectrometer performance includes the photon count-rate capability and spatial resolution.

Keywords: X-ray imaging crystal spectrometer, segmented position sensitive detector, TDC, KSTAR, ITER
PACS: 52.70.La

INTRODUCTION

An X-ray crystal spectrometer has been designed and fabricated for the Korea Superconducting Tokamak Advanced Research (KSTAR) device [1-3]. The spectrometer is based on imaging property, which utilizes the astigmatism of a spherical mirror [4] so that this type of spectrometer is called an X-ray imaging crystal spectrometer (XICS). The XICS provides essential measurements of the profiles of the ion and electron temperatures, toroidal rotation velocity, impurity charge-state distributions, and ionization equilibrium. The proof-of-principle experiment of the XICS has been successfully demonstrated the imaging properties and fundamental principles of the spectrometer. However, the photon count-rate capability was not satisfied due to the fact that the time-to-digital converter (TDC) based delay-line readout electronics are limited only to 400 kHz. Hence, improved concept with a multi-segmented detector and corresponding number of electronics are proposed to overcome the low photon count-rate of the current XICS. Recently, a two-segmented position sensitive detector and supporting electronics have been successfully showed the improvement of the photon count-rate capability of the XICS. The fabricated two-segmented position sensitive detector is a multi-wire proportional counter which is well proven and the most robust detector to survive under the severe radiation environments includes the ITER device and future burning plasmas. The proposed advanced XICS utilizes an eight-segmented position sensitive multi-wire proportional

CP988, *Burning Plasma Diagnostics*, edited by F. P. Orsitto, G. Gorini, E. Sindoni, and M. Tardocchi

counter and 8 TDC supporting electronics to increase the spectrometer performance includes the photon count rate capability and spatial resolution. The current development status of the advanced X-ray imaging crystal spectrometer is discussed.

DEVELOPMENT OF ADVANCED XICS

Eight Segmented Detector and Supporting Electronics

A schematic layout of the advanced XICS utilizes an eight-segmented position sensitive multi-wire proportional counter and eight TDC based delay-line readout electronics is shown in Fig. 1. Figure 2 shows an inside view of the fabricated two-segmented detector and F1 chip based TDC.

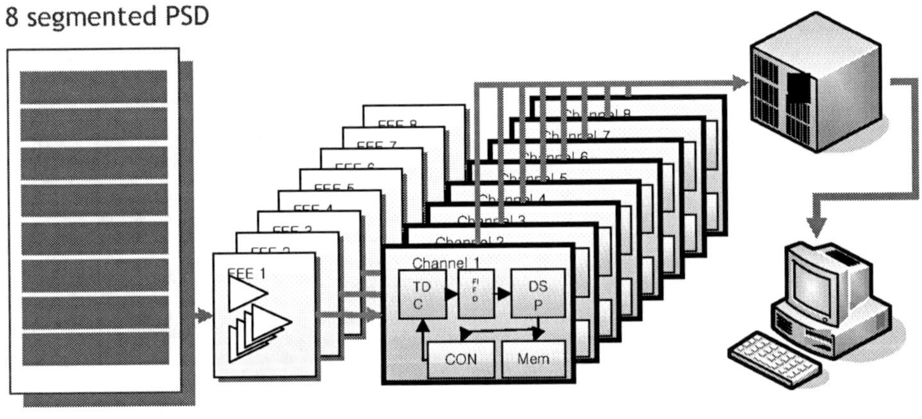

FIGURE 1. A schematic view of the eight-segmented detector and supporing electronics.

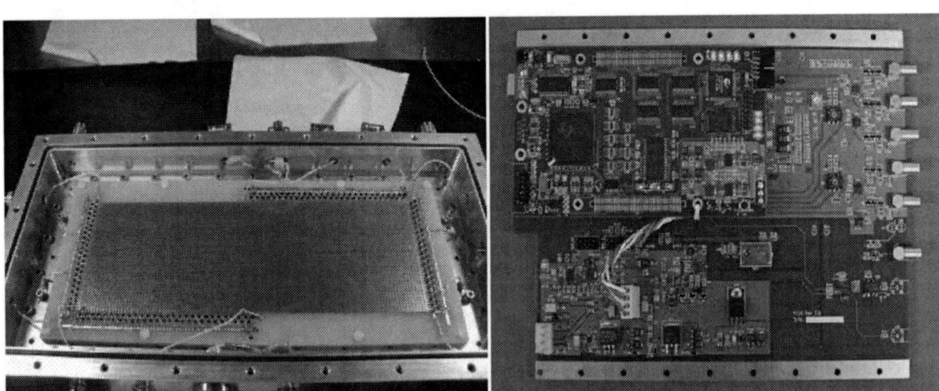

FIGURE 2. An inside view of the two-segmented detector and F1 chip based TDC.

As shown in Fig. 2 the inside printed circuit board (PCB) of the two-segmented detector is divided into two equal parts so that the performance of the detector is increased twice as compared to previous un-segmented one. Note that the measured spatial resolution of the two-segmented detector is about 0.3 mm which is much improved as compared to the previous result [3]. The improvement is due to the fact that the distance between anode and cathode PCBs was reduced. The proposed detector for the advanced XICS is an eight-segmented position sensitive multi-wire proportional counter, whose PCB is divided into eight equal parts so that eight identical segmented detectors are mounted on a single detector body. The photon count-rate capability is increased by a factor of 8 because each detector has its own data acquisition electronics. The TDC shown in Fig. 2 is comprised of a F1 chip and readout electronic module is equipped with an USB 2.0 communication port, which can be easily expanded to a multi-segmented version at low cost [5].

Vacuum Brazing Technique

The position sensitive multi-wire proportional counters usually have an X-ray entrance beryllium window. This window can be easily attached on the detector by a typical epoxy-bonding technique. However, the epoxy may react with the quenching gas inside of the detector and this effect degrades the detector performances. Therefore, a vacuum brazing method between the Be window and detector chamber made of Al was developed. The developed vacuum brazing technique will be very beneficial for the detector not only increasing the performances but also more safe under the burning plasma environments. Figure 3 shows a vacuum test and SEM image for the sample piece of the Be window after brazing. The base pressure was reached to 9.8×10^{-7} Torr and the SEM image showed good results so that the developed vacuum brazing technique is successful to the small sample Be window.

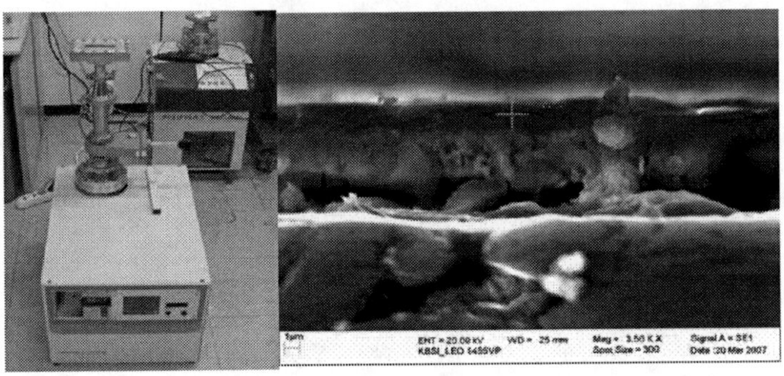

FIGURE 3. A vacuum test and SEM image of the sample Be window after brazing.

X-ray Tube Development

A novel X-ray tube for calibration of the advanced XICS has been developed [6]. The Al anode in the X-ray tube and typical example of the spectrum includes characteristic peaks and Bremsstrahlung continuum from the Al and Cu anodes measured a Si(Li) detector are shown in Fig. 4.

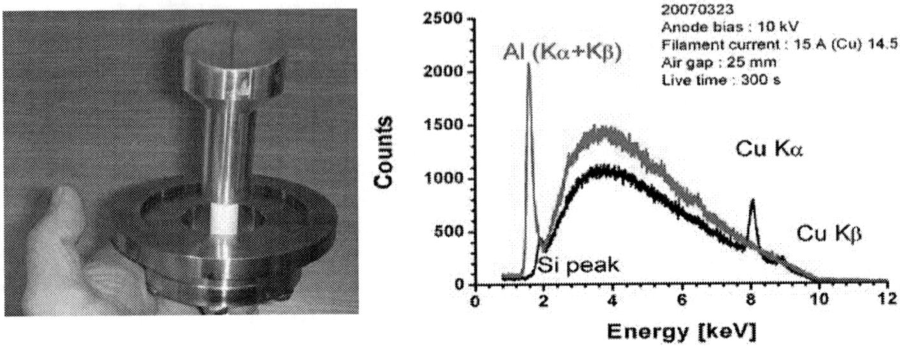

FIGURE 4. Al anode in the X-ray tube, and typical spectrum from the Al and Cu anodes.

SUMMARY

The advanced X-ray imaging crystal spectrometer has been designed using an eight-segmented position multi-wire proportional count and eight TDC based delay-line readout electronics. A vacuum brazing technique for the Be window on the detector and a novel X-ray tube for calibration of the XICS have been developed for future reference.

ACKNOWLEDGMENTS

This work was supported by the Korea Research Council of Fundamental Science and Technology under Contract No. C-research-2006-08-NFRC, and the Korea Ministry of Science and Technology under KSTAR project contracts.

REFERENCES

1. S. G. Lee, J. G. Bak, M. Bitter, M. K. Moon, U. W. Nam, K. C. Jin, K. N. Kong and K. I. Seon, *Rev. Sci. Instrum.* **74**, 1997-2000 (2003).
2. S. G. Lee, J. G. Bak, M. Bitter, K. Hill, U. W. Nam, Y. J. Kim and M. K. Moon, *Rev. Sci. Instrum.* **75**, 3693-3695 (2004).
3. S. G. Lee, J. G. Bak, U. W. Nam, M. K. Moon and J. K. Cheon, *Rev. Sci. Instrum.* **78**, 063504-1 - 4 (2007).
4. M. Bitter, K. Hill, A. L. Roquermore, P. Beiersdorfer, S. M. Kahn, S. R. Ellot and B. Fraenkel, *Rev. Sci. Instrum.* **70**, 292-295 (1999).
5. U. W. Nam, S. G. Lee, J. G. Bak, M. K. Moon and J. K. Cheon, submitted to *Rev. Sci. Instrum.* (2007).
6. J. G. Bak and S. G. Lee, to be published in *Plasma and Fusion Research* (2007).

Development of a soft-X ray detector for energy resolved 2D imaging by means of a Gas Pixel Detector with highly integrated microelectronics

D. Pacella [*], A. Romano[*], G. Pizzicaroli[*], L. Gabellieri [*], R. Bellazzini[%], A. Brez [%]

**) Associazione Euratom-ENEA sulla Fusione, Frascati, Italy*
%) Istituto Nazionale Fisica Nucleare, Pisa, Italy

Abstract. Soft-X ray 2-D imaging on ITER is not considered yet. We propose a new approach, based on a gas detector with a gas electron multiplier (GEM) as amplifying structure and with a two-dimensional readout fully integrated with the front end electronics, through an ASIC developed on purpose. The concept has been already tested by means of a prototype, with 128 pixels, carried out in Frascati in collaboration with INFN-Pisa and tested on FTU in 2001 and NSTX in 2002-2004. Thanks to the photon counting mode, it provides 2-D imaging with high time resolution (sub millisecond), high sensitivity and signal to noise ratio. Its capability of energy discrimination allows the acquisition of pictures in X-ray energy bands or to perform a spectral scan in the full energy interval. We propose the realisation of such kind a detector with a readout microchip (ASIC) equipped with 105600 hexagonal pixels arranged at 70 μm pitch in a 300 x 352 honeycomb matrix, corresponding to an active area of 2.1 x 2.1 cm^2 , with a pixel density of 240 pixels/ mm^2. Each pixel is connected to a charge sensitive amplifier followed by a discriminator of pulse amplitude and counter. The chip integrates more than 16.5 million transistors and it is subdivided in 64 identical clusters, to be read independently each other. An important part of the work will be also the design of the whole detector to fulfil all the constraints and requirements as plasma diagnostic in a tokamak machine. Since the detector has high and controllable intrinsic gain, it works well even at very low photon energy, ranging from 0.2 keV to 10 keV (X-VUV region). This range appears therefore particularly suitable for ITER to monitor the outer part of the plasma. In particular pedestal physics, edge modes, localization and effects of additional heating, boundary plasma control etc.. The capability of this proposed detector to work in this energy range is further valuable because solid state detectors are not favorite at low energy because of the lack of intrinsic gain.

Keywords: Soft X-ray imaging, gas detector, magnetic plasma physics
PACS: 52.55Hc , 52.70La , 87.59-E

INTRODUCTION

Soft-X ray 2-D plasma imaging is still an unexplored domain in magnetic fusion experiments. We propose a new camera, based on a gas detector with a gas electron multiplier (GEM) as amplifying structure and with a two-dimensional readout fully integrated with the front end electronics, through an ASIC developed on purpose. The concept (fig. 1a) has been already tested by means of a prototype [1], with 128 pixels, carried out in Frascati in collaboration with INFN-Pisa and tested on tokamak FTU (Italy) in 2001 and NSTX (USA) in 2002-2004. Thanks to the photon counting mode, it provides 2-D imaging with high time resolution (sub millisecond), high sensitivity and signal to noise ratio. Its capability of energy discrimination allows the acquisition

CP988, *Burning Plasma Diagnostics*, edited by F. P. Orsitto, G. Gorini, E. Sindoni, and M. Tardocchi

of pictures in X-ray energy bands or to perform a spectral scan and a better contrast as well. Single pixel can be addressed and programmed. Since the detector has high and controllable intrinsic gain, it works well even at very low photon energy, ranging from 0.2 keV to 10 keV (X-VUV region). Preliminary experiments performed at NSTX (fig. 1b) revealed many attractive features of the direct plasma imaging [2] and moreover of the energy discrimination: constraints for magnetic reconstructions in the core, plasma shaping, fast MHD activity detection, temperature map, impurity dynamics studies and so on.

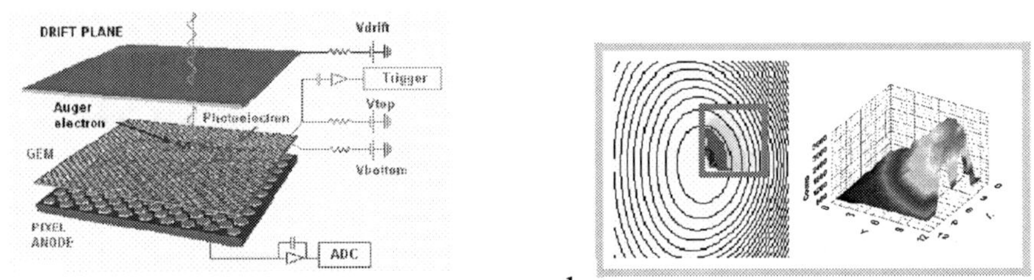

a b

FIGURE 1. Scheme of principle of a Gas Pixel Detector with Gem as amplifier (a). Example of an X-ray image (color plot superimposed to EFIT reconstructions or in a 3D plot) obtained with the prototype PIXCS-128 at NSTX (b)

GAS PIXEL DETECTOR FOR X-RAY ASTROPHYSICS

After this very successful prototype with 128 pixels, were the soft X-ray imaging capability at very high flux rate has been demonstrated, a way toward high pixel density gas detectors has been undertaken by the group of R. Bellazzini, at INFN-Pisa. In Micro Pattern Gas Detectors (MPGD) when the pixel size is below 100 μm and the number of pixels is large (above 1000) it is virtually impossible to use the conventional PCB read-out approach to bring the signal charge from the individual pixel to the external electronics chain. For this reason a custom CMOS array of 2101 active pixels with 80 μm pitch, directly used as the charge collecting anode of a GEM amplifying structure, has been developed and built [3]. Each charge collecting pad, hexagonally shaped, realized using the top metal layer of a deep submicron VLSI technology (0.35 μm) is individually connected to a full electronics chain (pre-amplifier, shaping-amplifier, sample & hold, multiplexer) which is built immediately below it by using the remaining five active layers. The GEM and the drift electrode window are assembled directly over the chip so the ASIC itself becomes the pixelized anode of a Micro Pattern Gas Detector. With this approach, for the first time, gas detectors have reached the level of integration and resolution typical of solid state pixel detectors. This approach has been used for an Astronomical X-Ray Polarimetry.
Three ASIC generations of increasing size, reduced pitch and improved functionality have been realized (fig. 2a). The last technological achievement is the ASIC with 105600 hexagonal pixels (PIXIE 105K) arranged at 50 μm pitch in a 300x 352

honeycomb matrix, carried out in technology 0.18 μm, corresponding to an active area of 15x15 mm^2 with a pixel density of 470/ mm^2 (fig. 2b). Each pixel is connected to a charge-sensitive amplifier followed by a shaping circuit (peaking time 3-10 μs externally adjustable) and a sample&hold multiplexer circuit. The chip integrates more than 16.5 million transistors and it is subdivided in 16 clusters of 6600 pixels (22 rows of 300 pixels) each one with an independent differential analog buffer. The read-out clock for each cluster is up to 10 MHz, with a total framing rate of 1.5 KHz. The pixel noise is 50 electrons ENC and the full-scale linear range of 30000 electrons (dynamic linear range of 500). Direct access to pixel content is possible, in alternative to the serial read-out of the cluster. The fill fraction ratio of metal area to active area) is 92%. Data are directly transferred form the mother board (approximate dimensions 10cm x 6 cm) to the host computer via USB cable.

a b

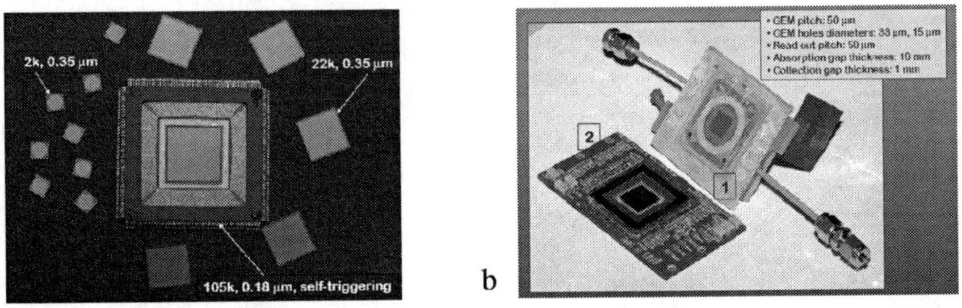

FIGURE 2. Three generations of ASIC (a). PIXIE 105K (b): upper part of the gas detector (1) and readout ASIC bonded on the mother board (2)

PROPOSAL OF AN IMAGING CAMERA FOR MAGNETIC PLASMAS

Following the very encouraging and promising results obtained at NSTX with the prototype PIXCS-128 we propose to develop a high density Gas Pixel Detector, similar to the last one developed for the astrophysics applications (PIXIE 105K) and previously described, but working in single photon counting with adjustable thresholds of discriminators, like the prototype PIXCS-128.

We can develop a similar ASIC, but with a different analog and digital electronics: charge sensitive preamplifier, pulse shaper, discriminator with adjustable thresholds, and counters. The pixel size is 50-100 μm, active area 20mm x 20mm, with the same sensitivity (50 electrons ENC) and dynamic range (500) of the ASIC already built for astrophysics (PIXIE 105K). The read-out architecture can be designed to have a full framing rate of a few KHz, or higher for a region of interest, with the capability to program the thresholds of the discriminators individually or for miniclusters.

The upper part of the detector will be carried out in a modular form (fig. 3a) in order to work with single or double GEM, depending on the energy photon bands (2-35 KeV or 0.2-2 KeV respectively), with different entrance windows (ultra thin for the lower band) and spacers that define the thickness of the conversion region, and the transfer and collecting gap. The camera will work with different gas mixtures depending on the required energy range, spatial resolution and total counting rate. All these components (entrance windows, GEM foils, spacers and read-out ASICs) have to be easily replaced to change the working configuration or to substitute them in case of damage. The GEM hole pitch is 50 μm, but the real physical processes limiting the spatial resolution are the range of the photoelectron following the photon absorption and the cross spread of the electrons migrating toward the GEM in the conversion gap. The combination of these effects produces a total spatial resolution of 100 μm at lower energy (up to a few KeV) and a bit higher at higher energy. The ASIC will be designed to have all the connections to the mother board along two sides, in order to place 4 ASICs in a matrix 2x2. It allows a larger active area of 40mm x 40mm. The detector and the mother board will be enclosed in a vacuum tight case, compatible with the high vacuum, with small Be windows in correspondence of the pinholes and filled with He or under a low vacuum, depending on the energy range of working (fig. 3b).

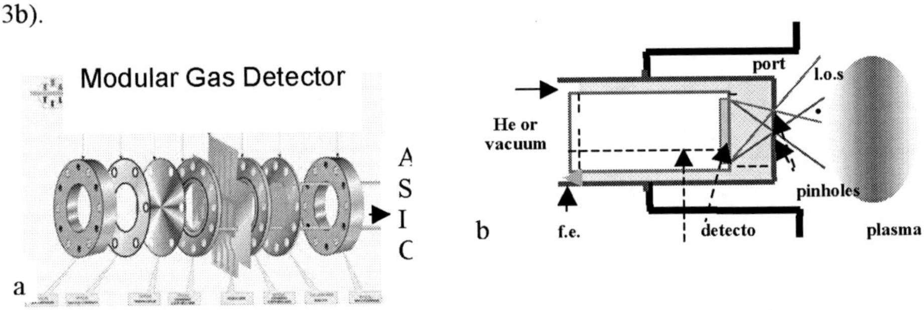

FIGURE 3 Design of the modular detector (a). Setup for a tokamak (b): the detector and the mother board (red) is enclosed in a high vacuum tight case (brown) to be inserted inside the port

CONCLUSIONS

This proposed device for energy resolved, high definition soft X-ray imaging will provide important contributions in a number of magnetic plasma physics issues: constraints for the inner reconstruction of magnetic surfaces and plasma current profile, recognition and localization of inner MHD instabilities, fluid effects (centrifugal forces). With adequate analysis tools, particle and impurity transport could be inferred (barriers, impurity pump-in/out and confinement times). Perturbative transport can be monitored also. Setting the detector at very low energy (0.2-2 KeV), the physics of pedestal can be studied and the monitoring of plasma position as well. Finally a very important perspective is the use of this detector for real time plasma control.

REFERENCES

[1] D. Pacella et al, Rev Sci. Instrum., vol 72, N2, 2001
[2] D. Pacella et al, Plasma Physics and Controlled Fusion 46(2004) 1075
[3] R. Bellazzini et al., Nuclear Instruments and Methods in Physics Research A535, 477484 (2004)

X-Ray Imaging Spectrometer for Recording Calibrated Time-Resolved K-Shell Spectra from Magnetically Confined Fusion Plasmas

John F. Seely[1], Uri Feldman[2], Szymon Suckewer[3], and Lawrence T. Hudson[4]

[1]*Naval Research Laboratory, Washington DC 20375 USA*
[2]*Artep Inc., 2922 Excelsior Spring Circle, Columbia Md 21042 USA*
[3]*Princeton University, Princeton NJ 08544 USA*
[4]*National Institute of Standards and Technology, Gaithersburg MD 20899 USA*

Abstract. Based on transmission crystal spectrometers recently optimized for recording high-resolution K-shell spectra from laser generated plasmas, an imaging spectrometer has been designed to record Mo and W K-shell spectra from magnetically confined plasmas. The spectrometer can be used to diagnose the electron energy distribution and transport of metal ions from the wall and divertor to the core plasma.

Keywords: x-ray spectroscopy, tokamak spectroscopy
PACS: 07.85.Nc; 52.70.La; 32.30.Rj; 39.30+w

INTRODUCTION

An imaging x-ray spectrometer has been designed for recording the space and time resolved K-shell spectra from magnetically confined plasmas. The spectrometer is based on spectrometers used to record hard x-ray spectra in the 10 keV to 100 keV range from plasmas generated by short-pulse laser irradiation. [1-6] The spectrometer can be optimized to record K-shell spectra from the wall and divertor elements such as Mo and W. The x-ray spectra are used to diagnose the hot electron distribution, for example generated by lower hybrid heating, and the transport of metal ions from the wall and divertor to the core plasma. In the case of Mo, the spectral resolution is sufficient to observe the spectral features from highly charged Mo ions from Ne-like to He-like in the 17.3 keV to 18.1 keV energy range as well as the nearby hard x-ray continuum.

The spectrometer implements a transmission crystal (Laué diffraction) and a 2D electronic x-ray detector operating with frame rates up to 30 Hz. In addition, a photodiode linear array records high-frequency emission fluctuations on a sub-microsecond timescale. The transmission crystal spectrometer can be adapted to observe the W L-shell emission in the 9 keV to 12 keV energy range and the W K-shell emission in the 60 keV to 70 keV range from future tokamaks such as ITER.

This spectrometer design combines the following features that are beneficial for tokamak diagnostics: (1) Imaging spectrometer provides spatial resolution with high

CP988, *Burning Plasma Diagnostics*, edited by F. P. Orsitto, G. Gorini, E. Sindoni, and M. Tardocchi
2008 American Institute of Physics 978-0-7354-0507-3/08/$23.00

throughput. (2) Linear, narrow instrument geometry is compatible with restricted viewing ports. (3) Proven ability to shield the detector against hard x-rays/gammas/neutrons (demonstrated at femtosecond laser facilities). (4) Robust quartz crystal is resistant to damage. (5) Spectrometer design is easily adapted for W L-shell and K-shell spectroscopy and hard x-ray continuum measurements (already achieved for laser-produced plasmas). (6) Proven 2D x-ray detector and photodiode detector techniques.

SPECTROMETER DESIGN

As shown in Fig. 1 (left), the spectrometer module consists of a cylindrically bent crystal, a slit, and a detector. The x rays from the source are diffracted by the crystal, converge through the slit, and diverge to the detector. Focusing occurs on the Rowland circle with diameter equal to the radius of curvature of the crystal. This symmetric-type spectrometer was originally developed for the characterization of medical x-ray sources. [7-9]

The spectrometer adapted to the Alcator C-Mod is illustrated in Fig. 1 (right). The spectrometer views the plasma through a vacuum window that transmits > 15 keV x-rays, for example Al with 0.5 mm thickness. Illustrated in Fig. 1(right) are 30 lines of sight through the plasma, with approximately 60 cm vertical extent, resulting in 2 cm spatial resolution in the vertical direction. Different lines of sight through the plasma are achieved by rotating the spectrometer about its center of mass by $\pm 11.5^{\circ}$ relative to the center of the plasma. When the detector operates at 30 Hz, a complete scan of the entire plasma is performed every 1 sec during the multi-second discharge.

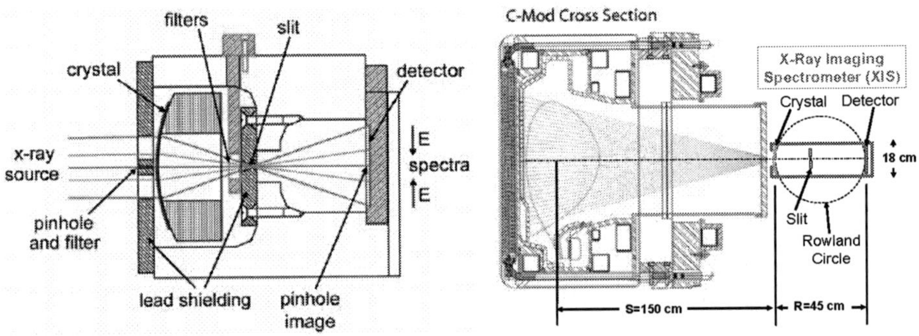

FIGURE 1. Left: Schematic of the spectrometer module. Right: The spectrometer mounted on the Alcator C-Mod tokamak indicating the field of view.

The spectrometer concept has been extensively tested by recording spectra from laboratory x-ray sources and from plasmas generated by high-intensity laser irradiation. Shown in Fig. 2 (left) is a spectrum from a Mo anode source. In this case the detector was placed behind the Rowland circle there the spectrometer's resolving power is enhanced, and the Kα and Kβ lines are well resolved. [10] Figure 2 (right) presents simulated Mo K-shell spectra calculated by the FLYCHK code [11] for

Alcator C-Mod conditions. The electron temperatures are 2 keV, 4 keV, and 6 keV, and a 10 % hot (20 keV) electron distribution is assumed. The spectral lines are broadened by the spectrometer's instrumental line width, and the Ne-like through He-like ionization stages are resolved. Absolute calibration of the spectrometer sensitivity, using the NIST x-ray calibration facility, provides a measure of the ion densities. This indicates the ability to study Mo ionization and transport in tokamak plasmas. Additional simulated spectra (not shown) demonstrate the ability to diagnose the hot electron distribution from the relative intensities of the spectral features.

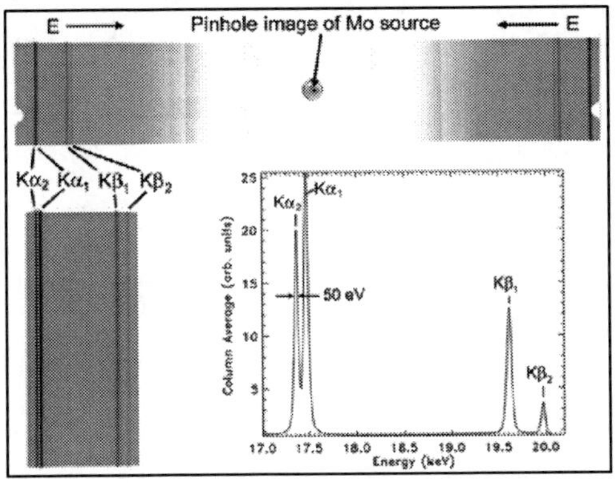

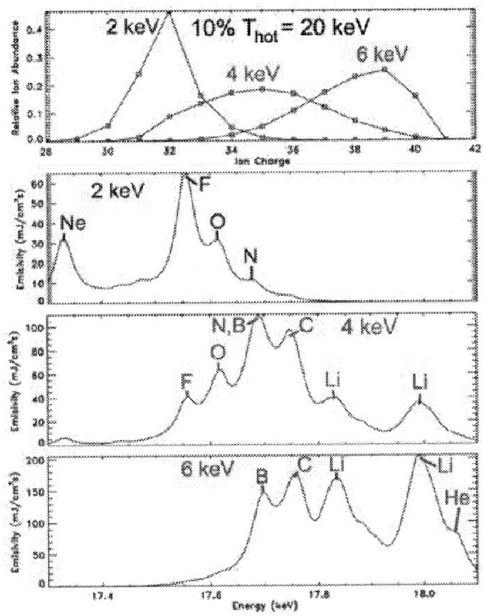

FIGURE 2. Left: Spectrum from a Mo anode source. Right: Calculated Mo spectrum indicating the resolution of the K-shell lines from the Ne-like through He-like ionization stages.

The spectrometer can be adapted for recording high-resolution W K-shell spectra in ITER. This is illustrated by the spectrum shown in Fig. 3 which was recorded using a W anode source. The W Kα and Kβ lines are well resolved in the second diffraction order.

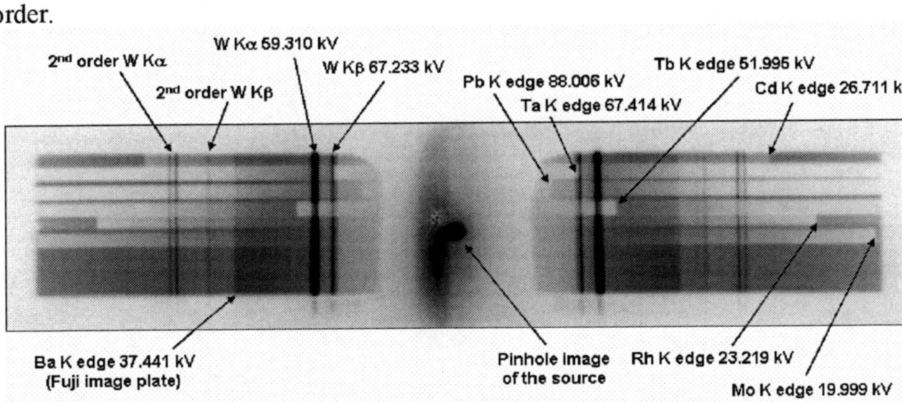

FIGURE 3. A W K-shell spectrum recorded using a W anode source. The K attenuation edges result from filters placed at the slit position.

SUMMARY

A transmission crystal spectrometer has been designed for recording the K-shell emissions from the wall and divertor materials such as Mo and W in magnetically confined plasmas. By resolving the Ne-like through He-like spectral features, the spectra can be used to diagnose the hot electron distribution and transport of wall and divertor material into the core plasma. The spectrometer provides coverage of the Alcator C-Mod plasma, for example, with 2 cm spatial resolution and 30 ms time resolution. The spectrometer can be adapted to record high-resolution W K-shell spectra from ITER. The spectrometer design is based on several spectrometers recently utilized to record K-shell spectra of the elements up to Au (70 keV) generated by short-pulse laser irradiation.

ACKNOWLEDGMENTS

We benefited from helpful discussions with Drs. Manfred Bitter, Kenneth Hill, John Rice, and Albert Henins.

REFERENCES

1. L. T. Hudson, A. Henins, R. D. Deslattes, J. F. Seely, G. E. Holland, R. Atkin, L. Martin, D. D. Meyerhofer, and C. Stoeckl, *Rev. Sci. Instrum.* **73**, 2270 (2002).
2. J. F. Seely, R. Doron, A. Bar-Shalom, L. T. Hudson, and C. Stoeckl, *J. Quant. Spectr. Rad. Trans.* **81**, 421 (2003).
3. J. F. Seely, C. A. Back, C. Constantin, R. W. Lee, H.-K. Chung, L. T. Hudson, C. I. Szabo, A. Henins, G. E. Holland, R. Atkin, and L. Martin, *J. Quant. Spectr. Rad. Trans.* **99**, 572 (2006).
4. L. T. Hudson, R. Atkin, C. A. Back, A. Henins, G. E. Holland, J. F. Seely, and C. Szabo, *Rad. Phys. Chem.* **75**, 1784 (2006).
5. C. I. Szabo, L. T. Hudson, A. Henins, G. E. Holland, R. Atkin, and J. F. Seely, *Rad. Phys. Chem.* **75**, 1824 (2006).
6. J. Seely, G. Holland, L. Hudson, C. Szabo, A. Henins, H.-S. Park, P. Patel, R. Tommasini, and J. M. Laming, *High Energy Density Phys.* **3**, 263 (2007).
7. R. D. Deslattes, J. C. Levin, M. D. Walker, and A. Henins, *Med. Phys.* **21**, 123 (1994).
8. C. T. Chantler, R. D. Deslattes, A. Henins, and L. T. Hudson, *Br. J. Radiol.* **69**, 636 (1996).
9. L. T. Hudson, R. D. Deslattes, A. Henins, C. T. Chantler, E. G. Kessler, and J. E. Schweppe, *Med. Phys.* **23**, 1659 (1996).
10. L. T. Hudson, A. Henins, J. F. Seely, G. E. Holland, in *15th International Conference on Atomic Processes in Plasmas*, edited by J. D. Gillaspy, J. J. Curry, and W. Wiese, AIP Conference Proceedings 926, Melville, New York: American Institute of Physics, 2007, pp. 34 - 41.
11. H.-K. Chung, M. H. Chen, W. L. Morgan, Y. Ralchenko, and R. W. Lee, *High Energy Density Phys.* **1**, 3 (2005).

An EUV narrow band imaging technique for diagnosing 10-30 keV ITER plasmas

U. Feldman*, J.F.Seely†, E. Landi* and Y. Ralchenko**

*Artep Inc. 2922 Excelsior Spring Court, Ellicott City, MD 21042 USA
†Space Science Division, Naval Research laboratory Washington DC 20375-5352 USA
**Atomic Physics Division, National Inst. of Standard and Technology, Gaithersburg, MD 20889

Abstract. In this paper we investigate the use of lines from tungsten (W) ions to measure the properties of W impurities in ITER plasmas. We calculate the intensities of the brightest spectral lines expected to be radiated by W ions with wavelengths longer than 45 Å, abundance of 10^{-4} and at densities of 10^{20} m^3. Using the calculated wavelengths and intensities we propose a concept for segmented multilayer-coated imaging telescopes that could be used to investigate the properties of W impurities as a function of time and space using spectral lines from highly ionized W.

Keywords: ITER plasmas – diagnostics
PACS: 50

INTRODUCTION

When deuterium and tritium fuse, two nuclei come together to form a helium nucleus and a high-energy neutron. This reaction occur spontaneously at a temperature of 10 keV and an electron density of 10^{20} m^{-3}. The stated objective of ITER is to demonstrate the scientific and technological feasibility of fusion energy for peaceful purposes. Once ignition starts, the ITER plasma temperature will be in the 10-30 keV range. The body of the confinement vessel will be made of stainless steel, the first material wall facing the high T plasmas will be made of Be and the front surfaces of the diverters will be of W. Trace amounts of W ions (e.g 10^{-4} or less) will be present as impurities in the mostly deuterium-tritium ITER plasma, and in this paper we propose to use them to measure the properties of impurities in ITER plasmas.

Usually the normal incidence reflectivity of a simple coating in the EUV is very small. However by depositing specially designed multilayer coatings on front surfaces of the optical elements the EUV normal incidence reflectivity at a narrow predetermined wavelength band can be significantly increased. As an example, the highest reflectivity from osmium coatings is $\simeq$1% at 190 Å and much less at 65 Å. Using multilayer coatings the reflectivity at these wavelengths were measured to be $\simeq$30%.

In this paper we also introduce a normal incidence multilayer coated instrument that is able to record lines from highly ionized W in the EUV and soft X-ray range to be used for the measurement of the physical properties of ITER plasmas.

CP988, *Burning Plasma Diagnostics*, edited by F. P. Orsitto, G. Gorini, E. Sindoni, and M. Tardocchi

TABLE 1. Strongest W lines formed in the 7-40 keV T range at $\lambda > 45$ Å (Intensities in phot cm^{-2}s^{-1}sr^{-1}). For each ion $T_{0.2}$ is the lowest temperature (in keV) at which the fractional abundance is 20%, T_{max} is the temperature (in keV) of maximum fractional abundance.

Ion	I.P. (keV)	$T_{0.2}$ (keV)	T_{max} (keV)	Wvl. (Å)	Intensity per unit EM 10 keV	20 keV	30 keV	40 keV
W^{58+}	6.0		7	53.353	1.56e-18	2.47e-21	-	-
W^{59+}	6.6	8	8	53.125	1.39e-17	7.63e-20	-	-
W^{60+}	6.6	9	9	65.185	5.10e-18	1.08e-19	1.88e-21	-
W^{61+}	6.7	10	10	73.245	1.01e-16	7.44e-18	2.48e-19	1.33e-20
W^{62+}	7.0	10	12	79.923	1.84e-16	5.09e-17	3.35e-18	2.75e-19
W^{63+}	7.1	11	15	77.605	9.19e-17	9.59e-17	1.28e-17	1.58e-18
W^{64+}	15.6	14	17	48.307	7.67e-20	4.84e-19	1.52e-19	3.25e-20
W^{65+}	15.9	20	23	52.108	9.23e-21	2.77e-19	1.82e-19	6.06e-20
W^{66+}	16.2	25	27	202.903	-	1.23e-19	1.84e-19	1.02e-19
W^{67+}	16.6	29	32	59.800	-	9.14e-21	2.99e-20	2.56e-20
W^{68+}	18.5	31	27	91.932	-	-	2.30e-21	3.17e-21
W^{69+}	18.9	39	47	45.567	-	4.18e-19	6.01e-18	1.25e-17
W^{70+}	19.4	50		60.127	-	8.67e-20	2.79e-18	9.37e-18
W^{71+}				63.981	-	4.54e-21	3.37e-19	1.87e-18

EUV W LINES AND DIAGNOSTIC APPLICATIONS

In order to search for highly ionized EUV tungsten lines in ITER-type plasmas we calculated the emissivity of tungsten ions as a function of temperature from a plasma with electron density of 10^{20} m^{-3} and abundance of 1×10^{-4} relative to deuterium + tritium. To do so we calculated the expected line intensities for W ions formed in the 7-40 keV temperature range, and checked their diagnostic applications.

Using the FLYCHK ([1]) code we calculated the tungsten ion abundances in the 7 to 40 keV range assuming an electron Maxwellian distribution and an electron density of 10^{20} m^{-3}. The most abundant ions in the 7-40 keV range are S-like (W^{58+}) to Be-like (W^{70+}) ions. In Table 1 we list the ionization potential (I.P.), the lowest temperature at which the ion abundance is 20%, and the temperature T_{max} of maximum abundance, of each ion. We also calculated the ion fraction abundances assuming that among the electron population with Maxwellian distribution, an additional 10% of non-thermal electrons with energy of 100 keV are also present, finding that the effect of the non-thermal electrons is small at $T = T_{max}$.

Using the FAC ([2]), the energy levels, radiative and collision rates for W^{58+} to W^{70+} were calculated in the 7-40 keV T range, in order to predict the emission that we expect to observe from ITER plasmas at $\lambda > 45$ Å range. Because of the limitations imposed by present day coating technology, the minimum wavelength needs to be 45 Å.

Pairs of lines from two different stages of ionization of W can be successfully used to measure the plasma temperature under the assumption of isothermal plasma. Therefore, line intensity ratios from adjacent stages of ionization can provide excellent temperature measurements. ITER plasmas also emit strong Bremsstrahlung continuum radiation, largely due to interactions between free electrons and deuterium and tritium. Once the

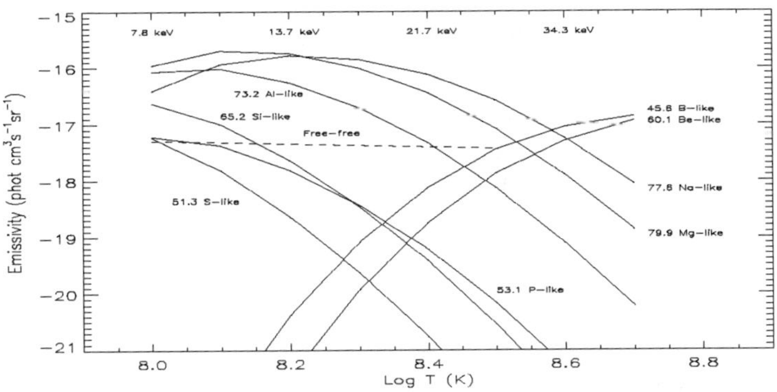

FIGURE 1. Emissivities of bright W lines and free-free radiation (dashed line) as a function of temperature.

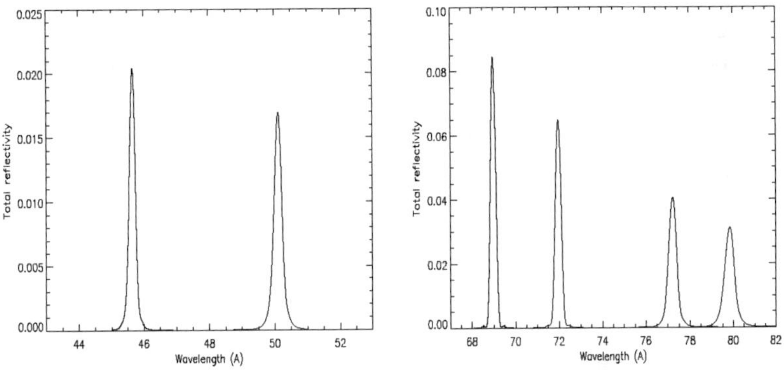

FIGURE 2. Reflectivity of an optical system consisting of two mirrors coated with $C-Co$ (left) and B_4C-La (right) multilayers. The leftmost curve in each panel is referred to as "1" in Table 2.

electron temperature is known, the comparison between line and free-free continuum intensities in the ITER plasma can provide a measurement of the W abundance as a function of spatial location and time in the ITER chamber. Examples of line emissivity for strong W lines and bremsstrahlung continuum are given in Figure 1.

RECORDING OPTICS

We envision a system where the radiation through a small opening in the ITER chamber is deflected by a gimbel mounted flat grazing incidence W mirror into a Cassegrain like telescope. The telescope in turn focuses the incident light on the detector. The telescope front and secondary mirrors are divided into n segments, each equipped with a multilayer coating specially designed to reflect a specific wavelength. A rotating mask

TABLE 2. Multilayer coating parameters.

Multilayer	Layers	Thickness (Å)	Interface (Å)
(a) curve 1	C	11.5	3.9
	Co	11.5	3.9
(b) curve 1	B4C	20.9	6.5
	La	13.9	6.5

with an opening identical in shape to that of a front mirror segment, designed to select the reflective coating associated with a desired wavelength, is placed in front of the telescope. A filter assembly is placed between the rotating mask and telescope to protect the system from undesired light at other wavelengths.

For wavelengths in the $45 < \lambda < 100$ Å, multilayer coatings can be designed to have high normal-incidence reflectance over a narrow region centered on a selected wavelength. We calculated normal-incidence reflectances in two wavelength regions near 45 Å and 75 Å using the parameters listed in Table 2. Examples of the results are shown in Fig. 2.

We assume that radiation will be reflected from the ITER chamber to the telescope by a grazing incidence mirror. Several segmented filters made of C, Ti and Zr could be placed in front of the first telescope mirror in order to prevent long-wavelength emission from reaching the detector. It is expected that the transmission of these filters will be in the 30% range.

SUMMARY

In this work we have studied how trace amounts (10^{-4}) of W atoms drifted from divertors into the ITER chambers can be used to measure the temperature and W abundance in ITER plasmas. We calculated the expected emission of all lines from W ions present in ITER plasmas with wavelength larger than 45 Å. We demonstrated how these lines can be used for diagnostic purposes by using an instrument specifically designed to observe W lines at wavelengths longer than 45 Å. We describe such an optical system and discuss the multilayer coatings necessary to enhance its sensitivity in small spectral ranges around selected W lines.

ACKNOWLEDGMENTS

The work of E. Landi is supported by the NNG06EA14I, NNH06CD24C and other NASA grants.

REFERENCES

1. H.-K. Chung and M.H. Chen and W.L. Morgan and Y. Ralchenko and R.W. Lee, *High Energy Density Physics*, **1**, 3 (2005)
2. Gu. M.F. 2003, *The Astrophysical Journal*, **582**, 1241

The CXRS diagnostic for ITER and the CXRS-Pilot Experiment on TEXTOR

R.J.E. Jaspers[1], M.G. von Hellermann[1], E. Delabie[1], J.E. Boom[1], A.J.H. Donné[1], W. Biel[2], N.C. Hawkes[3], O. Neubauer[2], M. Di Maio[2], S. Sadakov[2], F. Klinkhamer[4], B. Snijders[4], A. Hogenbirk[5]

[1]*FOM-Institute for Plasma Physics Rijnhuizen, Association EURATOM-FOM, member of Trilateral Euregio Cluster, PO Box 1207, 3430 BE Nieuwegein, The Netherlands*
[2]*Institut für Energieforschung - Plasmaphysik, Forschungszentrum Jülich Gmbh, Association EURATOM-FZJ, member of Trilateral Euregio Cluster, 52425 Jülich, Germany*
[3] *Euratom/UKAEA Fusion Association, Culham Science Centre, Abingdon OX14 3DB, UK*
[4]*TNO Science & Industry, P.O. Box 155, 2600 AD Delft, The Netherlands*
[5]*Nuclear Research and Consultancy Group, P.O. Box 25,1755 ZG Petten, The Netherlands*

Abstract. Charge eXchange Recombination Spectroscopy (CXRS) will play a crucial role in the diagnosing of burning plasmas: items like helium ash, transport barriers, impurity content or fuel ratio can all be assessed with CXRS. In fact this is the only direct method to obtain information about the light impurity ions, such as temperature, concentrations and rotation, in the core of the plasma. Two of such systems are foreseen at ITER, one in the upper port plug and one on the equatorial level, both viewing the 100 keV diagnostic H-beam.

Challenges for this diagnostic development project are: constructing a mirror close to the plasma able to keep its reflectivity under the extreme bombardment by neutrons, erosion by fast neutrals and deposition of sputtered material, design of an optical labyrinth with well enough neutron shielding, developing spectrometers with a throughput about a factor 100 larger than on present day devices, with a similar or better dispersion, calibration and alignment techniques and last but not least, finding ways to obtain the physics relevant information from the measured complicated spectra in real time.

With all these issues in mind, an ITER-CXRS pilot experiment has been developed and became operational on the TEXTOR tokamak, to address some of the above aspects. The results of this show the feasibility and the sensitivity of the ITER system: the simulation code for ITER signal levels could be verified, the sensitivity to the current profile has been tested, the beam emission has been compared to beam attenuation calculations and used to obtain absolute impurity concentrations and a high throughput spectrometer became operational.

Keywords: Charge Exchange Recombination Spectroscopy, Beam Emission, ITER, TEXTOR
PACS: 52.70.Kz, 52.55.Fa, 52.20.Hv, 32.60.+i, 34.70.+e

THE ITER CXRS SYSTEM

The motivation for the CXRS diagnostic on ITER is the unique measurement of the core helium ash concentration, as well as the ability to provide profiles of the ion temperature, plasma rotation and low-Z impurity content. The measurement requirements of this system are summarized in table I. To achieve these, a 3.6 MW, 100 keV hydrogen diagnostic neutral beam (DNB) is foreseen. Two optical systems are planned: one periscope in the upper port 3 for the core observation (from ρ=0 to ρ=0.8) and one in the equatorial port for the edge (from ρ= 0.5 to ρ=1.0).

CP988, *Burning Plasma Diagnostics*, edited by F. P. Orsitto, G. Gorini, E. Sindoni, and M. Tardocchi

TABLE I: Measurement specification of the CXRS system for ITER, according to [1].

Parameter	Range	Time resolution	Space resolution	Accuracy
Helium density	1-20 %	100 ms	a/10	10%
Ion temperature	0.5-40 keV	100 ms	a/10	10%
Poloidal plasma rotation	1-50 km/s	10 ms	a/30	5 km/s – 30 %
Toroidal plasma rotation	1-200 km/s	10 ms	a/30	5 km/s – 30 %
Impurity conc.[1] : $Z \leq 10$	0.5-20 %	100 ms	a/10	20 %
$Z > 10$	0.01-0.3 %	100 ms	a/10	20 %

[1] presently foreseen is the diagnosis of the elements: H,D,He,Be,C, N, Ne, Ar

The analysis of the core (upper port) CXRS system starts with a simulation of the CXRS measurement process, aiming to determine the minimum sensitivity S required to achieve at the statistical error for the parameters as listed in table I:

$$S = E \times T \times Q.E. \times t_{\text{int}} \tag{1}$$

where E is the etendue E, T denotes the total optical transmission (mirror labyrinth, fibres and spectrometer), $Q.E.$ the quantum efficiency of the detector and t_{int} the required integration time. The most critical parameter is the core helium density. It is found that $E=1\ mm^2sr$ is necessary to arrive at the requirements (for T=0.05 and $Q.E.$=80 %). This provides the basic number for the spectrometer development [2].

The main components of the conceptual design of the core system are: (see Fig. 1):

- The first mirror; this 100 mm diameter mirror will be subject to high radiation levels, erosion by fast particles and deposition of sputtered material. Mirrors of mono crystalline Mo or with a Rh coating are presently under discussion.
- The retractable tube for 'easy' maintenance and replacement of the first mirror, mounted on the tube. The tube will also hold the shutter.
- The optical labyrinth to transport the light of the CXRS process with the required etendue and spatial resolution to the fibres at the end of the port plug.

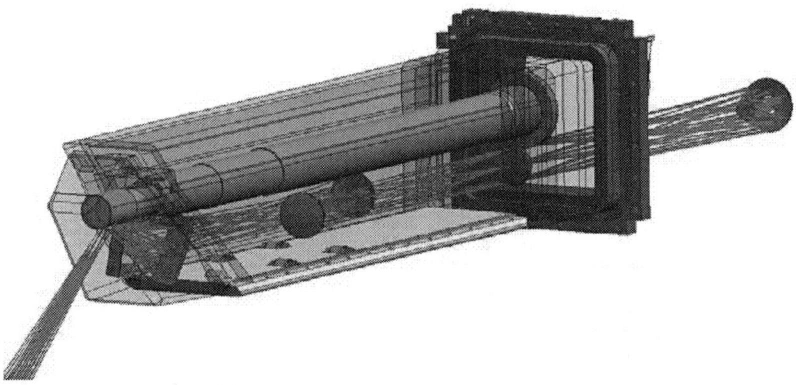

FIGURE 1. Overview of the conceptual design of the core CXRS system on ITER, located in the upper port plug number 3. The central tube and the optical labyrinth are shown.

- The fibre bundle to transport the light over tens of meters to the spectrometer room. Goal is to reduce the radiation level inside the port plug to such values that there is no need to use radiation-hard fibers
- The spectrometer and detection system; Several options are yet under study to arrive at a system measuring simultaneously four wavelength bands (1: around 465 nm for He/Be; 2: 525 nm for C/Ar/Ne; 3 656nm for H/D/T; 4: 567 nm for N) with E=1 mm^2sr, spectral resolution < 0.2 nm, time resolution = 10 ms, Q.E. > 80 %, spectrometer efficiency > 60 %.

THE TEXTOR PILOT CXRS SYSTEM

Several aspects in the design of the ITER CXRS system lack a firm scientific basis and additional experimental validation is desirable. For this specific purpose a CXRS system has been implemented on TEXTOR. This system has approximately the same geometry as on ITER, comprises of a spectrometer with characteristics close to the ITER requirements and ample experimental time on TEXTOR is available for specific diagnostics tests.

Potentially, the emission of the excited neutral beam particles can be used as a cross-calibration tool for absolute local impurity densities. Combining the measurement with the active CXRS signal, allows deducing the impurity concentration from the ratio of both intensities, without the need for a cumbersome absolute calibration [3,4]. This method will also avoid the accumulation of errors in the beam attenuation calculation. A test has been made on TEXTOR to compare three ways of deducing the helium density: from the electron density increase after a helium puff, from the calibrated CXRS signal in combination with a beam attenuation code, or from the ratio of CXRS to beam emission light. The result is shown in Figure 2.

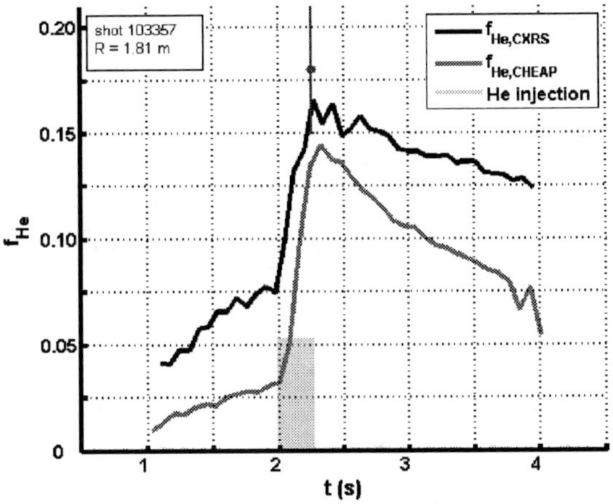

FIGURE 2. The helium concentration f_{He} determined in three ways: a) by the increase of the electron density: f_{He}=0.18±0.03 (blue data point), b) from the I(CXRS) and a beam attenuation code (CHEAP, red) and c) from the I(CXRS) and beam emission intensity (black).

The three methods give similar values, but an extensive assessment of the accuracy of the latter method has not been done yet, and some discrepancies are still observed.

The beam emission spectral feature is also used for the determination of the direction of the magnetic field. On the TEXTOR setup it is tested if this diagnostic information, usually denoted as MSE (Motional Stark Effect), can be obtained from a purely spectral measurement such as shown in the left graph of Figure 3. On the right hand side the intensity ratio derived from this spectrum is compared to the theoretical prediction. Whereas in the centre these values correspond reasonably well, more to the edge large discrepancies appear which are attributed to the polarization characteristics of the metallic mirror used in the periscope. An assessment of the value of this technique is still ongoing.

A third item to be tested on TEXTOR is the validation of the CXRS simulation code, used to predict the ITER performance [4]. In general sufficient agreement between the signal to noise from the code prediction and the TEXTOR data is obtained, as shown in Figure 4. However, an accurate model for the passive emission in the code is still lacking, but absolutely necessary for predicting the ITER performance. An effort in this direction has now been initiated.

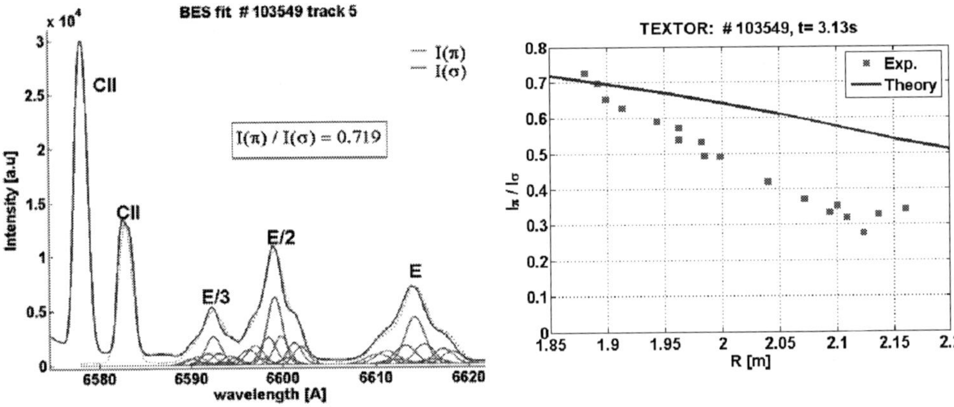

FIGURE 3. Typical beam emission of a TEXTOR discharge. The three multiplets are the D_{α}- lines of the beam emission. From this the intensity ratio is deduced. The result is shown on the right hand side: In the plasma centre (1.88 m) good agreement with the expected value is obtained. More to the edge large deviations appear, attributed to polarization characteristics of the metallic mirror in the periscope.

Finally, a spectrometer has been tested on TEXTOR, with an etendue of 0.75 mm^2sr and resolution of 0.25nm. This device [5] is used to record the spectrum of Figure 4. An assessment on the overall efficiency of the periscope optics, fiber transmission and spectrometer efficiency has been made, leading to a total transmission of 2% (with a spectrometer efficiency of 15 %). This shows that there is still room for improvement.

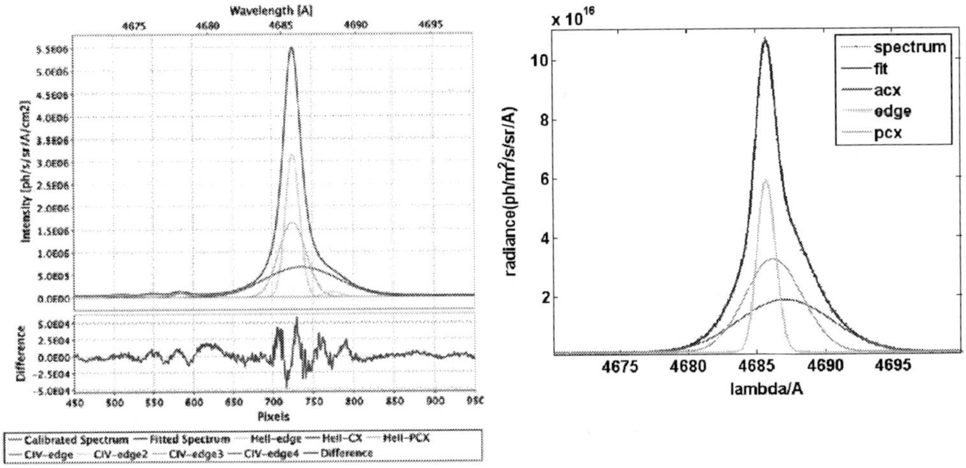

FIGURE 4. Left: measurement, Right: Simulation. Shown here is a typical example of a Helium spectrum of TEXTOR, compared with the results of the simulation code. In both cases the signal-to noise is comparable (S/N=35 at half width), but the passive emission is apparently not well modeled.

Future work on the TEXTOR CXRS system with respect to the ITER application includes: the investigation of the possibility to measure fast alpha particles with this setup, quantitative assessment of accuracies of the Beam emission and Motional Stark effect, simulation code validation and passive emission modeling and verification.

REFERENCES

1. A.J.H. Donné et al, Nucl. Fusion **47** (2007) S337
2. M.G. von Hellermann et al., Rev. Sci. Instrum. **77** (2006) 10F516
3. W. Mandl et al., Plasma Phys. Control. Fus. **35** (1993) 1373
4. M.G. von Hellermann et al., these proceedings
5. S. Tugarinov, private communcation , contact: tugar@triniti.ru

An Overview Of The Motional Stark Effect Diagnostic On DIII-D And Design Work For An ITER MSE

C.T. Holcomb. S.L. Allen, M.A. Makowski, R.J. Jayakumar, M.F. Gu, S. Lerner, K.L. Morris, J. Latkowski, J.M. Moller, W. Meyer, R. Ellis, R. Geer, D. Behne, R. Chipman*, P. Smith*, S. McClain*

Lawrence Livermore National Laboratory, 7000 East Avenue, Livermore, California, U.S.A. 94550
**College of Optical Sciences, The University of Arizona, 1630 East University Boulevard, Tucson, Arizona, U.S.A. 85721,*

Abstract. The advanced tokamak research program at DIII-D relies critically on the measurement of the current density profile. This was made possible by the development of a Motional Stark Effect (MSE) polarimeter that was first installed in 1992. Three major upgrades have since occurred, and improvements in our understanding of critical performance issues and calibration techniques are ongoing. In parallel with these improvements, we have drawn on our DIII-D experience to begin studies and design work for MSE on burning plasmas and ITER. This paper first reviews how Motional Stark Effect polarimetry (MSE) is used to determine the tokamak current profile. It uses the DIII-D MSE system as an example, and shows results from the latest upgrade that incorporates an array of channels from a new counter-Ip injected neutral beam. The various calibration techniques presently used are reviewed. High-leverage or unresolved issues affecting MSE performance and reliability in ITER are discussed. Next, we show a four-mirror collection optics design for the two ITER MSE views. Finally, we discuss measurements of the polarization properties of a few candidate mirrors for the ITER MSE. Work supported by the U.S. Department of Energy under W-7405-ENG-48.

Keywords: Motional Stark Effect, optical diagnostic, current profile, calibration, mirrors, ITER.
PACS: 52.55.Fa, 52.70.Ds,52.70.Kz,

INTRODUCTION

Motional Stark Effect polarimetry [1,2] is a standard measurement on the DIII-D tokamak that is used to constrain equilibrium reconstruction of the current profile. The use of MSE is anticipated on ITER as well. We are applying the experience gained making MSE successful on DIII-D to the new design challenges posed by an ITER-MSE. This paper is meant to provide a snapshot overview of the present state of our work on MSE for both experiments.

Briefly, MSE works by measuring the linear polarization state of a Stark component of Dα emission from a neutral beam. The σ components are polarized perpendicular to the Stark electric field (**v**x**B**) and the π components are parallel. A background plasma radial electric field may also contribute to the polarization state, and this must be accounted for. The DIII-D MSE has 65 channels viewing 2 neutral

CP988, *Burning Plasma Diagnostics*, edited by F. P. Orsitto, G. Gorini, E. Sindoni, and M. Tardocchi

beams from 5 arrays. Each view has either zero or one fold mirror in vacuum, followed by a vacuum window, refractive collection optics, and then the polarimeter. This consists of a dual photoelastic modulator made by Hinds Instruments and a fixed linear polarizer. This arrangement transfers the input polarization information by amplitude modulation in fiber optics to a remote lab where the signals are recovered using lock-in amplifiers or Fourier analysis. On DIII-D, the σ component is selected using narrowband filters. The five independent views allow evaluation of the Er component of the measurement with good spatial resolution (~1/2 cm in the edge, ~ 7 cm in the core).

MSE measurements are combined with magnetic measurements at the wall to do realtime equilibrium fitting using the EFIT code [3] for feedback with the plasma control system [4]. The EFITs also are used for between shot analysis to guide the experiment. While kinetic data is usually incorporated into fits for further analysis, experience has shown that the MSE and magnetics only EFITs are sufficient for constraining a wide variety of current profiles. These range from simple monotonic q profiles with $q_{min} \leq 1$, to advanced tokamak discharges with $q_{min} \geq 2$ and negative central shear varying from weak to strong (i.e. $q_0 \approx q_{min}$ to $q_0 >> q_{min}$). The resulting MSE EFIT q profiles are frequently compared to independent measurements of MHD activity, and are observed to match theoretical expectations, such as, for example, tearing modes located at mode rational surfaces.

Calibration and Reliability Issues For MSE

An ideal MSE diagnostic would employee collection optics that do not alter or distort the incoming linear polarization signal. In the language of Stokes vectors and Mueller matrices that describes partially polarized light, the Mueller matrix of such a system would be identity. This is never realized in practice, because one or more fold mirrors may be required to collect light around corners and refractive elements (vacuum window, lenses, photoelastic modulator) may be susceptible to Faraday rotation in the strong magnetic field. These items will change the signal polarization and should be minimized where possible, and calibrated out where they cannot. Changes to component surface coatings due to interaction with the plasma (erosion, deposition, thermal fatigue and cracking, neutron damage and darkening, etc.) require periodic recalibration.

On the DIII-D MSE, we have made design choices that attempt to minimize the starting polarization distortion. These include using at most one steering mirror with a custom dielectric coating that maintains the incident polarization to within a few percent, and using ~zero Verdet constant glass for lenses to minimize Faraday rotation.

We calibrate in situ a fitting function based on Mueller matrix modeling of the collection optics [5]. The constants in the function are determined by scanning a linear polarizer placed in the vessel at the neutral beam location during a vent. Another constant that accounts for Faraday rotation is determined using fixed, thin polarizers mounted in-vessel while the vacuum magnetic field is pulsed. Finally, we are investigating a technique that uses a fixed linear polarizer followed by a rotating quarter wave plate to inject time dependent elliptical polarization into the optical train.

Fourier analysis of the resulting signals is used to determine the system response to elliptical polarization.

In ITER, the neutron shielding requirement mandates more than one mirror in each system, and the first one or two mirrors may have to be metal rather than dielectric, which will have less ideal polarization preserving properties. Furthermore, the evolution of the front mirror calibration due to plasma interaction is expected to be far more rapid than in present experiments [6]. These create a need for a dependable calibration method that may be performed daily, or between shots. Worker vessel entry will be prohibited or extremely limited, so the calibration techniques used on DIII-D that require placing polarization references inside the tokamak will have to be done by remote insertion or by building these into the port plug. A better option that may allow frequent calibration is to fill the vessel with gas and fire the beam with known vacuum fields. The "beam-into-gas" calibration technique is presently limited to densities too low to produce statistically populated sublevels, so the resulting Stark spectrum is known to have different intensity and polarization properties than in the plasma case [7]. A full understanding of this spectrum is required before beam-into-gas calibration will be reliable.

ITER MSE Design Work and Mirror Studies

We have designed a collection optics train for the edge and core ITER MSE views. Each system uses four mirrors and a single low Verdet (SFL6) lens to bring light radially out of the port plug. Figure 1a shows the edge design in a Pro |E CAD model. The core design (not shown) uses the same design form, but the first two mirrors are set farther back in the port plug, whereas the edge design has these inside the blanket shield module. The design form is the result of trade offs between desired image quality, low polarization aberration, neutron shielding, and cost. We anticipate that the first two mirrors in the edge view will have to be metal because of their proximity to the plasma. For this reason the angle of incidence on these mirrors is kept high. The other mirrors may use dielectric coatings, and their design was chosen to satisfy neutron shielding. Compared to a preliminary MSE optics design [8], the present design has: two less mirrors, about a factor of 3 less polarization aberration, about a factor of three less projected cost due to smaller number and size of mirrors, a factor of ten improvement in the image resolution, and roughly equivalent light throughput and neutron shielding. The mirror and lens diameters (mm)/shapes for the edge design are as follows (in order from M1 to L1): 205/spherical, 230/cylindrical, 420/spherical, 250/anamorphic, 200/spherical. For the core design, they are: 270/anamorphic, 200/spherical, 250/spherical, 250/anamorphic, 200/conic.

Polarization ray tracing was part of the design optimization, and was carried out using Jones vectors with the Zemax code. For this analysis, we assume the first two mirrors have a uniform aluminum coating, and only these have any polarization aberration. In this case, a uniform input 45° linear polarization exits the system with some nonuniformity and finite ellipticity caused by a relative phase shift between S and P polarization of 2.5 to 3.2 degrees, but the pupil-averaged output angle is only rotated by 0.1° to 0.2°.

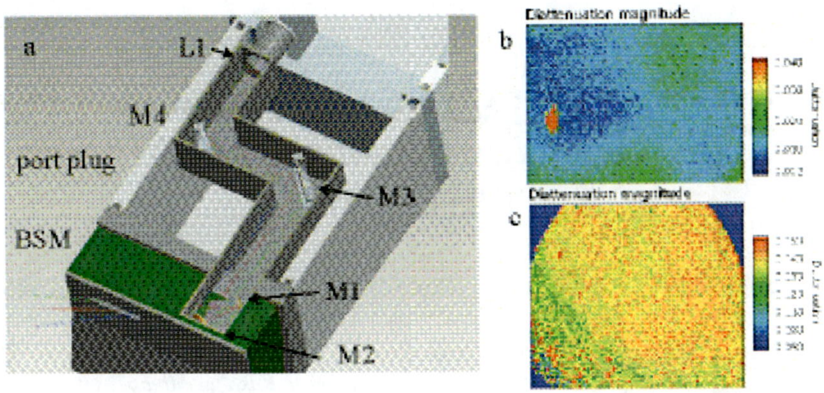

FIGURE 1. a) The LLNL 4b design, b) and c) Diattenuation images for dielectric and Rh mirrors

The next step in the design validation is to measure Mueller matrices of various first mirror surface materials as a function of wavelength and angle of incidence, and use these in a more complete polarization ray trace in Code V. These measurements will be made by a Mueller matrix angle of incidence imaging polarimeter [9]. Two example mirror candidates have already been checked using this technique for a limited number of angles. Figure 1b and c compares the diattenuation derived from Mueller images of a custom dielectric and a Rhodium mirror. These were both performed at the same wavelength (660 nm) and angle of incidence (60°), and it is clear that for this case Rhodium has significantly more polarization aberration than a dielectric mirror.

ACKNOWLEDGMENTS

This work was performed under the auspices of the U.S. Department of Energy under Contract W-7405-Eng-48.

REFERENCES

1. F.M. Levinton, et al. *Phys. Rev. Lett.* **63**, 2060 (1989).
2. D. Wroblewski et al., *Rev. Sci. Instrum.* **61**, 3552, (1990).
3 L. L. Lao et al., *Nucl. Fusion* **30**, 1035 (1990).
4. J.R. Ferron et al., *Nucl. Fusion* **38**, 1055 (1998).
5. M.A. Makowsi et al., *Rev. Sci. Instrum.* **76**, 023706 (2005).
6. "ITER MSE Diagnostic First Mirror Analysis", J.P. Allain, J.N. Brooks, A. Hassanein, Argonne National Laboratory, presented by Dave Johnson, USBPO Diagnostics Task Group Meeting, General Atomics, San Diego, CA, February 2007.
7. M.F. Gu et al., "Atomic Models for Motional Stark Effect Diagnostics", submitted *to J. Phys. B: At. Mol. Opt. Phys.*
8. A. Malaquias et al., *Rev. Sci. Instrum.* **75**, 3393 (2004).
9. J. Pezzaniti and R.A. Chipman, *Optical Engineering* **34**, 1558, (1995).

Spectroscopic Measurement System for ITER Divertor Plasma: Impurity Influx Monitor (divertor)

Tatsuo Sugie*, Hiroaki Ogawa*, Satoshi Kasai† and Yoshinori Kusama*

*Japan Atomic Energy Agency, 801-1 Mukoyama, Naka-shi, Ibaraki, 311-0193, Japan
†Nippon Advanced Technology, 3129-45 Muramatsu, Tokai-mura, Naka-gun, Ibaraki 319-1112, Japan

Abstract. The detailed design of the Impurity Influx Monitor (divertor) has been carried out to provide the measurement capability in the harsh environment such as higher irradiation levels of neutron, gamma-ray and particles than in present devices. The in-situ calibration system using a micro retro-reflector array has been developed to monitor the sensitivity change of the optical system due to the environmental effects. The optical alignment system for the Monitor has been developed by using a dedicated optics for alignment in the collection optics for measurement.

Keywords: Spectroscopy, Divertor, Impurity, Calibration, Alignment, ITER.
PACS: 52.70.-m, 28.52.-s, 52.55.Fa

INTRODUCTION

The main functions of the Impurity Influx Monitor (divertor) are to identify impurity species and to measure the two-dimensional distributions of the particle influx densities in the divertor plasmas by using spectroscopic techniques in the wavelength range of 200 - 1000 nm [1].

The optical components mounted close to the plasma will experience higher levels of radiation due to neutron, gamma ray and/or particles than in present devices. An additional potential damaging effect for the first mirrors is the formation of coatings of impurities due to deposition of material eroded from the first wall and/or divertor. Protective and mitigating measures are adopted in the design: for example, where possible the plasma is viewed through a small aperture, and baffle plates are inserted in front of the first mirror. In order to make the required measurements [2] it is important to know the sensitivity change of the optical systems due to the environmental effects. An in-situ sensitivity calibration system withstanding in the strong radiation field has been developed. In addition, the optical alignment system and the related developments of optical components are described in this paper.

OVERVIEW OF IMPURITY INFLUX MONITOR (DIVERTOR)

The divertor region will be observed from the divertor port, the equatorial port and the upper port as shown in Fig. 1. The light from the divertor region passes through a front end optics located near the plasma and goes through quartz windows and mirrors

CP988, *Burning Plasma Diagnostics*, edited by F. P. Orsitto, G. Gorini, E. Sindoni, and M. Tardocchi

on the port plug. It is then focused on a micro-lens array attached on a fiber array by a collecting optics. The fiber array guides the light to spectrometers. The spectrometers covering the wavelength range ($\lambda < 450$ nm) are installed just behind the bio-shield to minimize the transmission loss in fiber. The light with $\lambda \geq 450$ nm is guided by long optical fibers to the spectrometers located remotely in the diagnostic hall. The outer and inner divertor regions will be observed by a front end optics located on the bottom of the divertor cassette. In addition, lines of sight through the gap of 10 mm between the divertor cassettes will be provided. The detailed design has been carried out to provide the measurement capability in the harsh environment [3, 4, 5].

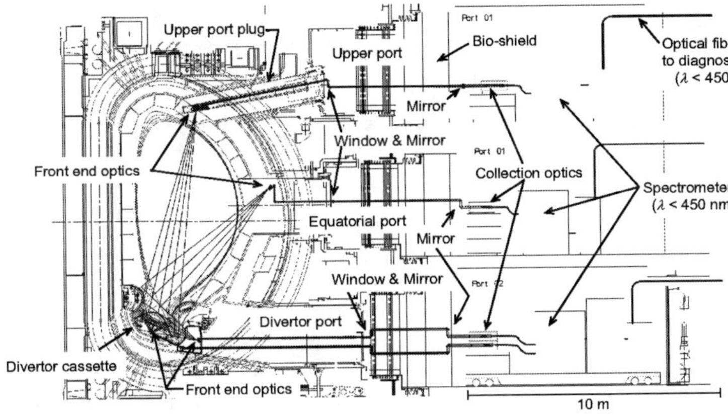

FIGURE 1. Arrangement of Impurity Influx Monitor (divertor) and typical lines of sight.

IN-SITU SENSITIVITY CALIBRATION

It is not feasible to install a light source in the vacuum chamber of ITER during operations. Therefore, an alternative method of calibration is needed. Here, an in-situ calibration system using a micro retro-reflector array has been developed as a possible method.

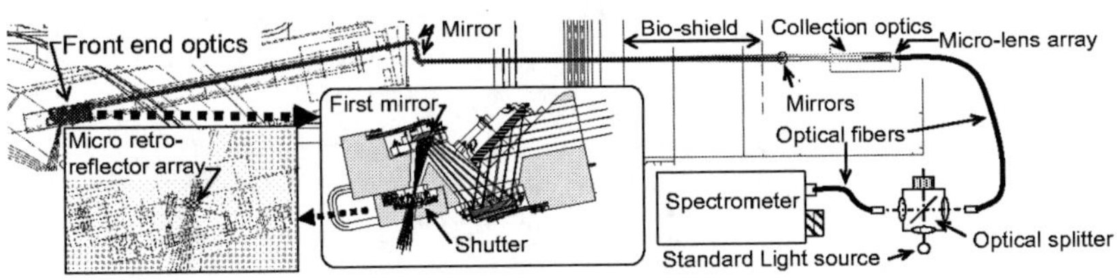

FIGURE 2. In-situ sensitivity calibration system for the upper port optics.

A standard light is set behind the bio-shield or in the diagnostic hall and the light is applied to the micro retro-reflector array mounted on a shutter through the same optics for plasma measurement with the collection optics as shown in Fig. 2. The reflected light is measured with a spectrometer to evaluate the sensitivity change of the optics.

During the plasma measurement, the micro retro-reflector array moves backward into the shelter to prevent the particle bombardment and deposition.

Ray-trace analysis of the calibration system was carried out by using the optical design software ZEMAX. For irradiance analysis, rays start from a micro lens (2.0 mm × 0.25 mm) of the micro-lens array and go through the collection optics. The rays are reflected by the micro retro-reflector array and come back to the micro-lens through the same optics. Figure 3(a) and 3(b) show the irradiance distribution on the aperture plane of the collection optics for the incident and returned light, respectively. Almost reflected light comes back into the collection lens and then focuses on the same area of the micro lens, as shown in Fig. 3(c).

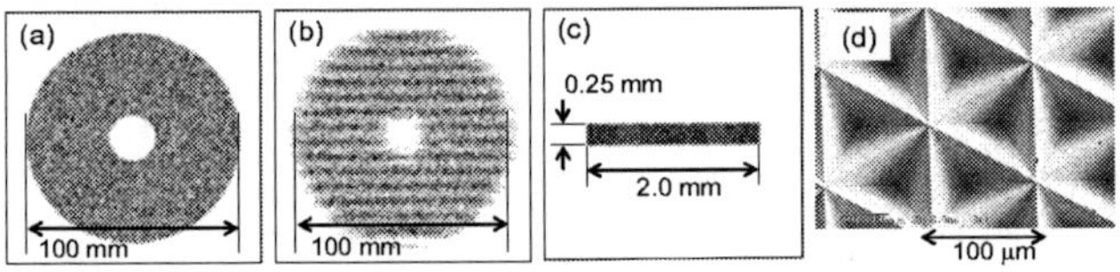

FIGURE 3. (a) Irradiance distribution on the aperture plane of the collection optics for the incident light and (b) returned light. (c) Irradiance distribution on the micro-lens for returned light. (d) Surface of the micro retro-reflector array observed by a scanning electron microscope.

Two types of micro retro-reflector array made of nickel were manufactured by electro-forming technique. One is coated with aluminum by vacuum evaporation and another is not coated. The picture of the surface observed by a scanning electron microscope is shown in Fig. 3(d). Figure 4(a) shows the measured spectral reflectivities and the reflectivities expected from ideal values of the materials.

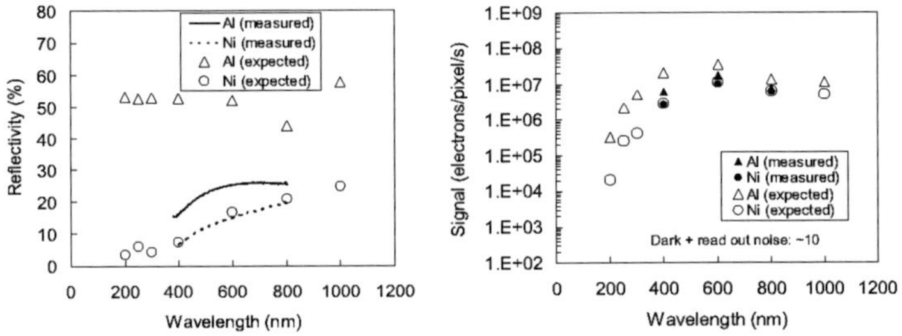

FIGURE 4. (a) Measured spectral reflectivities of micro retro-reflector arrays and the reflectivities expected from ideal values of the materials. (b) Estimated signals of the detector with measured reflectivities and expected values of micro retro-reflector arrays .

Detected signals were estimated for the calibration system shown in Fig. 2. Here, we assume that the first mirror in the front end optics is made of molybdenum and other mirrors are made of aluminum and lenses and windows are made of fused silica. A xenon lamp (HAMAMATSU: L7810) and a back-thinned type CCD (ANDOR DU741N) are assumed. The calculated results are shown in Fig. 4(b). From the figure, sufficient signals will be detected compared with the detector noise by this calibration system even if the micro retro-reflector array without aluminum coating is used.

OPTICAL ALIGNMENT

The Impurity Influx Monitor (divertor) has a long optical path, and more than two mirrors will be mounted on the port plug as shown in Fig. 1. The port plug will be expanded and distorted by baking, nuclear heating and electromagnetic force. It should be necessary to align the optics remotely. To realize this, optical fibers are leaded to appropriate points on each mirror or mirror holder and on the front end optics from the outside of the port. The images of the end of the fibers are formed on a CCD camera by a dedicated optics for alignment, which is installed in the dead area of the Cassegrain type collection optics for plasma measurement. The alignment is carried out by adjusting the angle of each mirror remotely so that the images on the CCD camera come on the decided positions. The light with the wavelength $\lambda > 800$ nm is used to prevent the large decrease of the transmissivity of the optical fiber due to neutron and gamma-ray irradiation. The induced losses of some fibers at 800 nm are 1.5 dB/m and 5 dB/m for the neutron fluence of 1×10^{22} n/m^2 and 5×10^{23} n/m^2, respectively [6]. The expected neutron fluences at the behind of the front end optics and at the just outside the port are less than about 1×10^{23} n/m^2 and 2×10^{20} n/m^2 for the ITER life, respectively [7]. It is expected that this system can be used in the radiation environment of ITER. From the ray-trace analysis, it has been confirmed that it will be possible to align each mirror and front end optics within about 2 mm.

CONCLUDING REMARKS AND FURTHER WORK

The in-situ calibration system using a micro retro-reflector array has been developed. Sufficient signal to noise ratio will be obtained in the wide wavelength range (200 nm – 1000 nm). The performance depends on the manufacturing accuracy of the micro retro-reflector array and the optics. Therefore, the test will be carried out with an optics simulating the real system to confirm the practical performance. In addition, neutron irradiation test should be required for the micro retro-reflector array.

ACKNOWLEDGMENTS

We would like to express our gratitude to Drs. A. Costley, C. Walker of the ITER International Organization and their colleagues for their kind suggestions.

REFERENCES

1. T. Sugie, H. Ogawa, T. Nishitani, et al., *Rev. Sci. Instrum.* **70,** 351-354 (1999).
2. A.J.H. Donne, A.E. Costley, R. Barnsley, et al., *Nucl. Fusion* **47,** S337–S384 (2007).
3. T. Sugie, H. Ogawa, A. Katsunuma, et al., "Design of divertor impurity monitoring system for ITER (II)", *JAERI-Tech* 98-047 (1998).
4. H. Ogawa, T. Sugie, A. Katsunuma and S. Kasai, "Design of Impurity Influx Monitor (Divertor) for ITER", *JAEA-Technology* 2006-015 (2006).
5. H. Ogawa, T. Sugie, S. Kasai, et al., "Development of Impurity Influx Monitor (Divertor) for ITER" to be published in *Plasma and Fusion Research.*
7. T. Kakuta, T. Shikama, T. Nishitani, et al., *J. Nucl. Mater.* **307-311,** 1277-1281 (2002)..
8. H. Iida, V. Khripunov, L. Petrizzi, et al., "Nuclear Analysis Report", *ITER_D_22F2ST* v2.0 (2004).

Design of a tomographic hard X-ray spectrometer for suprathermal electron studies with ECRH

S. Gnesin and S. Coda

Ecole Polytechnique Fédérale de Lausanne (EPFL), Centre de Recherches en Physique des Plasmas, Association Euratom-Confédération Suisse, CH-1015 Lausanne, Switzerland.

Abstract. Electron cyclotron resonance heating (ECRH) and current drive (ECCD), disruptive instability events and sawtooth activity have been demonstrated to produce suprathermal electrons in fusion devices. The importance of these phenomena for fusion reactors renders suprathermal electron generation and dynamics a key topic in the physics of burning plasmas. Here some significant results from the TCV tokamak are briefly reviewed and a preliminary design of a novel tomographic hard X-ray spectrometer proposed for TCV is discussed. The design is aided by simulations of tomographic reconstruction.

Keywords: Hard –X-rays, suprathermal electrons, tomography, spectrometer, ECRH, ECCD.
PACS: 52.70.La, 52.50.Sw, 52.55.Wq

Introduction

The physics of electron cyclotron current drive (ECCD)[1] is intimately tied with the dynamics of suprathermal electrons, which remains only partly understood in present-day magnetic-confinement devices. An increased understanding of the suprathermal electron dynamics can have direct implications for the first experimental fusion reactor currently under construction, ITER. Indeed, ECCD is planned to be used as a tool in ITER for MHD stability control[2]; additionally, sawtooth activity and disruptive instabilities, which are also accompanied by substantial suprathermal electron generation, are expected to occur in the ITER baseline regime. The development of increasingly advanced fast-electron diagnostics is therefore important in view of the next generation of burning-plasma devices.

Brief Review of the TCV Results

The TCV (Tokamak à Configuration Variable) plasma is routinely heated by means of ECRH; moreover ECCD is extensively used[3] to tailor the current profile for applications ranging from fully non-inductive operation to high-performance scenarios with electron internal transport barriers. With ECCD a suprathermal electron population is generated in an energy range from a few tens to a few hundred keV, resulting in hard X-ray (HXR) bremsstrahlung emission. Studies have been conducted on TCV with a single, multichordal spectroscopic HXR camera (on loan from Tore Supra intermittently between 1999 and 2005)[4] and with a high-field-side electron cyclotron emission (ECE) radiometer[5]. These studies have led in particular to the recognition of the role of the spatial transport of suprathermal electrons, at high power densities, in regulating the magnitude and spatial distribution of the driven current[3]. The temporal dynamics of the suprathermal population has been investigated in detail with non-perturbative pulsed ECCD experiments combined with coherent averaging of the ECE signals[6]. More recent measurements of oblique ECE have provided evidence

CP988, *Burning Plasma Diagnostics*, edited by F. P. Orsitto, G. Gorini, E. Sindoni, and M. Tardocchi

of distribution function asymmetries with ECCD[7]. Additionally, the absorption of third-harmonic ECRH on TCV has been shown to be significantly enhanced by the presence of a fast–electron tail[8]. Evidence of the generation and subsequent rapid cross-field transport of suprathermal electrons during sawtooth crashes has also emerged from ECE measurements[9].

Proposed tomographic spectroscopic system for TCV

Experimental evidence as well as theoretical predictions indicate that poloidal asymmetries exist in the HXR emissivity. A natural next step is therefore the development of tomographic HXR systems to investigate the full 2D emissivity distribution in the poloidal plane. For these reasons we have carried out a preliminary design of a new tomographic set of two to nine spectroscopic poloidal HXR cameras for the TCV geometry, with broad coverage of the poloidal cross section and very high spatial and temporal resolution. The proposed system is based on a multi-camera set of fan views which have been conceived to provide high flexibility and compactness, as required by the constraints of the TCV vessel environment. We envision using CdTe semiconductor diodes, which do not require bulky cooling or magnetic-shielding equipment and retain acceptable energy resolution (typically 5-10 keV)[4,10]. The HXR radiation is channeled to the detectors by a novel tungsten collimator configuration, adapted from the Soller collimator concept[11], and compatible with an extremely compact camera design. In this configuration the collimator plates are arranged in a fan and each plate defines the aperture and provides the necessary shielding against oblique illumination (see Fig. 1).The design of the collimator aperture can be adjusted for each individual detector in order to determine the étendue and the attendant chord spread in the plasma region. The spatial sampling properties of the system are determined by the typical chord separation in the plasma region, ~2cm (8% of the minor radius) in our design.

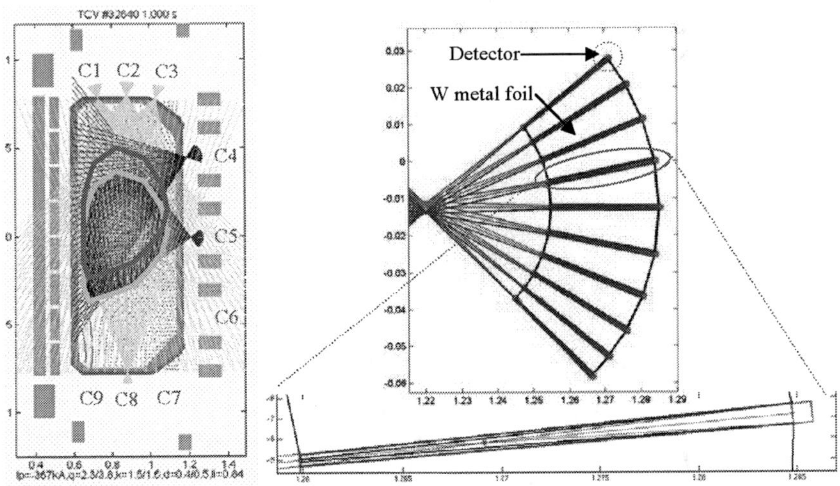

FIGURE 1. Chord fan coverage for the hard X-ray camera system in TCV (left); detail of the collimator system in camera 5 (right). Possible cameras are numbered from 1 to 9, corresponding to the nine ports seen in the figure, in clockwise order from the top left.

The detectors can be placed immediately behind the Soller plates, enabling them to be housed directly within the port opening. The extreme compactness and flexibility of this design should make it readily adaptable to other fusion devices.

Tomographic simulations and validation

The number and distribution of the cameras is determined by a compromise between the quality of tomographic reconstruction and cost, within the constraints set by the TCV port geometry. In order to base the choice on a quantitative foundation, we have performed simulations to compare imposed ("phantom") HXR emissivity patterns with the tomographic inversions resulting from the calculated detector signal distributions in different configurations. Different emission patterns have been used in the analysis: poloidally-symmetric emissivity profiles reconstructed from past, non-tomographic HXR measurements during ECRH in TCV, as well as periodic poloidal distributions with a poloidal wave number m, to simulate increasingly complex poloidal asymmetries. The tomographic reconstruction algorithm performs a line integration of the simulated photon emissivity g(R,z) along the i-detector's line of sight (L_i) in order to obtain the so called chord brightness $f_i=\int_{Li} g(R,z)dl$. This information is thus fed into the tomographic algorithm that generates the reconstructed plasma emissivity on a pixel grid using the minimum Fisher information method in order to condition the solution[12]. The simulations show the expected (see Fig. 2) increase in the quality of the tomographic reconstruction when cameras viewing the plasma from maximally separated directions are added. As a general rule, m+1 independent views are required to recover an emissivity pattern with wave number m. This is confirmed by Fig. 2 for the cases m=0, 1, and 2. The algorithm used in this study enables one to recover the main features (peak positions and relative intensity distribution) of the plasma emissivity; but the robustness in reconstructing emissive patterns with high spatial gradients must still be improved. Further work will be devoted to the upgrade and comparison of the present algorithm with alternative ones (maximum entropy and linear regularization). A final quantitative assessment of optimal camera orientation and distribution is expected to come in the near future.

Acknowledgment

This work was supported in part by the Swiss National Science Foundation.

References

1. N.J. Fisch, Rev. Mod. Phys. **59**, 175 (1987).
2. R.J. La Haye et al, Nucl. Fusion **46**, 451 (2006).
3. S. Coda et al, Nucl. Fusion. **43**, 1361 (2003).
4. Y. Peysson, S. Coda, F. Imbeaux, Nucl. Inst. Meth. A **458**, 269 (2001).
5. P. Blanchard et al, Plasma Phys. Control. Fusion **44**, 2231 (2002).
6. S. Coda et al, Plasma Phys. Control. Fusion **48**, B359 (2006).
7. T.P. Goodman et al, Proc. 34th EPS Conf. on Control. Fusion and Plasma Phys., Warsaw, Poland, 2007 (P2.147).
8. S. Alberti et al, Nucl. Fusion **42**, 42 (2002).
9. I. Klimanov et al, Plasma Phys. Control. Fusion **49**, L1 (2007).
10. T. Takahashi and S. Watanabe, IEEE Trans. Nucl. Sci. **48**, 950 (2001).
11. W. Soller, Phys. Rev. **24**, 158 (1924).

12. M. Anton et al, Plasma Phys. Control. Fusion **38,** 1849 (1996).

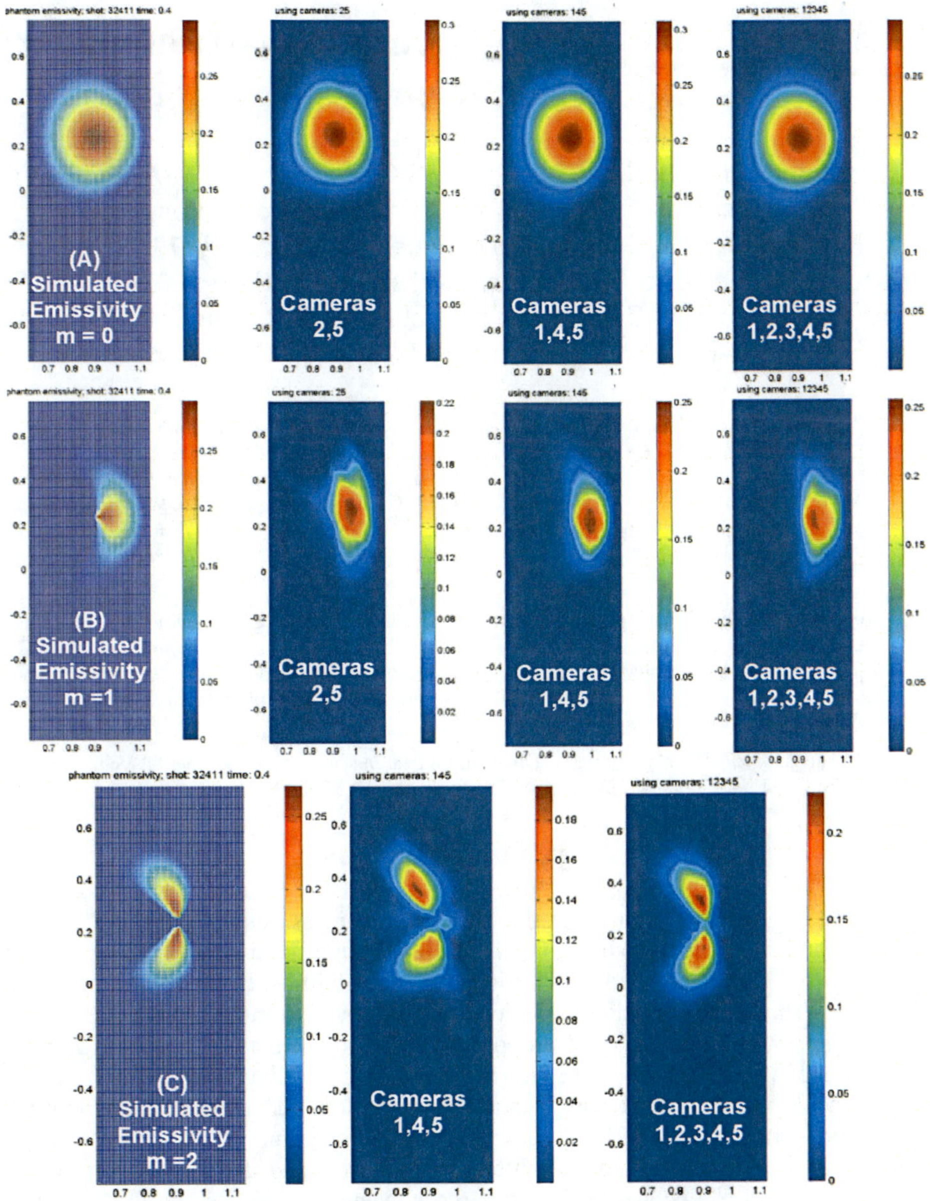

Figure 2. Tomographic reconstructions (2D contour plots in the poloidal plane) for three different postulated emissivity patterns: A) poloidally symmetric (m=0) emissivity; B) m=1 emissivity; C) m=2 emissivity. The leftmost figure in each row is the phantom, followed by reconstructions with different camera combinations. The quality of the reconstruction increases by adding cameras viewing the plasma from different directions.

Concept And Development Of Instruments For ITER Thermography

R.Reichle[1], J.M.Badie[2], C.Balorin[1], S.Carpentier[1], Y.Corre[1], M.Davi[1], E. Delchambre[1], C.Desgrange[1], F.Escourbiac[1], S.Fougerolle[1], J.L.Gardarein[1], E.Gauthier[1], S.Gicquel[3], D.Guilhem[1], D.Hernandez[2], M.Jouve[1], C.Le Niliot[4], Th.Loarer[1], A.Martin[3], S.Martinez[4], J.P. Martins[1], J.B. Migozzi[5], J.C.Patterlini[1], C.Pocheau[1], F.Rigollet[4], H.Roche[1], J.L.Sans[2], S.Salasca[1], J.M.Travere[1], C.Walker[3]

[1] *Ass. EURATOM-CEA, DSM/DRFC, CEA-Cadarache, 13108 St. Paul-lez-Durance, France*
[2] *PROMES CNRS 8521, Centre Felix Trombe, BP5, Odeillo, F-66125 Font Romeu, France*
[3] *ITER-IT, Joint Work Site, c/a CEA-Cadarache, 13108 St. Paul-lez-Durance, France, France*
[4] *Ecole Polytechnique Universitaire de Marseille, I.U.S.T.I, UMR CNRS N°6595,Technopôle de Château Gombert, 5 Rue Enrico Fermi, 13453 Marseille Cedex 13, France*
[5] *JBM Optique, 11 avenue de la division Leclerc, 92310 Sevres, France*

Abstract. We give here a short overview of the status of the development for ITER thermography as performed by the CEA-Cadarache and some of its collaboration partners. The topics that have been included in this synthesis are the status of the optical design, the role of multi-wavelength mesurements, multicolour pyroreflectometry, photothermal methods, and reflection simulations and measurements.

Keywords: ITER, IR-thermography, Divertor, Optical instruments, Experimental techniques
PACS: 07.57.Ty, 07.60.Vg, 42.30.-d, 42.79.-e, 52.70.-, 52.70.Kz

INTRODUCTION

ITER requires fast high resolution temperature measurements in the divertor and visible and thermal viewing of all plasma facing components ($800m^2$). These measurements shall be used in security loops to minimize the risk of damage to the machine. Tore Supra is using a conventional [1] and a fibre based [2] thermography system to perform this task. We propose for ITER to use these concepts together with elements of the JET wide-angle viewing thermography diagnostic [3] and add novel components to account for complications such as low and variable emissivity of the observed surfaces, reflections, inhomogeneous temperature distributions within the observation spots, strong bremsstrahlung from plasma and the degradation in time of the optical chain. First we present our most recent optical designs for divertor and midplane. Then we explain the proposed add-ons, such as multi-wavelength thermography [4,5], multicolour-pyroreflectometry [6,7], photo-thermal methods [8] and the role of reflection simulations [9] and measurements and discuss their eventual integration as measurement and/or calibration technique.

CP988, *Burning Plasma Diagnostics*, edited by F. P. Orsitto, G. Gorini, E. Sindoni, and M. Tardocchi

1. STATUS OF OPTICAL DESIGNS

1.1 Divertor thermography

A first version of the optical design for the divertor [10] has become obsolete due to a change of the divertor structure. The new design (fig. 1) has a different geometry but has maintained the design philosophy and the principle optical elements: a Cassegrain telescope close to the divertor port, a dividing mirror in the middle of the cassette and for each target a plane folding mirror and a concave first mirror. This system allows to see in one poloidal section the plasma-wetted parts of the carbon targets if suitable cut-outs of the lower tungsten armour tiles are accommodated. The optic allows a spatial resolution of 3 mm. The toroidal width of the viewing zone is 20 mm which is compatible with the free space between the individual bars of the finger like structures at the sides of the divertor dome through which the diagnostic looks onto the targets. The optical transport through the divertor port will either be by optical fibres [11] and/or a conventional optical relay connected to the Cassegrain telescope. In order to be compatible with the use of optical fibres, the numerical aperture of the system is 0.22 (F/2.2). With this high light throughput it is possible to fulfil the ITER requirements even with significant spectral resolution (30 wavelength intervals in the range of 2 - 6 μm) – on carbon targets. The present concept uses a cooled diagnostic box to hold the mirrors which is compatible with the maintenance of the cassettes in the vacuum vessel and the Hot Cells. More work is needed to assure simplicity and reliability for mounting it with the robotic tools. Shutters for calibration purposes are foreseen between the first mirror and the plasma inside the diagnostic box. Due to the impurity rich divertor environment first mirror cleaning is one of the two important maintenance issues, the other being the radiation hardness of the transparent elements [11] needed in the light transport form the divertor to the bioshield.

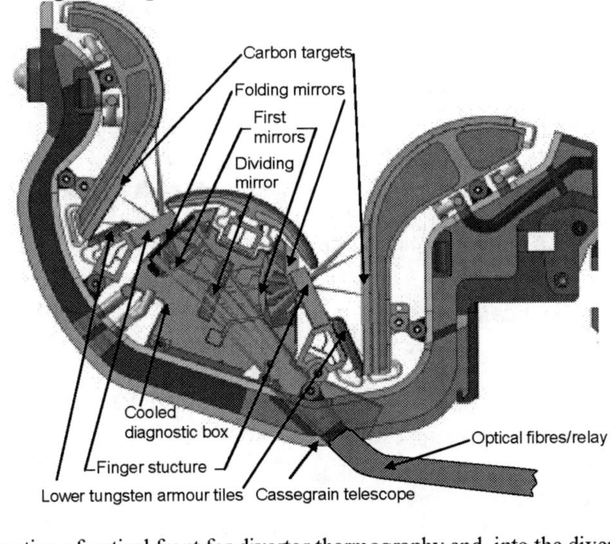

FIGURE 1. Integration of optical front for divertor thermography end into the divertor cassette.

1.2 Mid-plane Thermography

The mid-plane thermography system is integral part of the Wide Angle Viewing Visible IR diagnostic to be installed in 4 equatorial ports. Each of these ports is supposed to contain 3 individual viewing lines (fig. 2), one looking down into the divertor the other two tangentially left and right (fig. 3).

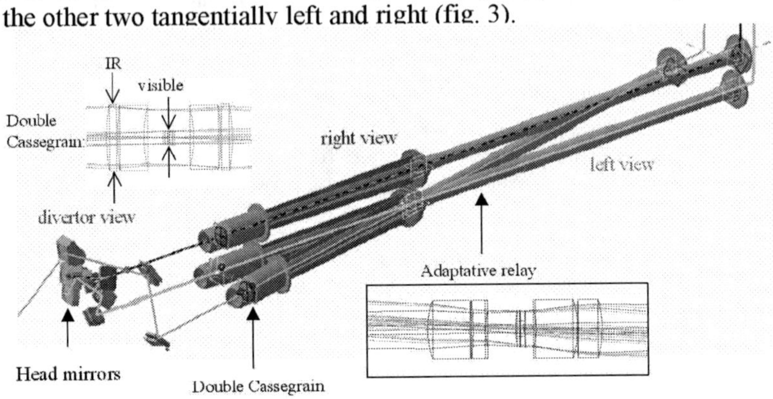

FIGURE 2: Principal of optical layout for Wide angle Viewing Visible IR diagnostic.

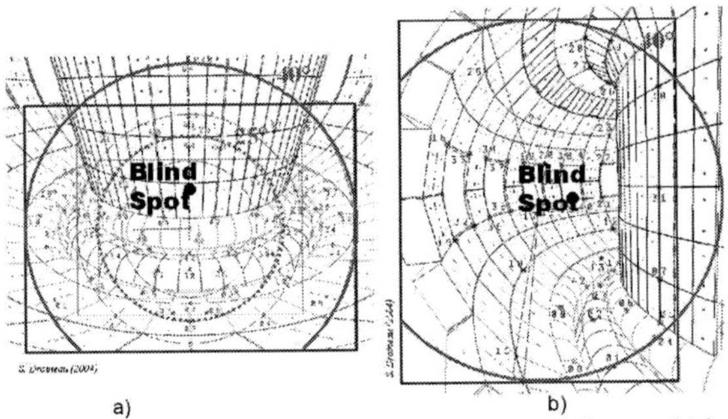

FIGURE 3: Viewing areas of mid-plane thermography system: a) divertor b) left-hand tangential view.

A generic optical design has been developed which integrates infrared and visible viewing capability in each of the 12 optical lines. Several optical solutions have been studied. The final choice is not yet done. It will be driven by 1) the available volume in the port plugs 2) the required optical performance (spatial resolution, coverage of a maximum of the total wall surface) 3) the feasibility and reliability of the solution. One of the solutions (fig. 2) could be to have a central element with two Cassegrain telescopes which are concentrically arranged within each other, the larger one (diameter 10 cm) for the IR, the smaller one (diameter 2 cm) for the visible light. Feasibility of manufacturing and optical performance which imply stringent alignment procedures still need to be validated. Simpler optical designs are currently under study. The other optical elements are head-mirrors close to the plasma, a conventional optical relay and an auto-adaptive optical relay which compensates relative movements between torus and bio-shield.

A performance analysis shows that with an optical system of an aperture of F/3.3 (numerical aperture 0.15), the required spatial resolution can be achieved. The midplane system can cover about 70% of the inner wall with a field of view per pixel between 3·3mm² and 7·7 mm². The integration of the optical line in the environment of the mid-plane port n° 3 is shown in figure 4. The main maintenance issue are the first mirrors which may be affected not only by deposition but also by erosion through charge-exchange particles.

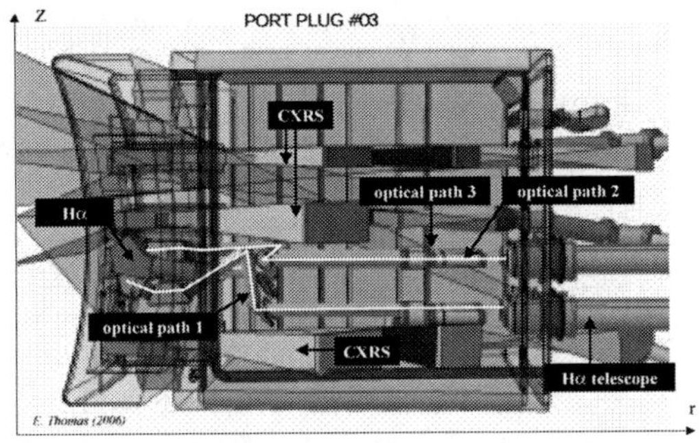

FIGURE 4: Port Plug 3 with optical paths for the three optical lines for the thermography system.

2. PASSIVE MULTI-WAVELENGTH MESUREMENTS

2.1 Two colour imaging

Both the divertor and midplane system shall be transparent over a certain spectral range in the IR. A range of 2-6 μm would be a good choice [10]. Within this range one can find cameras capable of taking simultaneously images at two wavelengths. The attraction of this approach would be the possibility to have images that would be easier to interpret than one colour images or isolated point like measurements. In the absence of important parasitic radiation (e.g. bremsstrahlung in the divertor – fig. 5) and if the emissivities (and the transmission of the optical chain) at both wavelengths are the same or known (section 3) temperatures can be deduced from the ratio of the spectral radiances at the two wavelengths.

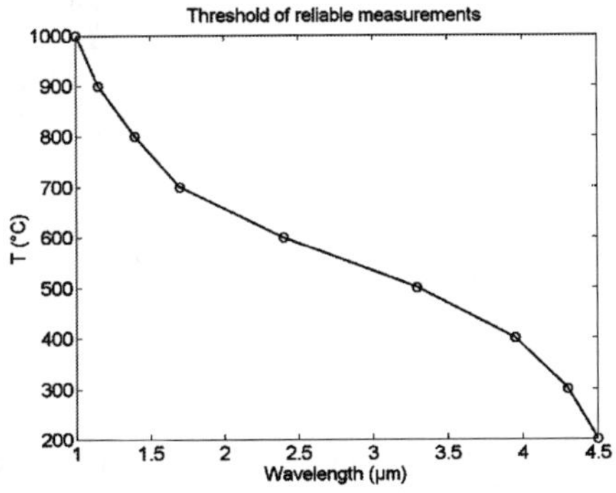

FIGURE 5: Approximate target temperature at which the bremsstrahlung variation between a 'high density' (peak density; peak temperature : $10^{22}/m^3$, 3 eV) and a 'normal density' ($2 \cdot 10^{20}/m^3$; 200eV) divertor scenario becomes insignificant for the temperature measurement.

2.2 High spectral resolution thermography

Multi wavelength thermography measurements can go further in the analysis bremsstrahlung, hot-spots and reflections than one or two wavelength measurements but imaging is more limited - e.g. by slit-type imaging with multiple filters or with reduced spatial resolution 2D imaging using optical fibre bundles and spectrally dispersing elements [4]. This method does not require changes to the optical front end - the spectral resolution is achieved further down the optical chain and can be done via beam splitter in parallel to higher quality one or two colour imaging.

Since the bremsstrahlung and the thermal radiation have different spectral radiance distributions a multi-wavelength measurement may allow to distinguish between the two contributions and thus lower significantly the threshold shown in fig. 5.

The non-uniformity of the target temperature due to micrometric hot-spots [12] is an explanation of the experimental fact that near-infra red measurements yield higher temperature values than mid-infrared measurements [13]. A recent theoretical investigation in the range 1-10 μm and laboratory tests in the range 0.9-1.6 μm [5] indicate that spectrally resolved thermography can allow to determine for a given measurement channel (e.g. the observation spot of a fibre) the upper part of a temperature distribution within its field of view. It may range from the highest temperature present to about 600°C below that maximum temperature (fig. 6). The lowest temperature of a population for which this method was tested experimentally was 400°C.

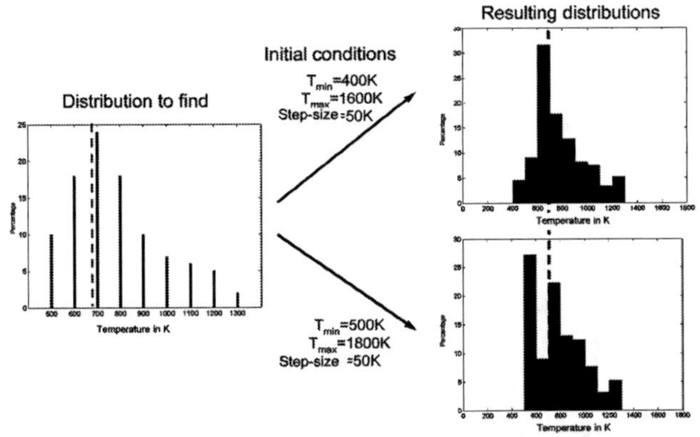

FIGURE 6: Example of partial reconstructions (only for T > broken line) of a temperature distribution function by inversion methods working on simulated spectral radiance distributions [5].

Multi-colour thermography does not distinguish reflections from emitted light, but it may indicate their presence as an additional peak in the temperature distribution.

3. MULTI-COLOUR PYROREFLECTOMETRY

Section 2.1 assumes that the emissivities at both wavelengths are identical – or after basic calibration - that their ratio is constant. Already the first assumption is not satisfied as laboratory measurements on clean tungsten samples have shown [14] even for wavelengths as close together as 1.3 μm and 1.55 μm. The presence of carbon deposits with wavelength dependent absorption [15] subjected to changes in time on metallic targets of interest will render the second assumption invalid. This situation requires normally additional methods with which one can determine the emissivity ratio of the wavelengths of interest in situ and during the experiments. These could be cavities in the target that can act as black body reference (e.g. inner wall blanket cut-outs for bolometers), or thermocouples in the targets. In critical places however, where such methods will not be applied, multi-colour pyro-reflectometry [6,7] could be envisaged to be used. It uses reflected low power laser light of several colours (at least 2) to determine the emissivity and passive thermography at the same colours for the temperature measurements (fig. 7).

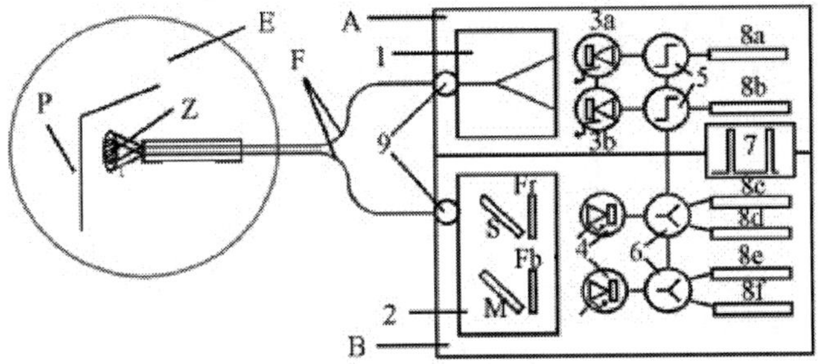

FIGURE 7: Schematic of two coulour pyroreflectometer [6]. Target (P) in vacuum vessel (E) observed and illuminated via fibres (F) with overlap area (Z) connected to measurement device (9). Pulsed (8a,8b) illumination circuit (A) with unification of output of two colour lasers (3a, 3b) through fibre junction (1). Circuit (B) with separation (2) of received light via semitransparent (S) and reflecting mirror (M) and filters (Fr,Fb) and measurement in photodetectors (4). Separation of signal periods measuring thermal light and signal periods measuring thermal light plus reflected light by gated (6) detection (8c-8d) according to pulse on/off sequence (7).

The assumption of the method is that the diffusivity factor (= ratio of bidirectional to hemispherical reflection) is constant for the wavelength of interest. This assumption is likely to be more readily satisfied than assumptions concerning the emissivity. The method worked with 10 kHz temporal resolution in the laboratory on different Tungsten samples for measurements within a cone of 30 degrees around the angle of specular reflection and for a temperature range from 700°C - 1500°C when using wavelengths of 1.3 and 1.55 µm [14]. To gain access to lower temperatures, longer working wavelengths are needed. But for a security application it could be sufficient to operate in the near infrared range, for which fibres [11] and lasers for illumination are readily available. The method has recently been successfully tested on tungsten targets over a distance of more than one meter in the electron test bed facility FE200 and verified against thermocouple measurements [16]. The measurement is sensitive to reflections and bremsstrahlung. This requires that these two main perturbations are small in relation with the temperature to measure – which may not be the case in the divertor, but rather on the antenna or zones hit by fast particles inside the main vacuum chamber. The method, when used in its three colour version [7] allows an error analysis. Due to the relatively low illumination power needed, large surface area coverage may be possible. For these reasons this method could become applicable mainly for the mid-plane. The next step is a tokamak test of the set-up in a reflective environment with parasitic light. One could envisage illuminating not only from the port used for the measurement but also from the other ports. In any case it will be necessary to assess the reflections (section 4) properly, for the measurement and for its perturbation.

4. REFLECTION SIMULATION AND MEASUREMENT

It is obvious that direct specular reflection on metal targets with a high degree of reflectivity can cause a severe overestimation of the danger to the target. Simple modelling using an integrating sphere approach [9] to estimate the effect of multiple and diffuse reflections for the metallic walls of ITER show that it may play an important role in certain conditions, in particular when the temperature of the inner (Be) wall shall be determined in the presence of strong radiation from the divertor (fig. 8). This has an impact on the usability of the pyroreflectometric measurements at low temperature as they can not distinguish between reflected and emitted light.

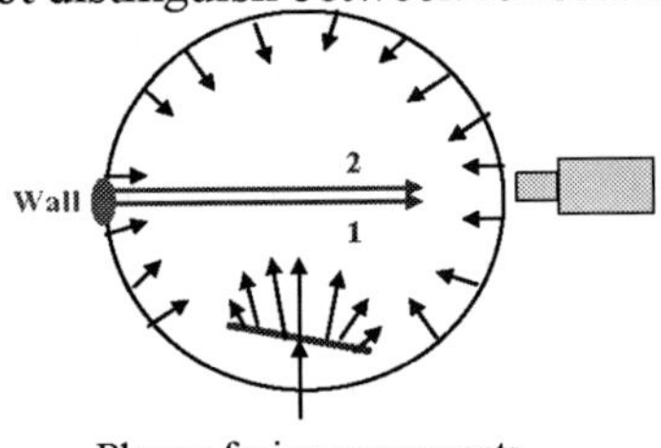

FIGURE 8: Part of the thermal light emission form the hot plasmafacing components is added after multiple specular and diffuse reflections (1) to the thermal emission of the first wall (2) in the view of the midplane cameras.

For finer modelling of the reflections Monte-Carlo ray-tracing codes need to be used, and in order to make such calculations relevant, dedicated in situ measurements of the reflectivity are needed – as presently under way in Tore Supra.

5. PHOTOTHERMAL METHODS

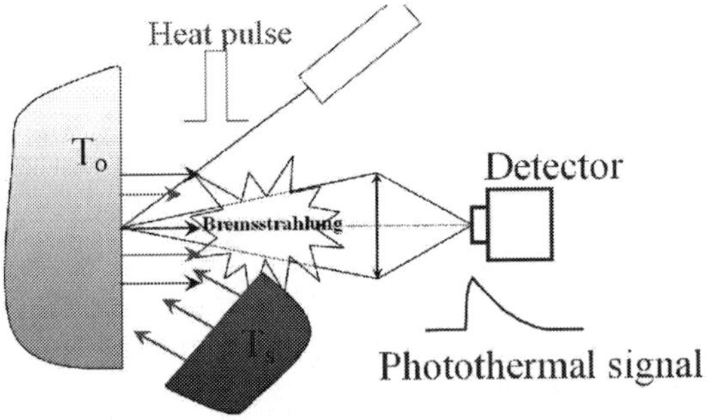

FIGURE 9: Schema of photothermal measurement using a laser as local heat source to measure the small temperature modulation independent from parasitic light from a hot target or bremsstrahlung.

Photo-thermal methods [8] (fig. 9) using relatively strong lasers as modulated heat-source may be an answer to measure surface temperature under conditions where the parasitic light (mainly bremsstrahlung) and the specular reflection from another hot target area are important, which is mainly expected in the divertor. One creates with these lasers a small oscillation ΔT of the temperature T (ΔT =10K at T=1000K) and measures at two wavelengths the amplitude of the radiance variation. The temperature derivative of the spectral radiance is – like the spectral radiance itself – a unique function of temperature. With the grey body hypothesis (or using additionally methods as described in section 3) we can deduce the temperature from the ratio of the amplitudes of the oscillations. This method is presently a localised one, but with methods like sweeping, stepping and/or multiple point illumination one could imagine to develop it to cover a certain area. In particular in the divertor area (see section 1) one could use a silica fibre bundle in the NIR range for heating via laser light and an optical relay transport of the IR light for measurement.

5. DISCUSSION AND CONCLUSION

The complexity of thermography in ITER requires a broader approach to measurement methods than previously used. Dedicated light simulations in the relevant are necessary. Thermocouples, calorimetry, black body structures (holes in the target), shutters and reference fibres will help to eliminate a number of unknowns. Additional benefits are expected from spectral analysis, and the combination of 2D viewing and localised precision measurements using active elements in the diagnostic as illumination and heating sources. In particular the application of these latter (active) methods also to the shutters which could have a black and hot part and a reflecting part should help to eliminate the uncertainties in the optical transmission chain. These options should be integrated in the hardware designs and the analysis software.

This work was partially supported by the European Communities. Views and opinions expressed do not necessarily reflect those of the European Commission.

REFERENCES

1. D. Guilhem, et al., Rev. Sci. Instr. Vol **70** (1999) 427
2. R. Reichle, D. Guilhem, R. Mitteau, J. C. Vallet. et al. Nucl. Fus. **43** (2003), 797
3. E. Gauthier, H. Roche, E. Thomas, et al. Proc. 24th SOFT (Warsaw, 11-15 Sep. 2006)
4. R. Reichle, et al. Rev. Sci. Instrum. **75**,10 (2004) 4129
5. S. Fougerolle et al. 15 Cong. Franç. de Therm., 29 Mai-1 Juin, Île de Embiez, France, (2007) 837
6. D. Hernandez et al., Rev. Sci. Instrum. **66** (1995) 5548
7. D. Hernandez , Rev. Sci. Instrum. **76** (2005) 024904
8. Th. Loarer, F. Brygo, E. Gauthier, et al. Journ. Nucl. Mat., **363-365** (2007) 1450
9. D. Guilhem, R.Reichle, H.Roche, QUIRT, **3**,2 (2006) 155
10. R. Reichle et al., Proc. 32nd EPS Conf. on Plasma Phys., ECA Vol **29C** (2005) P4.083
11. R. Reichle, B. Brichard, C. Pocheau et al. these proceeding
12. A. Herrmann, et al., Phys. Script.,Vol T**111** (2004) 98
13. E. Delchambre, R. Reichle, R. Mitteau et al., Journ. Nucl. Mat. **337-339** (2005) 1069
14. R. Reichle, B. Brichard, F. Escourbiac et al, Journ. Nucl. Mat., **363-365** (2007) 1466
15. A.H. Lettington, 'Application of Diamond-like (Hard Carbon) films' in: 'Diamond and Diamond like film coatings', Eds. R.E. Clausing et al. , NATO ASI serie B, 266, Plenum Press, New York (1991) 481
16. D. Hernandez, et al., submitted to Fusion Engineering and Design (2007)

Omegatron mass spectrometer in a hydrogen/deuterium detached plasma

A.Tonegawa, T.Nishijima, H.Ishioka, A.Nakanowatari, and K.Kawamura

Department of Physics, School of Science, Tokai University,
Kitakaname 1117, Hiratsuka 259-1292, Japan.
Phone/FAX: +81-463-58-1211/+81-463-50-2013 E-mail: tone@keyaki.cc.u-tokai.ac.jp

Abstract. Diagnostics of the relative concentrations of the atomic and molecular ions were carried out in hydrogen/deuterium plasma with hydrogen/deuterium gas puff in the linear plasma device, TPD-SheetIV. An omegatron ion mass spectrometer is used for analyzing ion species.

Keywords: Omegatron, detached plasma, divertor, hydrogen, deuterium.
PACS: 52.70.-m, 52.70.Nc, 52.75.-d

INTRODUCTIN

In magnetic fusion reactor, detached plasma is often formed in the divertor region. In this type of a low temperature plasma, the production of atomic and molecular ions has been predicted to greatly enhance the plasma recombination rate, that is, molecular assisted recombination (MAR) [1,2]. Therefore, diagnostics of atomic and molecular ion density of the detached plasma play an important role in the divertor's performance and modeling.

The atomic and molecular processes involved in the production and destruction of molecular ions have been studied for many years both theoretically and experimentally in various linear plasma divertor simulator device [3-5]. The role of molecular ions in the plasma, however, is still under discussion and various conclusions. Molecular processes have not been reported clearly for high-density plasma and for isotope effect of deuterium plasma.

In evaluating the effect of molecular ion in detached plasma, a reliable method of measuring the atomic and molecular ion concentration is necessary. In this paper, we describe the design and operation of an omegatron mass spectrometer for atomic and molecular ion concentration in detached plasma.

The omegatron mass spectrometer is a device invented to measure the charge-to mass ratio of ions in a high-vacuum system through the use of ion cyclotron resonance [6-8]. In an omegatron, a high-frequency electric field is set up normal to a uniform dc magnetic filed and ions are excited at their various ion cyclotron resonances, causing their orbital radii to increase; the current induced when they strike a collector in the omegatron is measured. This technique of ion cyclotron resonance has been applied to mass spectrometry and isotope separation.

CP988, *Burning Plasma Diagnostics*, edited by F. P. Orsitto, G. Gorini, E. Sindoni, and M. Tardocchi

EXPERIMENTAL APPARATUS AND MEASURING SYSTEM

The experiment was performed in the linear plasma device TPD-SheetIV, shown in Fig. 1 [8]. Ten rectangular magnetic coils formed a uniform magnetic field of 0.8 kG in the experimental region. The hydrogen plasma was generated at a hydrogen gas flow of 75 sccm, with a discharge current of 30-100 A. The neutral pressure P in the experimental region was controlled between 0.1 and 20 mTorr with a secondary gas feed. The sheet plasma flowed from the plasma source along the magnetic field to an endplate located about 1.0 m downstream. Electron density and electron temperature were measured using a planar Langmuir probe, which were located 3 cm in front of the endplate.

The basic layout of the omegatron spectrometer is shown in Fig.2. An "omegatron" mass analyzer, situated behind a small hole (Φ 0.5 mm) in the endplate with a differential pumping system, is used for analyzing ion species.[4] The omegatron mass analyzer is electrically isolating and is water-cooled to allow the entire housing to be inserted into the plasma column. The background pressure in the omegatron analyzer was kept low enough to allow a collisionless analysis to be performed. Therefore, the differential pumping was applied to maintain the pressure of around 10^{-5} Torr in the region of analyzer, which is reduced by a factor of 10^2 in comparison with that in a experimental region. Inside the diagnostic body of this mass analyzer, electrons are stripped away from the particle beam with a pair of electron repeller grids, which are typically dc-biased to $V_e \approx -150$ V. The resulting ion beam is then slowed by an ion repeller grid and an RF chamber which are dc biased to have slightly higher than the

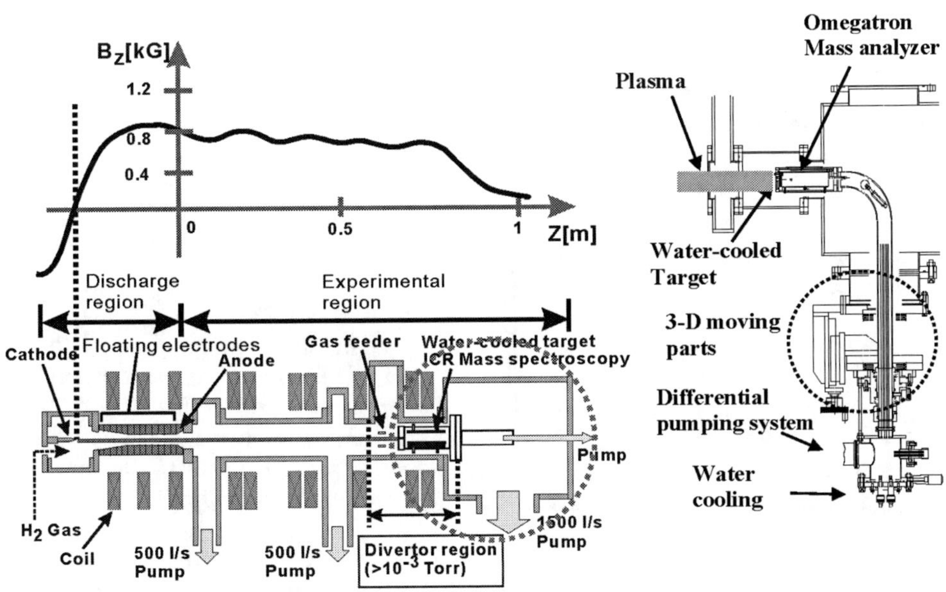

FIGURE 1. Schematic diagram of magnetized plasma simulator (TPD-SheetIV) and omegatron mass spectrometer.

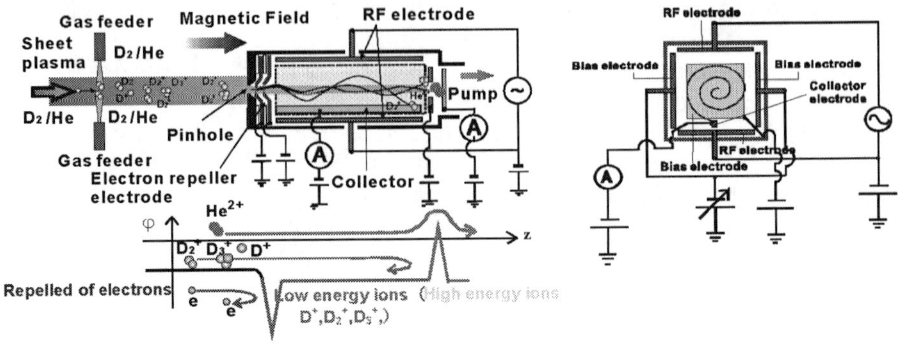

FIGURE 2. Schematic diagram of omegatron mass spectrometer showing the side view and front view.

plasma potential. The relative densities of the molecular and atomic ions were determined from the collector current of the mass analyzer. The peaks appear in the collector current of the analyzer when the frequency of the applied RF electric field is equal to the ion cyclotron frequency $f_c \equiv eB/2\pi m$. The current of an ion species j entering the collector should be proportional to the corresponding ion peak current $I_{p,j} \approx \frac{1}{4} e n_j c_{s,j} S$, where $c_{s,j} \approx \sqrt{k_B T_e / m_j}$ is the sound speed, k_B is the Boltzmann constant, and S is the area of the small hole in the endplate.

EXPERIMENTAL RESULTS

Mass spectroscopy is performed by applying an RF field to the parallel plate of the omegatron and measuring the resulting current on the collector plate. Figure 3 shows a typical mass spectrum at a discharge current I_d of 30 A and the P_{div} ~ 7.5 mTorr. From a typical plasma omegatron spectrum, it can be seen that the peaks correspond to H^+, H_2^+, and H_3^+. The peak identification is carried out by comparing the observed peak frequencies to the cyclotron frequency $f_c \equiv eB/2\pi m$ with obtain the mass-to-charge ratio m/e. The relative ion density of H^+, H_2^+, and H_3^+ should be roughly equal to the measured ratios of corresponding resonance peak heights.

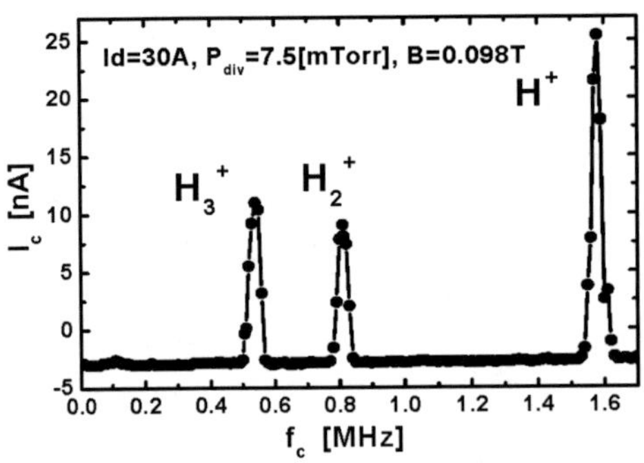

FIGURE 3. Typical mass spectrum showing collected current I as function of drive frequency.

Figure 4 shows the ion density ratio plotted against gas pressure P_{div} in the (a) hydrogen plasma and (b) deuterium plasma. The discharge current is 30 and 50 A. By defining the ion density ratio as the individual values of the ion densities divided by The sum of the total density, we can

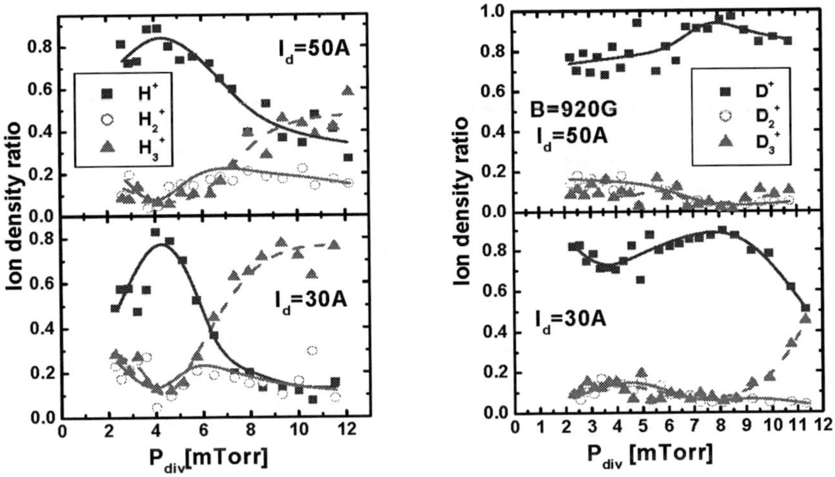

(a) (b)

FIGURE 4. The ion density ratio is plotted against gas pressure P_{div} in the (a)hydrogen plasma and (b)deuterium plasma.

Express the effect of the molecular processes on the hydrogen and deuterium plasma. At high density plasma and low gas pressure, the ion density ratio of H^+ is larger than that of H_2^+ or H_3^+. At low density plasma and high gas pressure, however, it is shown that the conversion H_2^+ into H_3^+ tends to increase and the plasma will contain H_3^+. The effects of D_3^+ in the deuterium plasma are smaller than that of H_3^+ in the hydrogen plasma for isotope effect.

ACKNOWLEDGMENTS

This work is part of a program supported by the LHD Joint Project, the National Institute for Fusion Science (NIFS04KOBP007) and is partially supported by a Grant-in-Aid No. 1803511 (2006-2007) from the Ministry of Education, Culture, Sports, Science and Technology of Japan.

REFERENCES

1. D. E. Post: J. Nucl. Mater. **220-222,** 143 (1995).
2. S. I. Krasheninnikov, A. Y. Pigarov and D. J. Sigmar: Phys. Lett. A **314,** 285(1996).
3. N. Ohno, N. Esumi, S. Takamura, S. I. Krasheninnikov and A. Y. Pigarov: Phy. Rev. Lett. **81,** 818(1998).
4. E. M. Hollmann and A. Yu. Pigarov: Phys. Plasma **9,** 4330(2002).
5. A. Tonegawa, M. Ono, Y. Morihira, H. Ogawa, T. Shibuya, K. Kawamura and K. Takayama: J. Nucl. Mater. **313-316**, 1046(2003).
6. H. Ogawa, K. Kumita, M. Ono, A. Tonegawa, T. Schibuya and K. Kawamura: J. Plasma Fusion Res. Ser. **6**, 453(2004).
7. C.J.Caldqwll-Nichols, R.-D.Pezhorn, S.Grunhage, M.Sirch, J.George, Fusion Eng.&Design, **58-59**,395(20001).
8. H.Yazawa, A.Tonegawa, M.Ono, T.Shibuya, K.Kawamura, Jpn.J.Appl.Phys.,**45**,8208(2006).

Behaviour of the ASDEX pressure gauge at high neutral gas pressure and applications for ITER

A. Scarabosio and G. Haas

Max-Planck-Institut für Plasmaphysik, EURATOM-Associaton, Boltzmannstr. 2, D-85748 Garching, Germany

Abstract. The ASDEX Pressure Gauge is, at present, the main candidate for in-vessel neutral pressure measurement in ITER. Although the APG output is found to saturate at around 15 Pa, below the ITER requirement of 20 Pa. We show, here, that with small modifications of the gauge geometry and potentials settings we can achieve satisfactory behaviour up to 30 Pa at 6T.

Keywords: ITER Diagnostic, Pressure gauges
PACS: 07.07.Df,52.70.m, 52.25.Ya

During plasma discharges in a fusion devices such as a tokamak, the neutral gas is continuously produced and destroyed due to the interaction between the plasma, the surrounding walls and the neutral gas itself. The knowledge of the neutral gas density or flux is not only required for plasma control but it also provides important information about plasma behaviour, plasma-wall interaction, gas recycling and divertor performance. To measure the pressure/flux of neutrals with sufficient temporal and spatial resolution, pressure gauges have to be installed inside the vessel. A suitable gauge has thus to work in presence of strong and variable (in magnitude and direction) magnetic field and strong background ionising radiation, and to provide reliable measurements for a pressure range variation of several orders of magnitude.

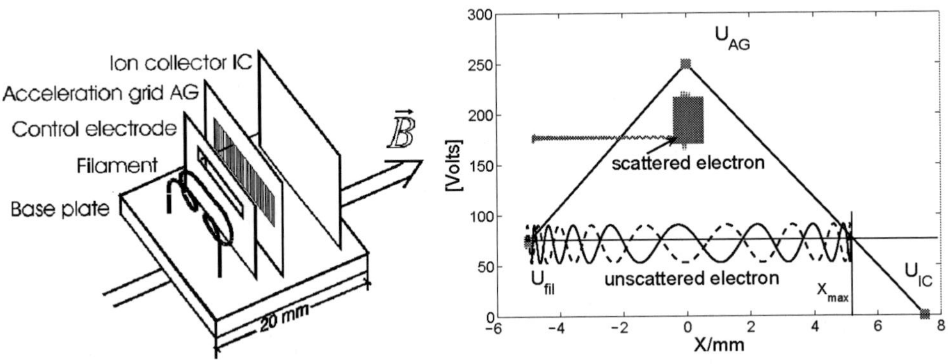

FIGURE 1. *Left: schematic of the APG representing the filament and electrode shape and arrangement. Right: the electrodes potential configuration forms a potential well for the electrons. Sketch (rescaled) trajectory of a collisionless electron returning to the filament (forward is full-line, backward dashed-line) and scattered electron collected at the acceleration grid are also shown.*

CP988, *Burning Plasma Diagnostics*, edited by F. P. Orsitto, G. Gorini, E. Sindoni, and M. Tardocchi

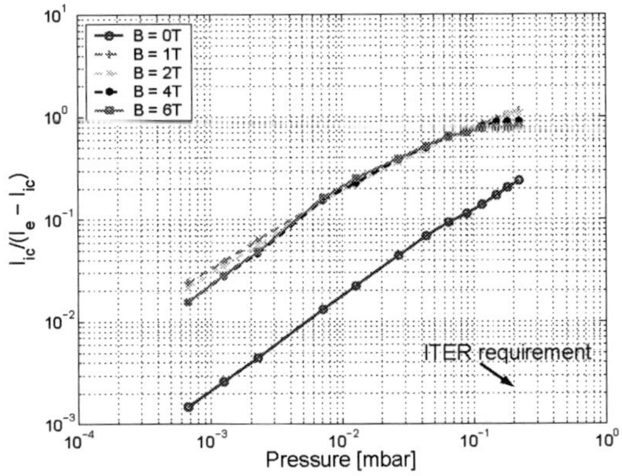

FIGURE 2. *Output of the standard APG as function of pressure and magnetic field.*

The so-called ASDEX Pressure Gauge or APG [1], a hot cathode ionisation gauge, copes well with all these requirements and has been used to provide neutral flux measurements in previous and present-day tokamaks and stellarators such as ASDEX, TEXT, ASDEX Upgrade, DIIID, JET and W7-AS for more than 25 years. This is achieved mainly by using a thick filament to stand $J \times B$ forces, linear geometry with the axis roughly parallel to magnetic field (see figure 1), feedback controlled electron current and by AC operation (using a control electrode) to reduce the noise and suppress DC background signals [1]. A schematic of the electric potential is drawn at the right of figure 1. The APGs are also foreseen as main in-vessel pressure gauges in ITER [2]. The aim of this work is to asses the reliability of the APG measurements close to the upper limit required by ITER [3]. Recent theoretical calculations [4] have suggested a high neutral pressure in the ITER divertor which leads to a requirement for the upper pressure range of 20Pa, much larger than what is typically observed in today's tokamaks.

The behaviour of the APG at low pressure is relatively well understood [1]. For magnetic fields larger than 1T, the APG sensitivity (output/pressure) increases by about a factor of 15 with respect to the $B = 0$ case. In the same pressure range, the APG behaves linearly (constant sensitivity) with p at $B = 0$ whereas it is not linear with $B \neq 0$.

In [1] was shown that the electron dynamic plays a crucial role. Because of the electric potential shape, the electrons are trapped (see sketch in figure 1) and some may eventually return to the filament due to their finite initial energy. These electrons contribute to electron flux in the ionisation region and thus to the ion production but they do not show up in the electron current, I_e, at the acceleration grid. This results in an effective increase of the normalised output, $I_{ic}/(I_e - I_{ic})$ where I_{ic} is the current at the ion collector and explains, at least partly, the increase in the sensitivity with B. We shall see that this dynamic is modified by collision at large neutral pressure.

Dedicated laboratory experiments have been performed with the standard gauge configuration in the medium and high pressure range up to 6 Tesla in pure hydrogen and the results are shown in figure 2. At $B = 0$ the gauge output increases monotonically for the full pressure range. For $B > 1T$ the sensitivity is weakly dependent on the magnetic field and the output saturates at about 15Pa at $B = 6T$, below the ITER requirement.

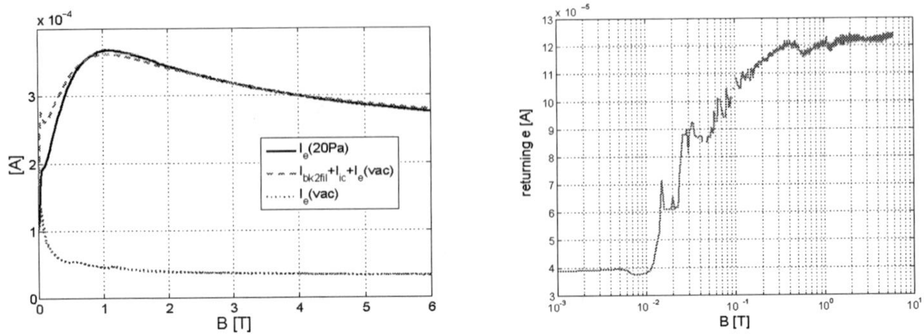

FIGURE 3. *Left: comparison of the electron current at 20 Pa with electron current in vacuum, returning electrons and secondary electrons from ionisation. Right: measurement of the electrons returning to the filament in high vacuum.*

In order to extend the available pressure range an effort was made to understand the reasons for the saturation. We note that the saturation occurs only for $B > 1T$, suggesting that the relevant mechanisms are related with the effects of the magnetic field and the accompanied increase of sensitivity. Ion losses (by scattering and/or recombination) and electron energy losses due to collisions with the neutral gas still play a small role at these pressures as suggested by the results at $B = 0$. The physics of the APG is best studied in experiments without feedback, at constant current in the filament with the magnetic field carefully varied. In this way we can consider constant the flux of electrons leaving the filament and study the effect of the magnetic field and pressure on the electrons. We observe an increase of the current to the acceleration grid with pressure (left of figure 3) that cannot be explained only by the production of secondary electrons from ionisation. Main candidates to explain this difference are the electrons lost to the filament in the low pressure regime. It is easy to realise that, even a single collision of the electron with the neutral gas has a great chance to prevent the electron from returning to the filament (figure 1). Any collision (elastic and inelastic) will substantially reduce the ratio of the parallel (to B) to perpendicular velocity, $v_{\parallel}/v_{\perp}$, leaving the electron with no sufficient parallel velocity to reach the filament again. The electron is thus trapped and will be eventually collected at the acceleration grid after a number of transits. Using cross-section data accounting for several elastic and inelastic processes we have calculated the probability for an electron to return to the filament as a function of pressure (figure 4) using the following expression for the probability of a free path λ_{free}, $P(\lambda_{free}) = e^{-\int_{x_{fil}}^{\lambda_{free}} dx/\lambda_{mfp}}$. Here we have $\lambda_{mfp} = 1/n_{gas}\sigma_{tot}$ with σ_{tot} the total cross-section [5] for electrons in a hydrogen gas. At p=1 Pa only a small fraction of the electron beam experiences a collision before returning to the filament whereas at p=20 Pa almost 90% collides at least once. We may thus expect that, at such pressure, most of the "returning" electrons are instead finally collected at the acceleration grid. We can measure the returning electrons in high vacuum by setting the electric potential of the ion collector higher ($U_{ic} = 259V$) than the potential of

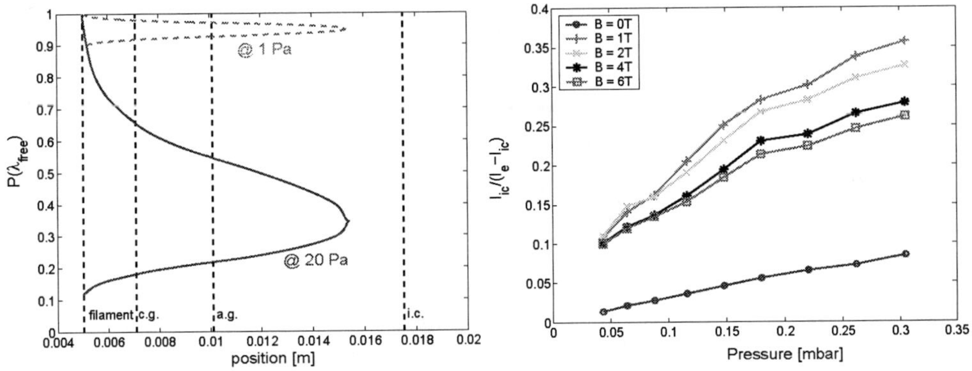

FIGURE 4. *Left: probability of a free path of an electron, λ_{free}, starting at the filament and moving along the gauge axis. Elastic and inelastic processes (excitation, vibration, rotation, attachment and ionisation) are considered. Right: output of the modified APG as function of pressure and magnetic field.*

accelerating grid in order to stop the oscillations of the electrons [1] and by assuming that the returning electron current is given by $I_{bk2fil} = (I_{ag} + I_{ic})_{U_{ic}=259V} - (I_{ag})_{U_{ic}=0V}$. The fraction of returning electrons (figure 3 right) increases with the magnetic field to a value comparable with the current at the acceleration grid. We thus compare the electron current at high pressure (20Pa) with the sum of the electron current in high vacuum, secondary electrons from ionisation and "returning electrons". We find that, except at very low field, $I_{ag}(20Pa) \sim I_{ag}(vac) + I_{secondary} + I_{bk2fil}$ (figure 3 left). It can be concluded that the increase of the electron current at high pressure (the part not accounted by secondary electrons) is due to the returning electrons. Consequently the sensitivity of the gauge decreases. One could say that the effect of collisions partially cancels the increase in the sensitivity due to the magnetic field.

It is reasonable to think that reducing the electron scattering may increase the upper pressure limit of the APG. This can be achieved by reducing the size of the acceleration volume and/or ionisation volume. To minimise the changes in the APG geometry we set $U_{cg} = U_{ag} = 270eV$ and reduce the ionisation volume to $2.5mm$. In this way, the control electrode becomes the effective acceleration grid. As a result, the probability for an electron to return to the filament is increased to 40% at 20Pa. We then explore the behaviour of this modified APG at high pressure. The APG output for pressure scans at different magnetic fields is displayed in figure 4. A monotonic output with sufficient slope is now obtained up to the maximum achievable pressure of 30Pa at $6T$. The monotonicity is achieved at the expenses of a lower sensitivity and, perhaps, to a stronger dependence on the magnetic field. In this pressure range the lower sensitivity does not cause any problem of signal-to-noise ratio. This result demonstrates that, with small modifications of the APG (reduction of the ionisation volume and increased control electrode potential), the upper pressure limit can be extended to satisfy the ITER requirements.

REFERENCES

1. Haas G., Bosch H.-S., In vessel Pressure Measurement in Nuclear Fusion Experiments with ASDEX Gauges, Vacuum 51 (1998), p. 39-46.
2. Mertens V., Haas G., In Vessel Neutral Pressure Measurement in ITER, Final Report on EFDA Contract 5-1349 Deliverable D 4.1, 2006
3. ITER Document, Plasma Facing Components and Operational Diagnostics, G 55 DDD 38 01-07-17 W, 2001
4. A.S. Kukushkin et al. IAEA-CN-149/IT/P1-16. IAEA 2006, Chengdu China.
5. http://consult.cern.ch/writeup/magboltz/

Critical Assessment of Pressure Gauges for ITER

Francisco L. Tabarés, David Tafalla and Isabel Garcia-Cortés

Laboratorio Nacional de Fusion. Asociacion Euratom/Ciemat. Av. Complutense 22. 28040 Madrid. Spain

Abstract. The density and flux of molecular species in ITER, largely dominated by the molecular form of the main plasma components and the He ash, is a valuable parameter of relevance not only for operation purposes but also for validating existing neutral particle models of direct implications in divertor performance. An accurate and spatially resolved monitoring of this parameter implies the proper selection of pressure gauges able to cope with the very unique and aggressive environment to be expected in a fusion reactor. To date, there is no standard gauge fulfilling all the requirements, which encompass high neutron and gamma fluxes, together with strong magnetic field and temperature excursions and dusty environment. In the present work, a review of the challenges to face in the measurement of neutral pressure in ITER, together with existing technologies and developments to be made in some of them for their application to the task is presented. Particular attention is paid to R&D needs of existing concepts with potential use in future designs.

Keywords: ITER Diagnostics, Pressure Gauges, Divertor
PACS: 07.07.Df, 52.70.m 52.25.Ya, and 52.55.Rk

INTRODUCTION

The measurement of neutral pressure in ITER, Pn, is required not only for operation purposes but also to provide unique insight into global particle behavior and divertor performance. At present, Monte Carlo simulations, as made through the improved version of B2-EIRENE code (1) indicate a high sensitivity of local values of Pn to divertor plasma parameters and a significant increase of its value close to the target plates has been deduced by the introduction of neutral elastic collisions into the code (2). Pressures up to 20 Pa (0.2 mbar) are predicted by the last calculations. Therefore, an accurate and spatially resolved monitoring of this parameter will provide a valuable benchmark for the validation of neutral modeling codes.

Although the field of pressure measurement, or any other parameter close linked to the flux of neutrals crossing a given element in the vacuum vessel, is quite old and many concepts have been put forward (3), no standard gauge presently exists which provides reliable data under the harsh scenario foreseen during plasma operation in ITER. Table I summarizes the main types of pressure sensors of wide use in vacuum applications to

CP988, *Burning Plasma Diagnostics*, edited by F. P. Orsitto, G. Gorini, E. Sindoni, and M. Tardocchi

date. As seen (even when radiation damage has not been included as selection parameter) all of the gauge concepts miss al least one of the requirements. It is also clear from the table that some concepts, not fully developed at the commercial level, remain still unexplored. Basically, the need of remotely sensing the gauge parameters (with distances of many tens of meter), motivated by the much higher sensitivity of the electronic components to the radiation and high stray field surroundings, precludes the use of some promising concepts, as the capacitance manometer for example. Another important source of incompatibility arises from the wide range of pressures to be measured in the divertor area, where the monitoring of neutral fluxes has the highest relevance. Present specifications (4) demand a range of 10^{-6} to 0.20 mbar to be covered. The fact that most of the existing concepts are meant to work either in the free molecular or viscous regimes, but not in both, will imply the use of more than one gauge type for the full diagnostic, therefore putting extra demand in calibration issues. Another aspect to be considered for the selection of existing concepts is the intrinsic dependence of the parameter directly monitored to actual parameter of interest. Thus, for example, ionization gauges (either B-A or Penning type) record the density of species present within the sensor head. For full thermalization of the gas with the sensor walls it can be shown that this parameter directly reflects the flux of particles crossing the aperture of the gauge. More important, the sensitivity of these manometers is highly species-dependent. Even when there is little difference to be expected between the main components in the gaseous mixture (D_2 and T_2), a much lower sensitivity for He, appearing as a byproduct of the fusion reaction, exists. For a typical concentration of ~10% of He in the mixture, however, the error can be masked by the 20% accuracy demanded in the measurements. However, this implies that reliable calibrations under the full operation regime of flammable gases must be performed. This will not be the case for capacitance manometers, which in addition provide direct information on true pressure, or any other type of gauge based in similar concept.

TABLE 1. Summary of pressure gauge concepts

Name	**Physical Principle**	**Measuring Range (mbar)**	**High T Operation**	**Magnetic Field Compat.**
Bayard-Alpert Gauge	Ionization of residual gas	10^{-10} to 10^{-2}	**Yes**	**No**
Penning gauge	Ionization of gas enhanced by B field		**Yes**	**B required!!**
Pirani/ Thermocouple gauge	Thermal conductivity of gas	10^{-4} to 1(up to 10^{3} with convectron)	**No**	**Yes**
Capacitance manometer	Deformation of a thin membrane	10^{-6} to 10^{3}	**<200 °C**	**Yes (W/o electronics)**
Other Concepts				
Piezo-gauge.	Force on piezo-electric crystal	1 to 10^{3}	??	??
Vibrating /rotating elements	Viscosity of gas	$<10^{-2}$ mbar?	??	??

FUSION-ORIENTED DEVELOPMENTS

The challenge of neutral pressure measurements in a fusion plasma has driven the development of new sensors since long. For ionization gauges, the inclusion of a modulated electrode together with a secondary electron repeller was proposed and worked out in the early sixties (5). In its present form, and with important new improvements, the concept has led to the birth of the Asdex Pressure gauge (6), at present the most tested system in real fusion environment. This gauge and its extension to high-pressure operation are widely described in these proceedings (7). Although it is by far the best-developed concept for ITER application to date, it has still to face some of the intrinsic problems associated to being an ionization gauge, and other gauges are presently foreseen as complimentary systems, either for calibration or redundancy.

Penning gauges offer their intrinsic compatibility with external magnetic fields, and fusion-devoted concepts have been successfully applied in some devices. Of particular relevance is the PDX-type Penning gauge (3) with reduced dimensions and using the confining magnetic field directly from the fusion device. Their extension to ITER, however, has to face its intrinsic non-linearity of the recorder signal (current in the gauge own plasma) with pressure, its low high pressure limit (<0.8mbar) and the fact that, providing a DC current output, it will be perturbed by the leakage currents induced by the background radiation. Also, igniting the discharge at low pressures may be problematic with no external source.

Of particular appeal is the capacitance manometer, which measure true pressure, independent of the type of gas, and has no hot elements facing the gas. To date, most of the commercially available models incorporate the RF circuitry into the sensor, which precludes its use in a fusion environment. Even those with bakable heads have a limited cable length, less than one meter, which extension is prevented by the required the accuracy of the balance in the capacitance-testing active circuit. Since the head itself is not sensitive to magnetic field (Inconel or ceramic membranes) and temperature stabilization, of critical impact on the low-pressure limit, can be solved keeping the head at the required temperature (up to 200 °C in some models), their use in Fusion has drawn the attention of many groups. A scheme particularly promising for their implementation in ITER was developed by the NBI group at JET (8). By replacing the conventional active circuit by passive elements in the capacitance-sensing circuit, they were able to insert 100 m long cables between then head and the conditioning electronics. This type of innovations, and the fact the capacitance manometers are of generalize use in radiation-dominated background scenarios, as in Fission Reactors (9), make their use very appealing as ITER diagnostic.

REQUIRED R&D ACTIVITIES

Although some concepts, as the APGs, are presently largely developed, there is still ample room for innovation. Particularly urgent is the design and testing of the gauges to compliment ionization manometers. Of particular appeal are capacitance and Pirani gauges. Very little work has been devoted to the latter regarding their implementation in ITER scenarios. Especially challenging is the adaptation of thermal conductivity-based ideas to the high T background present in the divertor cassettes during normal operation and the development of remote sensing electronics for capacitance manometers. Operation under dusty background, particularly conspicuous near the strike points, has to be included as a selection parameter. Figure 1 shows the scheme of integration of divertor diagnostics into the cassette. From the figure, it is obvious that the demand of port space for the wiring has to be considered for the final design, as redundancy, and hence a large number of heads, is mandatory for long-term reliable operation.

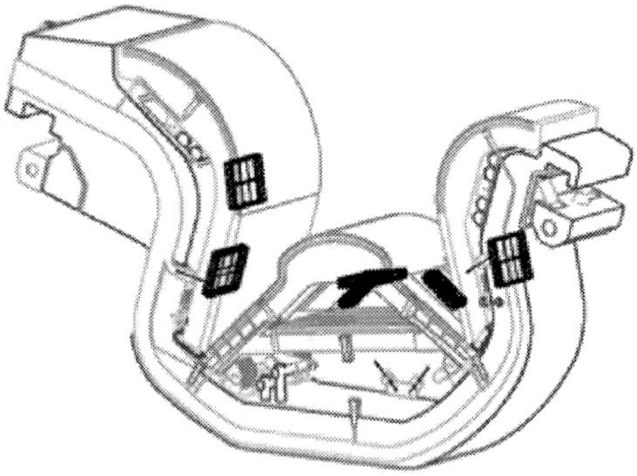

FIGURE 1. Integration of divertor diagnostics. Pick-up coils, bolometer micro cameras, Langmuir probe arrays and pressure gauges (small cylinders) are shown in their tentative location. From DD5.5N Annex.104. Div. Drawings

ACKNOWLEDGMENTS

Useful suggestions from the Task TW6-TPDS-DIADES, de.5 members are highly appreciated. Particular thanks are due to Dr. Günter Haas from IPP Garching, Stephane Vertanian, from CEA and Robert Pearce from the ITER team for very helpful suggestions.

REFERENCES

1.A.S. Kukushkin, H.D. Pacher et al. Nucl. Fusion **42** (2002) 14
2.A.S. Kukushkin, H.D. Pacher et al.. IAEA-CN-149/IT/P1-16. IAEA 2006. Chengdu China
3.H.F. Dylla. J. Vac. Sci. Technol.**20** (1982) 119
4.ITER Document, Plasma Facing Components and Operational Diagnostics, G55 DDD 38 01-07-17 W, 2001
5.J. Jaquinot, C. Deck and J. Godaert, Rev. Sci. Instrum. **43** (1972) 905 and references therein
6. G. Haas, H.S.Boch, Vacuum **51** (1998) 39
7. A. Scarabosio and G. Haas. These Proceedings (A-15)
8. A. Browne et al. "Remote sensing of a capacitance manometer pressure measurement head" Fus Technol. Proceedings.1996. Edited by C. Varandas and F. Serra. Elsevier Science. 1997
9. S. Vertanian, private communication

FUSION PRODUCTS

Neutron and Gamma-ray Measurements

Anatoly V. Krasilnikov[1,3], Mamiko Sasao[2], Yuri A. Kaschuck[3], Vasily G.Kiptily[4], Takeo Nishitani[5], Sergey V.Popovichev[4], Luciano Bertalot[6]

JET-EFDA Culham Science Centre OX14 3DB, Abingdon, UK

[1] RRC "Kurchatov Institute", Academician Kurchatov square1, Moscow, 123182, Russia
[2] Tohoku University, Sendai, Japan;
[3] SRC RF TRINITI, Troitsk, 142190, Russia;
[4] UKAEA, Abingdon, UK
[5] Japan Atomic Energy Agency, Tokai, Japan
[6] ITER Organization, Cadarach, France

Abstract. Due to high neutron and gamma-ray yields and large size plasmas many future fusion reactor plasma parameters such as fusion power, fusion power density, ion temperature, fuel mixture, fast ion energy and spatial distributions can be well measured by various fusion product diagnostics. Neutron diagnostics provide information on fusion reaction rate, which indicates how close is the plasma to the ultimate goal of nuclear fusion and fusion power distribution in the plasma core, which is crucial for optimization of plasma breakeven and burn. Depending on the plasma conditions neutron and gamma-ray diagnostics can provide important information, namely about dynamics of fast ion energy and spatial distributions during neutral beam injection, ion cyclotron heating and generated by fast ions MHD instabilities. The influence of the fast particle population on the 2-D neutron source profile was clearly demonstrated in JET experiments. 2-D neutron and gamma-ray source measurements could be important for driven plasma heating profile optimization in fusion reactors. To meat the measurement requirements in ITER the planned set of neutron and gamma ray diagnostics includes radial and vertical neutron and gamma cameras, neutron flux monitors, neutron activation systems and neutron spectrometers. The necessity of using massive radiation shielding strongly influences the diagnostic designs in fusion reactor, determines angular fields of view of neutron and gamma-ray cameras and spectrometers and gives rise to unavoidable difficulties in the absolute calibration. The development, testing in existing tokomaks and a possible engineering integration of neuron and gamma-ray diagnostic systems into ITER are presented.

Keywords: Fusion reactor, neutron and gamma-ray measurements.
PACS: 52.70.Nc

1. FUSION REACTIONS AND NEUTRON SPECTRUM

1.1. Cross-Sections and Reactivity of Fusion Reactions

Neutrons and charged fusion products are born in D-T fusion reaction

$$D + T \to {}^4He\ (3.5\ \text{MeV}) + n\ (14.1\ \text{MeV}) \qquad (1)$$

acting as an energy source of self-sustain "burning DT plasma", in D-D fusion reactions:

CP988, *Burning Plasma Diagnostics*, edited by F. P. Orsitto, G. Gorini, E. Sindoni, and M. Tardocchi
2008 American Institute of Physics 978-0-7354-0507-3/08/$23.00

$D + D \to t$ (1.0 MeV) + p (3.0 MeV) (2)
$D + D \to {}^3He$ (0.82 MeV) + n (2.45 MeV) (3)

and in the D-^{3}He fusion reaction:

$D + 3He \to 4He$ (3.7 MeV) + p (14.7 MeV) (4)

Cross-sections of these reactions are determined by tunneling probability through a potential barrier, composed by a repulsive long-range Coulomb potential and an attractive short-range nuclear potential. All these cross-sections have been measured experimentally down to the 10 keV [1]. According to ref. [2] the cross-section energy dependence can be described by equation (5) using introduced there parameters:

$$\sigma(E) = \frac{S(E)}{E}\exp\left(b / E^{0.5}\right) \tag{5}$$

$$S(E) = \frac{A_1 + E(A_2 + E(A_3 + E(A_4 + A_5)))}{1 + E(B_1 + E(B_2 + E(B_3 + EB_4)))}$$

The fusion reactivity is the product of the cross-section and the relative velocity averaged over the Maxwellian distribution and can be parameterized as a function of the ion temperature [2]:

$$\langle \sigma v \rangle = C_1 \theta \sqrt{\frac{\xi}{m_r c^2 T^3}} e^{-3\xi} \tag{6}$$

$$\theta = T \Big/ \left[1 - \frac{T(C_2 + T(C_4 + TC_6))}{1 + T(C_3 + T(C_5 + TC_7))}\right], \quad \xi = \left(\frac{B_G^{\ 2}}{4\theta}\right)^{1/3}, \qquad B_G = \pi\alpha Z_1 Z_2 \sqrt{2m_r c^2}$$

where B_G is the Gamov constant [3] and $m_r c^2$ is the reduced mass of the particles.

1.2. Thermal and Super-thermal Neutron Components

Very often the plasma contains super-thermal energetic ions originating from neutral beam injection (NBI) or accelerated by ion cyclotron range of frequency (ICRF) waves. In such a case, the actual reaction rate in a plasma of a unit volume, $Y_{total}(t)$, is comprised of three terms: a thermal term, $Y_{th}(t)$, a beam thermal term, $Y_{b\text{-}th}(t)$, and a beam-beam term, $Y_{b\text{-}b}(t)$,

$$Y_{Total} = Y_{th} + Y_{b-th} + Y_{b-b} \tag{7}$$

$$Y_{th} = n_i n_j \langle \sigma v \rangle_{Ti}, \quad Y_{b-th} = n_b n_i \langle \sigma v \rangle_{b-th}, \quad Y_{b-b} = \frac{1}{2} n_b^{\ 2} \langle \sigma v \rangle_{b-b},$$

$$\langle \sigma v \rangle_{12} = \iint f_1(\vec{v}_1) f_2(\vec{v}_2) \sigma(v_{rel}) v_{rel} d\vec{v}_1 d\vec{v}_2, \qquad v_{rel} = v_1 - v_2$$

Here, $n_i(n_j)$ and n_b denote the ion density of the i-th (j-th) species, and the beam particle density, and suffixes 1 and 2 in denote either thermal or beam particles, respectively.

In many today's DD and DT plasma experiments due to the increase of reaction cross-section with energy the dominant term in (7) is the beam-thermal term, $Y_{b\ th}$, even if the density of beam particles, n_b, is much smaller than n_i. Thus, the measurements of fusion neutron rate, neutron source profile and neutron energy distributions in present plasmas and at least in part of ITER experiments provide the information about fast ion behavior. This is in particular important for the energetic ion confinement study during: D and T NBI, ICRF heating at deuterium or tritium resonant frequencies and "triton burn-up", which measures the DT reaction rate produced by confined/ decelerated tritons originating from DD fusion reactions in deuterium plasma.

2. NEUTRON AND γ-RAY DIAGNOSTIC MISSION

Neutron diagnostics [4] provide information about fusion reaction rate, which indicates how close the plasma is to the ignition and burning required for making a power plant based on nuclear fusion. The neutron emission rate, which is directly related to the fusion power, can be determined by time-resolved emission monitors, which must be well calibrated in-situ, with support by activation systems and profile monitors with accuracy about 10%. The time-resolved neutron source profiles also provide useful information for transport analysis. The integral of the neutron rate from a shot is checked by the total yield, as measured by the activation system. Neutron diagnostics is also providing information about ion temperatures, fuel isotope ratio, fast ion behavior [5], degree of confinement, loss mechanism, *etc*. Velocity distributions and confinement properties of fast ions can be obtained from the neutron spectrometers and gamma ray measurement. The combination of several neutron measurement systems will provide the absolute fusion output and neutron fluence on the first wall. Fast time-resolved neutron emission rate and source profile measurements will play important roles in researches on self-heating burning plasma physics, and hence in the burning control of the device.

2.1. The time-resolved neutron emission rate measurements

The neutron emission rate will be used as a feedback parameter for fusion output control in ITER and future fusion reactor (neutron emission rate of 1 s-1 is equivalent to the fusion power of 1.06×10^{-12} W in the D-T plasma and 0.44×10^{-12} W in the D-D plasma). Feedback control by the neutron emission rate has been demonstrated for optimization of operation, such as avoidance of minor disruptions in JT-60U [6]. Neutron emission rate diagnostics have been employed for: real-time control in JET [7] and plasma control in TFTR [8].

Detectors must have a wide dynamic range, fast response time and be not sensitive to hard X-rays and γ-rays. BF_3 and ^{3}He proportional counters, ^{235}U and threshold ^{238}U fission chambers are the most commonly used neutron detectors having time response ~100 ns. Fission chambers operate in counting, Campbell and current modes. Semiconductor Si [9] and natural diamond (NDD) [10,11] detectors are also successfully used in counting mode for neutron flux monitoring in TFTR, JET (see

Figure 1) and JT-60U. 14 MeV neutron detector based on scintillating fibers [12] was used for time-resolved triton burn-up measurements at JT-60U.

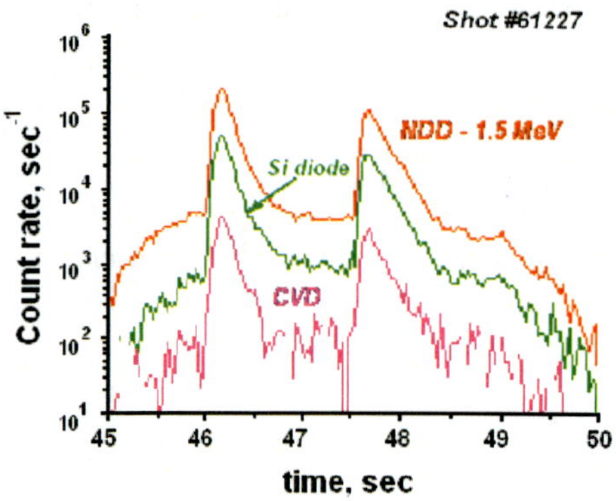

FIGURE 1. DT-neutron monitoring by NDD, CVD diamond and Si-diode in JET

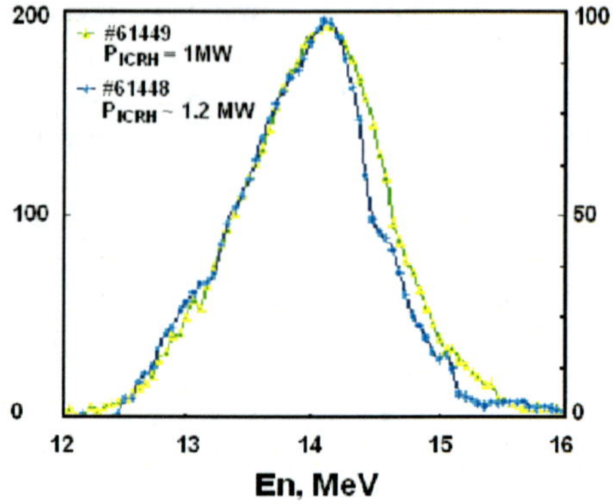

FIGURE 2 DT neutron spectra measured by Stilbene neutron spectrometer during T minority ICRH at JET

Absolute calibration is the most important problem in neutron emission rate measurements. This calibration is complicated because the neutron source is distributed in the plasma surrounded by many massive and complicated structures. Essential efforts have been devoted to absolute neutron calibrations at many tokamaks. The calibration has been performed by moving a neutron source along the known positions inside of the tokamak vacuum vessel. For absolute calibration of neutron measurements of JET and JT-60U deuterium plasmas, a ^{252}Cf neutron source was used [6]. At JET, later the activation technique is used for cross calibration. For calibration of the DT operation in TFTR a compact DT neutron generator was used [13].

2.2. Neutron Activation measurement

The neutron activation technique provides robust time-integrated measurements of the total neutron yield in a wide dynamic range. This enables accurate absolute calibration of fusion power and provides a flexible system for materials testing. In large tokamaks such as TFTR, JET, JT-60U the neutron activation systems consist of irradiation stations placed close to plasma, the γ-ray counting station and pneumatic operated sample transfer systems. To determine the relation between the neutron emission rate in the whole plasma and the neutron flux at the irradiation point the neutron Monte Carlo calculation with precise machine modeling should be applied.

At JET there is mirror symmetry of poloidal locations of irradiation ends with respect to horizontal mid-plane to correct the vertical plasma displacement influence on the total neutron yield. Sample foils transferred to the selected irradiation end via a "carousel" switching system before a plasma shot and returns to the γ-ray counting station by the pneumatic transfer system after the plasma shot.

Yields of DD and DT neutrons can be measured separately using several foils with different threshold energy reactions. From the ratio of DD and DT neutron yields the estimation of d/t fuel ratio in D-T plasmas can be performed. With the application of an unfolding codes the multifoil activation technique can provide a neutron energy distribution at the irradiation point [14], e.g. for test blanket modules experiments [15].

2.3. Neutron spectrometry

Spectrum of unscattered neutrons reflects the velocity distributions of reacting ions from the cone of measurements determined by collimator. For inhomogeneous Maxwellian plasma of ion temperature T_i this results in a Gaussian-shaped neutron energy spectrum with peak energy E_c, width $W \propto T_i$ and reaction rate $I_n \propto n_i n_j <\sigma v> T_i$ where n_i, n_j, and $<\sigma v> T_i$, are the local ion densities and reactivity, as determined in (6), respectively.

Plasmas with essential fast ion component due to auxiliary heating by NBI or ICRH will produce neutron spectra with non-Gaussian tails and Doppler energy shift relative to E_c. Fast confined α-particles can give rise to alpha knock-on neutron emission in DT plasmas with high-energy tail of the spectrum. Doppler shifts in the neutron spectrum could be also caused by rotations and anisotropy in the pitch angle distribution of the fuel populations.

Neutron spectrometry could provide measurement of the following plasma parameters: ion temperature, reaction rates $I_n \propto n_i n_j <\sigma v> T_i$, fusion powers for DD and DT reaction components, fuel ion densities in the core, plasma rotations, relative densities of super-thermal ions and their energy distributions, density and shape of fusion α-particle energy distribution.

Large and compact neutron spectrometers were essentially upgraded during recent years. Magnetic Proton Recoil (MPR) and Time of Flight for Optimized Rate (TOFOR) large neutron spectrometer [19] were successfully used in JET for detail high count rate neutron spectrometry. Natural diamond [11], stilbene [21] (see Fig.2) and NE-213 [22] compact neutron spectrometers also were successfully applied for neutron spectrometry at TFTR, JT-60U, JET and Tore Supra. Reached characteristics and performances of MPR and compact neutron spectrometers presented in Table 1.

TABLE 1. Characteristics and performance of neutron spectrometers

Spectro-meter	Useful for spectrometry sensitivity, count/(n/cm2)	Energy resolution (FWHM), %	Energy range, MeV	Dimensions W H L, cm3	Physics studied at JET	Application
MPR	5×10E-5	2.5	> 1	260×190 ×80	DT neutron spectrometry for: Ti(t), plasma toroidal rotation (t), fast ion energy (t) measurements with time resolution 50ms	JET
Natural Diamond Detector	2 × 10E-5	2	> 12.5	Ø1.6×3	Same, with time resolution 0.4s due to installation far (20m) from plasma	TFTR, JT-60U, JET
Stilbene (NE213)	10E-2– 10E-3	3	> 1	Ø4×35	Same, with time resolution 0.3s due usage slow analog electronics*	Tore-Supra JT-60U, JET

*Method based on digital signal processors showed excellent performance in spectrum measurement with pulse shape discrimination for neutrons and gammas using NE213 and stilbene spectrometers [22]

2.4. γ - rays measurements

Nuclear reaction γ-ray diagnostic is the powerful technique used for studies of the fast ion behavior in fusion devices. The emission of intense γ-ray lines is produced when fast ions (p, d, t, ^{3}He, and ^{4}He) react either with plasma fuel ions or with main plasma impurities, such as beryllium, boron, carbon, and oxygen.

γ-ray measurements were efficiently performed on the tokamaks Doublet-III [23], TFTR [24, 25], JET [26-28], and JT-60U [29] with the application of specially developed γ-ray spectrometers described in Table 2.

TABLE 2. Characteristics of the γ-ray spectrometers developed and used in fusion experiments

Tokamak	Doublet-III	TFTR	JT-60U	JET
Used γ-ray spectrometers	*NaI(Tl)* ∅5×5 cm, 1 - 10 MeV γs	Liquid fluorocarbon scintillator *NE226* ∅12.7×12.7 cm, up to 19 MeV γs	*NaI(Tl)* ∅5×5 inch 1-20 MeV γs	*Two BGO* (∅7.5×7.5 cm, 1-28 MeV (4% en.res.) *NaI(Tl)*(vert) ∅12.5×15 cm, 19 *CsI(Tl)* photo-diodes (∅20×15 mm) E_γ> 0.2 MeV

Intensity and 2-D source profile of γ-rays from the $^9Be(^3He,n\gamma)^{11}C$, $^9Be(^3He,p\gamma)^{11}B$ and $^{12}C(^3He,p\gamma)^{14}N$ reactions are using in JET as an indicator of the efficiency and spatial distribution of ICRF power deposition in ^{3}He-minority heating scheme [30,31].

To identify the existence in plasma the fusion α-particles with energies higher 1.7 and 4 MeV the measurements 4.44 and 3.21 MeV γ-rays from threshold reaction $^9Be(\alpha,n\gamma)^{12}C$ were performed [32] during and after 110 keV tritium NBI.

The influence of monotonic and non-monotonic q profiles onto 2 MeV ^{4}He ions injected into ^{4}He plasma and accelerated by ICRF waves at their third harmonic was also successfully studied [33] be means of γ-ray tomography.

2.5. Neutron/γ-ray source profile measurements

The neutron and γ-ray profile monitors are providing the measurements of the neutron and γ-ray emissivity over a poloidal cross-section of the plasma using line-integrated measurements performed by neutron and γ-ray detectors viewing the plasma along a number of lines-of-sight.

Being used inside neutron profile monitor, the compact neutron spectrometers is providing the information not only about the emissivity but also about the energy distribution of neutrons born at different magnetic surfaces of plasma. So, neutron source profile measurement is the principal method for the measurement fast deuteron and triton spatial and energy distributions, alpha-particle birth profile and ion temperature profile. Being absolutely calibrated the neutron source profile monitor could provide independent measurements of fusion power.

1D neutron source profiles were successfully measured in TFTR and JT-60U experiments. But in plasma with essential fast ion component the fusion neutron source and γ-ray profiles are not a functions of the magnetic surfaces due to non-uniformity of fast beam and ICRF-driven ion distributions on magnetic surfaces. In such cases as was demonstrated in varios JET experiments with NBI and ICRH (see

Fig.3) the 2D neutron and γ-ray source profile measurements with tomography reconstruction [17,18] are required to control the dynamics of spatial and energy distributions of fast confined ions.

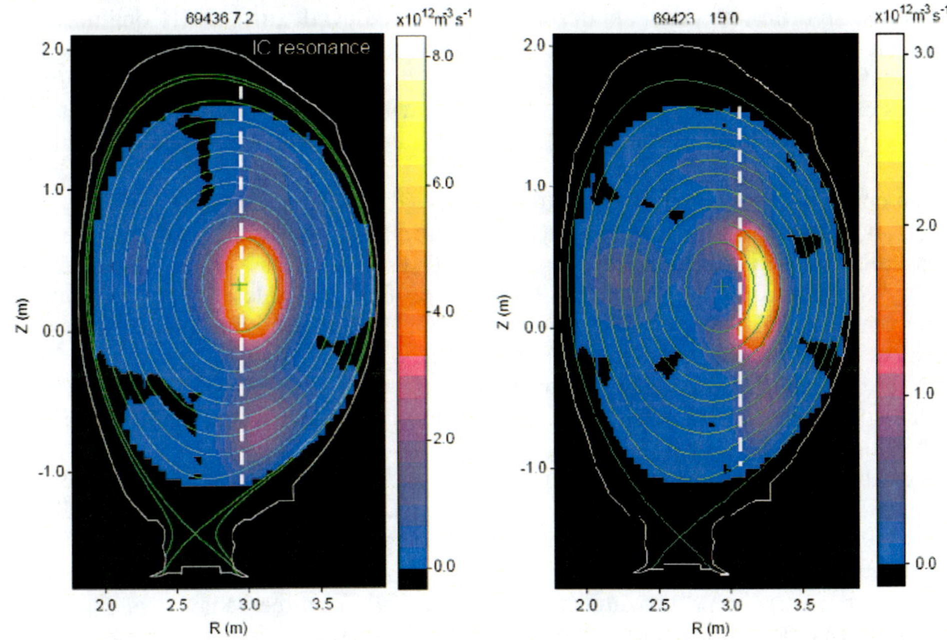

FIGURE 3. Tomography reconstruction of γ-ray emission profiles measured in JET during ^{3}He-minority ICRF heating discharges with central and low field side resonance layer positions.

3. NEUTRON AND γ-RAY DIAGNOSTICS ON ITER

ITER will be the first burning plasma experiment with collective behaviour of the alpha particles and other fast and thermal ions. ITER plasma parameters, such as fusion power, power density, ion temperature, fast ion energy and spatial distributions in the plasma core, can be well measured by means of neutron diagnostics [34-36]. The necessity to use a massive radiation shielding strongly influences the diagnostic designs, determines angular fields of view of the neutron cameras and spectrometers and gives rise to unavoidable difficulties in the absolute calibration [37, 38].

Neutron diagnostic systems under consideration and development for ITER include: radial (RNC) [39, 40] and vertical (VNC) [41] neutron cameras, in-vessel [42, 43], ex-vessel [44-46] and divertor [45] neutron flux monitors, neutron activation systems [47-49] and neutron spectrometers [10,11,19-22,50]. Arrangement of the neutron diagnostic techniques and lines of sight of RNC and VNC in ITER is described in [36].

The necessity of 2D neutron profile measurements in ITER arises from the fact that, due to fast ion components, the neutron source profile may not be a constant on magnetic surfaces, especially during: ICRH, NBI, sawtooth oscillations, AE modes and in advanced tokamak regimes with strongly negative magnetic shear. To measure the fusion power with required accuracy the 2D neutron source measurements will be necessary in all these cases.

Fast ions studies will be especially important in ITER for: optimization of ICRH absorption profile, effective application of ICRH induced drift, ICRH optimization from the point of view the establishment of the fast ion profile best for sawtooth and TAE activities stabilization, optimization NBI heating and current drive profiles, studies of the features of fast D or T spatial redistribution during TAE, IRE, sawteeth crash and other MHD instabilities and their confinement on different magnetic surfaces, etc.

Compact neutron spectrometers (diamond [10], stilbene [21] and NE-213 [22] detectors) placed inside the collimators of the RNC and VNC will provide the measurements of ion temperature in the range Ti>5 keV, fast deuteron and triton energy distribution and poloidal rotation profiles. Besides compact neutron spectrometers NDD [51], CVD diamond [52], NE-213 [53] and threshold ^{238}U fission chamber [54] flux monitors also will be used in RNC and VNC.

Neutron spectrometry using the magnetic proton recoil technique [19, 20] is also under consideration for n_T/n_D ratio, plasma toroidal rotation and fast ion energy distribution measurements.

Possible approaches to neutron knock-on tail measurements, which should provide information about the confined-alpha-particle density and energy distribution, include MPR [19] and bubble chamber neutron spectrometers [50].

Dedicated γ-ray profile camera with similar to neutron cameras arrangement and ^6LiH neutron filter could provide important information on the behavior of fusion α-particles in ITER by means of spectroscopy of γ-rays born in reactions $D(t,\gamma)^5He$ and $^9Be(\alpha,n\ \gamma)^{12}C$ [28]. Candidate detectors are single crystal spectrometers with heavy scintillators that allow measurement of γ-rays in the energy range from 1–30 MeV.

A strategy for the absolute calibration of ITER neutron diagnostics [36-38] includes: absolute calibration of all detectors at the manufacturer, calibration on site in a purpose built Neutron Test Aria and several different methods of absolute *in-situ* calibration of the fusion power and power density measurements.

The first method is based on the absolute *in-situ* calibration of the most sensitive RNC, VNC and NFM detectors after their installation on ITER, using a radionuclide neutron source and a DT neutron generator having a neutron output of about 10^{11} neutrons/s. The source will be moved inside the vacuum vessel in toroidal and poloidal directions. The most suitable period for in-situ calibration will be the end of hydrogen plasma phase. This method also involves a detailed MCNP analysis of the neutron fluxes and spectra in the RNC, VNC and NFM detector positions. The least sensitive detectors will be cross calibrated against more sensitive absolutely calibrated detectors using the plasma as the source.

The second method of the absolute calibration the RNC and VNC could be based on a characterization of the compact spectrometers and detailed MCNP calculations of the neutron flux and spectra in their locations. The compact spectrometers must be characterized on accelerator facilities and/or 2.5 and 14 MeV neutron generators in terms of absolute efficiency and neutron response function for different neutron and gamma energies. In such a way the calibration factor can be determined for all compact spectrometers. Gamma sources, 2.5 and 14 MeV neutron generators and/or

AmBe n/γ sources should be build in the detector blocks or should be periodically applied during maintenance for energy calibration and stability control of the compact spectrometers.

Another independent method of absolute calibration of the fusion power measurements will be based on the calibration of foil activation system. The advantages of this method are the intrinsic linearity and time stability. It requires essential MCNP calculations. Using activation foil materials with a range of threshold energies will increase the confidence of the MCNP calculations.

4. SUMMARY

The time-resolved neutron/γ-ray emission rate, source profile, spectroscopy measurements and neutron activation techniques are commonly used on large current tokamaks. Many years experience in application of well developed measuring systems together with essential recent progress with neutron spectrometer development, digital signal processing, tomography reconstruction, MCNP modeling etc. insures efficient operation of neutron and γ-ray diagnostics in ITER and other burning plasma installations.

The most of ITER neutron diagnostic systems have been selected, conceptually designed and integrated into the machine design. VNC and γ-ray camera integration is in progress. Methods of neutron spectrometry for the n_T/n_D ratio and for the energy and spatial distribution of confined α-particles are under study.

The combination of neutron emission rate and 2D neutron source profile measurement with neutron spectrometry, activation measurement and careful calibration experiments will provide the absolute fusion power, neutron/α source profile, neutron fluence on the first wall and ion temperature profile measurements.

Fast time-response measurements of neutron emission rate and 2D source profile can be used for burn control of the plasma in the presence of fast ions and MHD instabilities.

ACKNOWLEDGMENTS

This work was supported by Russian Agency of Atomic Energy, a Grant-in-Aid from the Ministry of Education, Culture, Sports, Science, and Technology of Japan, "Priority area of Advanced Burning Plasma Diagnostics" and United Kingdom Atomic Energy Authority (UKAEA) within the framework of the European Fusion Development Agreement (EFDA). The work leading to this article was funded by the European Atomic Energy Community and is subject to the provisions of the European Fusion Development Agreement.

REFERENCES

1. A. Krauss et al., *Nuclear Physics.*, A **465**, 150 (1987).
2. H-B. Bosch and G. M. Jale, *Nucear Fusion.*, 611-621 (1992).
3. G. Gamov, *Z. Phys*, **51**, 204 (1928).
4. O. N. Jarvis, *Plasma Phys. Control. Fusion.*, **36**, 209 (1994).

5. W. W. Heidbrink and G.J. Sadler, *Nuclear Fusion.*, **34**, 535 (1994).
6. T. Nishitani, H. Takeuchi, T. Kondoh, et al., *Rev. Sci. Instrum.*, **63**, 5270 (1992).
7. O. N. Jarvis, G. Sadler, P. van Bell and T. Elevant, *Rev. Sci. Instrum.*, **61**, 3172 (1990).
8. H. W. Hendel, R. W. Palladino, C. W. Barnes, et al., *Rev.Sci.Instrum.*, **61**, 1900 (1990).
9. S. Conroy, O. N. Jarvis, G. Sadler and G. B. Huxtable, *Nucl. Fusion.*, **28**, 2127 (1988).
10. A. V. Krasilnikov 1998 *Diagnostics for Experimental Thermonuclear Fusion Reactor 2* ed P.E. Stott *et al* (New York: Plenum) p 439.
11. A. V. Krasilnikov, et.al. *Nuc.r Instr. and Methods in Phys. Research* **A 476,** 500 (2002).
12. T. Nishitani, M. Hoek, H. Harano,et al., *Plasma Phys.Control.Fusion.*,**38**,355 (1996)
13. D. L. Jassby, C. W. Barnes, L. C. Johnson, et al., *Rev. Sci. Instrum.*, **66**, 891 (1995).
14. B. Esposito et. al., *Rev. Sci. Instrum.*, **70**, 1130 (1999).
15. M. J. Loughlin et. al., *Rev. Sci. Instrum.*, **70**, 1126 (1999).
16. M. Pillon, M. Martone, O. N. Jarvis, and J. Kalne, *Fusion Technol*,. **15**, 1420 (1989).
17. L. C. Ingesson et al., *Nucl. Fusion*, **38**, 1675 (1998).
18. J. Mlynar et al., *Plasma Phys. Control. Fusion*, **45**, 169 (2003).
19. L. Giacomelli, et. al., *Nuclear Fusion*, **45**, 1191 (2005).
20. H. Henriksson, et. al., *Plasma Phys. Control. Fusion*, **47**, 1–23 (2005).
21. Yu .A. Kaschuck *et al., 31st EPS PPCF Conf., London* (2004).
22. L. Bertalot et al., *Proc.32nd EPS Conf. on Plasma Phys.*, Tarragona, Spain, ECA (2005)
23. D. E. Newman, F.E. Cecil, *Nucl. Instrum. Methods*, **227,** 339 (1984).
24. F. E. Cecil, S.S. Medley, *Nucl. Instrum. Methods*, **271,** 628 (1988).
25. S. S. Medley et al., *Rev. Sci. Instrum.*, **61,** 3226 (1990) .
26. G. J. Sadler, et al., *Fusion Technology*, **18**, 556 (1990).
27. O. N. Jarvis et al., *Nuclear Fusion*, **36,** 1513 (1996).
28. V. G. Kiptily, et.al., *Plasma Phys.Control.Fusion*, **48,** R59-R82 (2006).
29. T. Kondoh et al., *J. Nucl. Mat.*, **241-243,** 564. (1997).
30. V. G. Kiptily, et al., *Nuclear Fusion*, **42,** 999 (2002).
31. M. J. Mantsinen et.al., *PRL*, **89,** 115004 (2002).
32. V. G. Kiptily, et al.. *Physical Review Letters*, **93** 115001 (2004).
33. V. G. Kiptily, et al., *Nuclear Fusion*, **45,** L21-L25 (2005).
34. L. C. Johnson, *et al* 1998 *Diagnostics for Experimental Thermonuclear Fusion Reactor 2* ed P.E. Stott *et al* (New York: Plenum) p 409.
35. M. Sasao,et al., *Plasma Phys.Control.Fusion.*, **46**, S107–S118 (2004).
36. A. V. Krasilnikov, et al., *Nuclear Fusion.*, **45**, 1503 (2005). ed P.E. Stott *et a*
37. G. J. Sadler, *et al* 1998 *Diagnostics for Experimental Thermonuclear Fusion Reactor 2* ed. P.E. Stott *et al* (New York: Plenum) p 501
38. Yu. A. Kaschuck 2003 ITER G XX ZZ 1 03-03-05 W0.1
39. F. B .Marcus, *et al* 1998 *Diagnostics for Experimental Thermonuclear Fusion Reactor 2,* ed P.E. Stott *et al.* (New York: Plenum) p 419
40. L. Petrizzi, *et al* 2004 Contract FU06 CT2003-00020(EFDA/02-1002)
41. A. V. Krasilnikov, *et al, Instrum. Exp. Tech.* **47,** 5 (2004).
42. T. Nishitani, *et al* 1998 *Diagnostics for Experimental Thermonuclear Fusion Reactor 2,* ed P.E. Stott *et al.* (New York: Plenum) p 491
43. M. Yamauchi, *et al, Rev. Sci. Instrum.* **74,** 1730 (2003).
44. C. W. Barnes and A. L. Roquemore, *Rev. Sci. Instrum.* **68,** 573 (1997).
45. A. V. Krasilnikov, *et al.,* "Progress in the development ofcompact in-plug neutron camera and NFM for ITER" *5th Meeting of ITPA TGD (St Petersburg, July 2003)*
46. K. Asai *et al, Rev. Sci. Instrum.* **75,** 3537 (2005).
47. C. W. Barnes, M. J. Loughlin and T. Nishitani, *Rev.Sci.Instrum.* **68,** 577 (1997).
48. T. Nishitani, K. Ebisawa, S. Kasai and C. I. Walker, *Rev.Sci. Instrum.* **74,** 1735 (2003).
49. Yu. A. Kaschuck *et al, Fusion Sci. Technol.* **43,** 1 (2003).
50. R. K. Fisher, S.V.Tsurillo and V.S.Zaveryaev, *Rev. Sci.Instrum.* **68,** 1103 (1997)
51. A. V. Krasilnikov *et al* 2002 *Advanced Diagnostics for Magnetic and Inertial Fusion* ed P.E. Stott *et al.* (New York: Kluwer Academic/Plenum) p 153
52. M. Angelone, *et al, Rev. Sci. Instrum.* **76,** 013506 (2005).
53. B. Esposito *et al, Rev. Sci. Instrum.* **75,** 3550 (2005).
54. I. N. Aristov *et al, Instrum. Exp. Tech.* **47,** 15 (2004).

Role of Fusion Product Measurements in Physics Understanding of a Burning Plasma

R.L. Boivin

General Atomics, P.O. Box 85608, San Diego, California 92186-5608, USA

Abstract. The roles of fusion products in diagnosing plasmas are quite diversified and rich in scientific content. From the early days to the new frontier of burning plasmas, the study of fusion products has allowed a deeper understanding of the physics found in these plasmas, spanning all configurations, in both magnetic and inertial confinement. In many early experiments, the measurement of plasma performance was based almost entirely on neutron yield measurements. Since then, the capability to measure fusion products has grown tremendously, and many new measuring techniques have been successfully tested and used in experiments. The first role is based on single-particle diagnosis, which can yield important information on performance, confinement, source and current profiles, plasma fluctuations, effects of ripples, MHD mode structure, *etc*. The second role, which has expanded tremendously in the last few years, includes the diagnosis of collective effects, which can yield additional information on instabilities, heating, and many others. This scientific capability is enhanced by the large number of fusion products (neutrons, charged particles, gammas) generated by the primary or secondary fusion processes such as the commonly studied, D+D, D+T and D+^{3}He reactions. This paper will review the applications of fusion product measurements used in the scientific study of plasma confinement and stability, and will describe the prospects and opportunities offered in burning plasma experiments such as ITER.

Keywords: Tokamak, diagnostics, fusion products, alpha particles, neutrons
PACS: 52.70.-m, 52.20.Dq, 52.55.Pi, 52.70.Nc

INTRODUCTION

The worldwide research effort in developing nuclear fusion as a viable source of energy has generated many exciting discoveries over the years. That excitement and everlasting hope for success could be, arguably, partly attributed to the fact that in virtually all magnetically confined fusion plasma experiments a sizable (but not yet sufficient) number of fusion events could be witnessed and measured. In early experiments, when deuterium was used as a fuel, production of fusion yields were in excess of 10^{10} events/sec- a staggering number by itself- while remaining well short of the $>\sim 10^{20}$ events/sec required for a sustainable burning plasma reactor. While the measurement of fusion yields by nuclear detection (*e.g.*, total neutron production) has always played a fundamental role in measuring the plasma performance, the field has grown into developing numerous techniques based on fusion products as a way to diagnostic many key plasma parameters.

The achievement of a sustained, fusion reactor will require two significant steps to be fulfilled for which fusion products (FP) diagnosis can be key in their optimization

CP988, *Burning Plasma Diagnostics*, edited by F. P. Orsitto, G. Gorini, E. Sindoni, and M. Tardocchi

and realization. First, sufficient plasma densities and temperatures have to be achieved in order to produce a sufficient number of fusion reactions. Second, in order to enable a self-heated plasma, the charged fusion products (*e.g.*, alpha particles) must be confined long enough to deposit their energy back to the plasma. These conditions thus require a careful analysis and diagnosis of production, transport, stability and heating of fast ions in the plasma, together with a proper monitoring of first wall components to avoid any short term material deterioration (*e.g.*, first wall damage).

This paper aims at reviewing and illustrating some key roles of diagnostic measurements based on fusion products detection.

FUSION REACTIONS

It is first important and interesting to review the basic and advanced fusion reactions relevant to magnetically confined plasmas. The steady progress in magnetic confinement experiments has allowed the study of fusion products from a number of reactions including:

$$D+D \rightarrow {}^{3}\text{He}\ (0.8\ \text{MeV}) + n\ (2.45\ \text{MeV})$$
$$D+D \rightarrow T\ \ (1\ \text{MeV}) + p\ (3\ \text{MeV})$$
$$D+T \rightarrow \alpha\ \ (3.5\ \text{MeV}) + n\ (14.1\ \text{MeV})$$
$$T+T \rightarrow \alpha + 2n\ (\text{total of } 11.3\ \text{MeV})$$

These basic reactions have been observed, and verified experimentally in many experimental devices. However, with new and planned experiments, conditions are now approaching burning plasma conditions for which the following aneutronic reactions are also becoming of interest.

$$D\ +\ {}^{3}\text{He} \rightarrow \alpha\ \ (3.6\ \text{MeV}) + p\ \ (14.7\ \text{MeV})$$
$$p + {}^{11}\text{B} \rightarrow 3a\ (\text{total of } 8.7\ \text{MeV})$$

Interestingly enough, the conditions found in a burning plasma experiment could approach the fusion conditions found in stars and help in verifying experimentally many of the conjectured reactions expected in stellar cores.

Study of confinement properties for fusion production optimization

Fusion product measurements made on early tokamaks and other confinement devices were based on the detection of the 2.45 MeV neutron. They showed the key importance of making accurate measurements in fusion performance, in part in trying to elucidate the reasons for the observed anomalous transport of heat and particles. At the same time, initial measurements were performed to determine the fusion yield of the D+^{3}He reaction, an aneutronic reaction, based on the direct collection of the 14.7 MeV proton product [1].

Later on, the fusion yield measurements, coupled to increasingly detailed determination of basic parameters such as electron density and temperature, enabled the study of the ion dynamics. An example of this application can be found from the Alcator

C-Mod tokamak where the ion temperature is determined from the measurement of the total neutron yield. In this particular case the information is also used to study the effects of the formation of an internal transport barrier on the main ion dynamics [2]. In many other applications or experiments, the inverse process has been applied. Modeling of the transport mechanisms, and derived profiles are constrained by the total measured neutron production (and/or profile).

One of the key measurements in diagnosing conditions necessary for optimal fusion production resides in the ability to resolve the source profile. This is usually accomplished by fielding many neutron detectors arranged in a fanning or parallel array. Examples of such systems have been fielded on TFTR [4], JET [7] and JT-60U [8]. They usually require considerable amount of shielding and an elaborate calibration process. Common results obtained from source profile measurements indicate a peaked emission centered and symmetric about the magnetic axis, as one would expect from simple confinement following flux surfaces. This would also imply a relatively isotropic and symmetric distribution of ions. Shown in Fig. 1 is an example of source profile measurements done on TFTR [4] in deuterium only plasma, in which the 2.45 MeV and 14.1 MeV neutrons were resolved. Source profile and subsequent energetic ion confinement can thus be inferred using the shape and magnitude of the emission.

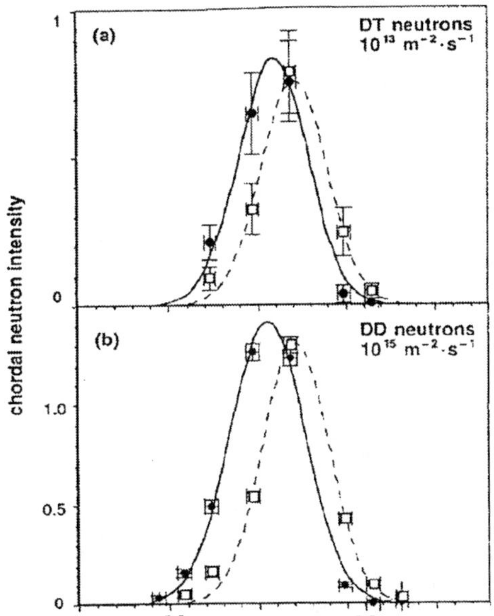

Fig. 1. Neutron profiles of (a) DT and (b) DD for two plasmas with different major radii on TFTR (in deuterium fill only). Solid line/circles are for plasma major radius of 2.45 m and dashed line/squares for 2.57 m plasma major radius. [Reprinted J.D. Strachan *et al.*, *Nucl. Fusion* **36**, 1189 (1996).]

In experiments where confinement and performance are sufficiently high, such as JET, TFTR and JT-60U, for example, measuring burnup as a tool to study energetic particle confinement have been quite illuminating. Burnup is defined as the fusion reaction in which a product of a primary reaction (*e.g.*, T, or ^{3}He from the DD reaction) becomes a reactant in the secondary reaction (*e.g.*, DT or D^3He). Initial measurements on TFTR using neutron activation techniques (from the 14.7 MeV neutrons) have shown a less than classical confinement of energetic ions [3]. However, later measurement using the TFTR neutron collimator [4] and activation measurements done at JET [5] and JT-60U [6] have shown much closer agreement with classical confinement.

However, an increasing number of deviations from centered and poloidally symmetric profiles have been observed, notably on JET, which are more readily resolvable given the 2-D tomographic capability of that system. In some instances, discharges with non-monotonic current profiles, such as current holes [9] showed a shift of the emission to the low field side, as illustrated in Fig. 2. Others, including rf heating schemes with tritium indicated a shift to the high field side [9]. Other techniques, based on the 2-D gamma ray detection following a secondary fusion reaction of high energy deuterium or alpha particles also showed unusual 2-D emission profiles which may be attributed to the generation, confinement (orbits) of these high energy particles [10]. These observations would support the need for the fielding of a 2-D tomographic system, even with partial coverage, in ITER, enabling increasing ability to optimize the main ion heating, stability and alpha heating in advanced scenarios, for example.

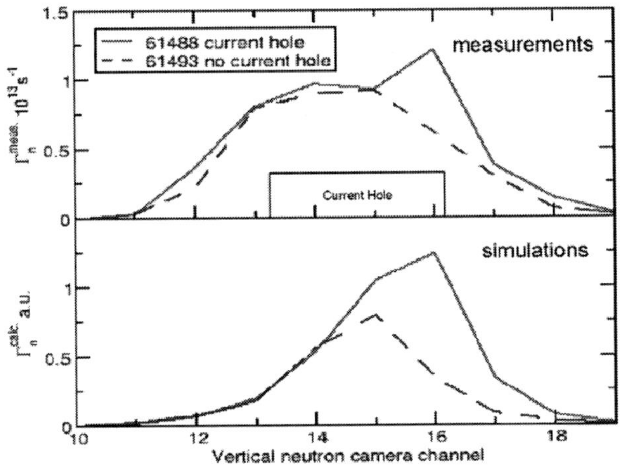

Fig. 2. Comparison of 14 MeV neutron profile for on-axis tritium beams (measurements – top box, simulation – bottom box) as measured on JET. An outward displacement of the neutron emission profile is clearly detectable with plasma with current hole (continuous line) (current hole shown in the inset – top box). Channel 15 corresponds to the magnetic axis. [Reprinted from A. Murari *et al.*, *Nucl. Fusion* **45**, S195 (2005).]

In addition, much information can also be gathered by measuring the energy spectrum of the produced neutrons. Finite temperature of the main ions (D, T, *etc.*) would manifest itself by a broadening of the neutron energy spectrum around the center-of-mass birth energy (2.45 MeV in D-D and 14.1 MeV in D-T). Furthermore, the presence of a separate energetic ion population (neutral beam ions, RF tail ions) would bring indicative wings to the spectrum. The full spectrum is dependent of the viewing angle of the detectors with respect to the plasma, in accordance with the pitch angle dependence of the fast ion population and the angular dependence of the fusion cross-section [11]. This information, usually unfolded by modeling, can be important in understanding the role of the different heating mechanisms and of transport on various classes of ions.

Study of confinement properties for alpha heating optimization

In seeking to reach conditions where burning plasmas can be sustained, it is important to study and characterize the confinement of energetic ions, such as the alpha particles produced in the DT reaction. However, since the production of a sufficient population of alpha particles (*e.g.*, in DT) is limited to TFTR and JET, it has long been realized that the 3 MeV proton and the 1 MeV triton, produced in one of the DD fusion branches, should possess very similar single particle behavior, since both gyroradius and slowing down time are very similar to the 3.5 MeV α. Collective effects, the effects of the fusion products as a sufficiently large population, can only be studied when they (or equivalent energetic ions) are in sufficient number to affect the plasma stability and/or confinement.

In the attempt to understand their behavior, new techniques were developed to study directly the confinement of charged fusion products (CFPs) (including energetic ions such as beam and RF tail ions) primarily through the measurement of particle flux to the first wall. The most comprehensive system has been fielded on TFTR using poloidally and toroidally displaced probes, fixed to the first wall, plus a radially scannable probe from the outer midplane [12,13]. With this system, the escaping flux is resolved in time, energy and pitch angle (angle sustained by the particle along the local magnetic field), allowing a full mapping of the particle orbits in physical and phase space. This technique has also been extended to other facilities.

In plasma conditions where the single particle confinement of energetic ions (*e.g.*, first-orbit) is partial, where passing particles are normally confined but not necessarily for all trapped particles, the diagnosis of escaping (*e.g.*, lost) flux can be extremely powerful to identify and quantify anomalous loss or diffusive mechanisms. The first group of diffusion or loss processes include deviations from non-asymmetric conditions which are externally imposed, such as the TF ripple, error fields or possibly resonant magnetic perturbations [more recently used for edge localized mode (ELM) control]. The potency of ripple diffusion, either through stochastic ripple diffusion or trapping has been long established and verified experimentally [14], see for example Fig. 3. The second group includes non-axisymmetric effects internally produced such as MHD activity [such as neoclassical tearing mode (NTM), fishbones, sawteeth, and others]. This group also includes Alfvén eigenmode activity, a category of modes which has the critical particularity of being driven by energetic ions which in turn can lead in their expulsion or loss. Such activity is a good example of collective effects. In presence of MHD activity the CFPs (and other energetic particles) which possess very high velocities (and short transit times), usually sample a "frozen" perturbed magnetic equilibrium and experience additional transport, especially when the mode is particularly strong and perturbations large compared to the particle's gyroradius or orbit shift.

A third category of perturbations, micro-turbulence, also internal in nature, has the potential for increased transport, in a process similar to the anomalous transport observed for thermal ions. However, experiments done on TFTR [15] has shown that this mechanism be largely inoperative for CFPs, an indication that large gyroradii particles (*i.e.*, CFPs) may be unaffected through averaging of micro-turbulence over a gyroradius and gyroperiod.

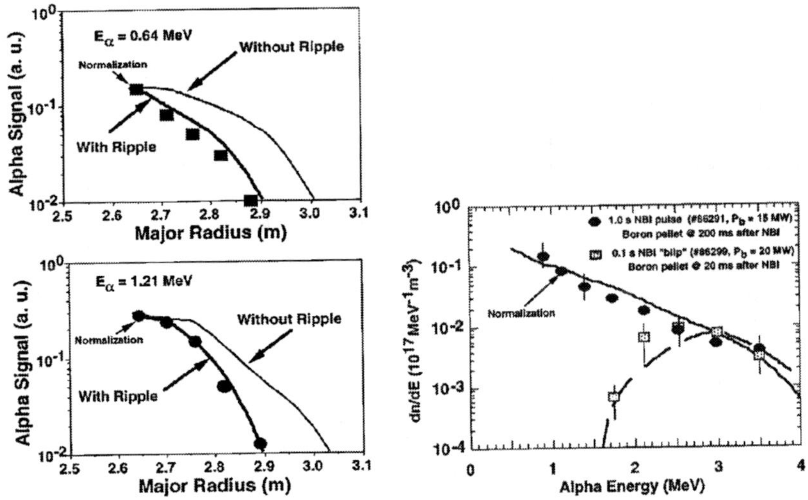

Fig. 3. Measurements of DT α populations as a function of major radius and energy in TFTR. The PCX measured radial profiles (left) agree well with modeling which includes stochastic ripple diffusion. The PCX measured alpha spectrum (right) shows a classic thermalization of alpha particles in the core of MHD quiescent discharges. [Reprinted from S.S. Medley *et al.*, *Nucl. Fusion* **38**, 1283 (1998).

Similar measurements of fast ion loss to the first wall will be more difficult in burning plasma experiments. The current profile and plasma size will prevent any first-orbit losses to the first wall of particles originating from the plasma core. To the first order, this means that any CFP transport mechanism acting on energetic particles will not be immediately visible on the first-wall. Since energetic particles drift away from a magnetic flux surface primarily in the toroidal direction, they will approach the outer wall surface very gradually (toroidal drift » poloidal drift » radial drift). They are likely to impact a leading edge (if present) located on the outer wall, urging the need for a complete IR camera coverage of the outer wall surface of a burning plasma experiment. Any direct detection and identification of lost particles will consequently be quite arduous, since the sensor would be required to be located very near the surface. In addition, broad trenches would be required to be cut in the blankets to allow escaping particle flux to reach the sensor. Similar arrangements were made necessary on TFTR as shown in Ref. [14]. The direct detection of ripple trapped losses has the opposite problem that the drifts are predominantly vertical and that for all purposes, a sensor would shadow itself (when including shielding and casing).

On the other hand, the measurement of confined CFP populations serves more than to validate, complement or replace loss measurements as described above. The primary goal of such measurement will reside in its ability to diagnose the temporal, spatial and phase space distribution of CFPs (*e.g.*, alphas) and their role in self-heating (deposit their full energy back to the plasma). The spatial and phase space (velocity distribution) information is important in understanding the drive mechanism of Alfvén eigenmode activity and in turn, how such activity affects the energetic ions themselves. Monitoring of the slowed down alpha particle population (also known as helium ash) is also a critical aspect in preventing large accumulation in the core, which would dilute the fuel and affect burning conditions.

A variety of techniques have been proposed and tested. They include spectroscopic techniques, such as alpha-CHERS [16]. A second category is based on scattering techniques, which at small angles, uses laser probing [17] or at large angles, microwaves [18]. Other techniques are based on nuclear detection of fuel ion fusion accelerated in the knock-on process [19]. Another category includes neutral particle detection [20], using neutral beams or pellet injection as a source of neutrals. An example of its application on TFTR is shown in Fig. 3. In this case, lithium pellets were injected neutralizing alpha particles which were then detected outside the torus. Energy and spatial distributions were obtained showing classical confinement, with TF ripple effects taken in account.

APPLICATIONS TO ITER

In ITER, a series of diagnostics are proposed for the measurements of fusion products. The selection of measurements and associated techniques corresponds to carefully studied requirements for control and physics understanding, coupled to severe space and environmental (*e.g.*, radiation) constraints [21,22]. The techniques proposed and their role is summarized in Table 1.

TABLE 1. Roles of Fusion Products Measurements, their Corresponding Measurement and Proposed Techniques for ITER [21,22].

Role	Parameter Measured	Proposed Technique
Fusion power	Neutron yield	Fission chambers (external and internal)
Alpha source profile	Neutron emission profile	Radial and vertical neutron cameras
Ion temperature	Neutron emission profile and spectrum	Radial and vertical neutron cameras, spectrometers
Isotopic ratio (n_D/n_T)	2.45 and 14.1 MeV neutron flux, energy spectrum	Neutron spectrometer
Loss of alpha particles	Heat flux to first wall	Infra-red cameras
Confinement of energetic particles (incl. alphas)	Energetic ion distribution function	Collective Thomson scattering
	Energy resolved Gamma emission profiles	Gamma cameras and spectrometer
	Ion knock-on tail (through neutron emission)	High energy neutron spectrometer

SUMMARY

With their ability to diagnose key features of burning plasmas, and because of their inherent compatibility with the environment (although some techniques are not), fusion product measurements are a fundamental part of the diagnostic set required for the optimization and viability of magnetically confined fusion reactors. Overall, measurements based on neutron and gamma emission are crucial in understanding and diagnosing conditions leading to optimal fusion production. When these measurements include profile and energy distribution, one can infer and optimize the composition, distribution and population of thermal and energetic ions.

Measurements based on the detection of confined charged fusion products (CFPs) are critical in diagnosing their important role in self-heating (through alpha particles for example). In addition to profile and energy distributions a key element of information is their phase space distribution, which influences the stability of the plasma and their sensitivity to loss mechanisms, either internally or externally driven. In addition, the detection techniques are sufficiently broad to include also the study of the confinement of the energetic fuel ions, which contribute significantly in the expected fusion production.

Finally, it is expected that the conditions expected on a DEMO reactor will require a reduced set of diagnostics, including fusion product measurements, while taking additional duties related to burn control. This set of diagnostics will require a dedicated vetting process on burning plasma experiments such as ITER.

ACKNOWLEDGMENT

Work supported by the U.S. Department of Energy under DE-FC02-04ER54698. We wish to acknowledge the valuable contributions of K.M. Young, W.W. Heidbrink, A. Murari, D.S. Darrow, R.K. Fisher, C.L. Fiore, M. Sasao, M. Van Zeeland in preparing this article.

REFERENCES

1. W. W. Heidbrink, *Nucl. Fusion* **24**, 636 (1984).
2. C. L. Fiore *et al.*, *Phys. Plasmas* **8**, 2023 (2001).
3. C. W. Barnes *et al.*, *Nucl. Fusion* **38**, 597 (1998).
4. J. D. Strachan *et al.*, *Nucl. Fusion* **36**, 1189 (1996).
5. S. Conroy *et al.*, *Nucl. Fusion* **28**, 2127 (1988).
6. T. Nishitani *et al.*, *Plasma Phys. Contr. Fusion* **38**, 355 (1996).
7. J. M. Adams *et al.*, *Nucl. Instrum. Methods Phys. Res. A* **239**, 277 (1993).
8. M. Ishikawa *et al.*, *Rev. Sci. Instrum.* **77**, 10E706 (2006).
9. A. Murari *et al.*, *Nucl. Fusion* **45**, S195 (2005).
10. V. G. Kiptily *et al.*, *Nucl. Fusion* **45**, L21 (2005).
11. W. Heidbrink, *Rev. Sci. Instrum.* **56**, 1098 (1985).
12. S. J. Zweben *et al.*, *Nucl. Fusion* **35**, 893 (1995).
13. R. L. Boivin *et al.*, *Nucl. Fusion* **33**, 449 (1993).
14. S.J. Zweben *et al.*, *Nucl. Fusion* **35**, 1445 (1995).
15. S. J. Zweben *et al.*, *Nucl. Fusion* **31**, 2219 (1991).
16. G. McKee *et al.*, *Phys. Rev. Lett.* **75**, 649 (1995).
17. T. Kondoh *et al.*, *Rev. Sci. Instrum.* **77**, 10E505 (2006).
18. S. B. Korsholm *et al.*, *Rev. Sci. Instrum.* **77**, 10E514 (2006).
19. R. K. Fisher *et al.*, *Nucl. Fusion* **34**, 1291 (1994).
20. S. S. Medley *et al.*, *Nucl. Fusion* **38**, 1283 (1998).
21. M. Sasao *et al.*, Plasma Phys. Contr. Fusion **46**, S107 (2004).
22. A. V. Krasilnikov *et al.*, Nucl. Fusion **45**, 1503 (2005).

Absolute Neutron Emission Measurement in Burning Plasma Experiments

Takeo Nishitani*, Masao Ishikawa*, Takashi Kondoh*, Yoshinori Kusama*, Keisuke Asai† and Mmamiko Sasao¶

**Japan Atomic Energy Agency, Naka, Ibaraki-ken, 311-0193, JAPAN*
†Nagoya University, Chikusa-ku, Nagoya, 464-8603, JAPAN
¶Tohoku University, Aoba-ku, Sendai, 980-8579, JAPAN

Abstract. The absolute measurement of neutron emission rate from the whole plasma is a very important diagnostics as a fusion power monitor in fusion experimental devices with D-T or D-T operations. Here measurement techniques of time-resolved and time-integrated absolute neutron emission on the present tokamaks and ITER are reviewed. In the present tokamaks, fission chamber installed outside the vacuum vessel are very popular in the absolute neutron emission rate measurement. As well as conventional neutron monitors installed outside the vacuum vessel, in-vessel neutron monitors using micro fission chamber are proposed for the absolute neutron emission rate measurement in ITER. The neutron activation system provides time-integrated measurements of the total neutron yield with high accuracy by using well known neutron reaction cross sections, which is useful to maintain a robust measure of fusion energy production with stability and wide dynamic range. The calibration of the relation between the neutron emission rate in the whole plasma and the output of neutron monitor is the most important issue in the absolute neutron emission rate measurements. The calibration of the neutron detectors has been performed by moving a neutron source such as a ^{252}Cf neutron or a small accelerator-based neutron generator. For the calibration in ITER, the neutron generator with neutron emission rate of 10^{11}s^{-1} or stronger is required to obtain high accuracy.

Keywords: Neutron Emission Rate, Neutron Monitor, Transport Analysis, Fission Chamber, Calibration, Burning Plasma Diagnostics, ITER.
PACS: 52.70.-m, 28.52.-s, 52.55.Fa

INTRODUCTION

The absolute measurement of neutron emission rate from the whole plasma is a very important diagnostics as a fusion power monitor in fusion experimental devices with D-T or D-T operations. A neutron emission rate of 1 s^{-1} is equivalent to the fusion power of 2.82×10^{-12} W in the D-T plasma and 1.06×10^{-12} W in the D-D plasma, respectively. So the neutron emission rate will be used as a feedback parameter for the burning control in ITER or other burning plasma devices. On JT-60U, a feedback control by the neutron emission rate was already demonstrated for the operation optimization such as minor disruption avoidance or keeping steady state plasma condition, where the neutral beam (NB) power was modulated as an actuator [1]. Now a feedback control by the neutron emission rate is popular technique in other tokamaks [2]. The absolute neutron emission rate is important parameter even in the transport analysis of D-D or D-T plasmas. The neutron emission rate is analyzed by a transport

CP988, *Burning Plasma Diagnostics*, edited by F. P. Orsitto, G. Gorini, E. Sindoni, and M. Tardocchi

code by using measured basic parameters such as ion and electron temperatures, and the electron density. Comparison of the calculated neutron emission rate with the measured one permits refinement of the input basic parameters and justifies the analysis results. These analyses provide the fraction of neutron emission process i.e. the thermonuclear, beam-plasma and beam-beam components.

This paper reviews measurement techniques of time-resolved and time-integrated absolute neutron emission on the present tokamaks and the applications. Also plan of the absolute neutron emission on ITER is presented.

TIME RESOLVED MEASUREMENT OF NEUTRON EMISSION

Time-resolved Neutron Emission Measurements in Present Devices

In fusion experiments with additional heating such as a NB injection, the neutron emission rate changes 10^4-10^6 in the time scale of fast ion slowing down time (~100 ms). Therefore, the detector of the neutron emission rate has to have a wide dynamic range and fast response time without noises from hard X-ray and gamma-rays. For this requirement, a combination of fast response neutron detector operated in pulse counting mode with different sensitivity or a detector operated in DC like mode such as the current mode is suitable. A BF_3 proportional counter, a ^{3}He proportional counter and a fission chamber are the most popular as the neutron detector. In the present tokamak experiments, fission chamber are employed for the neutron emission rate measurement. The fission chamber is a ionization chamber with electrodes coated with fissile material such as ^{235}U or ^{238}U, where the ionization is caused by a fission fragment with the kinetic energy of 50-100 MeV. Because the energy of background hard X-ray or gamma-ray is less than 10 MeV in the fusion devices even in ITER, the fission chamber can eliminate hard X-rays or gamma-rays by conventional pulse height discrimination technique. Typical pulse width is less than 100 ns. The fission chamber can be operated simultaneously in pulse counting mode, Campbelling (Mean Square Voltage) mode [3,4] and current mode. BF_3 proportional counter, a ^{3}He proportional counter utilize ^{10}B(n,α)^{7}Li and ^{3}He(n,p)T reactions, respectively, whose Q values are 2.78 MeV and 0.77 MeV, respectively. Because sensitivities of those proportional counters are about 100 times higher than that of same size ^{235}U fission chamber, BF_3 and ^{3}He proportional counters are useful for low neutron yield experiments. In the use of those detectors, the gamma-ray shielding should be considered. Recently, a CVD diamond detector [5] was tested as a 14 MeV neutron detector during the Trace Tritium Experiment in JET, which showed good linearity to the neutron emission rate measured with fission chambers. However, the dynamic range is limited because it is operating in pulse counting mode.

Figure 1 shows the schematics of the JT-60U neutron monitor using ^{235}U and ^{238}U fission chambers [6]. The detectors are all long cylinders in geometry. The ^{235}U detector is surrounded 50 mm thick polyethylene of moderator, and 1 mm thick cadmium of thermal neutron shield, whose neutron sensitivity is estimated to be almost constant in the energy range of 0.55 eV to 2.5 MeV. The ^{238}U detector is surrounded by 50 mm thick lead of gamma shield only. The detectors were oriented

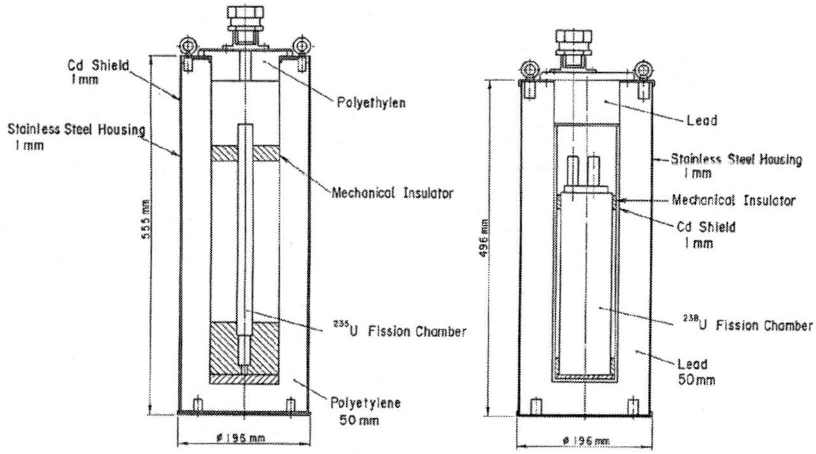

FIGURE 1. Schematic diagram of the JT-60U neutron monitor using ^{235}U and ^{238}U fission chambers [5].

vertically and placed just beyond the outer diameter of the toroidal field coils to reduce the effects of the magnetic fields. Those fission chambers are operated in both pulse counting mode and Campbelling mode. The former is suitable for low count rate less than 10^6 cps and the latter is for counts higher than 10^5 cps. Three pairs of ^{235}U and ^{238}U fission chambers are located on the torus mid-plane in JT-60U. Similar neutron mission rate diagnostics have been employed in JET [7] and TFTR [8].

In the present tokamaks with D-D or D-T plasma, the neutron emission rate is one of the most important parameter of the plasma operation. The neutron mission rate is commonly used as a feedback control parameter aiming at steady-state operation. Figure 2 shows the example of the feedback control of the neutron emission rate in JET [2], where active control of an internal transport barrier was demonstrated by using the feedback control of the neutron emission rate with NB power combined with the feedback control of the normalized electron temperature gradient with ICFH power.

The neutron emission rate is also important parameter in the transport analysis of D-D or D-T plasmas. The neutron emission rate is analyzed by a transport code such as TRANSP [9], by using measured basic parameters such as ion and electron temperatures, and the electron density. Comparison of the calculated neutron emission rate with the measured one permits refinement of the input basic parameters and justifies the analysis results. Total neutron emission rate consists of the thermonuclear, beam-plasma and beam-beam components. These analyses provide the fraction of each component. Figure 3 shows the measured and analyzed neutron emission rate in the JET D-T experiment with NB and ICRH [10]. This discharge achieved the peak neutron emission rate of 5.8×10^{18} s^{-1}, which corresponded to the fusion power of 16 MW. Calculated values agreed well to the measured neutron emission rate. It was found that fractions of the thermal-thermal reactions, the beam-thermal and beam-beam reactions are 57 %, 42 % and 1 %, respectively at 13.3 s.

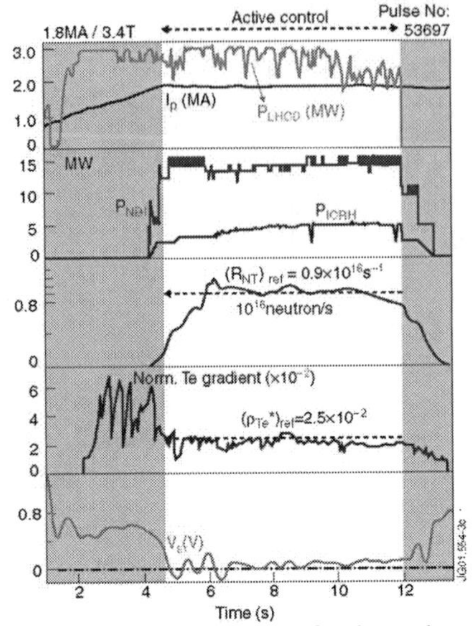

FIGURE 2. Active control of an internal transport barrier using the feedback control of the neutron emission rate with NB power combined with the feedback control of the normalized electron temperature gradient with ICFH power [2].

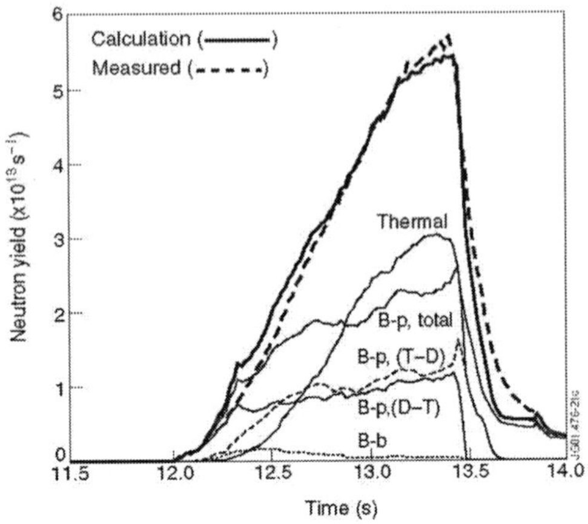

FIGURE 3. Measured and analyzed neutron emission rate in the JET D-T experiment with NB and ICRH [10].

Time-resolved Neutron Emission Measurements in ITER

In ITER, conventional neutron monitors installed outside the vacuum vessel and in-vessel neutron monitors using micro-fission chamber are proposed for the neutron emission rate measurement. Also the neutron emission rate can be derived from the neutron emission profile measurements employing a tomographic inversion technique.

Because ITER has a thick blanket and vacuum vessel, neutron detectors outside the vacuum vessel may not measure the neutron emission with sufficient accuracy. The in-vessel neutron monitor using micro-fission chambers, which are pencil size gas counters with fissile material inside shown in Fig.4, is proposed. In the beginning of the design work, micro fission chamber will be installed between the vacuum vessel and the shielding blanket module at several poloidal angles in order to reduce the effect of the plasma position and neutron emission profile on the total neutron emission rate [11]. Recently, it was found by neutron Monte Carlo calculations that the average outputs of the micro-fission chambers at two poloidal locations on the outboard vacuum vessel surface is insensitive to the changes in the plasma position and the neutron source profile [12]. Figure 5 shows the location of the micro-fission chambers and the cabling route on the vacuum vessel surface. The sheath of the micro fission chamber will be welded to the vacuum vessel which is cooled by water coolant. A heat transfer analysis indicates that the micro-fission chambers can be cooled to be under the limit of the operation temperature by the heat conduction to the vacuum vessel [13].

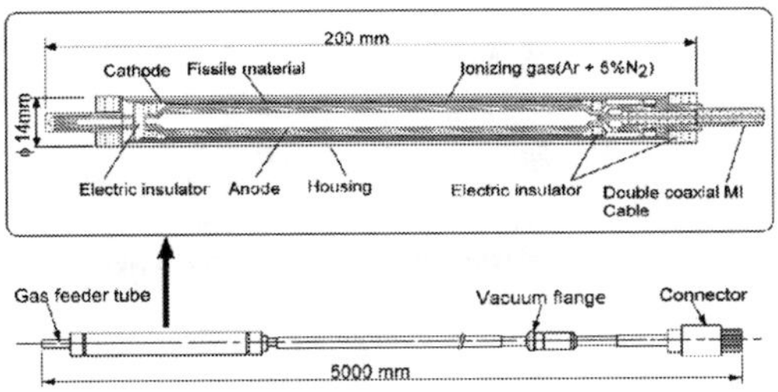

FIGURE 4. Schematic diagram of the micro-fission Chamber [14].

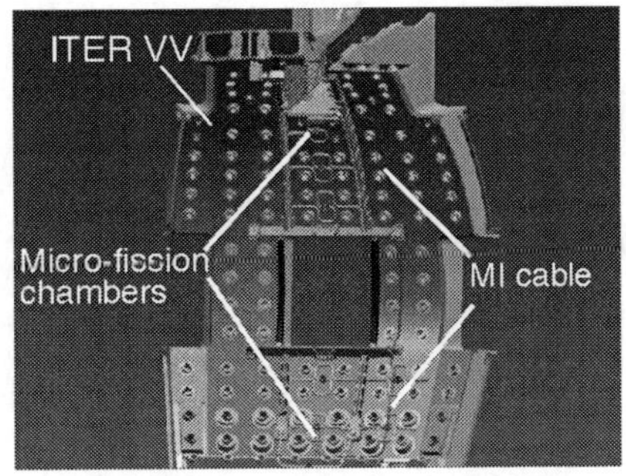

FIGURE 5. Location of the micro-fission chambers and the cabling route on the surface of the ITER vacuum vessel.

In the iv-vessel neutron monitor, micro-fission chambers must be operated in the high vacuum, high magnetic field and high radiation circumstances inside the vacuum vessel. A prototype micro-fission chamber has been developed for the ITER application in order to check the availability under the ITER condition such as high vacuum, neutron and gamma radiation, high temperature, and mechanical vibration [14]. The prototype micro-fission chamber has been tested at JT-60U in the high toroidal magnetic field, where it is operating without any trouble [15]. By those tests, a good prospect of the applicability of the micro fission chamber to ITER has been obtained.

A conventional neutron flux monitor outside the vacuum vessel is also considered in ITER. However, a neutron flux outside the vacuum vessel is not sufficient to get required accuracy. A neutron flux monitor is proposed by using normal size fission chambers to be installed in the equatorial port of ITER [16]. A moderator material for ^{235}U fission chamber has been investigated to get flat energy response of detectors. Here graphite and/or beryllium are employed as moderator, which materials are stable in ITER relevant temperature in an equatorial port as shown in Fig. 6. Based on the nuetronics calculations, a fission chamber with 200 mg- several g of ^{235}U is adopted for the neutron flux monitor. Three detectors are mounted in a stainless steel housing with moderation material. Three flux monitor modules will be installed in equatorial ports; the others are for D-D and calibration operation, and another is for D-T operation. The module for the D-D operation and the calibration are installed just outside the port plug in the equatorial port. The modules for the D-T operation are installed just behind the shielding block in the port.

Alternative neutron flux monitor to be installed inside a divertor cassette is proposed (see Fig. 7) [17]. This neutron flux monitor is based on the combined application of two ^{238}U and two ^{235}U fission chambers. The amount of ^{235}U and ^{238}U fissile material will be in the range of a few milligrams–few grams to provide an adequate overlap of the linear operation ranges and allow for the fission chamber cross calibration to one another. The ^{235}U fission chamber will be surrounded by a water moderator to provide a flat energy response. The ^{238}U fission chamber will be surrounded by a B_4C thermal neutron shield in order to prevent Plutonium breeding

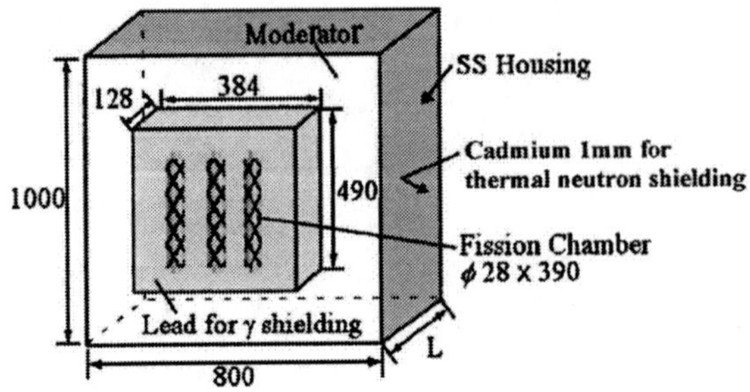

FIGURE 6. Schematic view of the neutron flux monitor module to be installed in the equatorial port [17].

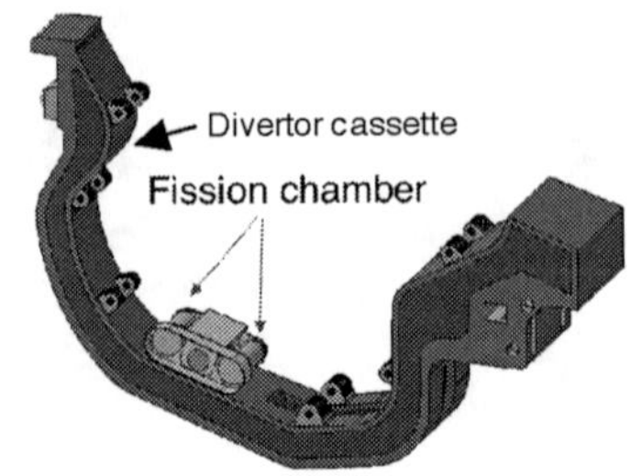

FIGURE 7. Schematic view of the divertor neutron flux monitor.

through the thermal neutron absorption reactions by ^{238}U. 'Blank' detectors will be included in the detector module to identify gamma-rays and noises.

TIME INTEGRATED MEASUREMENT OF NEUTRON EMISSION

The time-integrated measurements of the total neutron yield can be provided by a neutron activation technique with high accuracy based on well-known neutron reaction cross sections. Metal or other solid sample will transferred an irradiation point installed near the plasma by a pneumatic operated sample transfer system before plasma discharge, and will return to the counting station, where induce activities of samples will be measured. If the relationship between the neutron flux at the irradiation point and the neutron emission rate in the whole plasma is well known, the time-integrated neutron emission can be derived form the induced activities. JET [18] and TFTR [19], have multiple irradiation positions with different poloidal locations. JT-60U [20] has a single position of the irradiation end in a horizontal port. Also ITER will employ such neutron activation system [21], where irradiation points will be installed in upper ports, equatorial ports and divertor ports. The coaxial pneumatic transport tubes will have 2 cm outer diameter and 1.1 cm inner diameter and will carry 1cm diameter capsules. Helium will be uses as a pneumatic propellant to avoid the activation of the propellant gas. Candidate materials are; Fe, Al and Ti foils that appear to allow for reasonable sensitivity without creating too intense source. Si could be useful at low power. These samples can provide the neutron source strength of 10^{17}-10^{21} n/s with accuracy of 10 % in every few 100 seconds.

The primary goals of the neutron activation system are to maintain a robust relative measure of fusion energy production with stability and wide dynamic range and to enable an accurate absolute calibration of fusion power. Further more, information on the neutron energy spectrum can be obtained by using with several foils with different threshold energy reactions. Fluxes of D-D and D-T neutrons are easily measured with such multi-foil technique. From the ratio of D-D and D-T neutron fluxes, a triton burn-up ratio in D-D plasmas and a deuterium/ tritium fuel ratio (n_d/n_t) in D-T plasmas can be derived. A sophisticated multi foil activation technique using an unfolding code such as SAND-II can provide a neutron spectrum at the irradiation point, which is useful for valuable test of the neutron transport methodology from the fusion neutronics point of view. Also similar multi-foil activation system will be planed in the ITER Test Blanket Module (TBM) [22] in order to measure the neutron spectra in TBM, because the tritium breeding performance is sensitive to the neutron spectrum.

ITER plans other innovative neutron irradiation system employing a flowing water activation based on the $^{16}O(n,p)^{16}N$ reaction with reasonable temporal resolution. The neutron activation system with water flow [23] consists of a closed water loop laid between the first wall/blanket region and a remote low-background counting station. By measuring the gamma rays from ^{16}N with a gamma ray detector placed beside the water loop, it is possible to determine a D-T neutron yield continuously. The activated water will stay in a reservoir tank to diminish the activity and will be transferred to the first wall/blanket region again. The cross-section of the $^{16}O(n,p)^{16}N$ has a high threshold energy of 10.24 MeV. Therefore, the neutron activation is sensitive for almost non-scattered neutrons. A ^{16}N nucleus emits 6.13 (68.8%) and 7.12 (4.7%) MeV gamma-rays with a half-life of 7.13 seconds. The irradiation end should be as near to the plasma as possible. The proof of principle experiment of the flowing water activation measurement was carried out at the FNS facility [24] of JAEA, which predicts the flowing water activation system can measure the neutron source strength of 10^{16}-10^{21} n/s with the time resolution of 100 ms or less on ITER.

CALIBRATION OF THE NEUTRON EMISSION MEASUSREMENT

The absolute calibration of the relation between the neutron emission rate in the whole plasma and the output of neutron monitor is the most important issue in the absolute neutron emission rate measurements. The calibration is rather difficult because the neutron source is distributed in the plasma which is surrounded with many complicated structures such as first wall, shielding blankets in case of ITER, vacuum vessel, poloidal and toroidal coils. Many efforts and time are devoted to the calibration at many tokamaks [25]. The calibration of the neutron detectors has been performed by moving a neutron source such as a ^{252}Cf neutron source for D-D operation [6]. For D-T operations, ^{252}Cf neutron source is not suitable, because the average energy of 2.1 MeV is much lower than 14 MeV. A compact D-T neutron generator was used for the D-T operation of TFTR [8]. In ITER, also *in-situ* calibration using compact D-T neutron generator with a neutron intensity of 10^{11} s^{-1} or stronger is planed for the in-vessel and ex-vessel neutron monitors [26].

In the time-integrated neutron mission measurement, direct calibration using a neutron source is very difficult because the intensity of the available neutron source is too weak to activate the irradiation sample sufficiently. If the irradiation point is installed near the plasma, the relation between the neutron emission in the whole plasma and the induced activity of the sample can be calibrated by a neutron Monte Carlo calculation with a precise modeling of the machine.

SUMMARY

Time-resolved neutron emission is already used as a key parameter for the control. As well as conventional neutron monitors installed outside the vacuum vessel, in-vessel neutron monitors using micro fission chamber are proposed for the absolute neutron emission rate measurement in ITER. The neutron activation system provides time-integrated measurements of the total neutron yield with high accuracy by using well known neutron reaction cross sections, which is useful to maintain a robust measure of fusion energy production with stability and wide dynamic range. ITER plans conventional solid activation measurement using pneumatic sample transfer system and a flowing water activation system. The calibration of the relation between the neutron emission rate in the whole plasma and the output of neutron monitor is the most important issue in the absolute neutron emission rate measurements. ITER plans in-situ calibration using the neutron generator with neutron the intensity of $10^{11}s^{-1}$ or stronger.

ACKNOWLEDGMENTS

A part of this report was prepared as an account of work undertaken within the framework of ITER Transition Arrangement (ITA).

REFERENCES

1. Y. Neyatani, T. Fukuda, T. Nishitani, et al., *Fusion Eng. Design* **36**, 429-433 (1997).
2. D. Mazon, , X. Litaudon, D. Moreau et al., *Plasma Phys. Control. Fusion* **44**, 1087-1107 (2002).
3. Y. Endo, T. Ito and E. Seki, *IEEE Trans. Nucl. Sci.* **NS-29**, 714-716 (1982).
4. A.C. England, H.W. Hendel and E.B. Nieschmidt, *Rev. Sci. Instrum.* **57**, 1754-1756 (1986).
5. A. Murari, L. Bertalot, S. Conroy, et al., *Nucl. Fusion* **45**, S195-S202 (2005).
6. T. Nishitani, H. Takeuchi, T. Kondoh, et al., *Rev. Sci. Instrum.* **63**, 5270-5278 (1992).
7. O.N. Jarvis, *Rev. Sci. Instrum.* **63**, 4511-4516 (1992).
8. D.L. Jassby, Cris W. Barnes, L.C. Johnson, et al., *Rev. Sci. Instrum.* **66**, 891-893 (1995).
9. R. V. Budny, M. G. Bell, H. Biglari, et al., *Nucl Fusion* **32**, 429-447 (1992).
10. O.N. Jarvis and S. Conroy, *Plasma Phys. Control. Fusion* **44**, 1611-1687 (2002).
11. T. Nishitani, L.C. Johnson, K. Ebisawa, et al., *Rev. Sci. Instrum.* **70**, 1141-1144 (1999).
12. T. Nishiani, K. Ebisawa, C. Walker and S. Kasai, "Design of in-vessel neutron monitor using micro fission chambers", JAERI- Tech 2001-066, Japan Atomic Energy Research Institute, 2001.
13. T. Nishitani, M. Yamauchi, M. Izumi, et al., *Fusion Eng. Design* in Press.
14. M. Yamauchi, T. Nishiani, K. Ochiai, et al., *Rev. Sci. Instrum.* **74**, 1730-1734 (2003).
15. T. Hayashi, T. Nishitani and M. Ishikawa, *Rev. Sci. Instrum.* **75**, 3575-3577 (2004).
16. A.V. Krasilnikov, M. Sasao, Yu.A. Kaschuck, et al., *Nucl. Fusion* **45**, 1503-1509 (2005).
17. K. Asai, T. Iguchi, K. Watanabe et al., *Rev. Sci. Instrum.* **75**, 3537-3539 (2004).
18. M. Pillon, M. Martone, O. N. Jarvis, and J. Källne, *Fusion Technol.* **15**, 1420-1429 (1989).
19. C. W. Barnes, A. R. Larson, and A. L. Roquemore, *Fusion Technol.* **30**, 63-72 (1996).
20. M. Hoek, T. Nishitani, Y. Ikeda, A. Morioka, *Rev. Sci. Instrum.* **63**, 885-887 (1995).
21. C. W. Barnes, M. J. Loughlin, and T. Nishitani, *Rev. Sci. Instrum.* **68**, 577-580 (1997).
22. M. Enoeda, M. Akiba, S. Tanaka, et al, *Fusion Eng. Des.* **81**, 415-424 (2006).
23. T. Nishitani, K. Ebisawa, S. Kasai, *Rev. Sci. Instrum.* **74**, 1734-1738 (2003).
24. J. Kaneko, Y. Uno, T. Nishitani, et al., *Rev. Sci. Instrum.* **72**, 809-813 (2001).
25. J.D. Strachan, J.M. Adams, Cris W. Barnes, et al., *Rev. Sci. Instrum.* **61**, 3501-3504 (1990).
26. K. Asai, T. Iguchi, T. Nishitani, et al., *Fusion Eng. Des.* **81**, 1497-1502 (2006).

Neutron Emission Profile And Neutron Spectrum Measurements At JET: Status And Plans

S. Popovichev[1], L. Bertalot[2], F. Belli[6], G. Bonheure[3], S. Conroy[4], A. Hjalmarsson[4], G. Ericsson[4], B. Esposito[6], J. Källne[4], V. Kiptily[1], D. Marocco[6], A. Murari[5], M. Riva[6], B. Syme[1], M. Tsalas[7], M. Weiszflog[4] and JET EFDA contributors *

JET-EFDA Culham Science Centre, OX14 3DB, Abingdon, UK
[1]EURATOM/UKAEA Fusion Association, Culham Science Centre, Abingdon, UK
[2]ITER-JWS, Cadarache, 13108, St Paul lez Durance, France
[3]Association Euratom-Belgian state, Partners in the TEC, ERM-KMS, Brussels, Belgium
[4]INF, Uppsala University, EURATOM-VR Association, Uppsala, Sweden
[5]Consorzio RFX-Associazione EURATOM-ENEA per la Fusione, I-35127 Padova, Italy
[6]Associazione EURATOM/ENEA sulla Fusione, Frascati, Rome, Italy
[7] Association EURATOM-Hellenic Republic, NCSR "Demokritos", Agia Paraskevi Attica, Greece,
** See the Appendix of M.L. Watkins et al., Fusion Energy 2006 (Proc. 21st Int. Conf. Chengdu, IAEA, (2006)*

Abstract. In large tokamak devices, such as JET, TFTR, JT-60U and ITER, a neutron emission profile monitor is considered as one of the principal diagnostic tools, firstly in providing information to characterize the fusion source in terms of its location, shape, intensity and secondly in developing an understanding of fast ion physics in tokamak plasmas. The JET neutron profile monitor is one of the most advanced such instruments currently operating. It enables 2-D imaging of the non-circular plasma neutron source through a tomographic deconvolution of line-integrated measurements. Historically it was usually assumed that the fusion neutron source is a function of the magnetic flux surfaces (MFS), but recent JET results have clearly demonstrated the influence of non-uniform fast particle populations on the neutron emission profile.

Neutron energy spectroscopy also carries valuable information about fast ion behaviour and the fusion and nuclear reactions taking place in the plasma. Spectrum measurements of neutrons will be particularly important for the up-coming JET experiments with a Be wall. The neutron spectroscopy capabilities at JET were recently enhanced by several instruments: a new time-of-flight spectrometer optimized for high count rates (TOFOR), compact neutron spectrometers based on organic scintillator (NE213) and an upgrade of the Magnetic Proton Recoil system (MPRu). In recent JET campaigns, a prototype digital pulse shape discriminator (DPSD) has been successfully tested on one central chord of the JET neutron profile monitor and spectral information has also been obtained. Neutron spectrometry results obtained during the last JET experimental campaigns illustrate the measurement capabilities and indicate their potential use on ITER. Several projects are in progress to further develop such diagnostic capabilities and these are briefly summarized.

Keywords: Plasma diagnostics, neutrons, Tokamak.
PACS: 52.70.-m, 52.55.Fa, 29.30.Hs

CP988, *Burning Plasma Diagnostics*, edited by F. P. Orsitto, G. Gorini, E. Sindoni, and M. Tardocchi
2008 American Institute of Physics 978-0-7354-0507-3/08/$23.00

INTRODUCTION

Diagnosis of the plasma neutron source is of major importance for characterization of the fusion burning process and for determination of the performance of next step burning plasma devices, such as ITER. JET is the machine closest to the ITER parameters and it provides a unique tritium operation capability making it a good test bed for development of the ITER-relevant plasma diagnostics. The neutron emission from the JET tokamak is investigated using an extensive set of diagnostics, permitting the instantaneous neutron yield, the radial profile of the neutron emission and neutron energy spectra to be studied.

The JET neutron profile monitor is a unique instrument among the diagnostic systems available at large fusion devices. Section 1 of this paper briefly reviews the design of this device and presents some recent results. The necessity of 2-D neutron emission measurements arises from the fact that in the presence of fast ion component, contours of equal neutron emissivity are not necessarily coincident with magnetic flux surfaces (MFS). At JET, this disconjunction has been observed in recent DD and trace tritium experiments during periods of ion cyclotron resonance heating, neutral beam injection, sawtooth oscillations and other MHD instabilities. To improve the performance capabilities of the neutron profile measurements, the replacement of analogue pulse shape discrimination electronics with new digital systems has been started under a JET enhancement programme.

Significant attention is also paid on JET to developing new diagnostics for energetic particle studies. Neutron energy spectroscopy carries valuable information about fast ion behaviour and the fusion and other nuclear reactions taking place in the plasma. Recently, the major neutron spectroscopy development efforts at JET have been focused on the time-of-flight spectrometer TOFOR and on two different kinds of proton-recoil spectrometers: the Magnetic proton recoil (MPR) spectrometer and the organic liquid scintillator NE213. Neutron spectrometry results obtained lately at JET are provided in Section 2 to illustrate the measurement capabilities and indicate their potential use on ITER. Several projects are in progress to further develop such diagnostic capabilities and these are briefly summarized in Section 3.

1. NEUTRON EMISSION PROFILE MEASUREMENTS AT JET.

The primary function of a neutron profile monitor is to measure the neutron emissivity profile over a poloidal plasma cross-section using line-integrated measurements along a number of chords (lines-of-sight). Knowledge of the absolute detection efficiency of the system enables one to obtain the total instantaneous neutron yield from the tokamak, and this can independently supplement the results derived from primary methods of yield measurement, such as neutron activation or neutron flux monitors. Therefore a well-designed neutron profile monitor is potentially the most important among the other neutron diagnostics. The JET neutron emission profile monitor is one of the most advanced such instruments currently operating.

The JET Neutron Profile Monitor. The upgraded neutron profile monitor was installed in the mid 90's and since then has been used routinely, often playing a vital role in the physics understanding of fast ion population effects in the plasma and of the impact of different heating scenarios. The instrument was well described in several papers [1, 2], therefore only a short description will be given here.

The JET neutron emission profile monitor consists of two concrete shields ("cameras") which each providing a fan-shaped array of collimators. One camera views the plasma horizontally (10 channels), and the other vertically (9 channels). In each camera, the collimation can be remotely set up by the use of two pairs of rotatable steel cylinders with a choice of two aperture sizes. This permits one to tune the count rates in the detectors by a factor of 20 [3], so the detectors are able to cope with a neutron emission rate up to 10^{19} s^{-1}, making the instrument equally suitable for both DD and DT regimes of operation. The plasma coverage by neutron profile arrays is adequate for 2D tomography, although the spatial resolution is rather poor (~8 cm at the plasma centre for a collimator aperture of 10mm by 10mm when neighboring channels are 15–20 cm apart). Each channel is equipped with a set of three different detectors: i) a NE213 liquid scintillator with Pulse Shape Discrimination (PSD) electronics to measure 2.5 MeV, 14 MeV neutrons and gamma-rays simultaneously; ii) a Bicron BC418 plastic scintillator for 14 MeV neutron detection only with very low sensitivity to gamma radiation; iii) CsI(Tl) photodiodes known as the 'Fast Electron Bremsstrahlung (FEB)' detectors, for spatial measurements of Hard X rays and gamma emission with energies from 0.2 up to 6 MeV.

As mentioned above, each NE213 detector channel is equipped with quite complex analogue signal processing electronics with pulse shape discrimination capabilities to allow neutrons to be separated from gamma-rays and to provide the necessary energy discrimination, so that DD neutrons and DT neutrons within set energy windows can be recorded separately. The Bicron scintillators are positioned in front of the NE213 scintillators and record neutrons with energy above 8.5 MeV, i.e. DT neutrons only. Because of their small size they are relatively insensitive to γ-rays with $E\gamma < 10$MeV and the PSD complications are not required. The CsI(Tl) gamma detectors allow the JET neutron profile monitor to also act as a gamma-ray profile monitor as is illustrated in [4].

The absolute efficiencies and light outputs response of the two neutron detectors were measured using the Harwell Van de Graaff accelerator [1]. Since all of the detectors view the inner vessel wall of the JET vacuum vessel, extensive Monte Carlo neutron transport calculations are required to evaluate scattering and attenuation effects. This has been done and the profile monitor provides an independent, absolute measurement of the instantaneous neutron yield that agrees with measurements obtained from the neutron activation system to well within the estimated combined errors of about 10-15 % [2, 5].

Neutron Profile Measurements at JET. The JET neutron profile monitor was the key diagnostic for the Trace Tritium Experimental (TTE) campaign (with the introduction of trace amount (<5%) of tritium into D plasmas) in 2003, with over 80% of the experiments relying on its profiles of 14 MeV neutrons. The power of this instrument is demonstrated below by reference to some results recently obtained.

Many JET discharges employ auxiliary ICRH and NBI heating which can lead to anisotropy of fast ion velocity distributions. In such cases, the magnetic flux surfaces can no longer be assumed to be contours of constant neutron emissivity. For example, neutron emission spatial asymmetries have been clearly observed by the vertical camera in the TTE ICRH experiment with fundamental heating of tritium [6]. In this case, the high-energy tritons reacting with the deuterium bulk plasma induce a neutron emission profile which peaks off axis, close to the T cyclotron resonance layer [5, 6].

The effect of the current-hole, a region of negligible current in the core of the plasma with strong reverse shear, on the fast particles distribution was investigated in experiments using NB injected tritons as test particles and monitoring their spatial and temporal evolution via their 14 MeV neutron emission, as observed by the neutron cameras [7]. In the case of on-axis injection into a monotonic current profile, the emission was peaked on the plasma axis. For injection into a current-hole plasma, there was a clear shift in the emission in the radial outward direction [6, 8].

The radial distribution of NB-injected fast ions can be assessed using the neutron emission profile in cases where the thermal neutron yield is negligible. Dedicated experiments were carried out in the 2003 TTE campaign using short tritium beam blips (~300ms) with both on-axis and off-axis trajectories injected into deuterium plasmas to provide a basis for the validation of NB fast ion models. The measured 14MeV-neutron emission showed peaked or flat profiles corresponding to on-axis or off-axis beam injection respectively, in good agreement with TRANSP Monte Carlo simulations for both cases [8].

During 2003 TTE campaign, the neutron profile monitor was an indispensable tool for studies of tritium transport [10], fuel ratio [11], and different heating scenario [7,8,9]. As mentioned above, it also complements the activation system and neutron flux monitors providing an independent estimate of the absolute neutron yield from the plasma. The total 14 MeV neutron yields derived from the silicon diodes and the neutron profile measurements during TTE in 2003 were in an agreement with the profile monitor estimate to within ~8% [5].

Tomographic inversion of the line-integrated measurements has been also carried out to obtain a 2D mapping of the neutron emission strength. It should be noted that the number of lines of sight in the JET neutron profile is well short of what usually is considered needed for reliable tomography reconstruction. Several tomography techniques, including those used successfully for X-ray imaging, have been tested and used at JET for the purpose of neutron and gamma tomography giving clear indications of asymmetries in the poloidal neutron profiles [12, 13].

2. NEUTRON SPECTRUM MEASUREMENTS AT JET.

At JET, neutron spectrometry has been used since its early days and always rated as an essential tool for obtaining valuable information on the fuel ion temperature and composition and, in the case the plasma is subjected to auxiliary heating, the intensity and velocity distribution of the fuel ion populations for both DD and DT plasmas. JET plasma operating conditions can vary considerably in ion temperature and density, resulting in a large range of variation in both the rate and the energy distribution of the neutron emission. Since individual spectrometers have characteristic efficiencies and energy resolutions and operate within a more limited dynamic range, it is almost impossible to design a single neutron spectrometer which can satisfy all the JET operational demands. The result has been that a range of neutron spectrometers have been deployed at JET, for operation both in DD and DT plasmas. An early account of

JET neutron spectrometry is given in the review article [14]. Developments of neutron spectrometers have continued at JET and three systems will be described below: a new time-of-flight spectrometer optimized for high count rates (TOFOR), an upgrade of the Magnetic Proton Recoil system (MPRu) and compact neutron spectrometers based on an organic scintillator (NE213).

The JET TOFOR Neutron Spectrometer. The Time-of-Flight (TOF) technique for neutron spectrum measurements in fusion plasma studies is among those that can offer high energy and good time resolution. It is based on the measurements of the times of correlated neutron scattering events in a 'start' and a 'stop' detector, placed on a constant TOF sphere. In this configuration the flight time recorded is independent of the scattering angle and defines the energy of the incident neutron. The first such instrument for 2.5 MeV neutrons was installed at JET in the early 90s and had an energy resolution of about 5% and an efficiency ~5.10^{-2} cm^2 [15]. It performed satisfactory with count rates of up to a few kHz only requiring data integration times of several hundreds ms to provide good statistics. Such time resolution is not sufficient to study rapid plasma phenomena. Therefore the recent efforts at JET were aimed at improving the time and spectral characteristics of the neutron measurements at JET.

A new time-of-flight (TOFOR) neutron spectrometer was designed and installed at JET presently as part of the JET Enhancement programme. It has been optimized for high count rate measurements of 2.45 MeV neutrons. It is able to cope with the DD neutron rates from 10^{17} n/s (expected JET maximum) to below 10^{14} n/s (subject to minimum statistical accuracy required and hence time resolution). The instrument's design is described in [16, 17]. In short, it consists of a "start" detector, segmented in five layers to increase its rate capabilities, and 32 "stop" detectors forming a ring around the instrumental axis along the defining neutron collimator. Each of the "start" scintillators in the stack of five is equipped with 3 PM tubes to improve light collection. For the first time in fusion neutron TOF spectrometry, an innovative digital data acquisition (DAQ), based on time digitizing PC boards, has been especially developed, which contributes to raising the count rate capability limit of TOFOR to a projected level of 0.5 MHz [17]. The control and monitoring system, containing laser and LED test pulses and radioactive sources, has been implemented to allow setting of working points. Finally, the response of TOFOR to neutrons has been described with the help of detailed geometrical modelling and neutron transport calculations.

The TOFOR is now fully incorporated into the JET diagnostic and data acquisition systems and provides data to the present experimental campaigns. The maximum count rate achieved was 39 kHz at a neutron rate of 1.10^{16} n/s, where the observed "random-to-true" signal level was determined to 4.2% in the peak channel. From this one can project that TOFOR could be operated up to 400 kHz for a maximum DD neutron rate at JET of $1 \cdot 10^{17}$ n/s with an expected value of accidental events of ~43%. The attained energy resolution (E_n/E_n) is 7.4% only while it has very good flux efficiency of 0.12 cm^2. This allows DD spectral measurements to be performed with high statistical accuracy. The TOFOR capabilities and the quality of the neutron energy spectra that can be obtained are reported in [18, 19] where a discussion of neutron spectrum analysis for NBI and RF heated plasmas can also be found. An assessment of the impact of the scattered flux on the neutron spectrum has been recently carried out leading to improvement of the methods for studying TOFOR data [20]. It was also stressed that this has strong implications for the possibility of measuring 2.5-MeV neutrons on ITER.

The Magnetic Proton Recoil (MPR) Neutron Spectrometer. The original magnetic proton recoil (MPR) spectrometer was designed for diagnosis of the 14-MeV neutron emission in high-power DT fusion experiments. The principle and components of the MPR technique are described in detail in [21]. It works on the principle of converting the incoming neutron flux from the plasma into recoil protons which are further momentum analyzed in the magnetic field. It was installed at JET in 1996 and successfully used in the first deuterium-tritium (DTE1) campaign in 1997 and in the TTE campaign in 2003 [22]. It has a flexible setting for energy resolution within a range from 2.5% up to 10% with a flux efficiency of $\sim 10^{-4}$ cm^2. The MPR design also permits an accurate absolute calibration in both energy and flux efficiency, except for a profile shape factor derived from neutron profile monitor, making the instrument capable of providing an almost independent measurement of the neutron yield.

To explore its full potential, this device has been recently upgraded (MPRu JET enhancement project), permitting measurements for a wide range of plasma conditions, from relatively low-power DD plasmas (giving count rates of kHz for 10^{16} DD n/s) to DT plasmas of the highest power (10 MHz for 10^{19} DT n/s). The emphasis of the upgrade work has been to achieve a high level of background immunity in all measurement scenarios. New possibilities have been added by the combined use of phoswich detectors and state-of-the-art digital data acquisition electronics, making measurements of 14-MeV neutrons from DT plasmas virtually background-free, i.e. only the counting statistics will set the limit for the instrumental sensitivity. For DD plasma operations, the MPRu will deliver data of high absolute accuracy but moderate time resolution which will be complementary to the TOFOR data of high time resolution. In addition, with its narrow line width and high immunity to background, the MPRu can efficiently separate the scattered and direct neutron flux, thereby providing important data in the "intermediate" energy region 4 – 11 MeV, useful for benchmarking of neutron transport codes. These properties make the MPRu technique an interesting option for the next step fusion devices, such as ITER.

JET Compact Neutron Spectrometers. From the ITER engineering demands it is highly desirable to minimize the interfaces of any diagnostic. In particular, so called "compact" neutron spectrometers are considered as possible candidates for neutron flux and spectra measurements. In response to this, several candidate detectors have been subjected to tests at JET, with the focus mainly on spectrometers based on the NE213 organic proton recoil scintillator. Measurements with NE213 scintillators (5cm by 5cm) have been carried out at JET during the 2002 DD and 2003 TTE campaigns [23]. An energy resolution of 2% at 14 MeV and 4% at 2.5 MeV neutron energy has been achieved. Neutron spectra measurements have been performed for different plasma scenarios and the contribution of various neutron production mechanisms have been evaluated using unfolding techniques based on the Bayesian method. Simultaneous measurements of both 2.5 and 14 MeV neutron spectra have been conducted for the ohmic plasma with trace tritium gas puff. The unfolding analysis requires a detailed knowledge of the detector response function. In this respect the JET NE213 scintillator was fully characterized at the Physikalisch-

Technische Bundesanstalt accelerator facility. Different methods can be used for unfolding of neutron spectra from measured pulse height spectra. Therefore particular attention was paid to validation of several unfolding codes (summary is given in [24]).

3. NEUTRON DIAGNOSTICS DEVELOPMENT AND TESTS AT JET.

The development and improvement of neutron measuring techniques are among the main objectives of the EFDA JET programme. Several important upgrades are foreseen in the near future. With respect to the neutron profile monitor it is planned to replace the conventional analog pulse shape discriminators with a state-of-the-art Digital Pulse Shape Discrimination acquisition system [25]. Preliminary tests of such an approach have been conducted during the TTE campaign with a NE213 spectrometer and recently with a NE213 detector of the one central channel of the JET neutron profile monitor. It has been shown that such an upgrade, together with the *pile-up* resolving procedure proposed recently [26], will significantly improve the dynamic range in energy and count rate for both diagnostics. In addition, a new digital data collection system will allow a deconvolution of spectrum information from all neutron profile NE213 detectors since in fact they are acting as low-resolution spectrometers. No doubt it will be a tremendous step forward in establishing of key phenomena of fusion fast particles.

Essential upgrades are under consideration for TOFOR and MPRu neutron spectrometers as well. For TOFOR it is proposed, firstly, to install a new "start" detector to improve light collection, and secondly, to upgrade its data acquisition system to use electronics modules with simultaneous capability to measure time and pulse height. The combined effect of these upgrades would lead to significant improvement in the ability to reject random (false) coincidences and also give the possibility to correct for timing walk on an event-by-event basis. As was mentioned above, MPRu could provide an independent method to determine the total neutron flux, and thereby the fusion power, based on a neutron detection system combining a high-performance spectrometer with the neutron emission profile monitor. The method has been successfully tested in full power DT plasmas and trace T conditions at JET [27]. The installation of a high count-rate neutron flux detector in the collimated MPRu neutron "beam line" shall provide an increase in the time-resolution of the system in both full and trace T fusion power measurements and, furthermore, provide an opportunity to study the applicability of the method in D operations.

It is worthwhile to mention here that some other neutron diagnostic developments and R&D, with strong relevance to ITER, are now in progress at JET. There is current work on improving of flux spectral characteristics of the natural and CVD diamond detectors, Stilbene scintillators and an upgrade of the JET γ-ray cameras.

SUMMARY

In the last years a significant achievements have been made at JET in respect of development of burning plasma diagnostics. The data obtained by means of JET neutron profile monitor provide essential information for investigating fast particle physics. In particular, the vertical camera played a key role to perform these studies and the importance of this diagnostic tool for ITER is confirmed. Significant progress was also achieved on neutron spectroscopy and tomography. In addition, the upgraded Magnetic Proton Recoil (MPRu) spectrometer, together with the new time-of-flight at optimized rate (TOFOR) spectrometer, represents a significant step forward of the JET neutron emission spectroscopy capability permitting detailed studies of neutron energy spectra. Considerable attention at JET was also devoted to developing and testing of some other ITER-relevant neutron detectors, such as compact neutron spectrometers and radiation hard detectors, along with new electronic technology.

ACKNOWLEDGMENTS

This work was funded partly by the United Kingdom Engineering and Physical Sciences Research Council and by the European Communities under the contract of Association between EURATOM and UKAEA. The views and opinions expressed herein do not necessarily reflect those of the European Commission. This work was carried out within the framework of the European Fusion Development Agreement.

REFERENCES

1. M. Adams et al., *Nucl. Instrum .Methods,* A 329 (1993) 277-290.
2. O. N. Jarvis et al., *Fusion Eng. Design*, 59 (1997), 34-35.
3. M.J. Loughlin et al., *Rev. Sci. Instrum.*, 70, 1124 (1999).
4. V. Kiptily et al., *these proceedings*
5. S. Popovichev et al., *Proc. 31st EPS Conf. on Plasma Physics,* London,UK, ECA, Vol. 28G (2004).
6. P.U. Lamalle et al., *Proc. 31st EPS Conf. on Plasma Physics,* London, UK, ECA, Vol. 28G (2004).
7. N.C. Hawkes et al., *Plasma Phys. Control. Fusion*, 47 (2005), 1475.
8. I. Jenkins et al., *Proc. 32nd EPS Conf. on Plasma Physics*, Tarragona, Spain, ECA, (2005).
9. I. Jenkins et al., *Proc. 34st EPS Conf. on Plasma Physics,* Warsaw, Poland, ECA, (2007).
10. K. D. Zastrow et al., *Plasma Phys. Control. Fusion*, 46, B255 (2004).
11. G. Bonheure et al., *Nucl. Fusion*, 46, 725 (2006).
12. L.C Ingesson et al., *Nucl. Fusion*, 38 (1998) 1675.
13. G. Bonheure et al., *Proc. 32nd EPS Conf. on Plasma Phys*. Tarragona, Spain, ECA, (2005).
14. O. N. Jarvis, *Nucl. Instrum .Methods,* A 476 (2002) 474-484.
15. T. Elevant, *Rev. Sci. Instrum.*, 63 (1992) 4586.
16. A. Hjalmarsson et al., *Rev. Sci. Instrum.*, 74 (2003), 1750.
17. M. Gatu Johnson et al., *submitted for publication in Nucl. Instr. Meth. A*, September 2007.
18. L. Giacomelli et al., *Proc. 34st EPS Conference on Plasma Physics,* Warsaw, Poland, ECA, (2007).
19. C. Hellesen et al., *Proc. 34st EPS Conference on Plasma Physics,* Warsaw, Poland, ECA, (2007).
20. M. Gatu Johnson et al., *these proceedings*
21. J. Källne et al., *Rev. Sci. Instrum.*, 70 (1990) 1181.
22. G. Ericsson et al., *Rev. Sci. Instrum.*, 72 (2001) 759.
23. L. Bertalot et al., *Proc. 32nd EPS Conf. on Plasma Phys*. Tarragona, Spain, ECA (2005).
24. E. Andersson Sunden et al., *these proceedings*
25. B. Esposito et al., *Nucl. Instrum. Methods,* A 518 (2004) 626.
26. F. Belli et al., *to be published in Nucl. Instrum. Method*
27. H. Sjöstrand et al., *these proceeding (to be published in Nucl. Fusion)*

Fast ion JET diagnostics: confinement and losses

V.G. Kiptily,[1] D. Borba,[2] F.E.Cecil,[3] M. Cecconello,[4] D. Darrow,[5] V. Goloborod'ko,[6,7] K. Hill,[5] T. Johnson,[8] A. Murari,[9] F. Nabais,[2] S.D. Pinches,[1] M. Reich,[10] S.E. Sharapov,[1] V. Yavorskij,[6,7] I.N. Chugunov,[11] D.B. Gin,[11] G. Gorini,[12] A.E. Shevelev,[11] D.B. Syme,[1] V. Zoita [13] and JET-EFDA contributors [*]

JET-EFDA, Culham Science Centre, OX14 3DB, Abingdon, UK

[1] *Euratom / UKAEA Fusion Association, Culham Science Centre, Abingdon, Oxon, United Kingdom*

[2] *Euratom/IST Fusion Association, Centro de Fusão Nuclear, 1049-001 Lisboa, Portugal*

[3] *Colorado School of Mines, Golden, CO, USA*

[4] *Association EURATOM –VR, Royal Institute of Technology KTH, Stockholm, Sweden*

[5] *Princeton Plasma Physics Lab, Princeton NJ, USA*

[6] *Euratom/OEAW Association, Institute for Theoretical Physics, University of Innsbruck, Austria*

[7] *Institute for Nuclear Research, Kiev, Ukraine*

[8] *Euratom -VR Association, Alfvén Laboratory, Royal Institute of Technology, 10044 Stockholm, Sweden*

[9] *Consorzio RFX - Associazione Euratom-Enea sulla Fusione, I-35127 Padova, Italy*

[10] *Euratom / MPI für Plasmaphysik Association, Garching, Germany*

[11] *Ioffe Physico-Thechnical Institute, 194021 St. Petersburg, Russia*

[12] *Istituto di Fisica del Plasma, EURATOM-ENEA-CNR Association, Milan, Italy*

[13] *Euratom-MedC Association, National Institute for Laser, Plasma and Radiation Physics, Bucharest, Romania*

[*] *See the Appendix of M.L.Watkins et al., Fusion Energy 2006 (Proc. 21st Int. Conf. Chengdu, 2006) IAEA, (2006)*

Abstract. A study of magnetically confined fast ions in tokamaks plays an important role in burning plasma research. To reach ignition and steady burning of a reactor plasma an adequate confinement of energetic ions produced by NBI heating, accelerated with ICRF and born in fusion reactions is essential to provide efficient heating of the bulk plasma. Thus, investigation of the fast ion behaviour is an immediate task for present-day large machines, such as JET, in order to understand the main mechanisms of slowing down, redistribution and losses, and to develop optimal plasma scenarios. Today's JET has an enhanced suite of fast ion diagnostics both of confined and lost ions that enable to significantly contribute to this important area of research. Fast ion populations of *p, d, t*, ^3He and ^4He, made with ICRF, NBI, and fusion reactions have been investigated in experiments on JET with sophisticated diagnostics in conventional and shear-reversed plasmas, exploring a wide range of effects. This paper will introduce to the JET fast-ion diagnostic techniques and will give an overview of recent observations. A synergy of the unique diagnostic set was utilised in JET, and studies of the response of fast ions to MHD modes (e.g. tornado modes, sawtooth crashes), fast ^3He-ions behaviour in shear-reversed plasmas are impressive examples of that. Some results on fast ion losses in JET experiments with various levels of the toroidal field ripple will be demonstrated.

Keywords: tokamak, burning plasmas, fast ions, diagnostics
PACS: 52.55.Fa, 52.55.Pi, 52.70.La

The nuclear reaction $D(T,n)^4He$ between deuterium and tritium is the most promising source of energy in a thermonuclear fusion reactor. The power for the self-

CP988, *Burning Plasma Diagnostics*, edited by F. P. Orsitto, G. Gorini, E. Sindoni, and M. Tardocchi
2008 American Institute of Physics 978-0-7354-0507-3/08/$23.00

sustained *DT*-plasma burn is provided by the α particles (^4He-ions), which are born with an average energy of 3.5 MeV and transfer the energy to the thermal plasma during their slowing down. To reach ignition and steady burning of a reactor plasma an adequate confinement of α particles and energetic ions produced by NBI heating, accelerated with ICRF is essential to provide efficient heating of the bulk plasma. Thus, investigation of the fast ion behaviour is an immediate task for present-day large machines, such as JET, in order to understand the main mechanisms of slowing down, redistribution and losses, and to develop optimal plasma scenarios.

Nuclear reaction γ-ray diagnosis is one of the important techniques used on the JET tokamak for studying confined fast ions [1-5]. The intense γ-ray emission is produced in JET plasmas when fast ions (ICRF-driven ions, fusion products, NBI-injected ions) react either with fuel ions or with the main plasma impurities such as carbon and beryllium. Gamma-ray energy spectra are recorded with collimated spectrometers, while the γ-ray emission spatial 2D-profiles are measured with the JET neutron/gamma profile monitor [5,6]. Together, these provide information on the spatial distribution of fast ions and fast ion tail-temperature.

On JET, the α-particle diagnostic technique is based on the nuclear reaction $^9Be(\alpha,n\gamma)^{12}C$ between confined α-particles and the beryllium impurity typically present in the plasma. The γ-radiation due to the reaction $^9Be(\alpha,n\gamma)^{12}C$ has been observed for the first time in JET experiments with the third harmonic ICRF heating of ^4He beam ions in a ^4He plasma [3,5,7] and in *D*-plasmas with short *T*-NBI blips for α-particles [4].

During recent experiments, JET γ-ray diagnostics provided some of the most interesting results on fast-ion and α-particle physics [4,5]. These experiments on JET showed that the diagnostics work well, but a further improvement of the diagnostics is needed for the measurements in the high neutron yield *DD*- and *DT*-discharges to demonstrate the capabilities of γ-ray diagnostics for burning plasma physics in future reactors.

Recently, two fast ion loss diagnostic devices we have been installed. These consist of a set of thin Faraday foil collectors and a scintillator probe. Thin Faraday foil detectors, in which ions that are lost from a fusion plasma are detected as current to ground in a metallic foil near the plasma boundary, have been used to investigate ion losses on NSTX, DIII-D and JET [8]. Similarly, scintillation detectors have been widely employed to study ion losses on TFTR, NSTX, and other machines. The present devices are intended to study lost ions in general and *DT*-fusion product alpha particles in particular during the upcoming JET campaigns. The design and construction of a fast ion loss diagnostic for JET may have application to the International Thermonuclear Experimental Reactor (ITER).

EXPERIMENTAL EQUIPMENT

Gamma-ray spectrometers

Gamma-ray energy spectra are measured on JET with three independent devices, one with a quasi-tangential, and two with vertical line of sight, through the plasma centre. Two of these devices are a calibrated bismuth germanate (BGO) scintillation detectors with a diameter of 75 mm and a height of 75 mm. One of them is located in a well-shielded bunker, which views the plasma tangentially. In order to reduce the neutron flux and the gamma-ray background, the front collimator is filled to a depth of 500 mm with polythene. Behind the scintillation detector, there is an additional 500 mm long dump of polythene and a 1000-mm long steel plug. The detector's line of sight lies in a horizontal plane about 30 cm below the plasma magnetic axis. Another BGO-spectrometer is viewing the plasma vertically through the centre (R≈2.95 m). The γ-ray spectra are continuously recorded in all JET discharges over the energy range 1-28 MeV, with energy resolution of about 4% at 10 MeV. The third device for the gamma-ray energy spectrum measurements is a NaI(Tl) scintillation detector with a diameter of 125 mm and a height of 125 mm, viewing the plasma vertically through the centre (R≈2.95 m).

For identification of the fast ions, which exist in the plasma and produce the observed γ-ray emission, and in order to assess the effective tail temperatures of these fast ions, the γ-ray spectrum modelling code, GAMMOD [3] is used. This code is based on the known nuclear reaction cross-sections and it contains information on about hundred γ-ray transitions in the final nuclei of the low-Z impurity reactions. It also includes the γ-ray response function of the BGO-spectrometer. A Maxwellian energy distribution is used to describe the line-of-sight averaged tail of ICRH-accelerated ions. The GAMMOD code analysis gives the effective tail temperatures, the fast ion concentrations and the contribution to the neutron yield from the fast particle-induced reactions.

Measurements in high performance discharges with NaI(Tl) and BGO detectors showed the necessity to upgrade the data acquisition system (DAQ) in order to improve the energy and temporal resolution of the measurements in high performance *dt*- and *dd*-discharges. Recently, a modern DAQ system for γ-ray diagnostics based on PCI transient recorder was installed. It is used for digitising the incoming signal at a sampling rate of 25 MHz with an amplitude resolution of 14-bit at maximum average pulse rate up to 1MHz [9]. For a precise pulse height analysis (PHA) at high counting rates, sophisticated software has been developed. It allows avoidance of the pile-up effect and gain instability, which lead to a distortion of γ-ray spectra at high-count rates with fast rate variations. The main limitation for further improving of counting rate capabilities with NaI(Tl) and BGO-detectors is the rather long scintillation decay-times, of 250 ns and 300 ns, respectively. It is planned to replace the existing detectors by modern fast high-Z heavy scintillators $LaBr_3(Ce)$, known as "BriLanCe" and $Lu_{1.8}Y_{0.2}SiO5(Ce)$ - "LYSO". These new scintillators have short decay times, of 16 ns and 40 ns, and high photons yields. Their striking properties open a possibility to extend the counting rate limit beyond the 5 MHz, and at the same time to improve the energy resolution for γ-ray spectrometry in the range 2 - 30 MeV.

Gamma-ray imaging of fast ions

Spatial profiles of the γ-ray emission in the energy range E_γ>1 MeV have been measured in JET using the Gamma Cameras (Fig.1), which have ten horizontal and nine vertical collimated lines of sight [5,6]. Each collimator corresponds to a poloidal-viewing extent at the centre of plasma of about 10 cm. The detectors array is comprised of 19 *CsI(Tl)* photo-diodes (10 mm × 10 mm × 15 mm). The data acquisition system accommodates the γ-ray count-rate measurements in four independently adjustable energy windows. This allows allocating γ-ray peaks for a given fast ion population in specific windows to count them separately. The Monte-Carlo calculations of the detector response function show that the size of the *CsI(Tl)* scintillator is suitable for the γ-ray measurements in the MeV-range. It is found that a noticeable contribution to the γ-ray spectrum with energy $E_\gamma >> 2m_e c^2$ is coming from the so-called double - and single -escape peaks, which is resulting from the high probability of electron-positron pair generation in the detector. The double-escape peak efficiency of this detector is about 3% that in this case is sufficient for reliable counting.

Experimental data obtained for the 19 lines of sight allows tomographic reconstruction for deriving the local gamma-ray emissivity in a poloidal cross-section. In the tomographic reconstructions the full geometry of the collimators and detector efficiencies are taken into account, but small effects of attenuation and scattering of the γ-rays are neglected. It is assumed that the distribution of the low-Z impurities is uniform in the plasma core (a region, where 50% of neutrons are produced) as confirmed by atomic spectroscopy measurements. For the tomographic reconstruction a constrained optimisation method [10] is used, which was successfully applied earlier to soft x-ray, bolometer and γ-ray measurements on JET [6].

Lost ions diagnostics

The Faraday cup array detects the current of fast ions at multiple poloidal locations, with a dynamic range from 10 nA/cm^2 to 10 mA^2 at a temporal resolution of 1 ms. The detectable range of α-particle energies is about 1–5 MeV. The energy resolution for 3.5 MeV α particles is estimated to be about 15%–50%. The array consists of nine detectors spread over five poloidal locations (Z) between 22 cm and 80 cm below the midplane. Radially, the detectors are equally spaced on three locations between 25 and 85 mm behind the adjacent poloidal limiter. Each detector consists of at least four 75×25 mm^2 Ni foils (2.5 μm in eight of the detectors and 1.0 μm in the ninth), which are separated by insulating mica foils. Depending on its energy, a particle can pass through a certain number of foils before it is stopped in one foil, thus causing a current signal. The detection of the temporal evolution of the current signals in all foils in the radially and poloidally distributed detectors will allow a map of particle energies at different locations.

The scintillator probe allows the detection of ions with a pitch angle between 30° and 85° (5% resolution) and a gyroradius between 30 and 140 mm (15% resolution). It is located about 28 cm below the midplane. The underlying principle of scintillator measurements is the emission of light by a scintillating material after a particle strikes this material. Selection criteria for the particles that hit the scintillator are introduced by using a set of collimators within the magnetic field of JET. An optical arrangement within the scintillator probe is used to transfer the light emitted by the scintillator towards a charge-coupled device CCD) camera and a photomultiplier array through a coherent fibre bundle.

DIAGNOSIS OF FAST IONS IN JET

Alpha-particle confinement and losses

In the JET experiment with the 3rd harmonic ICRH heating of a ^{4}He neutral beam ($\omega=3\omega_{4He}$) in a ^{4}He plasma [7], the γ-radiation due to the reaction $^{9}Be(\alpha,n\gamma)^{12}C$ has been observed for the first time [3,5,7]. It was found that the ICRH waves accelerate the ^{4}He beam ions up to the energies in excess of 2 MeV. The presence of the 4.44-MeV peak in γ-ray spectra is evidence for the existence of alphas with energies that exceed 1.7 MeV. These measurements also indicate a surprisingly high level of the γ-ray emission from the reaction $^{12}C(d,p\gamma)^{13}C$ meaning that some ICRH power was absorbed by deuterons at the third harmonic *D*-resonance, which coincides with the 3rd harmonic ^{4}He-resonance. The fast *D*-ions give rise to 3.09-MeV γ-rays from the reaction $^{12}C(d,p\gamma)^{13}C$. A peak at 3.09 MeV in recorded spectra reflects the presence in the plasma of fast *D*-ions with energies exceeding 0.5 MeV.

In these experiments the relaxation of the distribution functions of fast ^{4}He-ions, accelerated with 3rd harmonic ICRH of ^{4}He-beam and D-minority ions, accelerated due to the parasitic ICRH absorption have been simultaneously studied during slowing down in ^{4}He-plasma (2.2T/ 2MA, 51 MHz). The γ-ray decay-time for both ions was measured during notches in the ICRH power when the accelerated ion source was strongly reduced. The γ-ray spectra with 100-ms integration time were recorded by means of a NaI-detector, which observes the plasma core through the vertical collimator. The data is obtained in discharges with monotonic q-profiles only. A clear correlation $\tau^{He}_{\gamma} \sim 0.5\tau^{D}_{\gamma}$ is observed. This is the first direct confirmation of the Spitzer slowing down dependence for different types of ions. An effect of the A/Z^2-factor for the classical slowing-down time is observed in these experiments.

In the Gamma Cameras, special energy windows, containing 3.09 MeV and 4.44 MeV peaks with their single and double escape satellites, were set up to measure the spatial profiles of the γ-ray emission from *D*- and ^{4}He-ions. The ability to separate γ-rays with the Gamma Cameras from different energy bands has allowed the study of the fast ion behaviour in an advanced regime plasma discharge with Alfvén cascades and sawtooth instabilities [5].

The first γ-ray measurements of fusion-born α particles were carried out in JET 'trace tritium' discharges [4], i.e. in deuterium plasmas with a small population of tritium Neutral Beam Injection (NBI) fast ions. The γ-ray emission from the nuclear reaction $^{9}Be(\alpha,n\gamma)^{12}C$ is used to measure changes in the density of the fast α particles with energy $E_\alpha > 1.7$ MeV in the post-NBI period. It was demonstrated in these experiments how the γ-ray spectrometry of the interaction between α's and Be impurity in plasmas could be used in future magnetic fusion machines to obtain essential information on the slowing down and confinement of the fast α-particles.

Another possibility to study fusion α-particles is the $D(^{3}He,p)^{4}He$ reaction, which yields 3.7-MeV alphas and 15-MeV protons. As a rule D-^{3}He experiments are carried out with the ICRF heating of the ^{3}He-minority ($\omega=\omega_{3He}$) in DD-plasmas. In the recent D-^{3}He JET experiments α-particle confinement was studied, and for the first time α-particle losses have been measured. The reaction $D(^{3}He,\gamma)^{5}Li$, which is a weak branch (10^{-5}) of the fusion reaction $D(^{3}He,p)^{4}He$ giving rise to fusion alphas, was identified in the plasma by γ-ray spectrometers. Nuclear reactions with main impurities, $^{12}C(^{3}He,p\gamma)^{14}N$, $^{9}Be(^{3}He,n\gamma)^{11}C$ and $^{9}Be(^{3}He,p\gamma)^{11}B$, have been detected as well. The spectrum analysis with GAMMOD code shows that the energetic ^{3}He-tail has an effective temperature in excess of 300 keV. Gamma-ray images of energetic ^{3}He-ions ($E_{3He} > 0.9$ MeV) were recorded with 2-D γ-ray cameras.

Fast ion losses were studied with the scintillator probe. Distribution of fusion α-particles born in the plasma with this type of heating is rather anisotropic, and has a broad energy spectrum. According to the calculated α-particle spectrum, most of alphas in the plasma have energy in range of 2-5.5MeV, and that is consistent with energy range of maximum losses. The modelling of orbits of the lost α-particles shows that most of energetic ions detected by the probe originates from the plasma region, which is coincided with the hot spot in the measured γ-ray emission profile. Thus, with a high probability, one can suppose that the drift losses of the trapped fusion α-particles were measured in these experiments.

A decay-time both for ^{3}He-ions and lost fusion alphas were measured during notches in the ICRH power when the accelerated ion source was strongly reduced or switched off at all. Decay times of intensities of 3.4-MeV α-particle losses (τ_L) and γ-rays (τ_γ) from ^{3}He-reactions were derived. It is found that $\tau_\gamma > \tau_L$ that is consistent with guess that change of the losses are faster than fusion α-particle source strength.

Observation of MHD effects

In JET experiments on fast H-ion redistribution carried out with low-density plasmas and high ICRH power, large populations of highly energetic fast ions were built up and a variety of instabilities were destabilised [11,12]. Two types of fast ions were identified by the γ-ray diagnostics: fast H-ions with an energetic tail in excess of 4.5 MeV due to ICRH tuned to the fundamental H-resonance ($\omega=\omega_H$) in the centre of the plasmas and D-ions accelerated with second harmonic heating whose energies exceeded 0.5 MeV. Fast ion losses associated to MHD activity were detected with scintillator probe. Three types of losses were identified, one associated with sawtooth crashes, one associated with TAE and high frequency fishbones and a third group that seems to correlate with tornado modes. When tornado modes became unstable, a significant increase in the number of losses was measured and a change in both the energy and pitch angle of the lost ions was observed. In addition, it was also observed that the number of losses increased extensively with the ICRH power.

Most of JET pulses with long period (typically fast particle stabilised) sawtooth activity show a strong correlation between the time when particle losses are detected by the scintillator probe and start of the sawtooth crash.

POSSIBLE IMPLEMENTATION OF γ-DIAGNOSTICS IN ITER

In ITER the γ-ray diagnostics could perform the same functions as on JET. The main difference is a severe neutron/gamma background that requires special measures to maximise the signal-to-background noise ratio. For the fusion α-particle measurements the 2-D γ-ray cameras could be used, however, the γ-ray detectors should be protected against harsh neutron emission with special neutron attenuators. The feasibility of the γ-ray measurements depends on effective the neutron suppression. It is important to avoid carbon-containing materials in the neutron attenuator because inelastic scattering of neutrons with energy exceeding 5 MeV, $^{12}C(n,n'\gamma)^{12}C$, leads to the unwelcome background of 4.44-MeV γ rays. A more convenient neutron attenuator is ^{6}LiH. It is compact, effective and rather transparent for γ-rays. It does not produce interfering γ-rays in the high-energy range. A 30-cm sample of the ^{6}LiH-filter reduces 2.4-MeV neutron flux ~900 times and the 15-MeV neutron flux ~30 times [13].

First experiments dedicated to the using ^{6}LiH for the γ-ray background reduction by means of neutron flux suppression have been carried out [14]. The test took place during JET experiments with *DD*-plasmas. A sample of the neutron attenuator (∅30mm x 300 mm) was installed in the 2-m vertical line-of-sight collimator used to view the plasma from above for fusion γ-ray spectrometry. Gamma-ray spectra recorded both with *BGO*- and *NaI(Tl)*-detectors were used for analysis of the γ-ray background. It was found that γ-ray background is mainly induced by neutrons due to radiative capture, (n,γ) and inelastic-scattering reactions, (n,n'γ). This neutron-induced gamma-background is produced in a) tokamak construction materials and b) detector itself. Only the last source of the background could be reduced with ^{6}LiH-attenuator plugged in the collimator. Assessments of the reduction obtained from change intensity of γ-ray lines give a factor of 100. It is found that this type of γ-ray background lies in the energy range below 3 MeV. A small reduction (factor of 2) of the spectra was found in the energy range above 3 MeV, which is defined by γ-ray transparency of the ^{6}LiH-attenuator. JET experiments with ^{6}LiH show that the attenuator is very effective in reduction of γ-ray background caused by *DD*-fusion neutrons, and results are consistent with neutron-generator measurements. Further tests in *DT*-discharges are needed to prove the full capability of ^{6}LiH-attenuators as an indispensable component of the γ-ray diagnostic system in ITER. A sophisticated neutron attenuation system is being developed for the γ-ray cameras to allow the 2-D γ-ray measurement in the high performance JET discharges. Severe design constraints limited the attainable parameters of the upgraded γ-ray cameras, and a solution based on the use of pure light water was developed for the neutron attenuators. Lithium hydride (natural isotopic composition) will be used in another upgraded diagnostics: vertical and tangential γ-ray spectrometers.

If the main impurity in ITER plasmas is beryllium, the slowed down deuterons will give rise to gammas, 2.87 MeV (^{10}B) and 3.37 MeV (^{10}Be) from the $^{9}Be(d,n\gamma)^{10}B$ and $^{9}Be(d,p\gamma)^{10}Be$ reactions, whereas the fusion alphas with energies around the 2-MeV resonance in the $^{9}Be(\alpha,n\gamma)^{12}C$ reaction with produce 4.44-MeV gammas. Furthermore, the source of fusion α-particles can be obtained by measuring the 17-MeV gammas from the $D(T,\gamma)^{5}He$ reaction, which is a 10^{-5} branch of the $D(T,n)^{4}He$ reaction. Using γ-ray spectrometers in every channel of the ITER neutron cameras, α-particle slowing down profiles could be measured with the technique successfully tested in JET Trace Tritium Experiments [4]. Simultaneous measurements of the NBI power deposition and α-particle slowing down profiles are very important for the optimisation of different plasma scenarios and understanding of the fast-ion confinement physics.

ACKNOWLEDGMENTS

This work was funded partly by the United Kingdom Engineering and Physical Sciences Research Council and by the European Atomic Energy Community under the contract of Association between EURATOM and UKAEA and is subject to the provisions of the European Fusion Development Agreement. Also this work was supported in part by US DoE Contract DE-AC02-76-CH03073 and government of the Russian Federation. The views and opinions expressed herein do not necessarily reflect those of the European Commission. This work was carried out within the framework of the European Fusion Development Agreement.

REFERENCES

1. G. J. Sadler et al Fusion Technology **18** (1990) 556.
2. O. N. Jarvis et al Nucl. Fusion **36** (1996) 1513.
3. V.G. Kiptily et al Nuclear Fusion **42** (2002) 999.
4. V.G. Kiptily et al Phys. Rev. Letters **93** (2004) 115001
5. V.G. Kiptily et al Nuclear Fusion **45** (2005) L21.
6. J. M. Adams et al Nucl. Instrum. Methods Phys. Res. **A239** (1993) 277.
7. M.J. Mantsinen et al., Phys. Rev. Letters **88** (2002) 105002.
8. D. Darrow et al. Rev. Sci. Instrum. 77 (2006) 10E701
9. V.G. Kiptily et al., 33 EPS Conf. on Plasma Phys., Rome, 19 - 23 June 2006, P1.077.
10. L.C. Ingesson et al., Nucl. Fusion **38** (1998)1675
11. M. Reich et al., 34 EPS Conf. on Plasma Phys., Warsaw, July 2-6, 2007, O-4.016.
12. F. Nabais et al., 34 EPS Conf. on Plasma Phys., Warsaw, July 2-6, 2007, P-1.123
13. V.G. Kiptily et al., Technical Physics **43** (1998) 471
14. I.N. Chugunov et al. to be published in Instruments and Experimental Techniques (Pribory i Tekhnika Eksperimenta, in Russian)

The ITER Radial Neutron Camera Detection System

D. Marocco[1], F. Belli[1], G. Bonheure[2], B. Esposito[1], Y. Kaschuck[3], L. Petrizzi[1], M. Riva[1] and JET EFDA contributors*

JET-EFDA, Culham Science Centre, OX14 3DB, Abingdon, UK
[1]Associazione ENEA-Euratom sulla Fusione, C.R. Frascati, C.P. 65, I-00044 Frascati (Roma), Italy
[2]Laboratory for Plasma Physics, Association "Euratom-Belgian State", Royal Military Academy, Avenue dela Renaissance, 30, B-1000 Brussels, Belgium
[3]TRINITI, Troitsk 142190, Moscow Region, Russian Federation
** See the Appendix of M.L.Watkins et al., Fusion Energy 2006 (Proc. 21st Int. Conf. Chengdu, 2006) IAEA, (2006)*

Abstract. A multichannel neutron detection system (Radial Neutron Camera, *RNC*) will be installed on the *ITER* equatorial port plug 1 for total neutron source strength, neutron emissivity/ion temperature profiles and n_t/n_d ratio measurements [1]. The system is composed by two fan shaped collimating structures: an ex-vessel structure, looking at the plasma core, containing tree sets of 12 collimators (each set lying on a different toroidal plane), and an in-vessel structure, containing 9 collimators, for plasma edge coverage. The *RNC* detecting system will work in a harsh environment (neutron flux up to $10^8 - 10^9$ n/cm^2s, magnetic field >0.5 T for in-vessel detectors), should provide both counting and spectrometric information and should be flexible enough to cover the high neutron flux dynamic range expected during the different *ITER* operation phases. *ENEA* has been involved in several activities related to *RNC* design and optimization [2,3]. In the present paper the up-to-date design and the neutron emissivity reconstruction capabilities of the *RNC* will be described. Different options for detectors suitable for spectrometry and counting (e.g. scintillators and diamonds) focusing on the implications in terms of overall *RNC* performance will be discussed. The increase of the *RNC* capabilities offered by the use of new digital data acquisition systems will be also addressed.

Keywords: ITER, radial neutron camera.

PACS: 28.52.-Lf, 52.70.-m, 29.30.HS

1. RNC LAYOUT

A *3D MCNP* model of the *ITER RNC* and a Measurement Simulation Software Tool (*MSST*), performing asymmetric Abel inversion of the *RNC* integrated data, have been developed in *ENEA* [2,3]; these tools were jointly used to refine *RNC* layout and check whether it satisfies the *ITER* measurement requirements for neutron (*n*) emissivity profile (10 % accuracy), spatial resolution (*SR*) (*a/10* with *a*= minor radius), time resolution (1 *ms*) and total neutron strength (10% accuracy) [1].

A sketch of the *RNC* lines of sight is shown in Fig.1a: the 3 set of 12 ex-vessel channels will be located on different toroidal planes (1° separation), embedded in a concrete shielding block anchored to the port plug; upper, lower and central (*in9*) in-vessel channels will lay in the port plug, each on a different toroidal plane; in-vessel detectors will be accommodated inside removable cassettes outside the machine vacuum in order to allow substitution/repair. Two possible sets of collimator diameters (Ø) are presently foreseen: 1 *cm* ex-vessel & 2 *cm* in-vessel and 2 *cm* ex-vessel & 4

CP988, *Burning Plasma Diagnostics*, edited by F. P. Orsitto, G. Gorini, E. Sindoni, and M. Tardocchi
2008 American Institute of Physics 978-0-7354-0507-3/08/$23.00

cm in-vessel. Considering a 1 *cm* thick *NE213* scintillator as detector (see section 2), the first set appears more appropriate for full power *DT* operation (and will be discussed in the present paper) and the second one for full power *DD* operation.

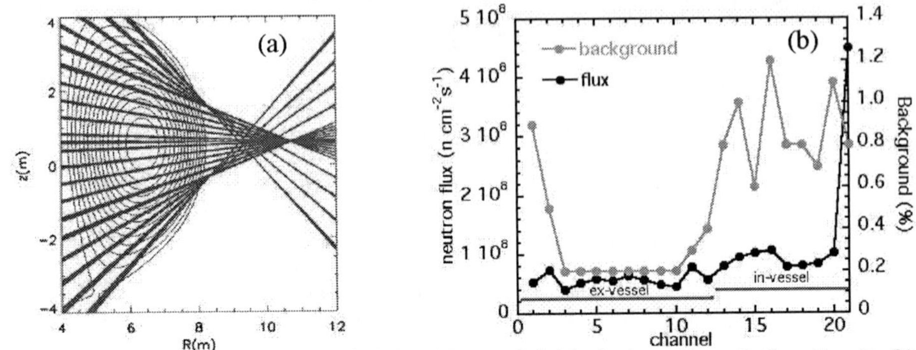

FIGURE 1. (a) Sketch of *RNC* lines of sight with actual field of views (1cm & 2cm Ø set); (b) expected 14 *MeV n* fluxes and background at *RNC* detectors (*MCNP* calculations, *ITER* scenario 2 [3]).

MCNP results for 14 *MeV n* fluxes and background (scattered *n* in the range 1-12.8 *MeV*) at the detectors positions are reported in Fig.1b; a typical 14 *MeV n* spectrum for an ex-vessel channel is reported in Fig.2, where also the estimated 2.5 *MeV n* spectrum during *DT* operations is shown (evaluated scaling by ¼ the results obtained with *MCNP* for a pure *DD* plasma): the 2.5 *MeV* peak is above the background produced by 14 *MeV n* indicating the measurability of the n_d/n_t ratio. More detailed analysis is needed to assess such result for all *RNC* channels. *MCNP* also indicates the gamma (γ) contribution to the total flux to be 3-20% depending on the line of sight [2].

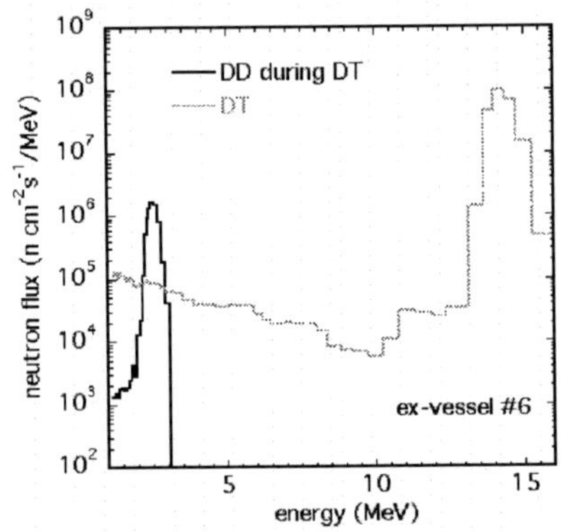

FIGURE 2. Comparison of *14 MeV* and *2.5 MeV n spectra* during *DT* operations for *RNC* ex-vessel line of sight #6.

The *RNC* neutron emissivity reconstruction capability has been analyzed with *MSST* considering background, counting statistics and random errors. Results for 1 cm thick *NE213* detector (see section 2) are reported in Fig.3a showing that 10% accuracy should be obtainable except that at the very plasma edge. *SR* has been investigated using double Gaussian emissivity test functions (separated by *a/10*) moved along the minor radius. With the standard *RNC* layout the *a/10* condition appears to be roughly achievable and a radial dependence of *SR* is observed [3]; the analysis also indicates that opposite tilting of two of the three ex-vessel channel sets should increase the achievable *SR* (Fig. 3b).

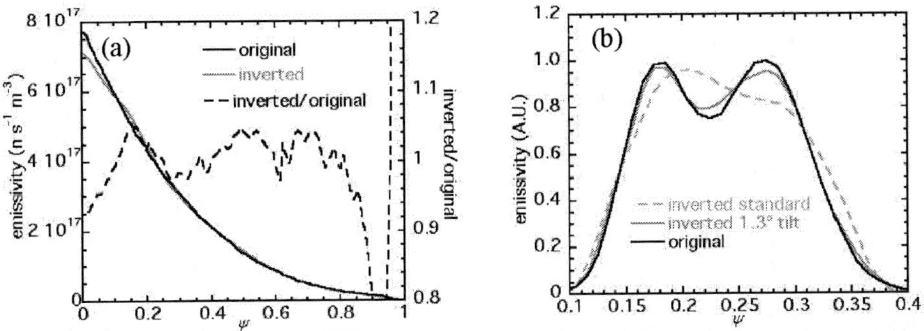

FIGURE 3. (a) comparison of *ITER* scenario 2 emissivity (original) and inverted emissivity with counting statistics error (*NE213* detector 10 *MeV* threshold, 1 *ms* time resolution), background and 5% random noise (the inverted/original ratio is also reported); (b) comparison between double gaussian emissivity test function (original, peak1 @ R=5.6 m, peak2 @ R=5.4 m, σ=7 cm), inverted emissivity with the standard *RNC* layout and inverted emissivity with opposite tilting (± 1.3°) of two of the tree ex-vessel detector sets. ψ= normalized poloidal magnetic flux.

2 DETECTORS AND ACQUISITION SYSTEM

Detectors with both flux and spectra measurement capability coupled to digital acquisition systems (*FPGA*-based, 200 *MSamples/s* sampling rate, 14-bit resolution, *PCI*-express data transfer to *PC*; see Fig.4a for data acquisition requirements) are presently foreseen for acquisition of *RNC* signals: liquid organic scintillators (such as *NE213*) and diamonds (natural (*ND*) or chemical vapor deposited (*CVD*)) are the main candidates as *RNC* detectors; a prototype of the digital acquisition system has been developed in *ENEA*-Frascati and tested both on the *JET* neutron profile monitor and at *PTB* (Physikalisch - Technische Bundesanstalt, Braunschweig, Germany) accelerator [4, 5, 6].

Liquid organic scintillators can provide a simultaneous measurement of the 2.5 *MeV* and 14 *MeV* *n* flux with a single detector unit: they have typical efficiency (with 1 *cm* thickness) of ~3% at 2.5 *MeV* (energy threshold (bias)=1 *MeV*), and ~1% at 14 *MeV* (bias=10 *MeV*). They are sensitive both to *n* and γ and provide pulse height spectra (*PHS*): pulse shape discrimination (*PSD*) and unfolding techniques are needed to determine *n* and γ count rates and spectra (measured energy resolution with digital systems <4%@ 2.5 *MeV* and <2% @14 *MeV* [6]). Diamond detection, being based on the ^{12}C(n, α) reaction (threshold ~8 *MeV*), is restricted to 14 *MeV* neutrons. Diamonds have lower intrinsic efficiency than liquid scintillators (~0.01% for 500 μm thickness), higher radiation resistance, low γ sensitivity and directly provide *n* spectra (~1% resolution @14 *MeV*). Expected *RNC* 14 *MeV* *n* count rates both with *NE213* and diamonds are reported in Fig.4b for *ITER* scenario 2.

The chosen digital acquisition system will provide several improvements compared to analog systems: handling of count rates > 1 *MHz* (foreseen in in-vessel channels with *NE213* (see Fig.4b)); storing of pulse data for off-line reprocessing; off-line pile-up elaboration; time resolved *n* pulse height spectra; real time control applications. Measurements performed on the *JET* neutron profile monitor with the *ENEA* digitizers have shown the possibility to follow the time evolution of high-energy neutron tails produced by *NBI* heating (Fig.4a) and to resolve with good accuracy pile-ups by fitting procedures (Fig.4b [5]).

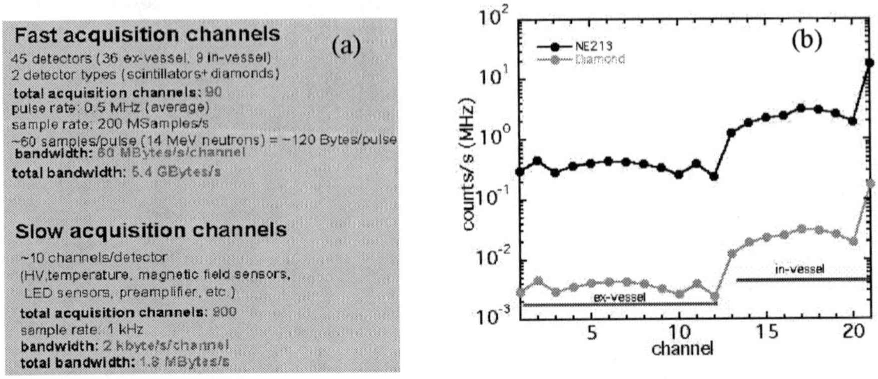

FIGURE 4. (a) *RNC* data acquisition requirements; (b) expected 14 MeV *n* count rates with *NE213* (1% efficiency) and diamonds (0.01% efficiency) for *ITER* scenario2 (*MCNP* calculations).

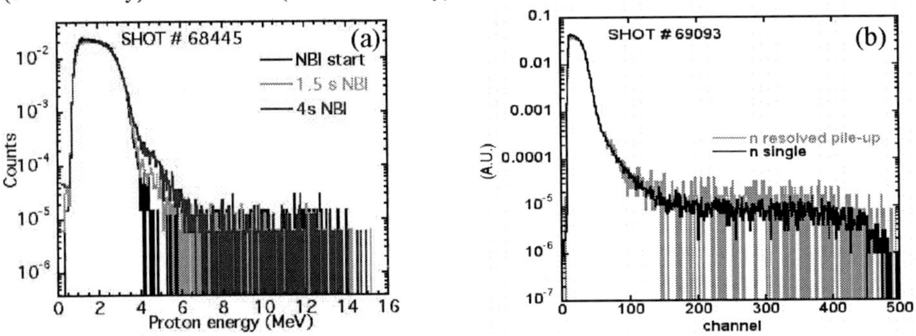

FIGURE 5. *n PHS* of plasma discharges obtained with digitizers coupled to *JET* neutron profile monitor scintillators: (a) time resolved *n PHS* of a *NBI* heated discharge (1 s time integration); (b) single and resolved pile-up *n PHS* (from [5]).

ACKNOWLEDGEMENT

The work leading to this article was funded by the European Atomic Energy Community and is subject to the provisions of the European Fusion Development Agreement.

REFERENCES

1. Project Integration Document PID, version 3.0, ITER_D_2234RH (2007).
2 L. Petrizzi et al., "Neutronic Design of the ITER Radial Neutron Camera", *Fusion Engineering and Design* (in press).
3 B. Esposito et al. "Final Report on the Design Analysis of the ITER Radial Neutron Camera", *Contract FU06-CT-2004-00172 (EFDA/04-1209)*.
4 D. Marocco et al. "A digital data acquisition system optimized for spectrometry with liquid scintillation detectors", *Proceedings of science* (http://pos.sissa.it/); FNDA 2006, Cape Town.
5 F. Belli et al. "A method for digital processing of pile-up events in organic scintillators", To be submitted to *Nuclear Instruments and Methods*.
6 B. Esposito et al., "Neutron measurements on Joint European Torus using an NE213 scintillator with digital pulse shape discrimination", *Review of scientific instruments*, 75 (10), 3550-3552, (2004).

Development of Fast Measurement System of Neutron Emission Profile Using a Digital Signal Processing Technique in JT-60U

M. Ishikawa*, T. Itoga†, T. Okuji†, M. Nakhostin†, K. Shinohara*, M. Baba†, T. Nishitani*

**Japan Atomic Energy Agency, Mukoyama 801-1 Naka, Ibaraki 311-0193, Japan*

†Cyclotron and Radioisotope Center, Tohoku University, Aoba, Sendai, Miyagi 980-8578, Japan

Abstract. Neutron emission profiles are routinely measured in JT-60U Tokamak. Stinbene neuron detectors (SNDs), which combine a Stilbene organic crystal scintillation detector (Stilbene detector) with an analog neutron-gamma pulse shape discrimination (PSD) circuit, have been used to measure neutron flux efficiently. Although the SND has many advantages as a neutron detector, the maximum count rate is limited up to ~ 1×10^5 counts/s due to the dead time of the analog PSD circuit. To overcome this issue, a digital signal processing (DSP) system using a Flash-ADC has been developed. In this system, anode signals from the photomultiplier of the Stilbene detector are fed to the Flash ADC and digitized. Then, the PSD between neutrons and gamma--rays are performed using software. The photomultiplier tube is also modified to suppress and correct gain fluctuation of the photomultiplier. The DSP system has been installed in the center channel of the vertical neutron collimator system in JT-60U and applied to measurements of neutron flux in JT-60U experiments. Neutron flux are successfully measured with cont rate up to ~ 1 $\times\ 10^6$ counts/s without the effect of pile up of detected pulses. The performance of the DSP system as a neutron detector is demonstrated.

Keywords: Neutron Measurement, Neutron Emission Profile, Neutron Detector, Digital Signal Processing

PACS: 29.40.Mc, 28.52.-s

INTRODUCTION

A measurement of neutron profile is of great importance for the knowledge of the profile of fusion power and the source of a-particles in burning plasmas and can also be used for plasma transport study in DD plasmas. Therefore, the neutron emission profile measurement is one of the key diagnostics in ITER [1, 2]. In JT-60U Tokamak [3], the neutron emission profile measurement has been performed using Stilbene neutron detectors (SNDs), which have been installed in the six channel neutron collimator system [4]. The SND combines a Stilbene organic crystal scintillation detector (Stilbene detector) with an analog neutron-gamma pulse shape discrimination (PSD) circuit for elimination of g-ray events. The SND has many advantages as a neutron detector, for example, excellent energy resolution and high gamma-ray suppression. So far, the neutron emission profile measurement has played an important role in analysis of fast ion transport due to MHD instabilities such as Aflvén Eigenmode in JT-60U [5]. However, the maximum count rate is limited to ~ 1×10^5 counts/s due to the dead time in the analog DSP circuit. Such a maximum count rate is not enough in ITER experiments. Because the maximum

CP988, *Burning Plasma Diagnostics*, edited by F. P. Orsitto, G. Gorini, E. Sindoni, and M. Tardocchi

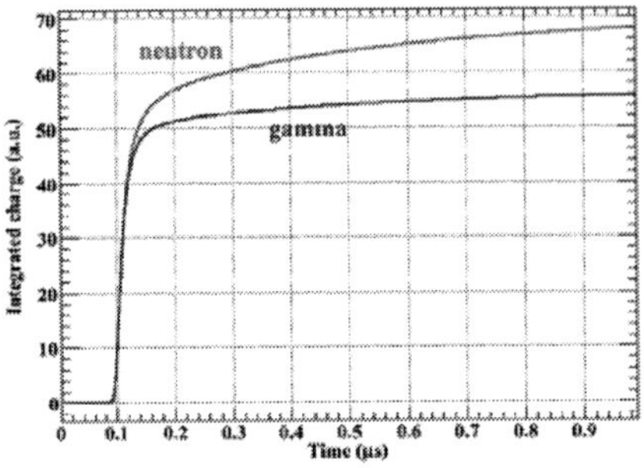

FIGURE 1. Integrated charge of a pulse produced by a neutron and a gamma-ray as a function of integration time

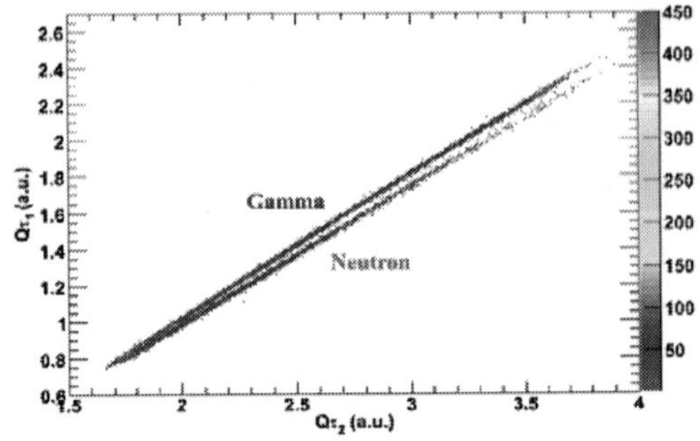

FIGURE 2. 3-D plots of integrated charge for the two integration intervals ($\tau_1 = 25$ ns, $\tau_2 = 60$ns)

count rate of neutron detectors of the neutron emission profile measurement in ITER is needed over 10^6 counts/s in order to meet the required 10 % accuracy. Therefore, improvement of the maximum count rate of neutron detectors is one of important issues for ITER. To overcome this issue, we have developed a digital signal processing (DSP) system for a Stilbene detector, which will replace the analog PSD system. Our new DSP system is based on a commercial Flash-ADC (Acquris DC282, 8 G sampling/s). By employing the DSP technique, PSD between neutrons and gamma-rays during a short period in the order of tens of nanoseconds, and even under the condition that two or more pulses overlap each other, will become possible. On the other hand, gain fluctuation of the photomultiplier under high counting rate is concerned. Since the gain fluctuation changes output signal, it affects neutron spectrum analysis and/or setting of threshold revel. In order to suppress and correct the gain fluctuation, the photomultiplier tube in the Stilbene detector is modified. After a performance test of the DSP system, It has been installed in the center channel of the vertical neutron emission profile monitor in JT-60U and measurements of neutron flux using the DSP system in JT-60U experiments has been started.

Experimental Setup

The DSP system is constructed from the Flash-ADC and a Personal Computer (PC). The data acquisition is carried out as follow; Anode signals of the photomultiplier of the Stilbene detector are directly connected to the Flash-ADC and digitized continuously. After the acquisition is completed, the data are transferred to the PC via PCI-bus. PSD between neutrons and gamma-rays is performed using a analyzing software in the PC. The software has been developed at Cyclotron and Radioisotope Center in Tohoku University, Japan. In order to discriminate between neutrons and gamma-rays efficiently, the charge comparison method is applied. The light pulse in the Stilbene detector comprise from fast scintillation component (3 ~ 5 ns) and slow scintillation component (~ 400 ns). Since the shapes of pulses produced by a neutron and a gamma-ray have different time function, integrated charges of pulses are different between a neutron and a gamma-ray as shown Fig.1. Compared with charges of pulses integrated for two different time interval, a neutron and gamma-ray can be discriminated.

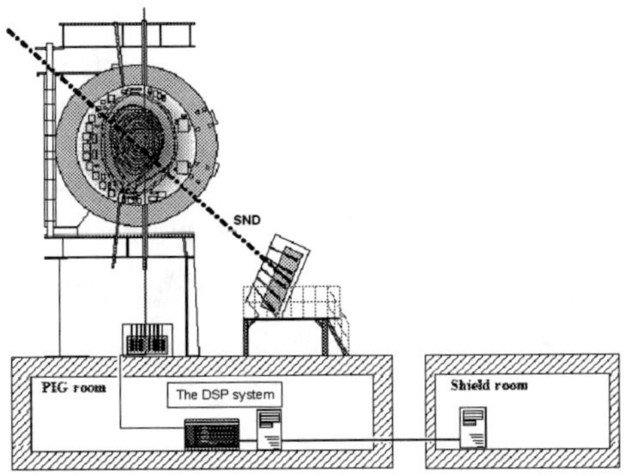

FIGURE 3. Schematic view of the neutron emission profile monitor

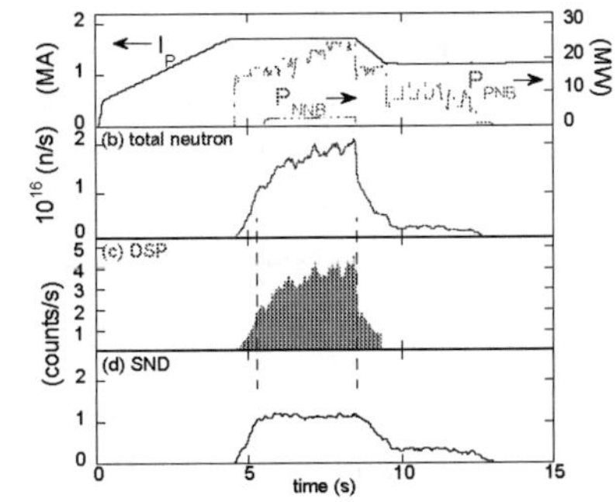

FIGURE 4. Time trace of (a) plasma current (Ip) and the injection power of neutral beam (P_{NB}), (b) the total neutron emission rate (Sn). (c) neutron flux measured by the DSP system and (d) by the SND.

3-D plot of charges integrated over two different time intervals for the Am-Be source is shown in Fig.2. Since the memory size of the Flash-ADC is limited, it is important that a neutron and a gamma-ray is discriminated efficiently, and in a short time. So, the optimization of two integration interval and the sampling rate is necessary. In order to optimize those, a performance test has been conducted using neutron + gamma-ray source Am-Be and ^{252}Cf. As a result, two integration intervals are selected at 25 ns and 60 ns, respectively, and the sampling late is selected at 200 MHz.

On the other hand, the gain fluctuation of the photomultiplier of the Stilbene detector is observed in the range of neutron flux over 10^5 ($n \cdot cm^{-2} \cdot s^{-1}$). The gain fluctuation changes an output signal and it affects not only neutron spectrum analysis but also the setting of threshold revel. The gain fluctuation is caused by small current in the voltage divider circuit of the photomultiplier tube in the Stilbene detector. In order to suppress it, we adopt new dividers circuit with higher divider current to make it insensitive to count rate change and to space charge effect. Further, a LED-optical fiber system for introducing the light of LED into a photomultiplier tube has been equipped to monitor the gain fluctuation quantitatively and enable the correction off line. By employing the new photomultiplier tube, it is demonstrated that the gain fluctuation is successfully suppressed and corrected in the neutron irradiation test conducted at Fusion Neutron Laboratory in Tohoku University.

Results and Discussion

The DSP system was installed in the center channel of the vertical neutron emission profile monitor in JT-60U as shown in Fig.3. The sight line of the center channel passes through almost the plasma center. A measurement of neutron flux using the DSP system is performed in JT-60U experiment. In order to investigate the performance of the neutron measurement using the DSP system, the measurement of neutron flux using the DSP system is compared with that using the existing SND (with the analog PSD circuit) in the center channel of the radial neutron emission profile monitor as shown in Fig.3. Because time trace of neutron flux in center channel of both neutron emission profile monitors is similar to that of a volume-averaged total neuron rate. Figure 4 shows the typical result of the neuron measurement using the DSP system. Time trace of plasma current (Ip) and beam injection power of the neutral beam (NB) are shown in Fig. 4 (a). Shown in Fig.4 (b) is the total neutron emission rate (Sn) measured by the Fission Chamber [6]. Sn increases with increasing power of the NB. The count rates of the Stilbene detector with the DSP system and the SND are shown in Fig.4 (c) and (d), respectively. The time trace of the count rate of the Stilbene detector with the DSP system is in agreement with that of Sn up to the $\sim 5 \times 10^5$ counts/s. On the other hand, the count rate of the SND is saturated at around 1.2×10^5 counts/s from 5.3 sec. to 8 sec. It is considered that the saturation is caused by the pile-up of pulses in the analog PSD circuit in the SND. So far, neutron flux has been successfully measured with the count rate up to $\sim 1 \times 10^6$ counts/s [7, 8]. Thus, the performance of the DSP system as a neutron detector is demonstrated.

The pulse height spectrum of neutrons measured with the DSP system is also investigated. It is found that though DD neutrons are dominant, about 1 % of neutrons in the energy range of a DT neutron are observed. Moreover, the pulse height spectrum in the energy range of a DT neutron are changed as the NB injection power changes. This satisfies that a measurement of neutron flux using the DSP system can be used as a neutron spectrometer.

SUMMARY

The DSP system using the Flash-ADC has been developed to improve the low maximum count rate limit of the SND. The optimization of two integration intervals and the sampling late is performed to discriminate between a neutron and a gamma-ray efficiently. The photomultiplier tube of the Stilbene detector is also modified to suppress and correct the gain fluctuation. Then, the DSP system has been installed in the center channel of vertical neutron collimator system and applied to measurements of neutron flux in JT-60U experiments. It is demonstrated that neutron flux has been successfully measured with the count rate up to $\sim 1 \times 10^6$ counts/s. Application to neutron spectrum measurement of the DSP system is also expected.

ACKNOWLEDGMENTS

This work is supported by Grant-in-Aid for Scientific Researches on Priority Areas from Ministry of Education, Culture, Sports, Science and Technology (No. 17044009)

REFERENCES

1. L. C. Johnson and Chris W. Barnes, Rev. Sci. Instrum., **70**, 1145-11148 (1999).
2. M. Sasao, et al., Nucl Fusion, **46**, S107-S118 (2004).
3. S. Ide, et al., Nucl. Fusion **45**, S48-S62 (2005).
4. M. Ishikawa, et al., Rev. Sci. Instrum., **73**,4237-4242 (2002).
5. M. Ishikawa, et al., Nucl. Fusion **45**,1474-1480 (2005).
6. T. Nishitani, et al., Rev. Sci. Instrum. **63**, 5270-5278 (1992).
7. M.Ishikawa, et al., Rev. Sci. Instrum., 77, 10E706-1 (2006).
8. T. Itoga, et al., Radiation Protection Dosimetry, 1-4 (2007).

Upgrade of the JET Gamma-Ray Cameras

S. Soare[1], V. Zoita[2], T. Craciunescu[2], M. Curuia[1], V. Kiptily[3], I. Lengar[4], A. Murari[5], P. Prior[3], M. Anghel[1], G. Bonheure[6], M. Constantin[1], E. David[1], T. Edlington[3], D. Falie[2], S. Griph[3], F. Le Guern[7], Y. Krivchenkov[3], M. Loughlin[3], A. Pantea[2], S. Popovichev[3], V. Riccardo[3], B. Syme[3], V. Thompson[3], I. Tiseanu[2] and JET EFDA contributors[8]

JET-EFDA Culham Science Centre, OX14 3DB, Abingdon, UK

[1]Association EURATOM-MEdC, National Institute for Cryogenics and Isotopic Technologies, Rm. Valcea, Romania
[2]Association EURATOM-MEdC, National Institute for Laser, Plasma and Radiation Physics, Bucharest, Romania
[3]Association EURATOM-UKAEA/JOC, Culham Science Centre, Abingdon, UK
[4]Association EURATOM-MHST, Jozef Stefan Institute, Ljubljana, Slovenia
[5]Association EURATOM-ENEA, RFX, Padova, Italy
[6]Association EURATOM-Belgian State, LPP-ERM/KMS, B 1000 Brussels, Belgium
[7]JET-CSU, Culham Science Centre, Abingdon, UK
[8]See the Appendix of M.L. Watkins et al., Fusion Energy 2006 (Proc. 21st Int. Conf. Chengdu, 2006) IAEA, (2006)

Abstract. The JET gamma-ray camera diagnostics have already provided valuable information on the gamma-ray imaging of fast ion in JET plasmas /1, 2/. The applicability of gamma-ray imaging to high performance deuterium and deuterium-tritium JET discharges is strongly dependent on the fulfilment of rather strict requirements for the characterisation of the neutron and gamma-ray radiation fields. These requirements have to be satisfied within very stringent boundary conditions for the design, such as the requirement of minimum impact on the co-existing neutron camera diagnostics. The JET Gamma-Ray Cameras (GRC) upgrade project deals with these issues with particular emphasis on the design of appropriate neutron/gamma-ray filters ("neutron attenuators"). Several design versions have been developed and evaluated for the JET GRC neutron attenuators at the conceptual design level. The main design parameter was the neutron attenuation factor. The two design solutions, that have been finally chosen and developed at the level of scheme design, consist of: **a)** one quasi-crescent shaped neutron attenuator (for the horizontal camera) and **b)** two quasi-trapezoid shaped neutron attenuators (for the vertical one). The second design solution has different attenuation lengths: a short version, to be used together with the horizontal attenuator for deuterium discharges, and a long version to be used for high performance deuterium and DT discharges. Various neutron-attenuating materials have been considered (lithium hydride with natural isotopic composition and ^{6}Li enriched, light and heavy water, polyethylene). Pure light water was finally chosen as the attenuating material for the JET gamma-ray cameras. The neutron attenuators will be steered in and out of the detector line-of-sight by means of an electro-pneumatic steering and control system. The MCNP code was used for neutron and gamma ray transport in order to evaluate the effect of the neutron attenuators on the neutron field of the JET GRC. The modelling was dedicated to the estimation of neutron and (plasma-emitted) gamma-ray attenuation, neutron-induced gamma-ray background and the neutron in-scattering impact on the neutron detectors due to the attenuator in the parking location. A numerical study of the gamma-ray detector (CsI(Tl)) was done by means of the IST Monte Carlo code. It provided preliminary results on the detector efficiency and response function.

CP988, *Burning Plasma Diagnostics*, edited by F. P. Orsitto, G. Gorini, E. Sindoni, and M. Tardocchi
2008 American Institute of Physics 978-0-7354-0507-3/08/$23.00

Keywords: tokamak, diagnostics, gamma-rays
PACS: 52.55.Fa, 52.70.La

1. INTRODUCTION

The JET gamma-ray camera diagnostics system (KN3 neutron/gamma-ray profile monitor) has already provided valuable information on the fast ion evolution in JET plasmas /1, 2/. The applicability of gamma-ray imaging diagnostics to high power deuterium and deuterium-tritium discharges is strongly dependent on the fulfilment of rather strict requirements for the control of the neutron and gamma-ray radiation fields. These requirements were augmented by the very hard design restrictions on JET (e.g., the requirement of minimum effects on the co-existing neutron camera diagnostics).

The main objective of the JET Enhancements (EP2) gamma-ray camera diagnostics upgrade is the design, construction and testing of neutrons attenuators for the two sub-systems of the KN3 gamma-ray imaging diagnostics: **a)** KN3 gamma-ray horizontal camera (KN3 HC) and **b)** KN3 gamma-ray vertical camera (KN3 VC). This diagnostics upgrade should make possible gamma-ray imaging measurements in high power deuterium JET pulses, and eventually in deuterium-tritium discharges.

Another objective of this upgrade project is to develop and test design solutions of relevance to ITER. Eventually, the JET KN3 gamma-ray cameras diagnostics upgrade should validate design solutions of interest for ITER.

2. JET GAMMA-RAY CAMERA NEUTRON ATTENUATORS

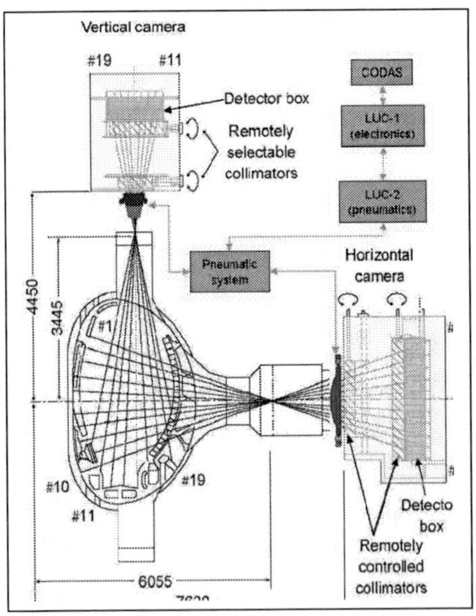

FIGURE 1.
neutron/gamma diagnostics with neutron attenuators and their

Several design versions were developed and evaluated for the JET gamma-ray camera neutron attenuators at the conceptual design level. The main design parameter was the neutron attenuation factor. The following design solutions were finally chosen and developed at the level of scheme design:

- One quasi-crescent shaped neutron attenuator for the horizontal camera
- Two quasi-trapezoid shaped neutron attenuators for the vertical camera, with different attenuation lengths: a short version, to be used together with the horizontal attenuator for deuterium discharges and a long version to be used for high performance deuterium and DT discharges.

The locations of the neutron attenuators are shown schematically in

Figure 1 together with the detector lines of sight of each of the two KN3 cameras. The components of the steering and control system are also shown (LUC, Local Unit Cubicle). The attenuators are placed within the KN3 diagnostics system in Octant 1 between the vacuum port and the collimator body (also called "radiation shield"), both in the case of the horizontal camera (HC) and vertical camera (VC) (Figure 1). The position of the neutron attenuators are is steered and controlled by a commercially available electro-pneumatic system and several additional custom-tailored parts (Figure 1).

The horizontal camera neutron attenuator is designed to function as a neutron filter when in working position (in the plane determined by the gamma-ray detectors lines of sight). To move the neutron attenuator to and from the working position two movements are required: first a 90° rotation (to the right when looking to plasma) and second a 630mm translation as shown in Figure 2.

The horizontal camera neutron attenuator (HC_NA) consists of a metal casing filled with the pure light water (as attenuating material) and a U-shaped profile that provides the structure with mechanical strength and connects with the steering and control system (Figure 2).

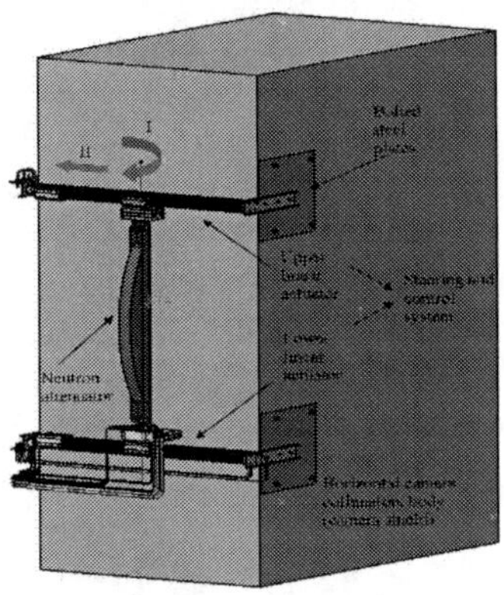

FIGURE 2. Horizontal Camera Neutron Attenuator in working

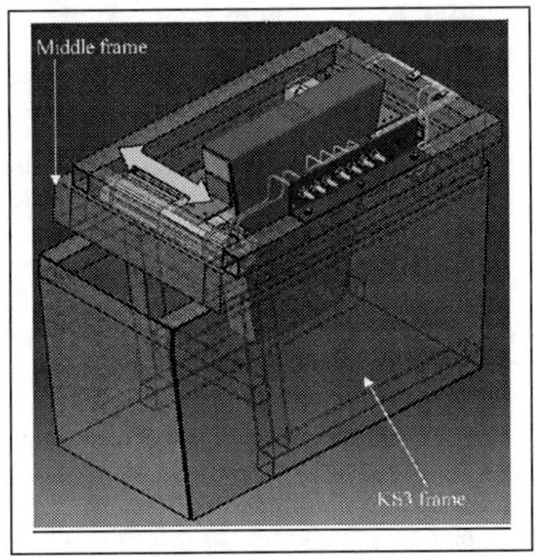

FIGURE 3. Vertical Camera Neutron Attenuator in working

The vertical camera neutron attenuator (VC_NA) is positioned on Octant 1, inside the KS3 optical diagnostics box (Figure 3). To move in and out of the working location the attenuator is translated 100mm by the steering and control electro-pneumatic system. Both vertical camera attenuator casings (short and long version) have a quasi-trapezoidal shape with internal reinforcements parallel to and between the lines of sight.

Both attenuators, of horizontal and vertical cameras, are situated in strong poloidal magnetic fields generated by the nearby poloidal and shaping coils. The vertical camera neutron attenuator is in a worse situation than that of the horizontal camera, due to larger torques from different directions. Finite Element Analysis was used to evaluate the casings deformation and stresses when subjected to torques larger than those thought to exist. Results suggest there is no cause of concern from this point of view.

3. RADIATION (NEUTRON AND PHOTON) ANALYSIS OF THE GAMMA-RAY CAMERA NEUTRON ATTENUATORS

Monte Carlo (MCNP, /3/) calculations of the neutron and photon transport were carried out for the vertical camera neutron attenuator and addressed to the following matters: attenuation factors for neutrons and plasma-emitted gamma-rays in the neutron attenuator; neutron induced gamma-rays in the neutron attenuator; relative level of the in-scattering of neutrons into the detector region from the attenuator in the parking position. The following results were obtained:

-attenuation factors (vertical camera neutron attenuator, short version) for 2.45 MeV neutrons: approximately 10^2 (short attenuator), and 10^4 (long attenuator)

-transmission of the plasma-emitted gamma-rays varies with photon energy from approximately 20% at 2 MeV to about 60% at 10 MeV

-level of the neutron in-scattering into the detector region from the vertical camera attenuator placed in parking position is estimated to be less than 3% of the overall in-scattering due to the collimator body (KN3 shield).

The influence of different CsI(Tl) gamma-ray detector geometries for an optimum detector response in the case of KN-3 gamma-ray imaging was analysed by means of the Integrated TIGER Series (ITS) Monte Carlo electron/photon radiation transport code /4/. The calculations were done for channel 15 (central channel) of the KN-3 vertical camera and they provided an evaluation of the global detector response. The response function of the CsI(Tl) detector was also calculated for the 1-10 MeV energy range using 256 energetic channels.

4. CONCLUSIONS

The design of the upgraded JET KN3 gamma-ray camera diagnostics has been developed up to the scheme design level. All issues regarding the impacts on adjacent systems ("interfaces") were addressed and adequate design solutions were developed. The analysis of the mechanical behaviour showed that the deformations were insignificant and will not present any causes of concern. The severe design constraints limited the attainable parameters of the upgraded gamma-ray cameras. The upgraded GRC diagnostics will work as a two-camera system for low power JET pulses and as a single (vertical) camera for high power deuterium and deuterium-tritium pulses. The addition of the neutron attenuators for the KN3 gamma-ray cameras has no significant impact on the co-existing neutron camera diagnostics.

ACKNOWLEDGEMENTS

The work leading to this article was funded by the European Atomic Energy Community and is subject to the provisions of the European Fusion Development Agreement.

REFERENCES

1 V. Kiptily et al., this conference

2 V. Kiptily et al., Plasma Phys. Control. Fusion, 48 (2006) R59–R82

3 Briesmeister J. F., Ed., "MCNP-A General Monte Carlo N-Particle Transport Code, Version 4C", LA-13709-M (April 2000)

4 Halbleib J.A. et al., "ITS version 3–The integrated TIGER series of Coupled Electron/Photon Monte Carlo Transport Code", Sandia National Laboratories.

Divertor Neutron Flux Monitor: Conceptual Design and Calibration

Kaschuck Yu.A., Batyunin A.V., Egorov O.G., Krasil'nikov A.V., Tsutskih A.Yu, Frunze V.V.

State Research Center of the Russian Federation, Troitsk Institute for Innovation and Fusion Research, Troitsk, Moscow oblast, 142190 Russia

Abstract. The operating conditions of a neutron diagnostic systems which are responsible for measuring of the total neutron yield in the ITER tokamak reactor are analyzed. Based on results of neutronic calculations and analysis of suitable methods for measuring the neutron yield, a concept of a system for neutron flux measurement in the divertor is proposed. The design for the neutron flux monitor located in the divertor cassette of the tokamak is selected in view of the requirements specified for the neutron diagnostic system of the ITER and its operating conditions. Several fission chambers with different sensitivities and radiator materials will be used for measurements. System is capable of neutron fluxes measuring over the entire dynamic range of the ITER neutron yield with an error of $\delta<10\%$ and a time resolution of 1 ms that are necessary for studying the physical phenomena of ignition and burning plasma. The problems of carrying out of the divertor neutron monitor efficiency calibration with the aim to measure the absolute value of the neutron yield in the ITER tokamak reactor are also discussed.

Keywords: ITER neutron diagnostic, neutron flux monitor.
PACS: 28.52.-s, 29.40.Cs

DNFM OPERATING CONDITIONS

The requirements specified for the ITER neutron yield measurements reactor are rather hard. The temporal resolution have to be less 1 ms and the error in determining the absolute neutron yield in the facility be $\leq 10\%$. These conditions impose a limitation on the minimum counting rate in a pulse mode of measurements (10^5 cps). If the duration of the fission chamber signal after preamplifier is $\leq$ 300 ns, the maximum counting rate reaches a value of $3x10^5$ cps at which the counts loss due to pulse pileup is $\leq 10\%$. Thus, the dynamic range of the fission chamber in which the required values of the time resolution and the measurement error are guaranteed is only 3. It should be noted that the requirement specified for the measurement error limits the dynamic range of the detector more rigidly than the requirement for the time resolution. Thus, for example, lowering of the time resolution to 2 ms increases the dynamic range to 6, whereas an increase in the measurement error to 20% expands the dynamic range to 10. The above results are an optimistic upper estimate, since they include only the error due to statistic and neglect the other factors [1].

CP988, *Burning Plasma Diagnostics*, edited by F. P. Orsitto, G. Gorini, E. Sindoni, and M. Tardocchi

From the above estimates, it follows that (1) four detectors, used in the pulse-counting mode, cover a dynamic range of ~100 ; for the ITER requirements to be met, fission chambers must operate mostly in the Campbell mode or in the current mode in order to cover the remaining range of neutron flux density, which is 10^5 , (2) the detectors must be installed in the area with the maximum neutron flux density, and (3) to provide a means for calibrating the absolute efficiency of the detectors, they must be located as close to the plasma as possible.

Neutron detectors installation at the first wall or placing them between the blanket modules seems to be impossible, because thermal loads at these places are extremely high. The best suitable site for the detectors of the neutron yield monitoring system is the divertor region. Results of MCNP calculations indicate that, at a maximum neutron yield of $2x10^{20}$ n/s, the neutron fluxes in the divertor are $2x10^{12}$ n/(cm^{2}s) for fusion neutrons and ~10^{13} n/(cm^{2}s) for fast neutrons with energies of 1–13 MeV.

DNFM CONCEPTUAL DESIGN

To expand the dynamic range of measurements, the detector unit for the divertor monitor of the neutron yield should be composed of four fission chambers, two of which have a ^{235}U radiator and the others radiators of ^{238}U. Two chambers with identical radiators are assembled in a common housing. These chambers differ in sensitivity by a factor of 10^3. To take into account the γ-ray background and electromagnetic noise, two chambers having an identical design and free from fissile materials will be located nearby. The chamber with a ^{235}U radiator is housed in a moderator block, in which water from the cooling system of the divertor assembly is used as a moderator. The typical thickness of the moderator is ~ 5 cm.

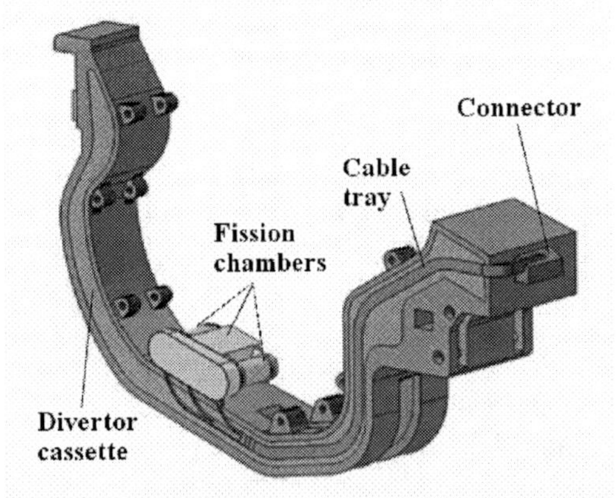

FIGURE 1. Integration of the neutron yield monitor into the divertor cassette.

The features of a fission chamber with a ^{238}U radiator were described in [2]. In this paper, it was shown that the radiator should be made of ultra-pure (≤99.999%) ^{238}U and that a boron shielding containing ^{10}B with an effective thickness of 1 g/cm^2 should be used to reduce uranium transmutation. A possible arrangement of the detector unit on the ITER divertor cassette is shown in Fig. 1.

CALIBRATION

The sensors of the neutron yield monitoring system may be located behind the blanket elements, inside either the divertor assembly or the plug of the horizontal port, and will thus be shielded from the fusion plasma. Therefore, it is necessary that absolute efficiency calibration be performed prior to beginning experiments. In all large tokamaks (TFTR, JET, JT-60U, and Tore Supra)[3], absolute efficiency calibration was carried out using a neutron source with the well-known characteristics (the spectrum, intensity, and anisotropy). The source is housed inside the vacuum vessel of the tokamak and moves in a toroidal direction along the plasma axis. A ^{252}Cf radionuclide neutron source is used for this purpose. To obtain acceptable value of the calibration error, a source with a yield of 10^{10}–10^{11} n/s is required for the ITER case.

The neutron spectra of all radionuclide sources differ from that of a thermonuclear facility. Therefore, a neutron generator based on a D–T reaction is an ideal neutron source for fusion plasma simulating. A compact neutron generator operating in both continuous and pulsed modes is currently considered as the main calibration source.

At low deuteron-accelerating voltages, neutrons produced by the fusion reaction D-T have an isotropic distribution. Nevertheless emission of the neutron generator has an essential anisotropy as a result of neutron scattering and absorption in its own construction. In addition, neutron yield may vary in the course of operation, thus introducing additional errors into calibration of the absolute sensitivity of neutron sensors. Therefore, radionuclide neutron sources characterized by a high isotropy of radiation should be also used during calibration.

Owing to the long-term stability of their neutron yield, radionuclide sources can be used to advantage in subsequent cross-calibration of new components added to the neutron yield monitoring system of the ITER. Analysis of the conditions in which the absolute efficiency of neutron yield monitor should be calibrated was based on the results of MNCP calculations. The model consists of the blanket's shielding modules and the vacuum chamber. The composition of materials is iron + water in the ratio of 50:50% vol. Calculations were done for a point source of 14-MeV neutrons, movable inside the vacuum vessel in a toroidal direction along the plasma axis with a step of 10°. The neutron energy spectra at the divertor level for several positions of the D–T neutron source at the axis plasma are shown in Fig. 2 a). The toroidal angles are shown with respect to the neutron yield monitor: an angle of 0° corresponds to the source located above the monitor, and an angle of 180° corresponds to the antipodal position of the source (behind the central column of the tokamak). At toroidal angles of ≥100°, the source lies beyond sight of the detector and the neutron spectrum consists of the albedo component only. Total neutron flux $\Sigma\Phi_n$ for fast neutrons (En > 1 MeV) and direct neutron flux $\Sigma\Phi_{14}$ (En = 14 MeV) per 1 source neutron at the DNFM position are shown in Fig. 2 b). The direct neutron flux varies between 6×10^{-7} n/(cm^2 s) per 1 source neutron at a toroidal angle of 0°. and zero at toroidal angles $|\theta| > 90°$. The contribution of scattered neutrons to the fluence is approximately 50% at $|\theta| < 90°$. For $|\theta| > 90°$, the neutron flux is governed solely by the albedo neutrons. Therefore, the source limited by angles $|\theta| < 120°$ will make the main contribution to the detector signals (98%).

The calibration time can be estimated assuming that the typical sensitivity of the detectors in the neutron yield monitoring system is ~1 cps/n/(cm^2 s)) for fission chambers based on ^{235}U and ~10^{-3} cps/n/(cm^2 s)) for fission chambers based on ^{238}U. A neutron generator with an intensity of 10^{11} n/s located at the plasma axis will produce in the divertor area a fast neutron flux varying from 1.2x10^5 n/(cm^2 s) at 0° to 6x10^2 n/(cm^2 s) at 180°. As a result, the count rate of the ^{238}U fission chamber will be ~0.6–100 cps. At an exposure time of 1000 s, the statistical error of measurements will be in the range of 0.4–4%. For the ^{235}U fission chamber, the count rate may be estimated using the conversion factor for fast neutrons into thermal neutrons, which is dependent on the moderator type and does not exceed 10^{-2}. The counting rate is expected to be ~10–10^3 cps, which guarantees a statistical error of measurements of better than 1%.

Fast neutron source used for calibration with 10^{10} n/s yield over 30 hours continuous irradiation will provide the total neutron yield 10^{15} neutrons. This value is by order of magnitude comparable to the neutron yield in the ITER operation in the deuterium phase, when the expected number of neutrons per discharge is ~(3–5)x10^{15}. If the mean detection efficiency is ε=10^{-9} counts per 1 source neutron, then, the counting of the ^{238}U and ^{235}U fission chambers will be 10^6 and 10^7, respectively, providing adequate statistics. As is shown in Fig. 2b), if a single DNFM is used, the efficiency strongly depends on the angle; it varies along the torus by a factor of 100 or more. This problem may be resolved by using several detectors uniformly distributed along the torus. Thus, the optimal solution is to use three detector systems located at an angle of 120° with one another, which will make it possible to essentially lower the angular dependence of the absolute efficiency of the DNFM.

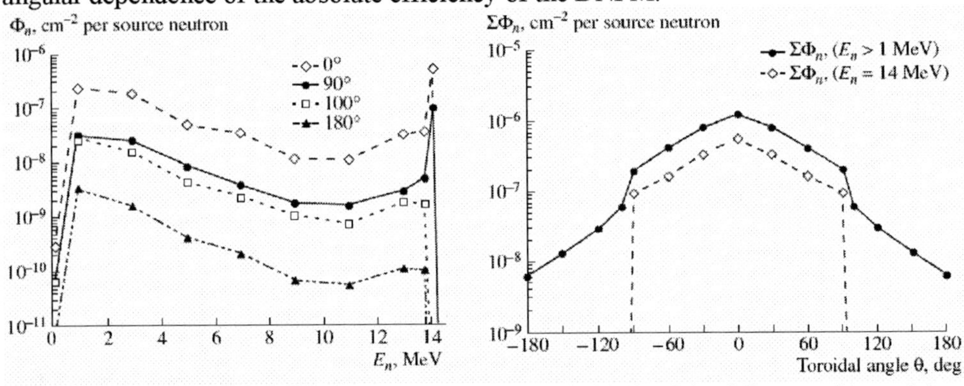

FIGURE 2. a) Neutron spectra per 1 source neutron at the DNFM level for different positions of the point source of *d–t* neutrons on the plasma axis. b) Total fast neutron flux (*En* > 1 MeV) and direct neutron flux (*En* = 14 MeV) per 1 source neutron at the DNFM level for a point source of *d–t* neutrons moving in a toroidal direction along the plasma axis inside the ITER vacuum vessel.

REFERENCES

1. Asai, K., Iguchi, T., Watanabe, K., et al., *Rev. Sci. Instrum.*, 2004, vol. 75, p. 3537.
2. Kashchuk Yu.A. and Frunze V.V., *Atomic Energy,* 2001, vol. 91, issue 3, p. 199.
3. J.D. Strachan et. al. *Rev. Sci. Instrum.*, 1990, vol. 61, No. 11, p. 3503

Prospects for High Resolution Neutron Spectroscopy on high power fusion devices in view of the recent diagnostic developments at JET

Göran Ericsson[1], E.Andersson Sundén[1], S.Conroy[1], M.Gatu Johnson[1], L.Giacomelli[1], C.Hellesen[1], A.Hjalmarsson[1], E.Ronchi[1], H.Sjöstrand[1], M.Weiszflog[1], J.Källne[2], G.Gorini[3], F.Ognissanto[3], M.Tardocchi[3], M.Angelone[4], S.Popovichev[5], and JET EFDA Contributors*

JET-EFDA, Culham Science Centre, OX14 3DB, Abingdon, UK
1) INF, Uppsala University, EURATOM-VR Association, Uppsala, Sweden
2) INF, Uppsala University, EURATOM-VR Association, Uppsala, Sweden
3) Istituto di Fisica del Plasma, EURATOM-ENEA-CNR Association, Milan, Italy
4) Association EURATOM-ENEA, Via E.Fermi, Frascati, Rome, Italy
5) EURATOM/UKAEA Fusion Association, Culham Science Centre, Abingdon, UK
**) See the Appendix of M.L.Watkins et al., Fusion Energy 2006 (Proc. 21st Int. Conf. Chengdu, 2006) IAEA, (2006)*

Abstract. An evaluation of three different candidate techniques for a 14-MeV High Resolution Neutron Spectrometer for a high power fusion device is presented. The performance is estimated for a modelled neutron emission for ITER plasma scenario 4. As performance indicators we use the estimated time-resolution achieved in measurements of three plasma parameters, namely, the ion temperature, the intensity of neutron emission due to neutral beam – thermal plasma interactions and the intensity of the so-called alpha knock-on neutron tail. It is found that only the MPR technique can deliver results on all three parameters with reasonable time resolution.

Keywords: Fusion, neutron, spectrometer, ITER
PACS: 52.70.–m 29.30.Hs 89.30.Jj 29.40.–

INTRODUCTION

Neutron diagnostics are an important tool in assessing the performance of high power fusion devices [1]. In particular, neutron spectrometers can supply detailed information on the state of fuel ions and confined alphas. Neutron monitors and so-called cameras are part of ITER's credited (i.e., budgeted) diagnostic systems, while a final decision on a High Resolution Neutron Spectrometer (HRNS) is still to be taken. In this contribution we evaluate three different techniques for a 14-MeV HRNS, namely, the Magnetic Proton Recoil (MPR), Time of Flight (ToF) and liquid scintillator (NE213). The evaluation is based on modelling of the neutron emission in a specific standard ITER scenario (S4). The techniques are evaluated based on their ability to measure three plasma parameters, pertaining to the ion temperature (T_i), the non-thermal (nt) fraction in Q (Q_{nt}/Q_{tot}) and the alpha number density (n_α).

MODELLED NEUTRON EMISSION

The information contained in the neutron emission is illustrated by model calculations for ITER scenario 4 (steady state, 9 MA, slight negative shear, P_{fus}=338 MW,

CP988, *Burning Plasma Diagnostics*, edited by F. P. Orsitto, G. Gorini, E. Sindoni, and M. Tardocchi
2008 American Institute of Physics 978-0-7354-0507-3/08/$23.00

P_{NB}=34 MW, Q=5) as shown in Fig. 1a. The calculation is for a HRNS in equatorial port cell 1, at about 18m from the centre of the machine, with a 300mm aperture in the First Wall (FW), tapering off to 10 cm^2 at the spectrometer. The neutron emission is dominated by a Gaussian-shaped component from the thermal plasma (green), with a weak component due to neutral beam - thermal plasma reactions (BP, red) and an even weaker component due to alpha knock-on neutrons (AKN, blue). The width of the thermal peak reflects the ion temperature, the width of the BP distribution is determined by the NB injection energy (1 MeV) and pitch angle, and the shape of the AKN is given by a model assuming classical slowing down of the fusion-born 3.5-MeV alphas. In Fig. 1b we show the situation for a hypothetical burning plasma, composed of only thermal and AKN components.

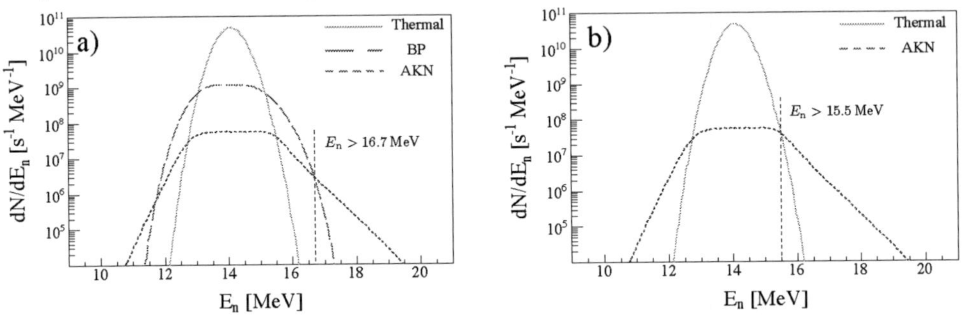

FIGURE 1. Examples of modelled neutron emission energy spectra at the ITER reference (radial) HRNS position for ITER reference scenario 4. a) Neutron energy spectrum divided into thermal (green), BP (red) and AKN (blue) components, b) same as in a) but without NB component (high-Q).

By measuring the intensities and shapes of the neutron emission components (thermal, BP and AKN), neutron spectroscopy can provide (line-integrated) information on a number of plasma parameters. Here we will restrict our study to the width of the thermal peak, pertaining to the ion temperature (T_i), the intensity of the heating component (I_{NB}), which determines the contribution to the fusion power from non-thermal (nt) sources (Q_{nt}/Q_{tot}), and the intensity of the AKN component (I_{AKN}), which gives information on alpha number density and plasma self-heating.

REQUIREMENTS ON HRNS

In order to access the full range of plasma parameters available to neutron spectroscopy (Fig. 1), the prime requirement on a HRNS is a very high sensitivity, meaning that weak components on the 10^{-4} level of the main emission peak must be discernible above any interfering background. Secondly, due to the weak intensity of the heating and alpha induced emissions, a very high count-rate of useful events is required to provide data of simultaneous good time resolution and accuracy; this demands an instrument of high count-rate capability. The energy bite of the spectrometer should cover the range 11 – 18 MeV. The instrumental resolution should be <5% (FWHM) although a poor resolution can to a certain degree be compensated by a well-known response function and high counting statistics. Operational stability (providing stable energy calibration) is a major concern, since even small fluctuations in the energy scale can induce large errors in the information extracted on the weak components.

INSTRUMENT CONSIDERATIONS

A number of neutron spectrometry techniques for fusion applications have been tested under quite different conditions over the years. During JET's full-scale DT campaign (DTE1) in 1997, instruments of the MPR, ToF (TANSY), thin-foil proton recoil (TPR - TANDEM) and (natural) diamond type were tested [2]. The MPR was put to the toughest test in terms of signal count rates (up to 610 kHz useful counts) and environmental conditions. More recently, the ToF technique has been tested under more challenging conditions with the TOFOR installation at JET, where signal rates of up to 40 kHz have been recorded, and rates up to 400 kHz are anticipated in conditions with full JET heating power. We thus consider the MPR and ToF techniques as proven under conditions approaching those on ITER, and we select them for the present study. In addition, the considerable interest in so-called compact spectrometers motivates including also an NE213 system. We also consider the diamond (in particular those produced with the Chemical Vapour Deposition technique) and TPR techniques as very interesting, although at present not fully proven, possibilities. The MPR technique is modelled after the one at JET, with a flux efficiency $1.5{\cdot}10^{-4}$ cm^2 and 4% (FWHM) energy resolution; this can represent all techniques with an ideal Gaussian response function. The hypothetical 14-MeV ToF instrument is modelled after JET's TOFOR spectrometer, with modifications to achieve 5% resolution. The response function, modelled with MCNP, exhibits low and high flight time tails due to multiple scattering events in the hydrogen and carbon nuclei of the instrument's scintillators. Another limitation is a rate-dependent background of random coincidences in the ToF spectrum. For the NE213 a response function given by the NRESP7 code is used [3], here with about 5% pulse height resolution. The high sensitivity of NE213 to stability variations (photomultiplier tube gain) is a major concern, but not included here.

PERFORMANCE POTENTIAL OF DIFFERENT TECHNIQUES

For the standard S4 ITER scenario, the count rates for the three instruments were 1 MHz for MPR, 400 kHz for ToF, and 1 MHz for NE213. The MPR rate is determined by the 300mm FW aperture, while for the ToF and NE213 techniques the count rate capability is the limiting factor. Perfect instrument stability was assumed. The extraction of plasma parameters was done with synthetic data analyzed using the so-called forward method, where three components (corresponding to the thermal, BP and AKN distributions), each with a free parameter corresponding to T_i, I_{NB} and I_{AKN}, are fitted to the data [4]. The total number of counts required to achieve the ITER specified accuracy level was determined and the corresponding time resolution given by the instrument's count rate. Results are presented in Table 1.

TABLE 1. Performance in terms of time resolution achieved for set ITER accuracy level of three neutron spectrometer techniques in the ITER reference position, for scenario S4 (Fig. 1a)

	Time resolution to achieve ITER accuracy in measurement of parameter; $T_i = 20$ keV, $I_{NB} = 6.9\%$, $I_{AKN} = 0.5\%$		
Technique (dE/E, FWHM)	**T_i (10%)**	**I_{NB} (10%)**	**I_{AKN} (20%)**
MPR (Gaussian, 4%)	<1 ms	20 ms	~2.5 s
ToF (MCNP, 5%)	10 ms	~1 s	(-)
NE213 (NRESP7, 5%)	80 ms	(5 s)	(-)

() Practically inaccessible due to response function, background and stability issues

For the burning plasma scenario (Fig. 1b), the time resolution in the I_{AKN} measurement for the MPR is 200 ms, due to the removed interference of the NB component.

DISCUSSION AND CONCLUSIONS

The results presented in Table 1 are obtained with idealized assumptions, i.e., no background and perfect stability. In reality, each instrument will have a sensitivity that is limited by both response function and background. For example, experience from JET has shown that the sensitivity set by background in the upgraded MPR instrument is at the level 10^{-5} to 10^{-6} of the peak amplitude. In fact, the MPR at JET has already observed the AKN distribution in less favourable conditions than those presented here [5]. The main limiting factors for the ToF technique, random coincidences and multiple scattering, are inherent in the method and already part of this analysis. For the NE213, the combination of background interference, non-ideal response function and stability limitation would in practice make measurements of the NB and AKN components inaccessible. In view of this we believe that only the MPR technique can deliver information on all three ITER relevant plasma parameters considered here. A suitable neutron diagnostic system for a high power fusion device, such as ITER, would consist of a 14-MeV MPR-type spectrometer for the highest yield conditions, complemented by a 2.5-MeV ToF instrument for the initial D phase of operations. Provisions should be taken for installing a suitable high-efficiency 14-MeV instrument for low-power conditions, e.g., of diamond or TPR type, to be decided on when experimental information for a more complete performance evaluation is available. We have found no interfacing issues that would prevent such a system on ITER.

ACKNOWLEDGEMENTS

The work leading to this article was funded by the European Atomic Energy Community and is subject to the provisions of the European Fusion Development Agreement

REFERENCES

1. A.V.Krasilnikov et al., Nucl. Fus. 45 (2005) 1246
2. O.N.Jarvis, NIM A 476 (2002) 474, and references therein
3. G.Dietze, H. Klein, Report PTB-ND-22, Braunschweig, Germany, 1982
4. E.Andersson Sundén et al., “Evaluation of spectral unfolding techniques ...”, these proceedings
5. J.Källne et al., Phys Rev Lett 85 (2000) 1246

The TOFOR Neutron Spectrometer For High-Performance Measurements of D Plasma Fuel Ion Properties

M.Gatu Johnson[1], L.Giacomelli[1], A.Hjalmarsson[1], M.Weiszflog[1], E.Andersson Sundén[1], S.Conroy[1], G.Ericsson[1], C.Hellesen[1], J.Källne[2], E.Ronchi[1], H. Sjöstrand[1], G.Gorini[3], M.Tardocchi[3] and JET EFDA contributors*

JET-EFDA, Culham Science Centre, OX14 3DB, Abingdon, UK
[1)] *Uppsala University, Department of Neutron Research, 75120 Uppsala, Sweden (EURATOM-VR)*
[2)] *Uppsala University, Department of Engineering Sciences, 75120 Uppsala, Sweden (EURATOM-VR)*
[3)] *Physics Department, Milano-Bicocca University, and Istituto di Fisica del Plasma del CNR, Milan, Italy (EURATOM-ENEA-CNR)*
*See the Appendix of M.L.Watkins et al., Fusion Energy 2006 (Proc. 21st Int. Conf. Chengdu, 2006) IAEA, (2006)

Abstract. The impact of scattered neutrons on the total flux reaching the TOFOR spectrometer at JET has been studied to allow for improvement of the data analysis. The scattered neutrons are demonstrated to contribute significantly to the flux. This will have implications for any neutron diagnostic on ITER.

Keywords: neutron spectrometer, time-of-flight, fusion diagnostics.
PACS: 52.70.-m 29.30.Hs 29.40.-n

INTRODUCTION

The TOFOR spectrometer [1,2,3] was designed, constructed and then installed in the JET roof laboratory for the purpose of measuring the 2.5-MeV neutron emission from $d+d \rightarrow {}^3He+n$ reactions. The collected neutron spectra can be analyzed using the tools of neutron emission spectroscopy (NES) to yield, e.g., information on ion temperature and the velocity states of the fuel ions.

The main design goal with TOFOR was to optimize the count rate achievable with the time-of-flight technique. TOFOR implements the proven constant time-of-flight sphere concept with the additional features of segmenting the in-beam detector in five layers to allow for higher count rate and using specially developed PC cards to make dead-time free recording of all detected events possible.

Detailed neutron transport modeling of the setup was performed to find the optimal balance between practicality, energy resolution and efficiency requirements. The energy resolution is 7.4% and the projected maximum count rate 0.5 MHz [4].

The analysis of neutron spectra measured with TOFOR is performed with the help of a response function R that relates the measured flight-times t_{tof} to the incident

CP988, *Burning Plasma Diagnostics*, edited by F. P. Orsitto, G. Gorini, E. Sindoni, and M. Tardocchi
2008 American Institute of Physics 978-0-7354-0507-3/08/$23.00

neutron energies E_n. R can be divided in two parts; the neutron response $R_n(t_{tof}, E_n)$ is based on the neutron transport modeling of the instrument and the pulse response $R_p(t_{tof}, E_n)$ includes the effects of proton energy loss in the detectors and signal processing in the electronics system on the measured time signals. R_n has been produced for the incident neutron energy interval E_n=1 to 7 MeV.

The accuracy in the analysis of TOFOR data depends on how well the response function describes the instrument and how well the analysis models used describe the measurement scenario. Fig.1 shows the result of NES analysis of TOFOR data from the NB heated interval from 65 to 72 s for JET discharge 70034. The fit represents the result of fitting a thermal and a high-energy NB component [5] to the spectrum.

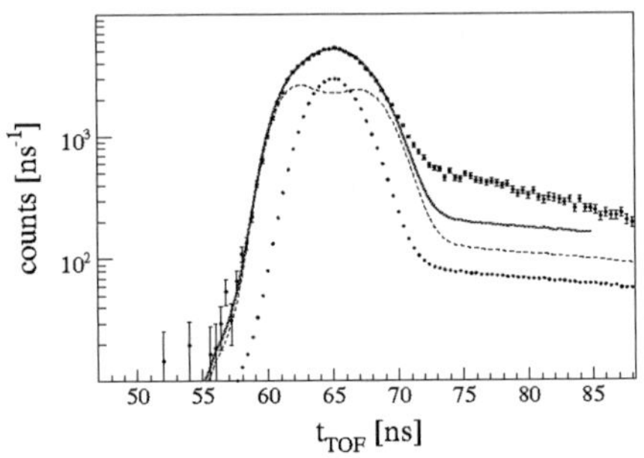

FIGURE 1. Fitted t_{tof} spectrum for NB heated JET discharge 70034. The measurement points represent the TOFOR counts integrated from 65 to 72 s, the solid curve the fitted sum of thermal (dotted) and HE-NB (dashed) components.

The neutron transport model used as the basis for the analysis in this case only takes into account the direct neutron flux from the plasma. As can be seen, the fit severely underestimates the contribution to the spectrum from neutrons on the high time side of the peak. This paper investigates the effect of the backscattered neutrons on the measured spectrum.

THE SCATTERED COMPONENT

TOFOR views the JET plasma vertically through a collimator in the roof laboratory floor. Direct neutrons emitted from the JET plasma in the upward direction as well as neutrons scattered on the torus walls and into the instrument sight line will reach the instrument. The characteristics of the scattered neutron flux have been investigated through MCNP simulations. The MCNP model used incorporates the geometry of the JET vessel, with the lower vertical port filled with inconel and a 1 cm layer of carbon added on the structure to account for the divertor tiles. The neutron source is a 1 m^2 square on the plasma axis with toroidal symmetry assumed.

To confirm the validity, the results from simulations for a uniform D plasma with a 2.5-keV temperature have been applied to Ohmic TOFOR data.

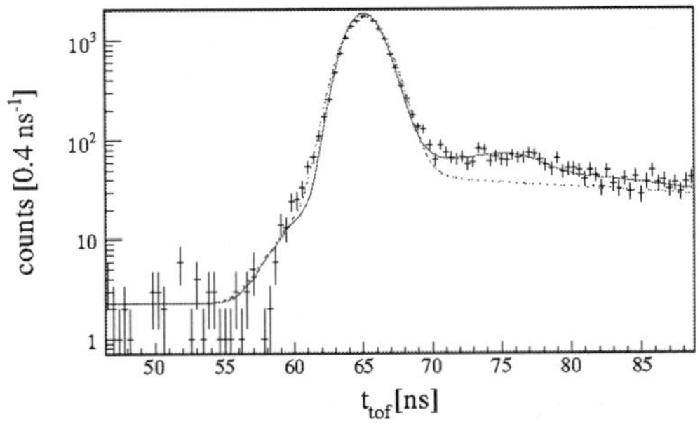

FIGURE 2. TOFOR t_{tof} spectrum for Ohmic parts of JET discharges 69550 to 70750 (summed). The solid curve represents a fit using the MCNP data for a 2.5-keV plasma. For comparison, a fit using a simple Gaussian component to describe the 2.5 MeV neutron peak is shown (dashed curve).

Fig.2 shows the MCNP result and, for comparison, a simple Gaussian, both folded with the response function and fitted to Ohmic TOFOR data. The spectrum is summed over JET discharges 69550 to 70750 with any plasma periods having auxiliary heating applied excluded. The MCNP result with the scattered flux taken into account matches the data better. TOFOR t_{tof} corresponding to 180° elastic scattering of 2.5-MeV neutrons on Fe and C are centered at 66.8 and 76.2 ns, while that for direct neutrons is centered at 65 ns. The C peak is visible in the spectrum; the Fe peak adds a broadening on the low energy side of the main peak. The broadening of the fitted simple Gaussian corresponds to a 4.5-keV plasma, i.e., warmer than expected from the Ohmic JET discharges. This is due to the Fe scattered component not being taken into account.

The fraction of indirect to total neutron flux integrated from 1.0 to 2.6 MeV in the simulation for a 2.5-keV plasma is 12%. This corresponds to an intensity ratio between the 2.5-MeV peak and the C peak in the spectrum of about 25.

The scattered component has been simulated for incident neutron energies in the interval 1.975-3.025 MeV to allow for inclusion of the component in NES analysis of NB heated JET discharges.

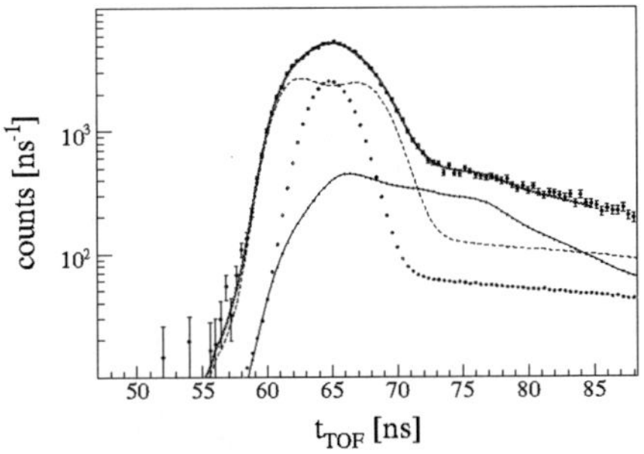

FIGURE 3. Fitted t_{tof} spectrum for NB heated JET discharge 70034 integrated from 65 to 72 s with the scattered neutron component included. The top solid curve represents the sum of a thermal (dotted), an HE-NB (dashed) and the scattered (solid broad) component.

Adding the scattered component in the fit for JET discharge 70034 (compare fig.1) yields the result shown in fig.3. The temperature of the thermal component comes out lower than for the fit in fig.1 as is to be expected. The broader spectrum of incident neutron energies in the NB heated case causes the scattered component to be smeared out compared to the Ohmic case (fig.2). The C peak is no longer clearly visible.

IMPLICATIONS FOR ITER

The demonstrated impact of the scattered flux on the neutron spectrum means that this component needs to be taken into account for ITER as discussed in [6]. It will have implications for the design of any neutron diagnostic and raises the issue of whether it will be possible to distinguish the 2.5-MeV signal in a DT plasma.

Using the JET MCNP model described above to simulate both 14 and 2.5-MeV fluxes from a 2.5-keV plasma source yields the result shown in fig.4. Here, the total 14-MeV rate is assumed to be 10^{18} and the 2.5-MeV rate 1/400 of that.

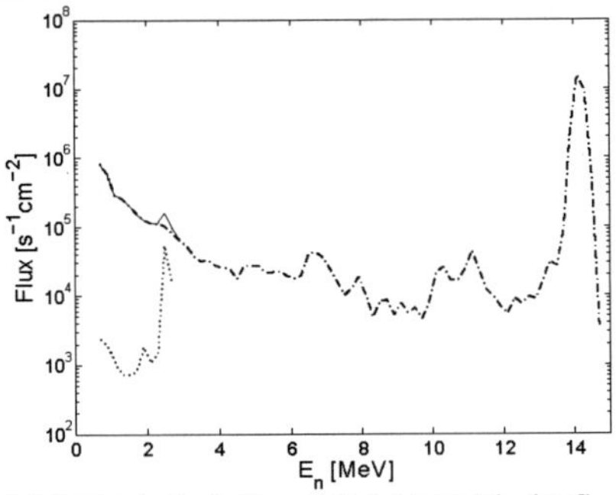

FIGURE 4. Simulated total 14-MeV (dash-dot) and 2.5-MeV (dash) fluxes for the TOFOR sight line from a JET DT plasma. Also shown is the sum (solid).

As can be seen, the 2.5-MeV peak is heavily obscured by the 14-MeV scattered flux. The collimation for TOFOR is essentially 20 m [4], which makes the results valid on an ITER relevant scale. To distinguish the 2.5-MeV peak would require diagnostics with both high energy resolution and ability to deal with a high DT flux.

ACKNOWLEDGEMENTS

The work leading to this article was funded by the European Atomic Energy Community and is subject to the provisions of the European Fusion Development Agreement.

REFERENCES

1. G. Gorini and J. Källne, Rev. Sci. Inst. 63 10 (1992) 4548-4550.
2. A. Hjalmarsson et al., Rev. Sci. Inst. 72 (2001) 841-844.
3. A. Hjalmarsson et al., Rev. Sci. Inst. 74 (2003) 1750-1752.
4. M. Gatu Johnson et al., "The 2.5-MeV neutron time-of-flight spectrometer TOFOR for experiments at JET", submitted for publication in Nucl. Instr. Meth. A, September 2007
5. C. Hellesen et al., "Measurement and analysis of the neutron emission from ICRH and NB heated JET D plasmas using the TOFOR spectrometer", UU-NF 07#13 Uppsala University, Neutron Physics Report ISSN 1401-6269 (August 2007), and to be published
6. P. Antozzi et al., Nucl. Instr. Meth. A 368 (1996) 457-472.

Evaluation of spectral unfolding techniques for neutron spectroscopy

Erik Andersson Sundén[1], S.Conroy[1], G.Ericsson[1], M.Gatu Johnson[1]
L.Giacomelli[1], C.Hellesen[1], A.Hjalmarsson[1], J.Källne[2], E.Ronchi[1],
H.Sjöstrand[1], M.Weiszflog[1], G.Gorini[3], M.Tardocchi[3] and JET EFDA contributors[*].

JET-EFDA Culham Science Centre OX14 3DB, Abingdon, UK

[1) Uppsala University, Department of Neutron Research, 75120 Uppsala, Sweden (EURATOM-VR Association)

[2) Uppsala University, Department of Engineering Sciences, 75120 Uppsala, Sweden (EURATOM-VR Association)

[3) Physics Department, Milano-Bicocca University, and Istituto di Fisica del Plasma del CNR, Milan, Italy (EURATOM-ENEA-CNR Association)

[*) See the Appendix of M.L.Watkins et al., Fusion Energy 2006 (Proc. 21st Int. Conf. Chengdu, 2006) IAEA, (2006)

Abstract. The precision of the JET installations of MAXED, GRAVEL and the L-curve version of MAXED has been evaluated by using synthetic neutron spectra. We have determined the number of counts needed for the detector systems NE213 and MPR to get an error below 10% of the MAXED unfolded neutron spectra is determined to be $\sim 10^6$ and $\sim 10^4$, respectively. For GRAVEL the same number is $\sim 10^7$ and $\sim 3 \cdot 10^4$ for NE213 and MPR, respectively.

Keywords: Unfolding, MAXED, GRAVEL, L-curve, MPR, NE213
PACS: 29.30.Hs, 29.40.-n, 52.70-m, 89.20.Ff, 89.30.Jj

INTRODUCTION

All neutron spectrometers measure physical quantities, e.g. light yield, time of flight, etc., which in turn are processed and interpreted as a neutron spectrum. Unfolding can be used for processing the measured data and obtaining the neutron spectrum. The advantage of the unfolding method is that the analysis can be totally unbiased. Several unfolding codes have been developed, e.g. MAXED [1], GRAVEL [2], etc. This paper evaluates how well these codes can unfold data with detector response functions (DRF) of two neutron diagnostic techniques, namely the Magnetic Proton Recoil (MPR) and of a liquid scintillator, NE213.

METHOD

We have setup a method to evaluate the performance of different unfolding techniques using synthetic neutron spectra of various shapes. In this paper, we limit our study to a Gaussian shaped neutron emission, corresponding to purely thermal

CP988, *Burning Plasma Diagnostics*, edited by F. P. Orsitto, G. Gorini, E. Sindoni, and M. Tardocchi
2008 American Institute of Physics 978-0-7354-0507-3/08/$23.00

plasma with an ion temperature of 20 keV. The DRFs in the study are those used at JET for analysis purposes. The target χ^2 used for MAXED is 1.1.

The synthetic neutron spectra are folded with a spectrometer response function resulting in an ideal instrument response without statistical fluctuations. Poisson errors are added to this to simulate a realistic measurement. The different unfolding codes are applied to these Poisson perturbed data, resulting in an unfolded spectrum. A quality value, q, is calculated for each unfolded spectrum

$$q = \frac{\sum_i |unf_i - synth_i|}{\sum_i synth_i} \tag{1}$$

where unf_i is the unfolded spectrum's i:th bin value and $synth_i$ is the same for the original synthetic spectrum. The quantity q shows how well a specific unfolding technique can recover the original neutron spectrum. This process is repeated many times and for each unfolded spectrum values of T and q are extracted. In this way we can assemble a frequency distribution of q and T, which are studied to evaluate the performance of the unfolding technique.

STARTING GUESSES OF UNFOLDING CODES

For a proper unbiased evaluation the unfolding should start with a uniform initial guess. If the starting guess is altered from uniform to peaked the result of the unfolding is changed, introducing a systematic error. This is illustrated in Fig. 1, where the unbiased analysis gives a distribution of T around the original synthetic data, while a peaked initial guess gives a distribution of T skewed towards lower values than the original. Thus, the results of the unfolding are dependent on the starting guess as shown in Fig. 1. This effect decreases with the number of counts.

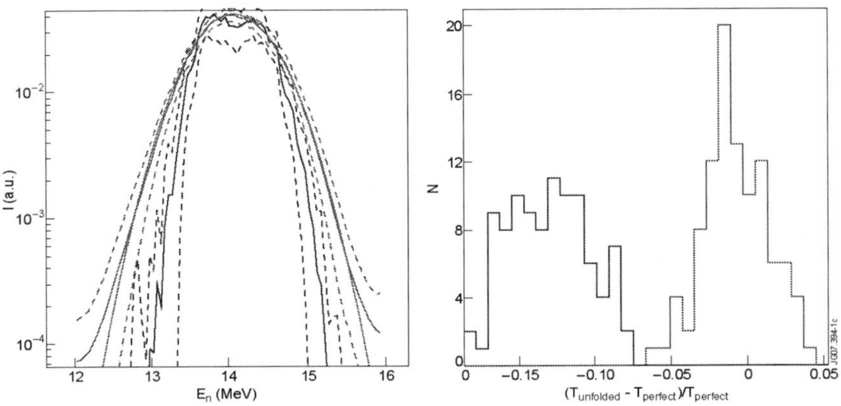

FIGURE 1. Left frame: Neutron spectra from an unfolding using MAXED on synthetic MPR data with 1000 number of counts. Green curve is the original synthetic spectrum, i.e. the correct answer. Solid blue and red curves are the averages of 100 runs (with standard deviations given by the dashed lines of the same color). Solid red curve is the result of unfolding with a uniform initial guess and solid blue curve corresponds to a Gaussian starting guess (of 1 keV). Right frame: The relative temperature

difference distribution of the unfolded and the synthetic spectrum using a uniform initial guess (red curve) and the same for blue curve, but for a Gaussian initial guess.

MAXED AND GRAVEL

Fig. 2 illustrates the performance of three different unfolding techniques for different number of counts. For MAXED, the synthetic data is reproduced with a q of about 10% with ~10^4 and ~10^6 counts for the MPR and NE213, respectively. For GRAVEL, the synthetic data is reproduced with a q of about 10% with ~$3{\cdot}10^4$ and ~10^7 counts for the MPR and NE213, respectively. A typical Gaussian DRF, like MPR, is easier to analyse for the unfolding codes.

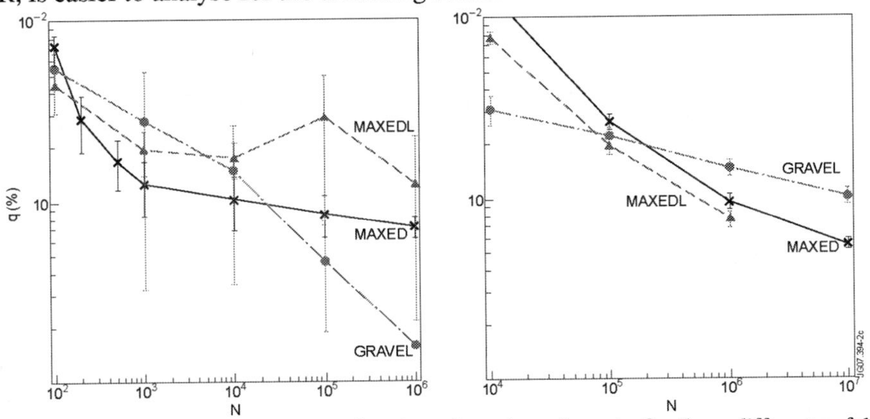

FIGURE 2. The quality value, q, as a function of number of counts for three different unfolding techniques, namely MAXED (black), GRAVEL (green) and MAXED-L (red). Left panel for MPR and right panel for NE213.

MAXED L-CURVE

An inherent problem of unfolding techniques is the risk of over fitting, i.e. using too low minimisation quantity, for example χ^2 in the case of MAXED. One way of dealing with such problems is to use the so-called L-curve method [3]. The L-curve implementation of MAXED (MAXED-L) is designed to find the best χ^2 for the problem and thus avoiding the risk of over fitting. To study this we have used ordinary MAXED to investigate if MAXED-L is finding the right χ^2.

By using the method described above and using different target χ^2 the corresponding q curves can be obtained as a function of the number of counts in the synthetic data (Fig. 3 left frame). The curve with the lowest q given a number of counts correspond to the best χ^2 to use. The best χ^2 to use for a given number of counts is plotted in Fig. 3 (right frame, solid line) together with the χ^2 calculated by MAXED-L. The unevenness of the curve is an effect of insufficient number of MAXED runs per data point in the analysis presented in the left frame of Fig. 3. MAXED-L seems to fail to find the best χ^2 of the problem, except for low number of counts.

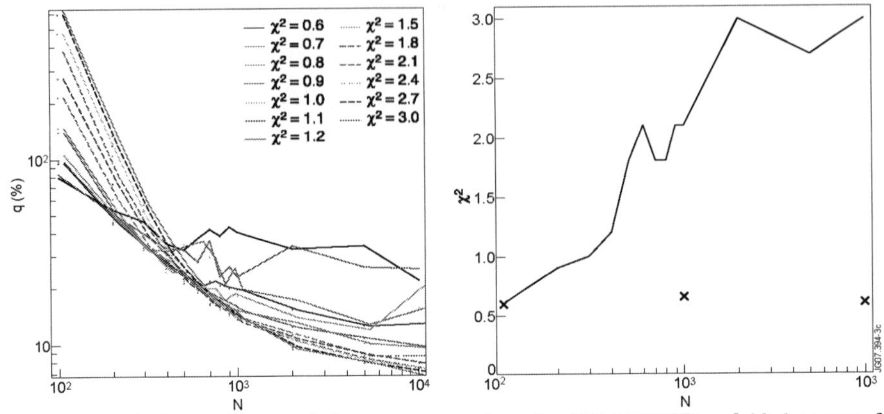

FIGURE 3. Left frame: Curves of the average q values for 100 MAXED unfolded spectra for a number of target χ^2 values as a function of the number of counts in the synthetic data. Right frame: The best χ^2 according to left frame as a function of number of counts (line) and the χ^2 suggested by MAXED-L (stars).

CONCLUSIONS AND OUTLOOK

The result of MAXED is dependent of the initial guess of the unfolding. This implies that it is important to show what starting guess is used when unfolding is done.

The NE213 detector system needs about 100 times more counts than the MPR system to achieve the same level of error.

The JET installation of MAXED-L seems to underestimate the best target χ^2 to use.

There are alternatives to MAXED and GRAVEL for finding the neutron spectra, e.g. MFR [4] and neural networks. MFR is still under development and therefore not included in this study. A future study would benefit of including, at least, these two unfolding techniques and using more DRFs of other instruments, e.g. diamonds or a hypothetical 14 MeV Time of Flight spectrometer.

ACKNOWLEDGMENTS

This work, supported by the European Atomic Energy Community under the contract of Association between EURATOM/VR(Sweden), was carried out within the framework of the European Fusion Development Agreement (EFDA Task on NES TW6-TPDS-DIASUP) and is subject to its provisions. The views and opinions expressed herein do not necessarily reflect those of the European Commission.

REFERENCES

1. Marcel Reginatto, Paul Goldhagen, Sonja Neumann, Nuclear Instruments and Methods A 476, Issues 1-2(2002) 242-246.
2. Reginatto M., Wiegel B., Zimbal A. UMG code package version 3.2. PTB. (2003)
3. P. Ch. Hansen, *Inverse Problems* **8** (1992) 849-872.
4. Jan Mlynar, J. M. Adamsa, L. Bertalot, S. Conroy, *Fusion Engineering and Design* **74** (2005) 781–786.

Fusion Power Measurement Using a Combined Neutron Spectrometer-Camera System at ITER

Henrik Sjöstrand[1], E.Andersson Sundén[1], S.Conroy[1], G.Ericsson[1], M.Gatu Johnson[1], L.Giacomelli[1], C.Hellesen[1], A.Hjalmarsson[1], J.Källne[2], E.Ronchi[1], M.Weiszflog[1] and JET EFDA contributors*.

JET-EFDA, Culham Science Centre, OX14 3DB, Abingdon, UK
1) INF, Uppsala University, EURATOM-VR Association, Uppsala, Sweden
2) Engineering Sciences, Uppsala University, Uppsala, Sweden
(Proc. 21st Int. Conf. Chengdu, 2006) IAEA, (2006)
** See the Appendix of M.L.Watkins et al., Fusion Energy 2006 See the Appendix of M.L.Watkins et al., Fusion Energy 2006 (Proc. 21st Int. Conf. Chengdu, 2006) IAEA, (2006)*

Abstract. A central task for fusion plasma diagnostics is to measure the 2.5 and 14 MeV neutron emission rate in order to determine the fusion power. A new method for determining the neutron yield has been developed at JET. It makes use of the magnetic proton recoil neutron spectrometer and a neutron camera and provides the neutron yield with small systematic errors. At ITER a similar system could operate if a high-resolution, high-performance neutron spectrometer similar to the MPR was installed. In this paper, we present how such system could be implemented and how well it would perform under different assumption of plasma scenarios and diagnostic capabilities. It is found that the systematic uncertainty for using such a system as an absolute calibration reference is as low as 3 % and hence it would be an excellent candidate for the calibration of neutron monitors such as fission chambers. It is also shown that the system could provide a 1 ms time resolved estimation of the neutron rate with a total uncertainty of 5 %.

Keywords: Fusion, diagnostics, neutron spectroscopy
PACS: 52.70.–m 29.30.Hs 89.30.Jj 29.40.– n

INTRODUCTION

The ultimate goal of fusion research is to produce fusion power and hence its determination will be an essential requirement at ITER. By separately measuring the 2.5 and 14 MeV neutron yield (Y) and neutron rate (R) at ITER the fusion energy and power can be determined. The fusion power measurement requirement for ITER is set to 10 % accuracy with 1 ms time resolution [1]. At JET a system consisting of a Magnetic Proton Recoil (MPR) High Resolution Neutron Spectrometer (HRNS) and a neutron camera has been used to measure the 14 MeV neutron yield independent of other neutron diagnostics [2] and this paper describes how a similar system could be used at ITER. The paper will concentrate on the ability to measure Y and the magnitudes of the systematic relative uncertainties that are associated to such measurements. The measured Y provides an absolute calibration for time resolved diagnostics such as fission chambers. The paper also investigates the accuracy of which R can be determined directly with the spectrometer-camera-system.

CP988, *Burning Plasma Diagnostics*, edited by F. P. Orsitto, G. Gorini, E. Sindoni, and M. Tardocchi
2008 American Institute of Physics 978-0-7354-0507-3/08/$23.00

METHOD

In order to measure Y the spectrometer's neutron flux F has to be determined and its relation (p) to Y has to be known. This is described in detail in Ref. [2], and simplified in Equation 1.

$$Y = \int_t R \,\partial t = \int_t F \cdot p^{-1} \,\partial t = \int_t \frac{C}{e} \cdot p^{-1} \,\partial t \qquad (1)$$

where p depends on the neutron emission profile (NEP), C is the spectrometer count rate, t is the integration time and e is its efficiency. Besides the terms given in Equation 1, material effects, such as scattering, attenuation and transmission have to be taken into account.

The uncertainties in this paper are relative and divided into systematic (λ), random (σ) and total (Δ), where $\Delta = \sqrt{\lambda^2 + \sigma^2}$. The uncertainty of Y and R, ΔY and ΔR, is given by the uncertainty of the parameters in Equation 1 combined with the uncertainties in the material effects. To evaluate these uncertainties two different ITER lines of sight (LOS) have been investigated. One is for the ITER HRNS reference position in equatorial port cell 1 and the other is for a hypothetical tangential LOS.

C and e have to be known in order to determine the flux at the spectrometer. As reference an MPR type spectrometer with a 4 % resolution setting was used. The systematic uncertainty of the efficiency, λ_e, for the MPR is 2.6 % [2] and similar or better performance is expected for a dedicated ITER spectrometer. ΔC is dominated by counting statistics, which gives σ_R.

In this work the NEP have a quasi-parabolic shape (see Equation 2) and is assumed to consist of 10 nested elliptical iso-emissivity contours (IECs)[1]. Each IEC has a centre in the poloidal plane and a neutron emission (NE).

$$\mathrm{NE}(r_n) = \mathrm{NE}_0 \left(1 - (r_n)^2\right)^{\alpha}, \qquad n = 1 \ldots 10 \qquad (2)$$

where n is the index of the IEC, r_n is the normalized distance to the plasma centre, NE_0 is the peak neutron emission and α is the peaking factor. In this work the radial position of the common centre has been fixed to 6.2 m. The height above the mid-plane (Z) of the centre was varied between 0 and 1.2 m and the peaking factor was varied between 0 and 10.

The λ_p depends on the NEP and its magnitude is determined by λ_Z, λ_{NE} and the uncertainty in the alignment of the LOS, where as σ_p depends on σ_Z and σ_{NE}. The propagation of the LOS and plasma position uncertainties was calculated by varying the centre of the NEP and evaluating how p changes with Z. The ITER requirement for the overall uncertainty in the determination of the NE in each IEC is 10 % [1]. λ_{NE} has still to be evaluated and is dependent on the characteristics of the ITER neutron camera. For this work three different scenarios have been examined with ΔNE of 2 %, 5 % and 10 %, where Δp was determined for a wide range of NEPs. This was done by reconstructing each IEC using ΔNE as the standard deviation and NE as the mean in a Gaussian random number generator and thereby producing a randomized NEP for which p was evaluated. This was repeated 1000 times with the spread in p giving Δp and hence ΔY.

In order to assess the material effects a MCNP [3] model of ITER and the MPR collimator have been set up to calculate the amount of scattered and transmitted[2]

[1] The ITER requirement for the spatial resolution is a/10, where a is the minor radius.

[2] Those neutrons penetrating the collimator

neutrons reaching the spectrometer. The amount of attenuation and scattering in the vacuum window has also been assessed. The model consists of a conical steel collimator with a radius of 17 mm at the foil and a radius of 150 mm at the first wall and a 5 mm vacuum window 5 m from the plasma edge. The neutron source in the model has full toroidal and poloidal coverage in order to correctly calculate the scattering contributions. MCNP point detectors (F5 tallies) were used [3].

RESULTS

The uncertainty for different parameters varies with changing plasma conditions. Figure 1a shows how ΔY varies as a function of α and ΔNE. The difference between using a radial and a tangential LOS is also illustrated. Figure 1b shows the Y dispersion in Z, $\partial Y / \partial Z$, as a function of Z and α, giving $\Delta Y = \partial Y / \partial Z \cdot \Delta Z$.

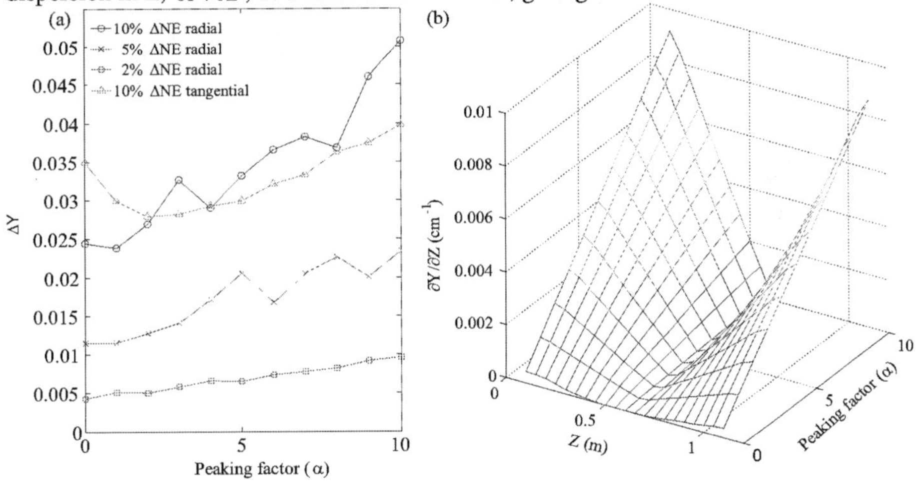

FIGURE 1. (a) ΔY as a function of α. **(b)** The $\partial Y_n / \partial Z$ as a function of Z and α.

A well-designed collimator should minimize the scattered and transmitted flux. It was found that a conical collimator where the angle and the size of the cone increase towards the plasma fill these criteria. In addition, the flux at surfaces close to the spectrometer should be minimized in order to reduce the scattered flux at the foil. It was established that scatters in the vacuum window gives a very small contribution (0.05 % of the direct flux), due to its distance from the plasma. This number would have to be re-evaluated if a vacuum window was placed close the plasma. Back-scattered neutrons (central column) is the only unavoidable scattering contribution.

A summary of the different contributions to λ_Y can be seen in Table 1 for a plasma (p1) with Z = 0.4, α = 2 and λ_{NE} = 5 % and for an extreme plasma (p2) with Z = 0, α = 5 and λ_{NE} = 10 %.

TABLE 1. Parameter sizes and their contribution to λ_Y

Contributor	Magnitude	λ_Y (p1)	λ_Y (p2)
Scattered neutrons	0.4 %	0.04 %	0.04 %
Transmitted neutrons	0.25 %	0.025 %	0.025 %
Attenuation (5 mm port)	8 %	0.4 %	0.4 %
λ_{NE} (see Fig.1a)	5 % / 10 %	1.3 %	3.2 %
λ_Z (see Fig. 1b)	1 cm	0.3 %	1.2 %

LOS alignment uncertainty	0.5 cm	0.1 %	0.4 %
Efficiency	$1.44 \cdot 10^{-5}$	2.6 %	2.6 %
SUM (in quadrature)	-	3.0 %	4.3 %

If the system is used to measure R a significant contribution to the total uncertainty comes from σ_C and σ_{NE} [4], which both are time dependent.

TABLE 2. Parameter sizes and their contribution to ΔR for a 400MW plasma

Time resolution	σ_{NE}	$\sigma_R(\sigma_{NE})$	$\sigma_R(\sigma_C)$	$\lambda_R=\lambda_Y$	$\Delta R=\sqrt{\sigma_R^2(\sigma_{NE})+\sigma_R^2(\sigma_C)+\lambda_R^2}$
1 ms	5 %	1.3 %	3.5 %	3.0 %	4.8 %
10 ms	2 %	0.5 %	1.1 %	3.0 %	3.2 %
100 ms	2 %	0.5 %	0.4 %	3.0 %	3.1 %

DISCUSSION AND CONCLUSION

The results given in Table 1 supports the idea that a HRNS with a broad central LOS combined with a neutron camera can determine Y with an accuracy (3.0 %) well below the ITER requirements. Consequently, the system could be used as an independent absolute calibration of the ITER neutron monitors. The uncertainty in the reconstructed NEP is an important contribution to λ_Y. Therefore, a research effort to evaluate the systematic uncertainties associated to the ITER camera should be pursued. λ_e is the dominating term for λ_Y and hence the choice of HRNS is very important. Thin foil HRNSs like the MPR have the advantage of e mainly depending on the well-known hydrogen cross section, which reduces the uncertainty.

ACKNOWLEDGEMENTS

The work leading to this article was funded by the European Atomic Energy Community and is subject to the provisions of the European Fusion Development Agreement

.REFERENCES

1. ITER Design Description Document N 55 DDD 2 01-07-05 W 0.3
2. H.Sjöstrand et al. 2006 UU-NF ISSN 1401-6269, http://www.inf.uu.se/Reports/Internal.html
3. MCNP manual, LA-UR-03-1987, Los Alamos
4. B. Esposito et al. ITPA Meeting-Moscow, 2006.04.11

Compact neutron spectometers and their performance under fusion-relevant conditions

A. Zimbal, M. Reginatto, H. Schuhmacher

Physikalisch-Technische Bundesanstalt, 38116 Braunschweig, Germany

Abstract. The Physikalisch-Technische Bundesanstalt (PTB) has many years of experience with the development, detailed characterization and use of organic liquid scintillation detectors. Such detectors, e.g. NE213, are routinely used as high resolution compact neutron spectrometers in many nuclear experiments. In recent years, a series of measurements with a carefully characterized detector was carried out by PTB in collaboration with ENEA and EFDA-JET at the Joint European Torus (JET). The results show that these detectors can be used as high resolution compact neutron spectrometers for diagnostic purposes in fusion research. The neutron energy spectrum can be reliably determined from the acquired pulse height spectrum provided the response function of the detector is well determined, a suitable unfolding method is applied and care is taken to ensure stable operation of the system during the measurement. We describe in this paper tests of the performance of the spectrometer carried out at the PTB accelerator. The irradiation facilities at PTB are well suited for such investigations since they allow testing of neutron and photon detectors over a wide range in energy and intensity. Among the reference neutron fields available, there are fields with energies of 2.5 MeV and 14 MeV, produced by means of D(d,n) and T(d,n) reactions. Detailed calculations of the energy spectra have been carried out, and they indicate that the widths of the peaks are comparable to those of thermal fusion plasmas with Maxwellian temperatures of the order of 1 keV. Measurements carried out with a compact neutron spectrometer under fusion-relevant conditions (i.e., using reference fields with energies of 2.5 MeV and 14 MeV and suitable ranges of count rates and statistics) are discussed in terms of the energy resolution that is achievable and the influence of the different irradiation conditions.

Keywords: neutron spectrometry, fusion diagnostics, scintillator, Bayesian methods.

PACS: 52.70.-m, 29.30.-h.

PTB ACCELERATOR FACILITY AND NE213 DETECTOR

The PTB accelerator facility can provide monoenergetic (24 keV < E_n < 19 MeV) neutron reference fields and fields with high intensity and broad energy distribution. These fields are produced with two ns-pulsed accelerators, a Van-de-Graaf (3.75 MV) and a cyclotron (E_p < 19 MeV, E_d < 13 MeV). The reference fields are completely characterized in terms of energy distribution and fluence. The characterization and test of neutron detectors is done in a low scatter experimental hall (24 m x 30 m x 14 m). For more information on the PTB facility, see reference [1].

The measurements of neutron spectra described in this paper were made with an NE213 liquid organic scintillator [2], which has a scintillator cell 5.08 cm in diameter and length coupled to a fast photomultiplier via a partially coated light guide to

CP988, *Burning Plasma Diagnostics*, edited by F. P. Orsitto, G. Gorini, E. Sindoni, and M. Tardocchi

minimize the geometrical dependence of the light transmission. The light collection system is designed to optimize the pulse height resolution. An LED is used to feed a signal into the light guide, and this signal is used to correct for count rate dependent changes in the gain of the photomultiplier. This is essential for applications where the neutron flux varies rapidly, e.g., fusion experiments. The photomultiplier signal from the anode and the 9th dynode are fed into different electronic chains for fast n/γ-separation by pulse-shape discrimination and energy determination using integration of the events. This results in separate pulse height spectra (PHS) for neutrons and photons. The ranges of neutron energies E_n and photon energies energy E_γ that can be measured with this spectrometer are 1.5 MeV < E_n < 20 MeV and 0.2 MeV < E_γ < 10 MeV. More details on the detector, its calibration and the determination of the detector response matrix can be found in references [3, 4].

COMPARISON OF MEASURED AND CALCULATED SPECTRA

We made measurements of neutron beams with energies ~2.5 MeV and ~14 MeV to test the energy resolution that is achievable with our spectrometer. The measured PHS, with total counts of the order of ~10^7, were unfolded using the maximum entropy method (the PTB code MAXED [5]) and the unfolded spectra were compared to calculations made with the TARGET code [6]. Details of the analysis can be found in reference [7]. The results are shown in Fig. 1. The measured mean energies derive directly from the response functions (i.e., no adjustments to the energy scale are made after measurements); at ~14 MeV the differences are of the order of only ~0.8 %. The high quality of the unfolded results reflects the careful characterization of the spectrometer and the good quality of the response matrix.

FIGURE 1. Calculated (thin lines) and measured (thick lines) energy distributions for neutrons using the reactions $d+d \rightarrow {}^3He+n$ (left) and $d+T \rightarrow {}^4He+n$ (right) at different detector positions. The neutron emission angles (measured relative to the incoming deuteron beam) are 106° (black) and 0° (grey) for the plot on the left, and 98° (black) and 0° (grey) for the plot on the right. The calculations have been normalized to match the total fluence of the measurement [7].

FULL WIDTH AT HALF MAXIMUM DETERMINATION AND MEASUREMENT STATISTICS

We analyzed 25 measurements of ~2.5 MeV neutrons produced by the reaction $d+d \rightarrow {}^3He+n$ (neutron emission angle of 106° relative to the incoming deuteron beam) with different statistics. The analysis of the data was done using two approaches [7]: (1) maximum entropy unfolding, and (2) a "two step method" in which Bayesian parameter estimation with a Gaussian model is followed by maximum entropy unfolding. Fig. 2 (left) shows one example of a measured spectrum derived using the "two step method"; the inset shows the measured PHS (histogram) as well as the PHS that results from folding the response matrix with the measured neutron spectrum (line). Fig. 2 (right) shows a plot of the full width at half maximum (FWHM) vs. the total number of counts in the PHS for different statistics, for both methods of analysis. The linear fit indicates that there is no trend with statistics when the "two step method" is used, and the scatter of the data points indicates that the uncertainty of the estimate decreases with better statistics, as expected.

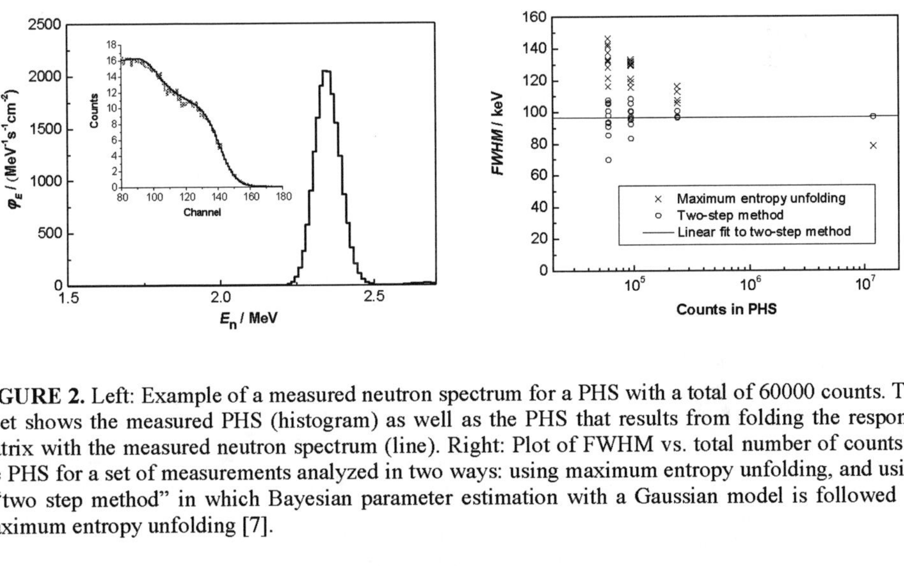

FIGURE 2. Left: Example of a measured neutron spectrum for a PHS with a total of 60000 counts. The inset shows the measured PHS (histogram) as well as the PHS that results from folding the response matrix with the measured neutron spectrum (line). Right: Plot of FWHM vs. total number of counts in the PHS for a set of measurements analyzed in two ways: using maximum entropy unfolding, and using a "two step method" in which Bayesian parameter estimation with a Gaussian model is followed by maximum entropy unfolding [7].

CONCLUSIONS

In recent years, a series of measurements with a carefully characterized NE213 spectrometer was carried out by PTB in collaboration with ENEA and EFDA-JET at the Joint European Torus (JET) [4,8,9,10,11]. The results indicated that these detectors can be used as high resolution compact neutron spectrometers for diagnostic purposes in fusion research. This work presents additional results from measurements made under controlled conditions at the PTB accelerator, where it is possible to look at the agreement between unfolded and calculated spectra and to study the effect of statistics on the quality of the result. The PTB facilities are well suited for these types of investigations, since neutron detectors can be tested over a wide range in energy and intensity. Neutron fields with energies of ~2.5 MeV and ~14 MeV are available, and detailed calculations of the energy spectra indicate that the widths of the peaks are comparable to those of thermal fusion plasmas with Maxwellian temperatures of the order of 1 keV. With well known detector parameters (response matrix), extremely good resolution can be achieved with NE213 spectrometers. With appropriate methods of data analysis, parameters like the FWHM can be determined with low uncertainty even in cases where the statistics of the measurements is not optimal (e.g., (96 ± 10) keV FWHM for a PHS with 60000 counts at E_n = 2.5 MeV).

ACKNOWLEDGMENTS

We thank Horst Klein, Ralf Nolte and our collaborators at ENEA and EFDA-JET, in particular Luciano Bertalot and Andrea Murari, for many valuable discussions. We also want to acknowledge the support of Angelika Kasper and of the members of the PTB accelerator team.

REFERENCES

1. R. Nolte, M.S. Allie, R. Böttger, F.D. Brooks, A. Buffler, V. Dangendorf, H. Friedrich, S. Guldbakke, H. Klein, J.P. Meulders, D. Schlegel, H. Schuhmacher and F.D. Smit, *Radiat. Prot. Dosim.* **110**, 97-102 (2004).
2. H. Klein and F.D. Brooks, "Scintillation Detectors for Fast Neutrons", in *Proceedings of "International Workshop on Fast Neutron Detectors and Applications"*, PoS(FNDA2006)097, http://pos.sissa.it//archive/conferences/025/097/FNDA2006_097.pdf.
3. H. Klein and S. Neumann, *Nucl. Instrum. Meth.* **A 476**, 132-142 (2002).
4. A. Zimbal, H. Klein, M. Reginatto, H. Schuhmacher, L. Bertalot, A. Murari and the JET-EFDA Contributors, "High Resolution Neutron Spectrometry with Liquid Scintillation Detectors for Fusion Applications", in *Proceedings of "International Workshop on Fast Neutron Detectors and Applications"*, PoS(FNDA2006)035, http://pos.sissa.it//archive/conferences/025/035/FNDA2006_035.pdf.
5. M. Reginatto, P. Goldhagen and S. Neumann, *Nucl. Instr. and Meth.* **A 476**, 242-246 (2002).
6. D. Schlegel, *Target User's Manual*, Report PTB-642-05-2 (2005).
7. M. Reginatto and A. Zimbal, "Bayesian and maximum entropy methods for fusion diagnostic measurements with compact neutron spectrometers", submitted for publication to *Review of Scientific Instruments*.
8. B. Esposito, L. Bertalot, D. Marocco, M. Riva, Y. Kaschuck, D. Skopintsev, A. Zimbal, M. Reginatto, H. Schuhmacher, J.M. Adams and A. Murari, *Review of Scientific Instruments* **75**, 3550-3552 (2004).
9. A. Zimbal, M. Reginatto, H. Schuhmacher, L. Bertalot, B. Esposito, F. Poli, J.M. Adams, S. Popovichev, V. Kiptily and A. Murari, *Review of Scientific Instruments* **75**, 3553-3555 (2004).
10. L. Bertalot, J.M. Adams, M. Angelone, S. Conroy, B. Esposito, Y. Kaschuck, H. Henriksson, P. Lamalle, D. Marocco, A. Murari, N. Hawkes, M. Pillon, S. Popovichev, M. Reginatto, M. Riva, H. Schuhmacher, D. Stork, K.D. Zastrow, A. Zimbal and the JET EDFA contributors, *Fusion Engineering and Design* **74**, 835-839 (2005).
11. L. Bertalot, S. Conroy, A. Murari, M. Reginatto, H. Schuhmacher, A. Zimbal and the JET EFDA contributors, "Neutron Energy Measurements of Trace Tritium Plasmas with NE213 Compact Spectrometer at JET", *Proceedings of the 32nd EPS Conference on Plasma Phys.*, Tarragona (Spain), 27 June 1 July 2005, ECA Vol. 29C, P 1.078 (2005).

Measurement of DT and DD components in neutron spectrum with a double-crystal time-of-flight spectrometer

K. Okada[a], K. Kondo[b], K. Ochiai[c], S. Sato[c], T. Nishitani[c], C. Konno[c], A. Okamoto[a], S. Kitajima[a], M. Sasao[a]

[a]*Tohoku University, Sendai 980-8579, Japan*
[b]*Osaka University, Yamadaoka 2-1, suita, 565-0871, Japan*
[c]*Japan Atomic Energy Agency, Tokai, Naka, Ibaraki 319-1195, Japan*

Abstract. To investigate the deuteron and triton density ratio in core plasmas, a new methodology with measurement of tritium (DT) and deuterium (DD) neutron count rate ratio using a double-crystal time-of-flight (TOF) spectrometer is proposed. Multi-discriminator electronic circuits for the first and second detectors are used in addition to the TOF technique. The optimum arrangement of the detectors and discrimination window were examined considering the relations between the geometrical arrangement and deposited energy using a Monte Carlo Code, PHITS (Particle and Heavy Ion Transport Code System). An experiment to verify the calculations was performed using DD neutrons from an accelerator.

Keywords: fuel ratio, TOF spectrometer, neutron spectrometer, PHITS code

PACS: 52.70.Nc

Introduction

To maintain stable tritium (DT) burning in ITER, it is essential to determine the deuteron and triton density ratio, n_d/n_t, in a core plasma for burning control [1]. The fuel ion density ratio is related to the deuterium (DD)/DT neutron emission ratio (S_{DD}/S_{DT}) and the Maxwellian reactivity $<\sigma v>$ ratio, which depend on the ion temperature [2], and is given by:

$$\frac{n_{\mathrm{d}}}{n_{\mathrm{t}}}=\frac{1}{2}\frac{S_{\mathrm{DD}}}{S_{\mathrm{DT}}}\frac{\langle\sigma v\rangle_{\mathrm{DT}}}{\langle\sigma v\rangle_{\mathrm{DD}}} \tag{1}$$

The DD/DT neutron emission ratio, obtained by measurement of DT and DD neutron spectra, and the ion temperature must be measured [3]. The required performance for measurement of fuel ion density ratio is $a/10$ in spatial resolution, 100 ms in time resolution, and 20% accuracy [4]. The required count rate is 250 cps.

Here, a double-crystal time-of-flight (TOF) method was studied as the means of measurement. However, TOF methods have a critical issue for measurement in ITER. The DT reaction will be dominant and the DT neutron emission rate is 200 times higher than that of DD neutrons under the condition $n_d/n_t = 1$, because $<\sigma v>_{DT}$ is 100 times higher than that of $<\sigma v>_{DD}$. These DT neutrons prevent the accurate

CP988, *Burning Plasma Diagnostics*, edited by F. P. Orsitto, G. Gorini, E. Sindoni, and M. Tardocchi

measurement of DD neutrons due to accidental coincidences. In addition, ITER is operated over a wide rage of fusion powers. The total neutron source strength changes from 10^{14} to 5×10^{20} neutrons/s [4]. Accordingly, the measurement instrument must have high count rate capability for high-power operation and must satisfy the required measurement accuracy in low-power operation.

Measurement methodology

In the method proposed here, DD and DT neutrons are measured separately with one system. The DT and DD events are identified using multi-discriminator electronic circuits for the fist (d1) and second detectors (d2). It is possible to prevent accidental coincidence because accidental events are separated in the circuit.

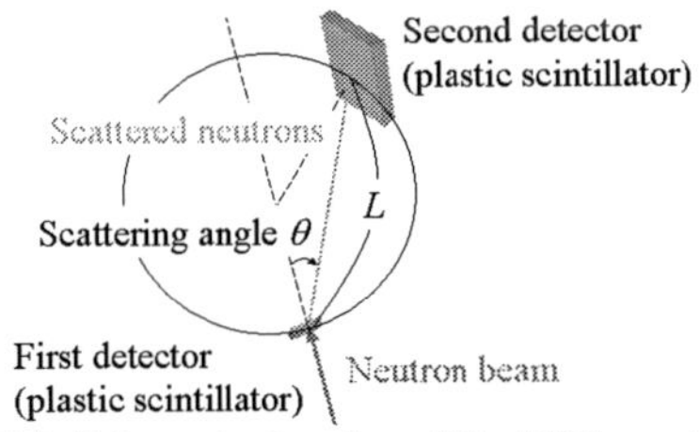

FIGURE 1. Schematic drawing of the TOF spectrometer.

Figure 1 shows a schematic drawing of the TOF spectrometer, which is placed in the line of sight of the plasma core to measure the fuel ion density ratio of the plasma core. The location must have sufficient distance to decrease the scattered energy-degraded DT neutron components [3]. Flight time is constant on the surface of the sphere [5]. The directly incident neutrons from the plasma enter d1, while d2 escapes from direct neutron bombardment. The major reaction in the plastic scintillator responsible for a neutron event is elastic scattering, ^{1}H(n, n')p.

The scattered neutron energy, E_s, and the proton energy in d1, E_p, are given by:

$$E_s = E\cos^2\theta \qquad\qquad E_p = E\sin^2\theta \tag{2}$$

Where E is incident neutron energy. The scattering angle is ranges from 0° to 90° because neutrons and protons have almost equal mass. As the scattering angle is limited by the geometrical arrangement, DD neutron events can be discriminated from most DT events by a pulse-height discriminator window in d1. The flight time is measured to obtained E_s, and hence the incident energy E.

Monte Carlo calculation with PHITS

To optimize the TOF geometry, the energy deposited in d1 is calculated with PHITS (Particle and Heavy Ion Transport Code System) [6]. The size of d1 is assumed to be $5\times10\times0.5$ cm, and that of d2 to be $30\times60\times2$ cm. The distance between d1 and d2 ranges from 40 to 200 cm and the angle ranges from 20° to 60°. The incident neutron energy from DD is 2.45 MeV and that from DT is 14.06 MeV. The radius of the incident neutron beam is assumed to be 1.0 cm. Typical results for incident DD neutrons are shown in Fig. 2, which shows the number of coincidence events between d1 and d2 on a map of deposited energy of each detector.

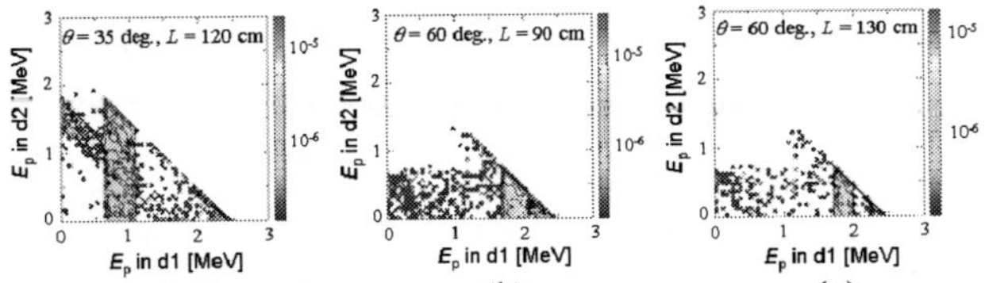

(a) (b) (c)

FIGURE 2. Coincidence events on a map of proton energy deposited in d1 (horizontal axis) and d2 (vertical axis) for incident DD neutrons. Energy bins for d1 and d2 are 0.05 MeV. (a) $\theta = 35°$, $L = 120$ cm. (b) $\theta = 60°$, $L = 90$ cm. (c) $\theta = 60°$, $L = 130$ cm.

The deposited energy in d1 is clearly grouped, while that in d2 is dispersed. To discriminate DD events with the pulse-height, deposited proton energy should be higher in d1. In the case of a large scattering angle, the deposited energy in d1 is high according to Eq (2). If scattering angle is too large, it is difficult to identify in d2 because the gap between the DT neutron and DD neutron energy is too small. Therefore, a suitable scattering angle is about 60°. Typical results in the case of incident DT are shown in Fig. 3b for the same condition of Fig. 2a. The energy range of (b) is the DD proton energy region, showing that the separation of DD events from DT events by a proton energy window is possible.

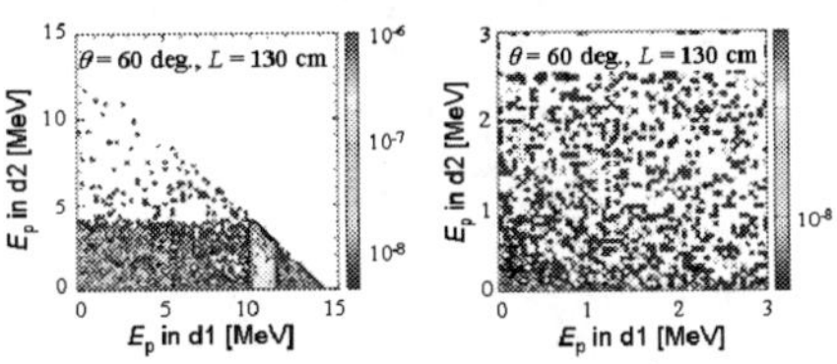

(a) (b)

FIGURE 3. Coincidence events on a d1-d2 deposited proton energy map for incident DT neutrons, when $\theta = 60°$, $L = 130$ cm. (a) Energy bins for d1 and d2 are 0.2 MeV. (b) Energy bins for d1 and d2 are 0.05 MeV.

Experiments and Results

An experiment to investigate the performance of a TOF spectrometer based on our method was performed at Japan Atomic Energy Agency (JAEA) Fusion Neutronics Source facility (FNS). Fig. 4 shows the measurement circuit. The incident neutron energy is 2.66 MeV. The neutron beam enters the TOF spectrometer, and deposited energy in d1 produces a pulse. If the pulse-height is within the window adjusted for DD by a Constant Fraction Differential Discriminator (CFDD/ORTEC583), a DD event is counted. The typical energy spectrum converted from time spectrum is shown in Figure5. FWHM of experimental result is 0.300 MeV (12.12 %), and the peak energy is 2.46 MeV. The peak counts which is calculated by the product of FWHM and peak value is 2.48×10^{-5} count/neutron. The measured spectrum was compared with PHITS calculation based on experimental geometry, and the results showed good agreement as seen in Fig. 5. The experimental results of other arrangement are shown in table1.

Peak counts in all results are order of 10^{-5} count/source-neutron.

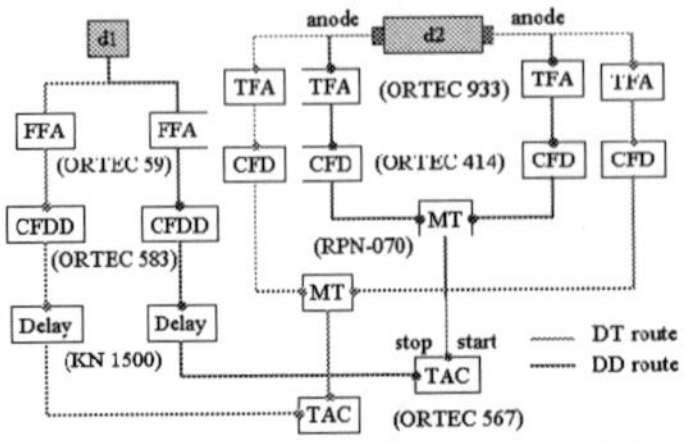

FIGURE 4. Measurement circuit. The blue line is the route of measurement of DD neutrons. The red line is the route of measurement of DT neutrons. In this experiment, only the blue line was used.

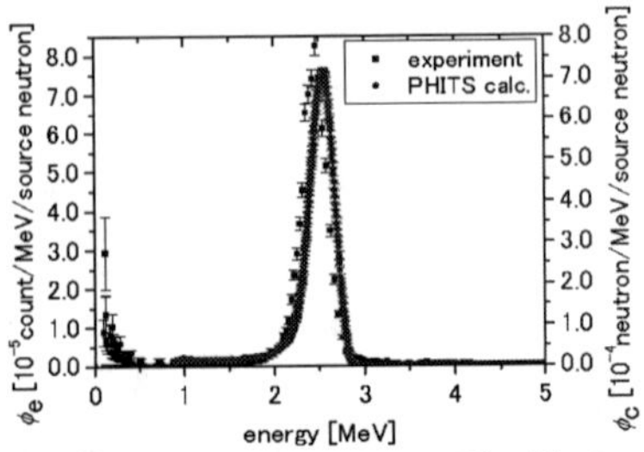

FIGURE 5. Energy spectrum. The black points are experimental results. The blue points are calculation results with PHITS. θ=60°, L=130 cm.

TABLE 1. Experimental result.

Arrangements	θ [degrees]	L [cm]	FWHM [MeV]	ϕ_{peak} [10^{-5} count/source-neutron]
No1	35	120	0.373	3.371
No2	50	150	0.281	1.114
No3	60	90	0.518	5.488
No4	60	130	0.300	2.477

A typical energy spectrum converted from the time spectrum is shown in Fig. 5. FWHM of experimental results is 0.300 MeV (12.12%), and the peak energy is 2.46 MeV. The peak count calculated as the product of FWHM and peak value is 2.48×10^{-5} counts/neutron. The measured spectrum was compared with PHITS calculation based on experimental geometry, and the results showed good agreement, as shown in Fig. 5. The experimental results of other arrangements are shown in Table 1. Peak counts in all results are of the order of 10^{-5} counts/source neutron.

The realistic upper limit of the count rate is of the order of 10^7 cps because the fluorescence decay time of the general plastic scintillator is a few ns. If the count rate for DT neutrons is 1.0×10^7 cps and measurement efficiency for DT neutrons is 10%, as determined by comparison between the experimental and the calculation results, incident DD neutron rate is of the order of 5×10^5 neutron/s. Then, by using 10 detectors as d2 with geometry No. 3, the 250 cps DD events will be obtained.

ACKNOWLEDGMENTS

This work was supported by a Grant-in-aid from the Japan Society for the Promotion of Science (1640177).

REFERENCES

1. ITER Physics Expert Groups on Diagnostics, *Nucl. Fusion* **39**, 2541 (1999).
2. H. S. Bosch and G. H. Hale, *Nucl. Fusion* **32**, 611 (1992).
3 K. Okada *et al., Rev. Sci. Instrum.* **77**, 10E726 (2006).
4. A. V. Krasilnikov *et al., Nucl. Fusion* **45**, 1503-1509 (2005).
5. M. Hoek *et al., Fusion Eng. Des.* **45**, 437-453 (1999).
6. K. Niita *et al., Radiat. Meas.* **41**, 1080-1090 (2006).

Modelling of γ-Ray and Neutral Particle Diagnostics for Fast Ion Examination in Tokamaks

K. Schoepf[1], V. Yavorskij[1,2], V. Goloborod'ko[1,2] and JET-EFDA contributors*

[1]*Association EURATOM-OEAW, Institute for Theoretical Physics, University of Innsbruck, Technikerstrasse 25, A-6020 Innsbruck, Austria*
[2]*Institute for Nuclear Research, Ukrainian Academy of Sciences, Kiev, Ukraine*
**See the Appendix of M.L.Watkins et al., Fusion Energy 2006 (Proc. 21st Int. Conf. Chengdu, 2006) IAEA, Vienna (2006)*

Abstract. Kinetic modelling of fast ions confined in tokamak plasmas is performed for the needs of gamma diagnostics of charged fusion products and RF heated fast ions as well as of the diagnostics of NBI ions via measurements of neutral particle fluxes.

Keywords: tokamak reactor, neutral beam, alpha particles, gamma diagnostics, charge exchange
PACS: 52.55.Fa, 52.65.-y, 52.65.Ff, 52.70.La, 52.40.Mj, 52.38.Ph, 52.55.Pi

INTRODUCTION

Studying the behaviour of fusion-born alpha particles produced by short tritium (deuterium) NBI pulses in deuterium (tritium) tokamak plasmas by means of γ-diagnostics looks to be a promising method for obtaining information on *MeV* charged fusion products in tokamaks [1,2]. We present an analysis of temporally evolving alpha distributions produced by short NBI pulses in a tokamak reactor and demonstrate the possibility of evaluating the loss rate of fusion alphas from measurements of γ-emission induced by alphas colliding with beryllium impurities. Further we define the requirements on γ-diagnostics, plasma parameters and on injection for attaining utmost information about fusion alphas. Similarly, NBI ions can be investigated, e.g. via γ-emission from $^{12}C(d, p\gamma)^{13}C$ reactions of NBI deuterons.

On the other hand, the confinement of NBI ions in tokamak plasmas can be examined by analyzing energetic neutral particle fluxes emitted from the plasma, which are mainly formed by recombination and charge exchange reactions of NBI generated ions. The effectiveness of neutral particle diagnostics was confirmed in recent JET experiments [3] where the influence of enhanced TF ripples on NBI deuterons had been investigated by measuring the neutral deuterium fluxes emitted from the plasma mid-plane using a neutral particle analyzer. Enlarged TF ripples were seen to reduce the D^0-fluxes, which quantitatively corresponds with the TF ripple induced decrease of trapped deuteron population. The measured reduction of D^0-fluxes gave valuable information on the NBI generated deuteron distribution function.

CP988, *Burning Plasma Diagnostics*, edited by F. P. Orsitto, G. Gorini, E. Sindoni, and M. Tardocchi

MODELLING OF γ-RAY DIAGNOSTICS OF FAST IONS

Basic equations

Evidently, the most important quantity for modelling the γ-emission induced by energetic ions in tokamak plasmas is the γ-emission rate per volume, $R^{\gamma}_{\eta i}$, associated with nuclear interactions of fast ions of species j with the η-species of impurity ions,

$$R^{\gamma}_{\eta j}(\mathbf{r},t)=\iint d\mathbf{V}_{\eta}d\mathbf{V}_{j}f_{\eta}(\mathbf{r},\mathbf{V}_{\eta})f_{j}(\mathbf{r},\mathbf{V}_{j},t)\sigma^{\gamma}_{\eta j}\left(\left|\mathbf{V}_{\eta}-\mathbf{V}_{j}\right|\right)\left|\mathbf{V}_{\eta}-\mathbf{V}_{j}\right| , \qquad (1)$$

where f_η and f_j are the corresponding distribution functions and $\sigma^{\gamma}_{\eta i}$ denotes the cross section of the specific j-η nuclear reaction resulting in γ-emission. Note that, due to the energy threshold of such nuclear reactions, only high-energy ions can induce γ-rays for measurements. For the reaction $^{9}Be(\alpha,n\gamma)^{12}C$ this is illustrated in Fig. 1. It is seen that significant γ-emission is effected only when fusion alphas with an energy $E>E_{cr}=1.6$MeV interact with Be ions. For the modelling of γ-emission induced by fast ions in reactor size tokamaks we can use the ion distribution function in drift approximation. This function is adequately determined by the bounce-averaged Fokker-Planck equation in 3D constants-of-motion (COM) space [4]

$$\partial_t f=\partial_{\mathbf{C}}\left(\mathbf{d}+\overleftrightarrow{\mathbf{D}}\partial_{\mathbf{C}}\right)f+S(t,\mathbf{C}), \qquad (2)$$

where $\mathbf{C}\equiv\mathbf{C}(\mathbf{r},\mathbf{V})$ is the 3D COM space of Lagrangian variables, $\mathbf{d}$ and $\mathbf{D}$ are coefficients describing convective and diffusive transport induced by Coulomb collisions and orbital motions, and S represents the bounce-averaged source term.

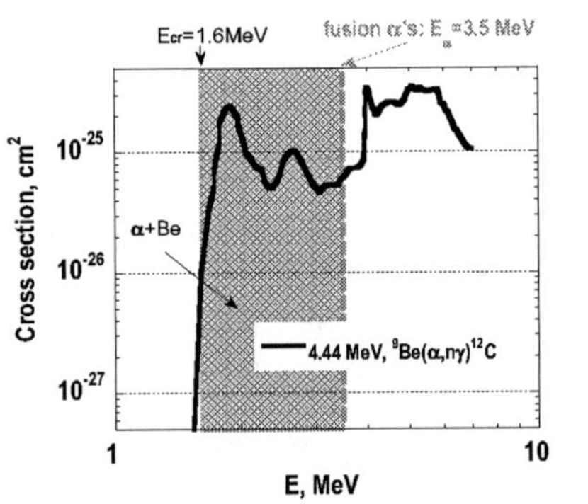

FIGURE 1. Cross Section of the Nuclear Reaction $^{9}Be(\alpha,n\gamma)^{12}C$. Only Partly Thermalised Fusion Alphas with $E>E_{cr}=1.6$MeV can produce γ-Rays.

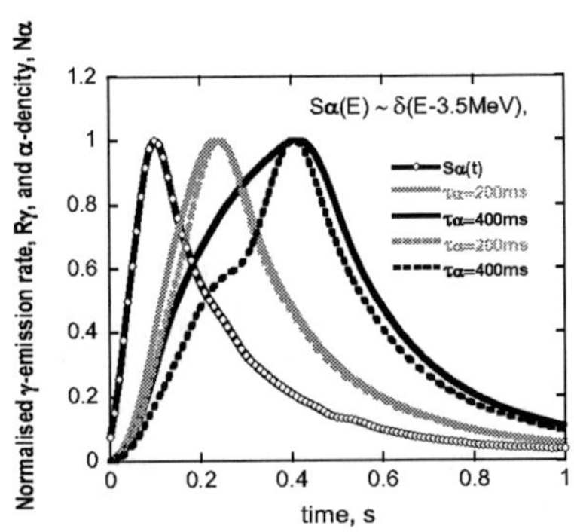

FIGURE 2. Modelled Evolution of γ-Emission Rate (---) and Fusion Alpha Density (—) in Typical JET-like TTE Plasmas for Different Alpha Slowing Down Times $\tau_\alpha=0.5\,\tau_S\ln(3.5/1.6)\cong 0.36\,\tau_S$ and $\tau_L=\infty$ (Absence of Alpha Loss).

Temporal Evolution of γ-Emission

For a qualitative interpretation, the analysis of γ-emission induced by fast ions with temporally evolving distributions, like fusion born alpha particles in JET TTE [1], is based here on a simplest 1D kinetic equation that accounts only for energy and time dependences of the fast ion distribution and considers ion loss in the loss time (τ_l) – approximation [2],

$$\partial_t f_j = 2E^{-1/2}\tau_s^{-1}\partial_E\left(E^{3/2}f_j\right) - f_j/\tau_l + S\left(E_j,t\right) \tag{3}$$

where τ_s~$T_e^{3/2}/n$ is the Spitzer slowing down time. Figure 2 compares the modelled evolutions of the γ-emission rate and of the fusion alpha density as actuated by a typical temporal shape of the DT fusion source term $S(t)$ in JET-like TTE plasmas for different alpha slowing down times τ_α=0.5$\tau_s$ln(3.5/1.6) ≅ 0.36τ_s and $\tau_l = \infty$ (absence of alpha loss). Obviously, the temporal behaviour of γ-emission accords with that of $N_\alpha(t)$. Notice that the γ-emission appears delayed in comparison to the fusion alpha source by a delay time τ_d~0.6τ_α~ 0.2τ_s determined by the alpha slowing down time. An important feature now is that the delay of alpha induced γ-emission reduces when the alpha confinement ($\tau_l < \tau_\alpha$) is substantially degraded. From the analysis above we thus conclude that measurements of the temporal evolution of γ–emission allows for evaluating the slowing down rates of MeV fusion alphas as well as their loss rates. High precision measurements of these rates, however, will require conditions with long Spitzer slowing down times and short-period fusion sources.

Poloidal Distributions of γ-Emission Induced by Fast Ions

Valuable information on the spatial distribution of energetic ions can be obtained from the poloidal distribution of fast ion induced γ-emission. For two ITER scenarios we present in Fig. 3 the results of 3D Fokker-Planck modelling of (R,Z)-profiles of 4.44MeV γ-emission induced by DT fusion alphas in $^{9}Be(\alpha,n\gamma)^{12}C$ interactions with 2.5% Be impurity. The spatial distributions of γ-emission are seen to well reproduce that of energetic fusion alphas (E>0.4MeV).

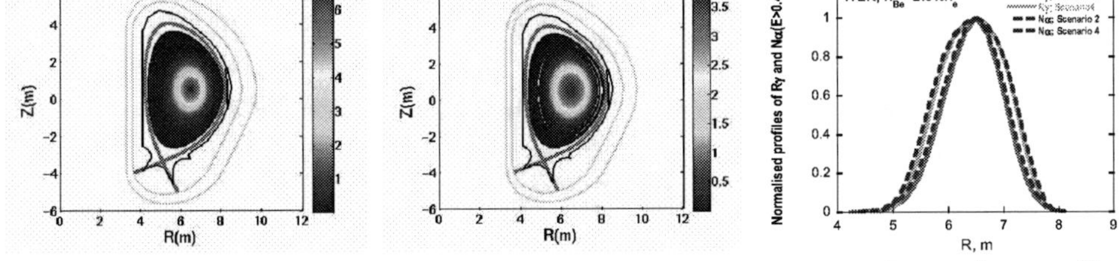

FIGURE 3. Modelled Profiles of Alpha Induced 4.44MeV γ-emission Rates R^γ (from $^{9}Be(\alpha,n\gamma)^{12}C$ Reactions) in $m^{-3}s^{-1}$ for the 2nd ITER Scenario (left) and the 4th ITER Scenario (centre). The Display on the Right Compares Mid-Plane Slices of R^γ with the Density N_α of DT Fusion Alphas with E>0.4MeV.

MODELLING OF NEUTRAL PARTICLE DIAGNOSTICS

The effectiveness of neutral particle diagnostics was confirmed in recent JET experiments [3] where the influence of enhanced TF ripples on NBI deuterons had been investigated by measuring the D^0-fluxes emitted from the plasma mid-plane using a neutral particle analyzer (NPA). Enlarged TF ripples were seen to reduce the D^0-fluxes, which was in quantitative correspondence with the ripple induced decrease of the trapped deuteron population. Employing a Fokker-Planck calculated 4-dim. beam ion distribution function in a Monte-Carlo code for the computation of neutral

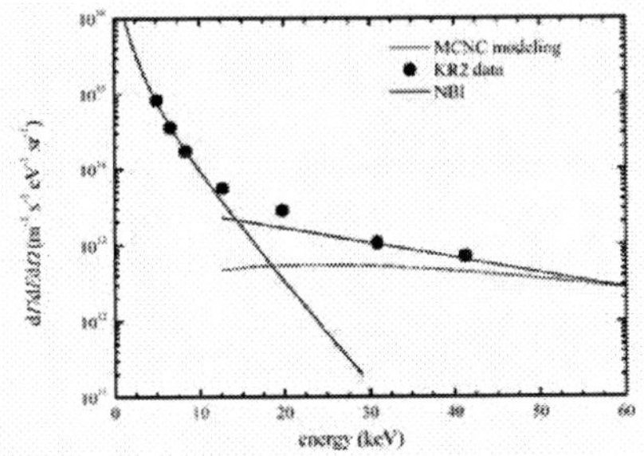

FIGURE 4. Modelled D^0 Fluxes in JET Ripple Experiments [3] as a Function of Energy

fluxes to the NPA detector, the modelled D^0-fluxes [3] are displayed in Fig. 4 for shot #69710 in JET ripple experiments. The interpretive modelling of ripple effects on NBI ions was seen to reasonably concur with neutral particle flux measurements.

SUMMARY

The analysis presented suggests that γ-diagnostics constitutes a promising method for studying the behaviour of *MeV* charged fusion products and NBI ions in tokamak plasmas. Measuring the delay time between the source of fast ions and the γ-emission induced by them allows to draw conclusions about the distribution and loss rates of energetic ions. Further was demonstrated that the confinement of fast ions in tokamaks can be examined by analyzing energetic neutral particle fluxes from the plasma.

ACKNOWLEDGMENTS

This work, carried out in the framework of EFDA, was supported by the European Communities under the contract of Association between EURATOM and OEAW.

REFERENCES

1. V.G. Kiptily, et al., *Phys. Rev. Lett.* **93**, 115001 (2004).
2. V.A. Yavorskij, et al., IAEA Fusion Energy Conference, Chengdu, China, 2006, TH/P6-7.
3. V.A. Yavorskij, et al., 34th EPS Plasma Physics Conference, Warsaw, 2007, P2.041.
4. V.Ya. Goloborod'ko, et al., *Nucl. Fusion* **35**, 1523 (1995).

Development of a He^0 Source for Confined Alpha Particle Measurement

N. Tanaka[a], M. Kisaki[a], K. Iwazaki[a], M. Kikuchi[a], A. Okamoto[a], T. Kobuchi[a], K. Shinto[a], S. Kitajima[a], K. Tsumori[b], O. Kaneko[b], M. Wada[c], M. Sasao[a]

[a]*Tohoku University, Sendai 980-8579, Japan*
[b]*National Institute for Fusion Science, Toki, Gifu 509-5292, Japan*
[c]*Doshisha University, Kyotanabe, Kyoto 610-0321, Japan*

Abstract. A probing He^0 beam for confined alpha particle measurement using a double charge exchange process is now under development. A proof of principle experiment for ground-state He^0 beam production will be performed on a test stand. Several methods are developed to measure the metastable fraction of a He^0 beam. A full-size strong-focusing He^+ source has been constructed and sufficient beam current was achieved with a beam size tolerable to be used on ITER.

Keywords: Alpha particle, Neutral beam, Negative beam, Helium beam
PACS: 52.70.Nc, 29.25.Lg, 52.55.Pi

INTRODUCTION

An alpha particle diagnostics method for next generation magnetically confined fusion reactors, such as ITER, has been suggested[1]. An energetic He^0 probe beam of 1-2 MeV is to be injected into plasma to neutralize alpha particles (Figure 1). To confirm the availability of the He^0 beam, we have been developing (1) a test stand for ground-state He^0 beam production[2-4], (2) several methods to measure the metastable fraction[5,6], and (3) a full-size ion source[7, 8].

A Test Stand for He^0 Beam Production

The fast He^0 beam should be produced from an accelerated He^- beam because of its high neutralization efficiency, and the He^- beam is produced from a He^+ beam through double charge exchange (CX) with alkali metal vapor (Fig. 2). To give an experimental proof, the beam transport system and He^- and He^0 beam producing processes are to be studied on a test stand. We have developed a high voltage station to apply a high voltage of 100-150 kV to accelerate the He^- beam. All of the components for He^- beam production are in the station, which is assisted by 6 insulators and covered by a rounded cover[2]. The system has a 90° bending magnet to separate He^+, He^0, and He^- beams produced in the CX-cell. The magnet prevents He^+ and He^0 beams entering the electrostatic accelerator and causing problems to the

CP988, *Burning Plasma Diagnostics*, edited by F. P. Orsitto, G. Gorini, E. Sindoni, and M. Tardocchi

accelerator. The bending magnet is also designed to focus the beam in the two transverse directions[3]. One of the focal points should be at the center of the CX-cell and the other should be at the entrance of the accelerator. The experimental results showed that the bending magnet has this lens effect for the He^+ beam[4].

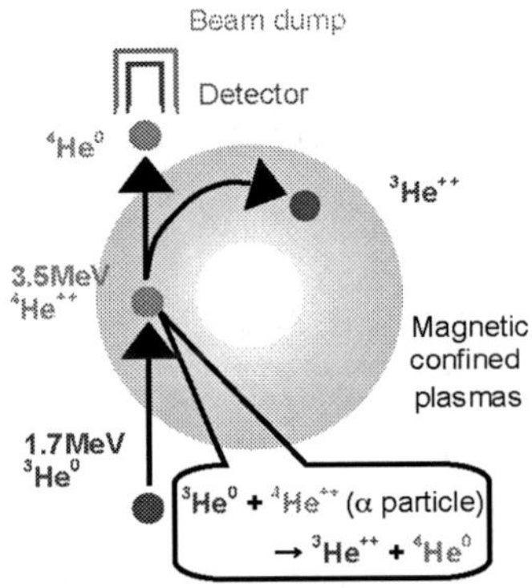

FIGURE 1. An alpha particle diagnostics method. The $^3He^0$ beam of 1-2 MeV is injected into the plasma and an alpha particles are neutralized and detected out of the confined plasmas.

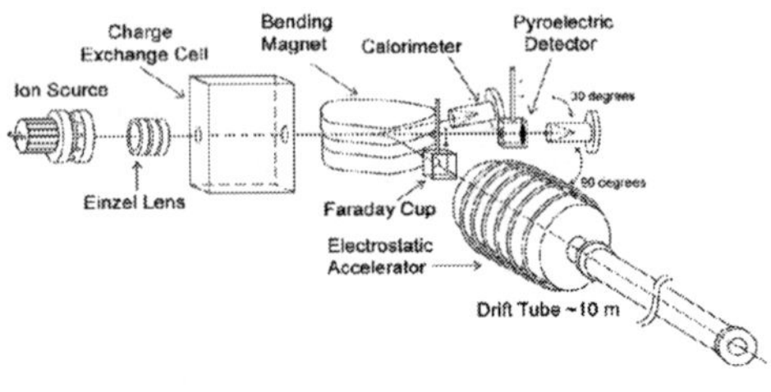

FIGURE 2. The test stand. The He^+ beam of 10-15 keV is extracted and enters a charge exchange cell filled with lithium vapor. He^+, He^0, and He^- beams are produced in the cell and separated by the bending magnet. The He^- beam is bent at an angle of –90° to the incident beam direction and accelerated up to 150 keV by an electrostatic accelerator. The fast He^- beam is neutralized by auto electron detachment or collisional detachment during flight in the long drift tube.

Production of a He^- Beam

A He^- beam has been produced in the test stand. The beam current intensities of He^+ and He^- beams are shown in Fig. 3. The beams of 10 keV were detected at -90° by the Faraday cup with a 6-mm aperture at each lithium temperature. The He^+ beam current intensity decreased with increasing He^- beam current intensity. The maximum He^- beam current intensity was almost 550 nA at the optimum lithium temperature of 590°C, and the He^+ current intensity at the room temperature (approx. 24°C) was 45 μA. The ratio of I_{He-}/I_{He+} was almost 1%. Here, I_{He+} is the intensity at room temperature and I_{He-} is that at the optimum temperature.

Measurement of Neutral Efficiency and Metastable Fraction

A fast $He^0(1s^2\ {}^1S_0)$ beam of 70-150 keV is to be produced by electron detachment in the drift tube. The attenuation equation in case of auto electron detachment is

$$I_{He^0} = I_{He^-0}(1 - e^{t/\tau}).$$

The converted He^- particles have three electron states. The existence rates of J=3/2 and 5/2 of He^- particles are 50 % for each state, and that of J=1/2 is negligible. The life times τ for each J=3/2 and J=5/2 state are 10 μs and 300 μs respectively. Fig. 4

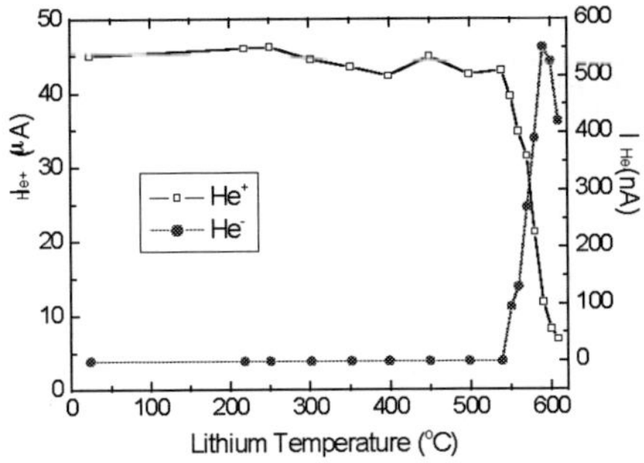

FIGURE 3. He^- beam and He^+ beam current intensities. The current intensity of the He^+ beam at room temperature is almost 45 μA. The optimum lithium temperature was 590°C and the maximum He^- beam current intensity was almost 550 nA. The beam current intensity ratio I_{He-}/I_{He+} was $\geq$ 1%.

shows estimated neutralization efficiencies *vs.* the beam energy at drift tube lengths of 1 m, 4 m, and 8 m. The neutral helium particle from the J=3/2 state of He^- particle is tens of a percentage, while that from the J=5/2 state is negligible.

The system for neutralization efficiency measurement is shown in Fig. 5. Several sets of detection apparatus composed of a pyroelectric detector[2] and a deflector are to be located at different points on the drift tube. The absolute current intensity of neutral helium beam can be measured by detecting both neutral and charged particles by the pyroelectric detector[2].

The neutral helium beam consists of two types of electron state. : *i.e.*, the ground state, He^0, and the metastable state He^*. Several methods have been developed to measure the metastable fraction of the He^0 beam. The metastable fraction can be determined by measuring the beam attenuation in a gas cell, or in a plasma. The former is unreliable due to attenuation by scattering, while this is much less problematic in the latter case. We have been developing Diagnostic Tool Assisted by Linear Plasma for Helium Atom beam (DTALPHA). The plasma is produced by a helicon wave. The required plasma density is about 10^{13} cm^{-3} and the length of the plasma is about 1 m. Another notable method using Laser Absorption Spectroscopy has also been developed and tested recently[5,6].

A full size He^+ ion source and the ITER probing beam

A strongly focusing high-intensity He^+ ion source has been designed and constructed[7, 8]. The ion source was installed with a beam diagnostic chamber on the Neutral Beam Injection (NBI) test stand at the National Institute for Fusion Science (NIFS). In the beam diagnostic chamber, the beam current was measured by using a Rogowski coil, and a current on a 2-mm carbon plate target, which is movable in the direction of the beam axis. The infrared radiation from the carbon plate was observed wiht an infrared camera to determine the beam size and the focusing property.

Acceleration currents of more than 3 A were achieved at an acceleration voltage of 20 kV. The measured beam diameter was about 30 mm at the focal point.

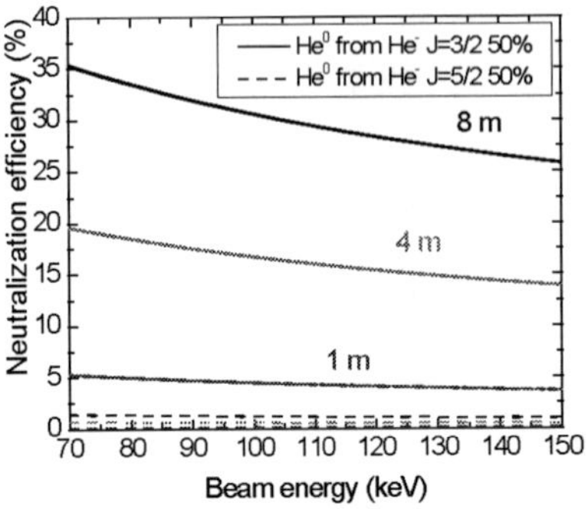

FIGURE 4. Estimated neutralization efficiencies for each drift tube length. The efficiency for the J=5/2 state is negligible.

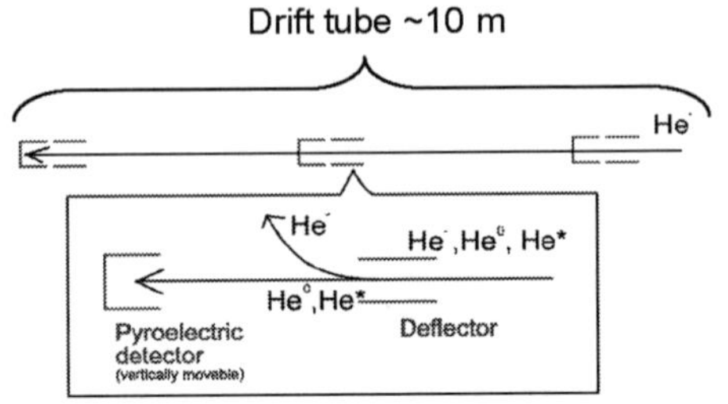

FIGURE 5. The neutralization efficiency measurement system. A pyroelectric detector and a deflector are located at several points in the drift tube.

The beam size is tolerable for injection into the charge exchange cell for the He^- beam production. The beam will then be accelerated up to 1-2 MeV by RFQ, and be neutralized during a flight of 20 m as described above, and injected into an ITER burning plasma.

ACKNOWLEDGMENTS

This work was supported by a Grant-in-aid from the Ministry of Education, Culture, Sports, Science, and Technology of Japan (priority area 442-16082201). This work was also partially supported by NIFS collaboration programs (NIFS05KCBB006 and NIFS06KCBB007).

REFERENCES

1. M. Sasao *et al.*, Rev. Sci. Instrum. **69**, 1063 (1998)
2. N. Tanaka *et al.*, Plasma Fusion Res. **2** S1105 (2007)
3. K. Shinto *et al.*, PAC conf. Proc. 2030 (2006)
4. N. Tanaka *et al.*, Rev. Sci. Instrum. (in press)
5. A. Okamoto *et al.*, Plasma and Fusion Research, **2** 1044 (2007)
6. A. Okamoto *et al.*, Plasma and Fusion Research, **2** 029 (2007)
7. K. Shinto *et al.*, Proc. of EPAC 06, pp 1726-1728 (2006)
8. M. Kisaki *et al.*, Rev. Sci. Instrum. (in press)

Helium measurements simulating alpha-particle diagnostics by the pellet charge exchange in Large Helical Device

T.Ozaki, P.Goncharov, E.Veschev[1)], S.Sudo, N.Tamura, TESPEL Group, High-Energy Particle Group and LHD Experimental Group

National Institute for Fusion Science, Toki, Gifu 509-5292, Japan
1) Graduate University for Advanced Studies, Hayama, Kanagawa, 240-0193, Japan

Abstract. It is very important to investigate the confinement of α particle, which is produced by the nuclear reaction in ITER or fusion reactor. The pellet charge exchange measurement (PCX) is one of the most powerful methods because it can directly provide the profile of α particle energy spectra in plasma. In Large Helical Device (LHD), it is possible to confine the high energetic particle (helium) accelerated by using the ion cyclotron resonance heating (ICH). Therefore it is possible to perform the simulation experiment of α particle diagnostic in LHD.

Keywords: Helium ion, Alpha particle, ICH, PCX, TESPEL, LHD
PACS: 52.70.Nc

I. INTRODUCTION

It is very important to investigate α particle heating mechanism in future fusion reactor because α particle has a main role to heat the fusion plasma. High-energy particles including α particle are emitted not only by the charge exchange but also by the MHD in the fusion reactor [1]. Their particles give damage to the plasma wall addition to create a poor plasma confinement. Decelerated α particle (or a helium ion) with the energy over 1 keV makes a bubble and gives a serious damage on the wall surface unlike hydrogen. Therefore the suitable method for measuring helium ion distribution should be established immediately.

It is very difficult to use spectroscopic methods or the passive charge exchange neutral particle method for helium ion. Helium ions are almost fully ionized except near peripheral region. A few helium atoms are escaped from plasmas by the double charge exchange reaction between the background helium neutral and the fully ionized helium ion, whose cross section is too small. Some techniques for measuring α particle have been proposed [2]. A pellet charge exchange measurement (PCX) is one of the most powerful candidates to obtain the spatial resolved helium energy spectrum. PCX had been performed in TFTR [3]. However it is not often utilized because there are a few opportunities to obtain nuclear reaction plasma. High-energy particle over 1 MeV can be generated and confined in LHD [4] by using ICH.

CP988, *Burning Plasma Diagnostics*, edited by F. P. Orsitto, G. Gorini, E. Sindoni, and M. Tardocchi

II. PELLET CHARGE EXCHANGE MEASUREMENT

The high-energy neutral particles, which are produced by the charge exchange between the injected Tracer Encapsulated Solid Pellet (TESPEL) [5] and the energetic ions, are observed in PCX. TESPEL is the impurity pellet as the polystyrene or the titanium etc. with the velocity of 400-500 m/s. TESPEL is ablated and produces the ablation cloud with several layers of different charge states around the traveling pellet in plasma. The pellet ablation cloud keeps the neutral or partially ionized states surrounding the pellet until fully ionization. Parts of injected particles into the pellet ablation cloud are escaped from the plasma due to the charge exchange reaction. It is 10^8 times larger than the background neutrals because the cloud density is high enough (10^{16} cm^{-3}). Therefore the measurement in the plasma is available because the double charge exchange can be expected if the high-Z material as the carbon or the lithium is used. The neutralization factor is very important in order to obtain energy spectrum in plasma. It is determined by the ratio of the recombination to the ionization. The precise calculation has been done under the LHD plasma parameter by Sergeev [6]. The neutralization factors for the hydrogen and the helium are 0.9 and 0.07, respectively if the polystyrene is used.

III. EXPERIMENTAL APPARATUS

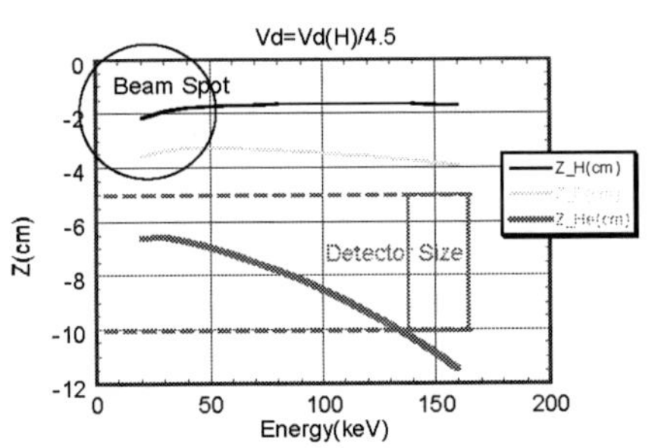

FIGURE 1. The beam spot on the detector. When the plate voltage of 1/4.5 against hydrogen setting voltage is chosen, the helium can be separated in lower energy

It is one of advantages in PCX to obtain the spatial information. When the compact neutral particle analyzer (CNPA) for measurement of the charge exchanged particle is installed just behind TESPEL trajectory, the time trace of the signal can be transferred to the information of the pellet position. The typical pellet velocity of 500 m/s makes the spatial resolution of 5 cm by the sampling time of 0.1 ms. Full trajectory of the pellet in plasma should be in the viewing cone of the CNPA by minimizing the angle between the sight line and the trajectory. The detection time of the particle corresponds to the position of the generated particle because the size of the cloud is same as the spatial resolution (5 cm). CNPA is a traditional E//B particle analyzers with a diamond-like carbon film as a stripping foil, the permanent magnet for the energy analysis of the particle and a condenser plate for the particle mass separation. The hydrogen with the energy range from 0.8 to 168 keV can be observed by 40 rectangular-shape channeltrons which is set on the position for the hydrogen measurement.

If the plate voltage is changed, the different mass as helium can be observed in principle. According to simple orbit calculation in CNPA, the beam spot of the helium is different from the channeltron array, which is adjusted to the hydrogen even if the plate voltage is tuned. Figure 1 shows the energy dependence of the position of the beam spots in hydrogen, deuteron and helium. Here we assume the single ionized helium ion after translation of the carbon film of helium. The spot size is assumed to be determined by the aperture size (2 mm) and the geometric configuration of the plasma and the detector. In low energy region, the spot size may be enlarged due to the scattering in the foil.

According to Fig. 1, the helium beam spots do not correspond to the detector array in higher energy channels when the plate voltage is adjusted to a low energy channel because the detector array position is adjusted to the proton. Now we continue accurate calculation for obtaining the detector efficiency of helium. The helium energy spectrum can be obtained because we are interested in lower energy helium spectra in LHD experiments.

IV. EXPERIMENTAL RESULTS

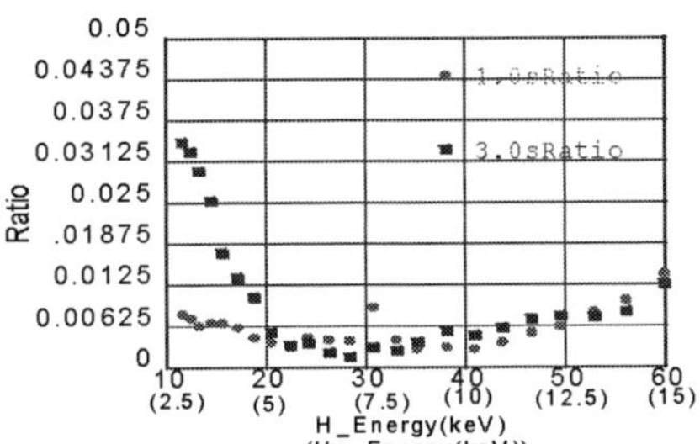

FIGURE 2. He/H ratios without and with ICH.
The ratio with ICH (3.0 s) is higher than the ratio without ICH (1.0 s) at lower energy region.

In order to obtain the high energetic helium ion, we use the higher harmonic ICH. At B=1.86 T, 38.47 MHz (1.9 MW), there is the resonance region of the helium-4 near the plasma center, although the electron is mainly heated in this heating mode. To confirm the helium acceleration, we compare the spectra of the helium and hydrogen by using CNPA with different plate voltages. Figure 2 shows the ratio between the signal of He and H in two similar shots. We must remember that most signals are hydrogen even if we set the plate voltage for helium. Therefore the ratio means the ratio between the scattered hydrogen and the real hydrogen. The large ratios at low and high energy regions are due to the scattering at the foil and the close trajectory of the hydrogen beams, respectively, not only with ICH but without ICH. The ICH is applied at 3.0 sec, but not applied at 1.0 sec. He/H ratio lower than the helium energy of 5 keV at 3.0 sec is obviously larger than at 1.0 sec. This means the low energy helium ion is accelerated by the higher harmonic ICH.

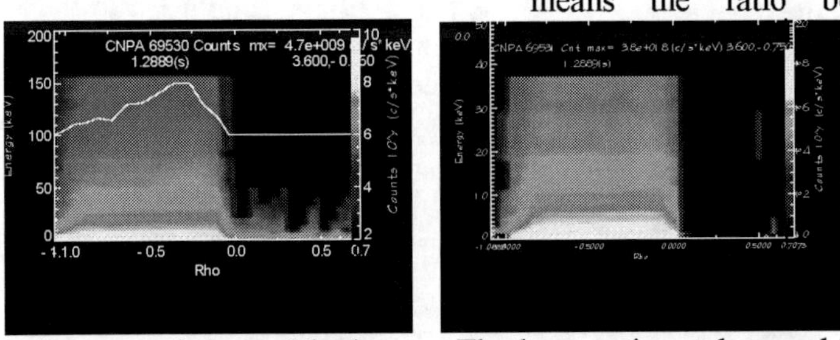

FIGURE 3. Radial energy profiles of H and He.
These profiles ((a) H, (b) He) are obtained by PCX. In PCX, the time history is corresponded to the radial information.

It is the final target to establish helium measurement through the PCX as mentioned in chapter I. The helium atom is surely observed by decreasing the plate voltage of the CNPA. The helium atom is generated by the charge exchange between the fully ionized helium and the partially ionized carbon ion in the TESPEL. Figure 3(a) and (b) show the energy spectra of hydrogen and helium spatial distribution measured by PCX in the helium plasma sustained by NBI. The time history is translated to the radial information. Figure 4 shows the He/(He+H) ratio on the detector observed by the PCX when the TESPEL is injected to the helium plasma. In the ratio, the scattering hydrogen contribution is considered by the combination between the calibration and the model calculation. In LHD, the hydrogen concentration is not small due to the hydrogen NBI in helium plasma. When the TESPEL is injected to the hydrogen plasma, the ratio is almost 0.1. In this helium plasma, the abundance of the low energy helium is large because the helium plasma has a Maxwellian distribution. Hydrogen mainly comes from the wall and the NBI. Therefore there is much hydrogen at the plasma edge. Hydrogen contribution is large at the high energy region because the high energetic hydrogen is provided by the NBI.

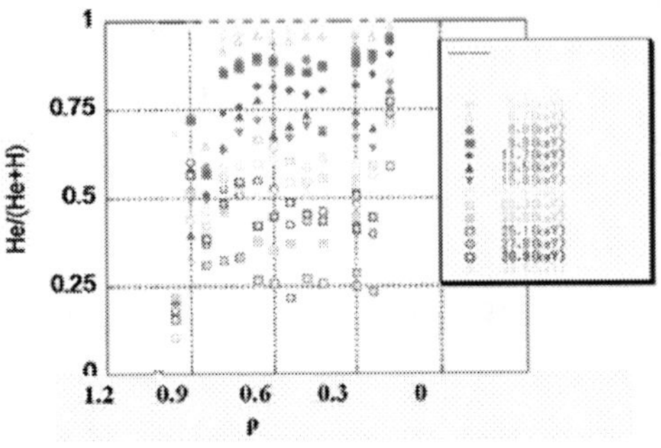

FIGURE 4. Calibrated He/(He+H) ratio.

V. SUMMARY

It is very important to establish the alpha profile measurement because the helium/alpha makes a bubble and gives a serious damage on the wall surface. The PCX is one of the techniques to obtain the energetic helium particle distribution in plasma. The helium acceleration by the higher harmonic ICH can be surely observed by the CNPA, which is the detector in PCX if the plate voltage is adjusted to the He orbit. By using this technique in PCX, we can obtain the He/(He+H) ratio in NBI plasma. The ratio profile between the hydrogen and the helium can be obtained. Through these experimental results, the PCX can be advanced as the alpha particle measurement method.

ACKNOWLEDGMENTS

This work was performed under NIFS07ULBB509, the grants aid No. 17540475 and 18035013.

REFERENCES

1. L. Chen,, Phys. Plasmas, 1 (1994) 1519-1522.
2. M. Sasao et al, Plasma Phys. Control. Fusion, 46 (2004) S107-S118.
3. R Fisher et al, Phys., Rev. Lett., 75 (1995) 846-849.
4. O. Motojima et al, Fusion Eng. Des., 20 (1993) 3-14.
5. S. Sudo et al., Proc. of 17th IAEA Fusion Energy Conference (Yokohama, Japan, 1998), 39 (1999) IAEA-CN -69/EXP1/18.
6. V. Sergeev, private communication.

Study of Escaping Alpha Particle Orbits for Detector Design on ITER

C. Carrander, M. Nishiura[a], M. Isobe[a], T. Hirouchi, D. Funaki, Y. Sato, S.Kitajima, A. Okamoto, K. Kobuchi, M. Sasao, and C. Walker[b]

Tohoku University, Sendai 980-8579, Japan
[a]National Institute for Fusion Science (NIFS), 332-6 Oroshi, Toki, Gifu 509-5292, Japan[b]ITER
[b]Cadarache Joint Work Site, 13108 St Paul-Les-Durance Cedex, France

Abstract. The placement of lost alpha particle detectors in ITER has been studied using a full gyro-motion orbit calculation code, in which the orbit is located in the plasma side of the first wall boundary. Detection rates are likely to be low with the detector is placed inside the blanket. Tilting the modules so that they interfere less gives a significantly higher probability of detection.

Keywords: Lost alpha particle, ITER, alpha particle diagnostics, orbit calculation

PACS: 52.70.Nc

Introduction

The detection of escaping alpha particles is important for the study of various types of collective phenomenon driven by alpha particles. In addition, alpha particle loss on the first wall is one of the problems for the safe operation of ITER [1].

Although ITER integration studies for some candidates for lost alpha particle measurement, such as camera imaging of scintillators on the first wall, scintillator probe, Faraday-cup, and imaging bolometer, are in progress, the question of whether it is realistic from an engineering viewpoint to detect these particles has been raised [1,2]. Figure 1 shows an example of detector placement in the blanket module.

To investigate escaping alpha particle orbits in ITER and determine whether it is possible to place detectors in such an arrangement that alpha particles from the plasma can be detected while maintaining reasonable detector survivability, an orbit calculation program of full gyro-motion under the ITER magnetic field configuration has been developed. It was checked that an orbit striking the detection point does not touch or intersect the first wall boundary. This article describes the poloidal and toroidal distributions of feasible detection location, and the feasible aperture direction of detectors. The usefulness of the tilting blanket modules is also discussed.

Calculation

The full gyro-orbit following calculation code calculates alpha particle orbits running on the ITER equilibrium field. The equilibrium field of ITER, *i.e.*, the

CP988, *Burning Plasma Diagnostics*, edited by F. P. Orsitto, G. Gorini, E. Sindoni, and M. Tardocchi

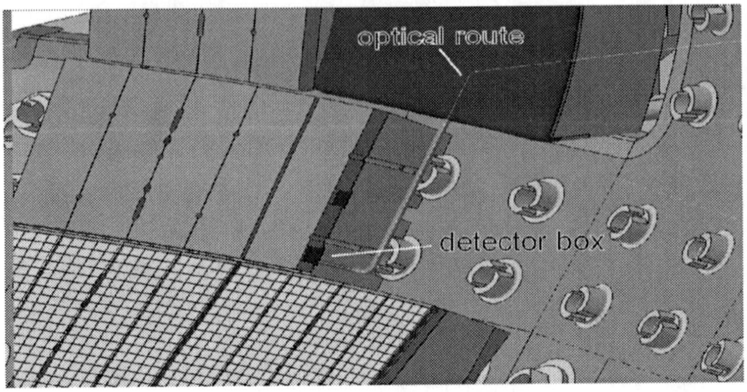

FIGURE 1. Cutout view of detector placement in the blanket module

poloidal flux function (R, Z) is first obtained from the EFIT equilibrium code and then the poloidal field components $B_r(=-(d\psi/dZ)/2\pi R)$ and $B_z(=(d\psi/dR)/2\pi R)$ are computed. The toroidal magnetic field is given as $B_{\phi}=B_0R_0/R$. Both plasma current I_p and B_{ϕ} are oriented in the clockwise direction because ITER accepts only this operation. The ion grad-B drift direction is downward in this case. Currently, the toroidal field ripple resulting from discrete toroidal coils is not considered in our code, *i.e.*, the equilibrium field is perfectly axisymmetric. Nevertheless, this approximation is sufficient for our purpose of detector design. Orbit calculations have been performed in two different equilibria. One has monotonic magnetic shear, with the parameters R_0=6.2 m, B_0=5.3 T, and I_p=15 MA. The other has weak reversed shear with parameters R_0=6.2 m, B_0=5.3 T, and I_p=9 MA.

The equations of motion were solved using the Runge-Kutta method[1], following the alpha particle backwards in time from the proposed detector location. For each step, the position of the particle was checked to ascertain that it was inside the first wall blanket. This was done by inserting the coordinates of the particle into the equations of the planes that describe the first wall blanket:

$$y = Ax + By + Cz + D \tag{1}$$

A result of zero corresponds to a point on the plane, while positive and negative results correspond to points inside and outside the torus, respectively. Fifteen such equations were used to describe the fifteen blanket modules in one segment of the torus, and positions outside this segment were then rotated toroidally to the corresponding positions assuming that all 18 segments were identical. The suggested placement for the detector was in a small cutout in one of the vertical slits of the blanket modules (see Fig. 1).

---

1 The Runge-Kutta 4-5 method ode45 in Matlab was used.

Calculation Results and Summary

In total, 600 positions[2] along the first wall were tested with the detector placed 1 cm behind the blanket module surface. For each of these positions, a number of different angles were tested, tilting the detector in the toroidal (phi) and poloidal (theta) directions, creating a cone of possible orbits. Figure 2 shows the distribution of sums of orbit length starting from the detection position to the wall surface in time inverse. This gave a total of 2,520 tested orbits per module, or 37,800 for the entire first wall segment.

Detection rates with orbit length longer than 10 m were low with untilted plates. Significantly less than 1% of the orbits were likely to have originated within the plasma.

Calculations were also run with the blanket modules tilted slightly from the original configuration shifting the plate corners a further 10 cm outwards (Fig.3), while the detectors were placed at the same coordinates as with untilted plates, *i.e.*, in front of the tilted blanket modules. As shown in Fig. 4, detection rates increased markedly. Detection rates differed strongly between different positions on the first wall, due to the differences in the magnetic field and geometrical conditions. Positions closer to the centre of the torus, *i.e.*, in the middle of the blanket modules, tended to give better detection rates.

The distance traveled in one rotation around the guiding center varied from a few millimeters to several decimeters. This must be kept in mind when designing the detector so the particles do not strike the housing around the detector before striking the detector itself. The best setup for the detector, regardless of position, is with the detector tilted slightly towards the incoming particles so they hit the detector straight on. Second, the detector should be tilted upwards until it is oriented parallel to the blanket module. A given orbit will, at the point where it is closest to the wall without hitting it, be parallel to the wall. Details will be described elsewhere.

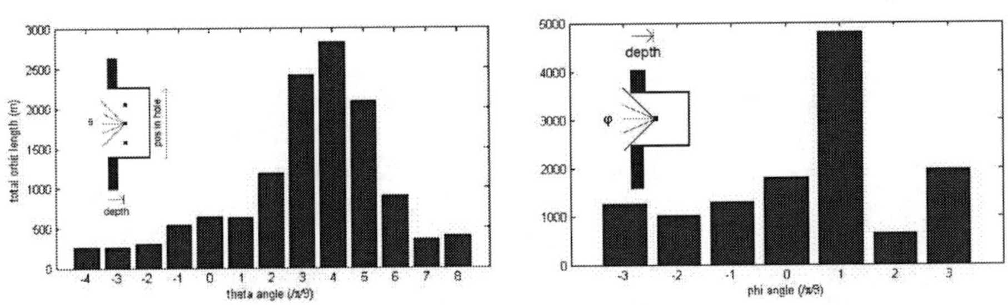

FIGURE 2. Poloidal (theta) and toroidal (phi) distribution of sums of orbit length starting from the detection position to the wall surface in time inverse.

---

2 40 per module, 15 modules.

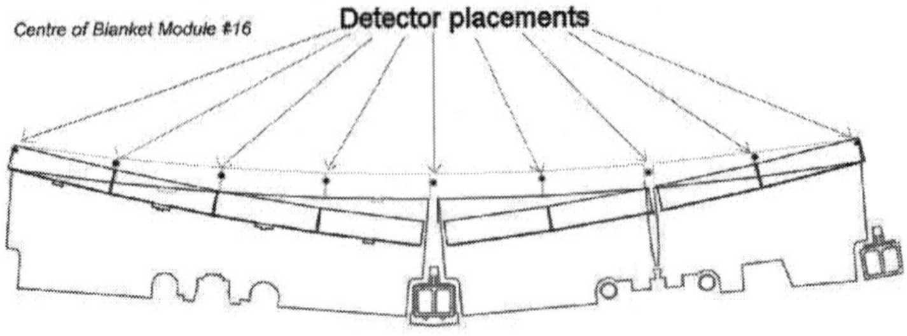

FIGURE 3. Detector placements. Note that the tested detector placements remain on the same coordinates when the blanket modules are tilted.

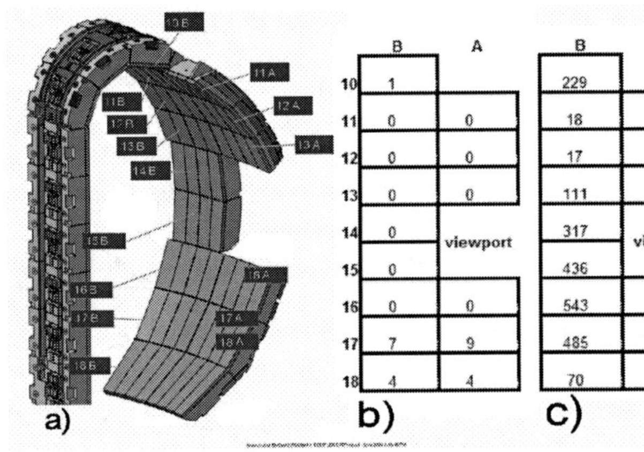

FIGURE 4. Detector placements. Note that the tested detector placements remain on the same coordinates when the blanket modules are tilted by 0.1 m behind.

ACKNOWLEDGMENTS

This work was supported by Grants-in-aid from the Japan Society for the Promotion of Science (1640177) and the Ministry of Education, Culture, Sports, Science, and Technology of Japan (priority area 442-16082201).

REFERENCES

1. M. Sasao *et al.*, *Plasma Phys. Control. Fusion.*, **46**, S107–S118 (2004).
2. M. Sasao *et al.*, in Proceedings of the 9th IAEA-TCM on Energetic Particles in Magnetic Confinement Systems, http://dg1.nifs.ac.jp/IAEATM-EP2005/program.html, OT11 (2005, Takayama).
3. B.J. Peterson *et al.*, in Proceedings of the 9th IAEA-TCM on Energetic Particles in Magnetic Confinement Systems, http://dg1.nifs.ac.jp/IAEATM-EP2005/program.html, P-24 (2005, Takayama).
4. C. Walker, "Accommodate B-09 Lost Alpha Faraday Cup Detectors," ITER_D_25Z8N4.

PLASMA FACING COMPONENTS, RADIATION EFFECTS AND DUST

Diagnostics for Dust Monitoring in Tokamak Environment

S. Rosanvallon[1], C. Grisolia[1], F. Onofri[2], J. Worms[1,2], S.H. Hong[1]

1 Association Euratom/CEA, DRFC/SIPP, 13108 St Paul lez Durance, France
2 IUSTI-CNRS, University of Provence, Technopole de Château Gomberi, 13453 Marseille, France

Abstract. During ITER lifetime, dusts and flakes will be produced due to the interaction of plasmas with the in-vessel materials or due to maintenance. They will be made of carbon, beryllium and tungsten and will be activated, tritiated and chemically reactive and toxic. Safety limits have been set in order to reduce dust hazards. Thus dust diagnostics and removal methods need to be developed for ITER within the constraints linked to magnetic field, radiation, vacuum and temperature. This paper reviews potential diagnostics to monitor the dust content using techniques already used for erosion or deposition monitoring or techniques specially developed for measuring dust in suspension.

Keywords: ITER, safety, dust, diagnostic, erosion, deposition
PACS: 28.52.Nh

INTRODUCTION

Experience feedback from existing fusion machines or from devices simulating plasma-wall interactions has shown that dusts and flakes in a Tokamak are produced by the interaction of the plasmas (operation and conditioning) with the Plasma Facing Components (PFCs) or during maintenance periods.

During operation, the dust formation is closely linked to the erosion of the PFC materials. The eroded materials tend to re-deposit and to form stratified layers made of mixed materials coming mainly from the plasma facing components. Due to the mechanical and thermal stresses applied to the layers, they can flake once they have reached a certain thickness. Dusts of different sizes are produced, depending on the stress levels and the local deposit structure (typically from some tens μm to few mm)[1].

In the edge of fusion plasmas, the growth of small particles from atomic or molecular precursors (which are released by physical or chemical sputtering) can also lead to dust creation. This mechanism starts with particles of nm size range that may nucleate and grow up to produce quasi spherical species with diameters exceeding 1 μm[2].

Plasma events such as Vertical Displacement Events (VDEs), Edge Localised Modes or disruptions can produce high heat load. Since the thermal conduction between the co-deposited layers and the bulk of the PFC is insufficient, the surface temperature can reach the boiling or sublimation point of the material. Dust particles are then created by either condensation and growth of the vaporized material, or

CP988, *Burning Plasma Diagnostics*, edited by F. P. Orsitto, G. Gorini, E. Sindoni, and M. Tardocchi

pressure-driven ejection of melt layer material or explosive brittle destruction by heating of gas bubbles [3]. In this case, significant amounts of small dust particles (< 1 μm) are produced and can agglomerate into fractal aggregates.

Unipolar arcs (present during plasma start-up or disruptions) can also produce large particles and droplets of molten metals. In this case, a large amount of energy is deposited at the metal surface leading to melting and vaporization of the material. The dust particles are of the μm size , spherical and composed of a single material.

The wall conditioning procedures involving He-glow discharge or reactive gases (carbonization, boronization, siliconization) also generates surface films, which are subject to the same degradation as the co-deposited films and are thus potential source of dust particles.

At last, particles can arise during maintenance activities due to the presence of air/humidity that enhances the flaking process or due to operations such as mechanical abrasion during component replacement.

DUSTS IN ITER: CHARACTERISTICS AND ISSUES

In ITER, dusts are foreseen to be made of Be, W, and C with the presence of metallic impurities. Due to the deuterium reaction, these dusts will be both activated and tritiated. Moreover, due to the presence of Be, the dusts will be chemically reactive and toxic.

In ITER, it is assumed that the dust diameters will be between 100 nm and 100μm. It is considered that particles larger than 100 μm will not be transported to the environment. Based on the experience feedback on carbon dusts from various tokamaks, it is also assumed that the dusts will have a mass-median diameter of 2 μm, except the dusts produced during disruptions that will have a diameter of 0.1 μm [4].

The amount of mobile dust in the Vacuum Vessel (VV) is then a source term in case of off-normal events and it has been taken into account in postulated accidental scenarii which could result in an environmental impact in case of dust spreading. The risk of hydrogen production at the hot divertor surface in case of steam ingress, and explosion hazard in presence of dusts, air and hydrogen have also been studied. In order to control the dust hazard, the current ITER strategy is based on a defense in depth approach designed to avoid failures, minimize, detect and measure the inventory with respect to safety limits and rely on mitigation systems such as injection of inert gas.

These safety limits have been set to 100 kg of W, 100 kg of Be and 200 kg of C mobilizable dusts inside the VV. This 1st set of limits is based on the radiological hazard in case of dust spreading, and on the chemical reactivity in case of a reaction of dust with steam (leading to the production of hydrogen). The 2nd set of limits is based only on the chemical reactivity, complete reactions occurring when the temperatures are higher than 400°C for Be or W, and 600°C for C. This leads to set the limits of dusts quantity on the hot surfaces of the divertor to 6 kg of W, 6 kg of Be and 6 kg of C [5].

As a consequence, dust diagnostics need thus to be developed for the ITER environment, considering the constraints linked to magnetic field, radiation, vacuum and temperature.

In this review, two categories of diagnostics will be considered:

- Diagnostics to improve the basic knowledge on dust and flakes behaviour (creation, transport, etc),
- Diagnostics related to safety (monitoring of the quantities inside the VV and on the hot surfaces of the divertor).

DIAGNOSTIC FOR BASIC STUDIES OF DUSTS AND FLAKES BEHAVIOUR

The direct measurement of the safety limits is very difficult, but might be determined indirectly from the knowledge on dust creation, transport, preferential accumulation areas, etc, coupled to computational fluid dynamics. Moreover, based on "shot by shot" data analysis, a possible link between the plasma operation scenarii and the dust production, transport and mobilization could be found.

Large numbers of techniques have been used, in particular by the low temperature dusty plasma community. Although these techniques have been proven to be very efficient in well-characterized laboratory conditions, their application to the ITER environment is not yet straightforward.

Some of these techniques are described in Table 1.

TABLE 1. Foreseen techniques to increase the basic knowledge on dust behaviour.

Light scattering [6, 21]	Principle: Analysis of the scattering properties (intensities) of dusts illuminated by a laser.
	Data obtained (analysis range): dust in suspension size and distribution (80nm to 1-5μm).
	Feasibility: Non invasive technique with extensive experience in existing tokamaks. But the interpretation of the results is difficult when the particles are non spherical. Only small particles are detected.
CCD cameras (visible or Infra Red) [7]	Principle: Observation of plasma and inside the VV.
	Data obtained (analysis range): Creation, transport and accumulation zones (10μm to ~1mm).
	Feasibility: Non invasive technique that enables 3D analysis if two cameras are used simultaneously. Quantitative data are not easy to obtain except if calibration is used (injection of calibrated dust).
Image and Tracking Velocimetry [8]	Principle: Superimposed flashing laser sheets are used to light the particles. The displacement of each particle is measured from one image to the second one.
	Data obtained (analysis range): 2D-3D Velocity and thus surface charge (μm size and smaller).
	Feasibility: This non invasive technique needs to be coupled to laser illumination and CCD cameras.
Langmuir probe [6]	Principle: Langmuir probes show a modified characteristic due to a non equal density of electrons and ions at infinity.
	Data obtained (analysis range): Presence of dusts, ion/electron densities and balance of electrostatic to gravity forces (100nm to 1-5μm).
	Feasibility: Invasive technique but relying on existing diagnostic in Tokamaks. But the interpretation of the measurement is difficult since many phenomena can disturb the signal. Only small particles are detected.
Electrostatic grids [9]	Principle: Particles landing on two interlocking grids create a short circuit.

	Data obtained (analysis range): Particle concentration and size (5 to 100μm).
	Feasibility: Quite simple system already installed in NSTX and AUG. But since it is an invasive technique, maintenance can represent an issue.
Quartz microbalance [10]	Principle: Evolution of the quartz resonance with the presence of layers.
	Data obtained (analysis range): Film deposition (100nm to μm).
	Feasibility: System already used in existing Tokamaks. But since it is invasive, maintenance can represent an issue. Moreover, in ITER, radiation effect on quartz is an issue.
Capacitive Diaphragm Microbalance [11]	Principle: accumulation of dusts, films creates deflexion of the diaphragm surface leading to capacitance change.
	Data obtained (analysis range): Film deposition (100nm to μm).
	Feasibility: Developed to be radiation resistant in comparison with quartz microbalance. But at the moment there is no experience in Tokamaks. Maintenance remains an issue.
Scanning tunnelling microscope [12]	Principle: Surface topography is measured by the monitoring of the level of the tunnel current between the microscope tip and the sample.
	Data obtained (analysis range): Size (Å to nm).
	Feasibility: This very precise technique is invasive (maintenance issue) and concerns only very small particles. Appears difficult to use in ITER but could be useful to characterize sampling of dusts..
Light extinction spectroscopy [13]	Principle: Particles in suspension are lightened by a light beam and the beam spectral transmission is measured after passing through the particulate medium.
	Data obtained (analysis range): Concentration and Size distribution (0.3 to 10μm).
	Feasibility: Easy measurement that needs mirror to reflect the light or to be embarked on a robot. Difficult interpretation due to the large range of size, shape and composition of the dusts.
Aerosol analyser [14]	Principle: The sampling is usually performed by sucking continuously a part of the suspension medium. The measurement is based on the detection of the light pulses scattered by the particles when they are passing through a laser beam.
	Data obtained (analysis range): Size distribution, density (0.3-20μm).
	Feasibility: Technique widely used for aerosol analysis but needs relevant sampling (Invasive? Line length?) of dusts in suspension.
Pump exhaust	Principle: Analysis of the dusts collected in the existing pump exhaust system.
	Data obtained (analysis range): Quantity exhausted (nm to 10 μm).
	Feasibility: The impact on availability of pumping system during dust analysis must be addressed. Relevancy probably limited on heavy dusts (W). Some drawbacks (spreading) if gas puffing is used.
Sampling via endoscope	Principle: Collection of dusts through robotised arm for external analysis.
	Data obtained (analysis range): Size distribution, quantity, composition, etc are measured ex-situ (50nm to mm).
	Feasibility: Extensive analysis made possible (needed as input data for other techniques). But it needs a robotised arm (not yet developed).

Looking at the technical feasibility of those various techniques, with the current state of the art, light scattering and light extinction, use of CCD cameras (with or without image and tracking velocimetry techniques), sampling via endoscope and capacitive diaphragm microbalance appear to be the most promising techniques that need to be further assessed.

All these assessed techniques provide rather local measurements, not directly applicable to the monitoring of the safety limits. A link between the local measurement(s) and the global inventory with reasonable models, carefully validated assumptions and arguments has to be found.

DIAGNOSTICS RELATED TO SAFETY

Diagnostics to Monitor the Dust Inventory Inside the VV

Concerning the limits inside the VV, the strategy proposed is based on the monitoring of the erosion source inside the VV, in order to determine the envelope quantity of materials that could potentially lead to dust creation. As a first approximation, 100% of the net eroded materials are considered as a potential source of dusts. In order to reduce the constraints on operation, this envelop value has to be decreased using the experience feedback from the H and D phases and/or more local measurements as presented in the previous section.

The order of magnitude to be monitored is preliminarily derived by considering a homogeneous erosion of the 700 m^2 of Be, 100 m^2 of W, 50 m^2 of C and a density divided by 2 for dust in comparison to bulk materials. The limits on dusts inside the VV would correspond to about 160 μm of eroded Be, 100 μm of eroded W, and 4000 μm of eroded C.

Potential techniques are listed in Table 2.

TABLE 2. Foreseen techniques to monitor the dust inventory inside the VV

Spectroscopic measurements (without dopant) [15]	Principle: Measurement of the impurity photon fluxes and providing that information on plasma edge condition are available, data on particle fluxes can be obtained.
	Data obtained (analysis range): Total amount of material eroded in the machine.
	Feasibility: Technique present in tokamaks but needs data (T_e, n_e, etc) and quite complex modelling to obtain erosion rate in particular in divertor region.
Spectroscopic measurements (with dopant) [11]	Principle: Dopants are implanted between PFC material mats and dopant signal is observed by spectrometry, indicating erosion of all the materials above.
	Data obtained (analysis range): Material eroded in areas where dedicated tiles are installed.
	Feasibility: Choice of dopants, manufacturing and impact on waste management still need to be addressed.
Gas physical adsorption [16]	Principle: Evolution of the physical adsorption of a neutral gas by the surfaces inside the VV, without and with dust/layers.
	Data obtained (analysis range): Variation of the active surface area.
	Feasibility: The feasibility is difficult to assess at this development stage but would enable an overall measurement.
Speckle interferometry [17]	Principle: Measurements of the interferences between the laser beam reflected by a mirror and by the surface to be analysed.
	Data obtained (analysis range): Eroded materials (resolution ~10μm).

	Feasibility: Non invasive and precise technique that would enable to integrate all the erosion sources (normal operation and conditioning). But need a reference surface to obtain relative values. Even if few cm^2 are scanned, it remains a local measurement. Moreover, the analysis of the tiles at glancing incidence can be an issue.
Laser Distance and Ranging [18]	Principle: Distance are measured via time-of-flight measurements for the reflection of ultra-short (<100ps) laser pulses from a surface.
	Data obtained (analysis range): Eroded materials (resolution ~100μm).
	Feasibility: This non invasive needs surface reference. Local measurement. Resolution could be improved by the use of femtosecond lasers.

Spectroscopic measurements and Speckle interferometry appear to be the most promising techniques to estimate the net erosion source in a Tokamak.

Diagnostics to Monitor the Dust Inventory on the Hot Surfaces of the Divertor

Since it is difficult to determine whether the co-deposited layers will remain stable or contribute to dust creation through flaking during operation or maintenance, the strategy proposed for the monitoring of the inventory on the hot surfaces of the divertor is based on a global measurement of the dusts and the co-deposited layer. Some of the techniques are described in Table 3.

TABLE 3. Foreseen techniques to monitor the dust inventory on the "hot surfaces" of the divertor

Infra Red (IR) thermography [19]	Principle: IR signal differs from a zone of pure erosion to a zone with co-deposited layers due to the low thermal conductivity and low adherence of dusts/layers.
	Data obtained (analysis range): Qualitative mapping of the eroded and deposited zones.
	Feasibility: Enables mapping during or in between plasmas, to be coupled to LIBS for thickness "calibration". Impact of metallic materials and analysis of the tiles at glancing incidence to be addressed.
Laser Induced Breakkdown Spectroscopy [20]	Principle: Analysis of the plasma emitted by the laser ablation gives the thickness of the co-deposited layers if a pre calibration of the ablation rate is available.
	Data obtained (analysis range): Coupled to the IR thermography, it provides a quantitative mapping (10 to 40 μm).
	Feasibility: Non invasive technique already tested in Tokamak. Pre calibration of the ablation rate still needs to be addressed.
Speckle interferometry [17]	Same as in Table 2 for codeposited materials (resolution ~10μm).
Laser Distance and Ranging [18]	Same as in Table 2 for codeposited materials (resolution ~100μm).

These techniques do not enable to separate the contribution of the deposits from dust at the moment so further R&D is needed in this direction. The management of the dust on the hot surfaces of the divertor can also rely on a regular removal or, at least, on displacing the dusts to areas not in direct contact with the plasma where the safety constraints are reduced.

CONCLUSIONS

This survey has shown that various diagnostic techniques can be foreseen for the dust monitoring in ITER, but no diagnostic solely provides directly a value that could be compared to the safety limits. Nevertheless, some promising techniques, already used in Tokamak but sometimes derived from their initial objectives, have been identified. Even if the environment will be more stringent in ITER than in the existing machines, the current developments are encouraging for their application in ITER.

The strategy for the dust management has also to rely on an improvement of the prediction (better understanding of the creation processes and modelling of the reactions involving dusts), and a reduction of the consequences (removal techniques and mitigation systems).

From a safety point of view, the monitoring of the 18 kg of dusts on the hot surfaces of the divertor remains the most challenging issue because of the accuracy needed and design constraints (divertor design and space allocated to dust diagnostic).

Moreover, these schemes have to integrate tritium, conditioning and dust management issues since there are highly interconnected (dust production, removal, impact on conditioning, etc).

ACKNOWLEDGMENTS

This work, supported by the European Communities under the contract of Association between EURATOM and several Associations, was carried out within the framework of the European Fusion Development Agreement. The views and opinions expressed herein do not necessarily reflect those of the European Commission. Authors would like to thanks the other contributors in particular: E. Gauthier (CEA), X. Courtois (CEA), G. Counsell (UKAEA), J. Winter (Bochum University), B. Annaratone (CNRS/Université de Provence), G. Maddaluno (ENEA) and P. Gasior (IPPLM).

REFERENCES

1. J. Winter, *Dust in fusion devices-a multi-faceted problem connecting high- and low-temperature plasma physics*, Journal of Nuclear Materials 283-287 (2000) 110-119
2. C. Arnas et al., *Characterisation of carbon dust produced in sputtering discharges and in the Tore Supra tokamak*, Journal of Nuclear Materials 353 (2006) 80–88
3. P. Sharpe et al., *A review of dust in fusion devices: Implications for safety and operational performance*, Fusion Engineering and Design 63-64 (2002) 153-/163
4. *Generic Site Safety Report – Volume III – Radiological and Energy source terms* – July 2004, G 84 RI 3 R0.2
5. S. Rosanvallon et al., *ITER safety limits related to dust*, 1st Workshop on the "Dust in Fusion Plasmas", Warsaw, July 8-10, 2007, to be published
6. B. M. Annaratone, W. Jacob, C. Arnas and G. E. Morfill, *Critical review of complex plasma (dusty plasma) diagnostics and manipulation techniques for the fusion community and other*, 2007, to be published
7. S. Hong et al., *Dusts observed by IR and visible cameras in Tore Supra and JET*, 1st Workshop on the "Dust in Fusion Plasmas", Warsaw, July 8-10, 2007, to be published
8. M Raffel, C. Willert, *Particle Image Velocimetry: A Practical Guide*, Springer, Tokyo, 1998.

9. C. Skinner et al., *Measuring and controlling surface dust in a tokamak,* 1st Workshop on the "Dust in Fusion Plasmas", Warsaw, July 8-10, 2007, to be published
10. H.G. Esser et al., *Quartz microbalance: a time resolved diagnostic to measure material deposition in JET,* Fusion Engineering and Design 66- 68 (2003) 855- 860
11. G.F Counsell, C.H. Wu, *In-situ detection and removal of Carbon debris - a challenge for the next step fusion device,* Physica Scripta. T91, 70-75, 2001
12. L.N. Khimchenko, *Tungsten and carbon fractal nanodust formation in ITER ELMs I Type simulation experiment,* 1st Workshop on the "Dust in Fusion Plasmas", Warsaw, July 8-10, 2007, to be published
13. J. Worms, F. Onofri, C. Grisolia and S. Rosanvallon, *Non invasive optical diagnostics for particles in suspension,* 1st Workshop on the "Dust in Fusion Plasmas", Warsaw, July 8-10, 2007, to be published
14. A.J. Armendariz, D Leith, *Concentration measurement and counting efficiency for the aerodynamic particle sizer 3320,* Aerosol Science 33 (2002) 133–148
15. J. Hogan et al., *Model for impurity generation, transport and deposition in the complex CIEL environment,* Journal of Nuclear Materials 363–365 (2007) 167–173
16. V.V. Filatov et al., *Dust Assessment (Phase 2),* ITER documentation, IDoMS.No. G 17 RE 90 01-06-13 F 1
17. P. Doré, E. Gauthier, *Speckle interferometry diagnostic for erosion/redeposition measurement in tokamaks,* Journal of Nuclear Materials 363–365 (2007) 1414–1419
18. C. Skinner, GA Gentile, M, Menon, R. Barry, *Flaking of co-deposited hydrogenated carbon layers on the TFTR limiter,* Nuclear Fusion, Vol. 39, No. 9, 1999
19. E. Gauthier et al., *Thermal behaviour of redeposited layer under high heat flux exposure,* Journal of Nuclear Materials 337–339 (2005) 960–964.
20. F. Le Guern et al., *First Results with Laser Induced Break-down Spectroscopy for Codeposited Layer Studies at JET,* 24th Symposium on Fusion Technology, Warsaw, Poland, 11 - 15 September 2006, to be published
21. E. Giovannozzi, C. Castaldo, G. Maddaluno, Evidence of dust in FTU from Thomson Scattering diagnostic measurements, 33rd EPS Conference on Plasma Phys. Rome, 19 - 23 June 2006 ECA Vol.30I, P-2.093 (2006)

On The Problem Of In-vessel Mirrors For Diagnostic Systems Of ITER

V.S. Voitsenya[a] and A. Litnovsky[b] for the SWG on First Mirrors of the ITPA Topical Group on Diagnostics

[a]*IPP, NSC Kharkov Institute of Physics and Technology, Kharkov 61108, Ukraine;* *voitseny@ipp.kharkov.ua;*

[b]*Institut für Energieforschung - Plasmaphysik, Forschungszentrum Jülich, Ass. EURATOM- FZ Jülich, D-52425 Jülich, Germany;* *a.litnovsky@fz-juelich.de*

Abstract. The present status of the investigations with ITER-candidate mirror materials and directed on solution of the in-vessel mirror problem, are presented in the paper. The current tasks in the R&D of diagnostic mirrors and outstanding questions are discussed.

Keywords: ITER Diagnostics, Mirrors, Sputter Erosion, Contaminant Deposition, .
PACS: 52.40.Hf; 52.70.-m; 52.70.Kz; 78; 78.68.+m

1. INTRODUCTION

The main goal of the International Thermonuclear Experimental Reactor (ITER) – is to provide a wide program of investigations devoted to obtaining the base data for development of the next step fusion device (DEMO). The planned scientific and technological programs require using of a great number of different methods of plasma diagnostics in ITER during every stage of a discharge. Among methods of plasma diagnostics being planned to be used, a large fraction belongs to optical and laser diagnostics in a wide spectral wavelength range, from a few nm up to a few mm [1].

Because of high level of gamma and neutron radiation, the systems of measurements of these diagnostics must be based not on a refractive optics (dielectric lens or prisms) but on a reflective optics, with the use of metal mirrors, at least for those components which are exposed close to a thermonuclear plasma. Some mirrors unavoidably have to be plasma-viewing component of the respective measuring schemes. The surface of these first mirrors (FM) will be subject of an impact of all kinds of radiation emanating by a fusion plasma: electromagnetic due to gammas and x-rays, VUV, UV, Visible radiation and corpuscular due to neutrons, charge exchange atoms and ions of a near-wall plasma

. The atom and ion fluxes will be composed of thermonuclear fuel particles (T^0, D^0, D^+, T^+), thermonuclear ashes (He^0, He^+, He^{++}), and of some impurities, mainly beryllium and carbon ions in different charge states [2]. The ratio between fluxes of radiation and particles on the mirror surface strongly depends on the mirror position inside the vacuum vessel of a fusion reactor.

This present paper is an attempt to give the short description of the main results obtained during about one decade of investigations in the field of FM problem and to discuss the questions that are not solved yet.

The paper consists of six parts. The 2nd part contains information about results obtained in simulation experiments on small-scale laboratory facilities. The 3rd part is

CP988, *Burning Plasma Diagnostics*, edited by F. P. Orsitto, G. Gorini, E. Sindoni, and M. Tardocchi

devoted to the description of main results obtained in experiments on behavior of in-vessel mirrors on present fusion devices. Part 4 concerns with the blistering phenomenon which is being discussed in connection with the FM problem. In the 5th part the questions are discussed which have to be subject of coming experiments, and the concluding remarks are presented in the 6th part.

2. RESULTS OF SIMULATION EXPERIMENTS

The majority of simulation experiments formerly were performed in the assumption that the main reason of FM degradation in ITER could be the development of micro-relief due to bombardment by charge exchange atoms (CXA). Correspondingly, the results obtained [3] can be applied to the situation when the sputter erosion of the mirror surface prevails over deposition of contaminants. The difficulties connected with necessity to clean mirror surface from contaminating deposit became clear much later and solution for mirror cleaning is yet to be found.

2.1. Mirrors Under Erosion-Dominated Conditions

According to the experimental data, the FMs have to be fabricated from single crystals (sc) of those metals that have a rather low sputtering yield combined with an acceptable reflectivity. The examples of such materials are molybdenum and tungsten, what was found after carrying the experiments with many mirror materials tested under long-term sputtering by ions of deuterium (or hydrogen) plasma: Be, Al, V, Ni, SS, Cu, Mo, Ta, W, and Rh film on different substrates [3]. The most of tested materials had a polycrystalline (pc) structure but SS, Mo and W mirrors were either machined from single crystals (sc) with different orientations of the main crystallographic axes relatively the mirror surface. The results for similar orientations (111) of sc mirrors and for the pc mirrors of the same materials are shown in Fig.1. For comparison, the behavior of one of the best Rh film mirrors on Cu substrate is presented as well. Along

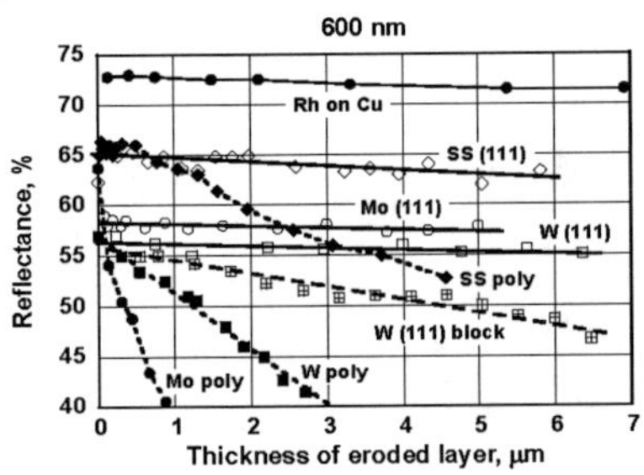

FIGURE 1. Dependence of reflectivity of poly- and mono-crystalline metals on the depth of layer eroded by long-term sputtering with deuterium plasma ions with wide energy distribution in comparison with similar dependence of the Rh film on Cu substrate mirror.

the X axis is shown the depth of layer sputtered due to bombardment with energy-distributed deuterium plasma ions. In Fig.1 the symbol "(111) block" denotes the data for mirror fabricated of pc tungsten with similar (111) orientation of crystalline plane for more than 95% of all individual grains. The reflectance of this mirror degrades significantly slower than the reflectance of a pc mirror but noticeably faster than the real sc mirror, where no degradation was found in the initial reflectance.

These data were provided using the DSM-2 device where nonelectrode plasma in deuterium and in hydrogen was produced by the non-contact Super High Frequency (SHF) discharge in electron cyclotron resonance (ECR) conditions, and the mirrors under the test were bombarded with ions having a wide energy distribution, ~50-1500 eV [3, 4]. The subsequent direct comparative test of sc and pc Mo mirrors in TEXTOR had confirmed advantageous properties of mono-crystal mirror materials [5].

It is evident from Fig. 1 that sc mirrors can maintain their optical properties being long-term ion sputtered up to the depth of several micrometers. The main reason of degradation of pc mirrors is development of a step structure because the grains with different orientations have different sputtering yields. The Rh film mirror behaves similar to sc mirrors in spite of the fact that a metal film has a pc structure, what is explained by the small size of the microcrystals (less than the wavelength of light) composed the film. However, the sputtering coefficient of Rh film is rather high and the film cannot be too thick (probably not thicker ~10 μm), therefore the mirror fabricated as Rh film on a high surface quality substrate (e.g., copper) may be used in positions where the CXA flux to the FMs in ITER is weakened for some reason (location of mirror behind an aperture, in the long duct, or as a second mirror).

It was observed in experiments with pc samples subject to bombardment with hydrogen (deuterium) ions of different energy and fluence that the ingrain roughness develops with different rates: some grains become very roughen but some ones continue to be smooth. For the stainless steel (SS) mirrors it was found that the ingrain roughness growth rate depends on the orientation of the every given grain face in relation to the mirror surface [6]. The (111) grains demonstrated the largest sputtering yield but at the same time their surface continued to be smooth much longer than the surfaces of surrounding grains with other crystallographic orientations. (To see such evident peculiarities of differently oriented grains became possible due to recent appearance of the newest technique EBSD (Electron Backscatter Diffraction) with high space resolution [7]). The pc steel used for fabrication of mirror samples in [6] belongs to the fcc group of metals where the (111) face has a highest density of atoms.

2.2. Mirrors Under Deposition-Dominated Conditions

In ITER divertor area the carbon-fiber composite material are planned to be used. As a consequence of this, carbon will be one of the main impurities in the core plasma and in the plasma out of the confinement volume. Therefore, deposition of carbon on the first and other mirrors will probably be unavoidable.

In laboratory experiments the effect of thin carbon film on SS and Mo mirrors was studied in detail [3] and it was found that the quite simple formulae by Hass [8] can be applicable within a good accuracy for the description of the optical properties of mirror coated with a half transparent thin C-film. For demonstration of the effect, in Fig. 2 are shown the dependences of reflectivity for two different contaminating films. Curve 1 is the dependence for the film of complicated composition that was found on the diagnostic window of the JT-60U tokamak and characteristics of which were provided in [9]; curve 2 is the result of calculation of reflectivity for the C-film deposited in an arc discharge between two graphite electrodes [10].

In reality, the deposits on mirrors in ITER, as well as in existing fusion devices with protecting graphite tiles, will have the characteristics somewhere between these two curves, so, even the quite thin contaminating carbon-based film (≥ 10 nm) will strongly modify the reflectance of in-vessel mirrors. Because of the wavelength dependence of the effect, not only the intensity of reflected signal will drop, but also its spectral characteristics will be changed as compared to that of the incident signal.

The deposition of the C-film causes also changes in the polarization angle of reflected light. An example of such an effect is shown in Fig.3. For the measurement of polarization angles depending on the C-film thickness for different angles of incidence, the zero-ellipsometer was used.

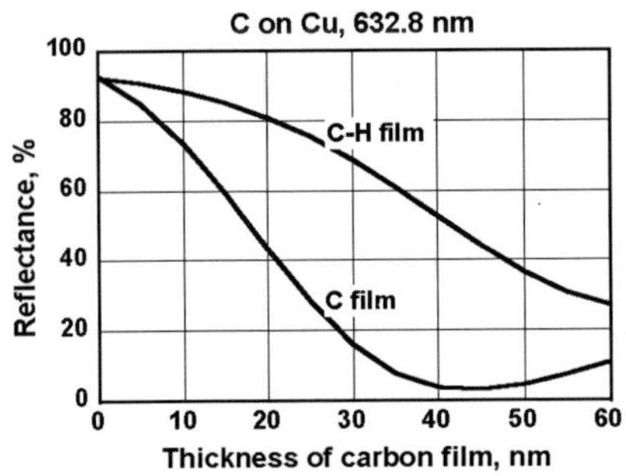

FIGURE 2. Calculated reflectivity values for hydrocarbon film found as deposition on the diagnostic window of JT-60U [9] and for the carbon film deposited in vacuum due to an arc discharge between two graphite electrodes [10].

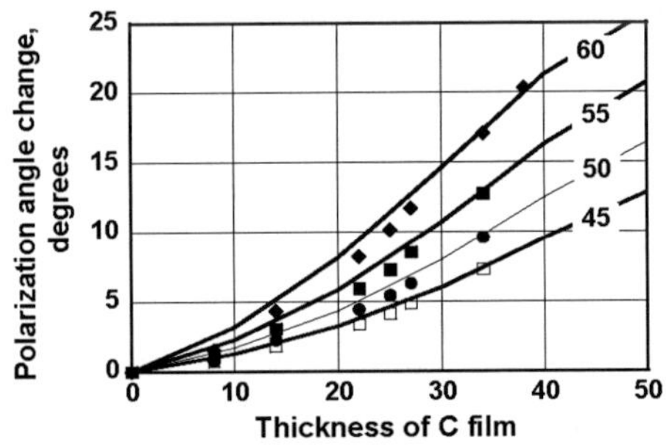

FIGURE 3. Calculated (lines and solid points) and experimental (open points) data measured by ellipsometry of the polarization angle rotation of the light at λ = 632.8 nm reflecting from a Mo mirror coated with different thicknesses of C film. The angle of incidence is shown near every curve [11].

The results of measurements (points) are compared with the results of calculations (lines) carried out by using the scheme described in [11]. It is seen that the effect of carbon film increases with increasing the film thickness and the incidence angle.

These data are very important for the e.g. MSE (Motion Stark Effect) diagnostics based on measurement of the polarization angle of the light radiated from a fixed plasma volume [12]. It is seen in Fig. 3 that there is a strong dependence of the effect on the incident angle. That is why to decrease of the influence of C-film on polarization angle change, the scheme of MSE measurements has to use several secondary mirrors and thus looks to be quite complicated.

3. MIRROR EXPERIMENTS ON EXISTING FUSION DEVICES

During last years an extensive experimental program on behavior of in-vessel mirrors was initiated at several fusion devices: JET [13, 14], LHD [15,16], T-10 [17], Tore Supra [16,18], TRIAM-1M [16], TEXTOR [19], TCV [16,20], DIII-D [21], HL-2A [22], HT-7, and ASDEX-U. The results obtained in these experiments have generally confirmed the findings made in the laboratory experiments but also demonstrated the necessity to change the emphasis of investigations directed to solution of the first mirror problem. An overview of the recent findings may be found in [23].

The degradation of optical properties of mirrors occurred both because of sputter erosion (JET [13], LHD, Tore Supra and TEXTOR) and due to growing the deposit containing predominantly carbon (LHD, T-10, TRIAM-1M, TEXTOR, TCV, DIII-D, HT-7). The erosion was observed for mirror samples fixed at the walls along the inner circumference in JET [13], and for samples closely located to the plasma (LHD, Tore Supra, TEXTOR); all these samples were fully open to the CXA flux and to glow discharge plasma used for conditioning. The main contribution into erosion is given by either working discharges (TEXTOR, JET [13], LHD) or glow discharge (Tore Supra).

The deposition of contaminants was observed for mirrors fixed at the walls along the outer circumference in JET [13], on the samples located in the divertor area and in the diagnostic port, far from the confinement volume in LHD, and on mirrors exposed in tokamaks T-10, TEXTOR, TCV, D-IIID, TRIAM-1M, HL-2A, HT-7.

However, it should be noted that in modern tokamaks the fraction of the inner surface protected with carbon-based materials (tiles, limiters) is noticeably greater than in ITER, and thus the role of carbon deposit can be too high in comparison with ITER conditions. Also, in ITER the working temperature of mirrors will be higher than in the majority of modern experiments in fusion devices.

Quite important for predicting behavior of mirrors in ITER are results of recent calculations of fluxes of ions and atoms of hydrogen isotopes and impurities contained in periphery plasma (He, C) on the FM surfaces [2]. According to results obtained, the carbon flux represents a significant part of the total particle flux to the mirrors. The first modeling of similar situation was provided in [24] and it was found that there is no chance to avoid the C-film deposition if the carbon ion portion exceeding ~4% of the deuterium ion flux, independently on the material exposed to such an ion flux. The physical reason of C-deposit growth is the low erosion rate of carbon with deuterium ions (of the order a few per cents) and low self-sputtering yield (less than unity).

4. IS THE BLISTERING PHENOMENON IMPORTANT?

During last years quite a few papers appear with observation of blistering phenomenon on W and in much less degree – on Mo surfaces, exposed to He^+ and H^+ (D^+) ions of rather low fixed energy (up to ≥5 eV) [25-30]. Using the EBSD method, the authors of [30] have shown that for a pc tungsten exposed to a high flux of D^+ ions with 38 eV energy, the (111) faces tend towards the appearance of blisters much stronger than the faces with any other orientation relatively the surface. In contrast to stainless steel, tungsten belongs to the bcc group of metals with the (110) face having the highest density, therefore, the result of [30] does not contradict the abovementioned result of [6] obtained for long-term sputtering of SS mirrors with H_3^+ ions at normal incidence angle.

Such "low-energy" blistering (LEB) is new phenomenon which probably has to be studied in more detail to understand its importance for FMs in ITER. For example, no blisters were observed in simulation experiments during long-term bombardment of both, sc and pc structures of W, Mo and SS with energy-distributed ions [3] (the small part of the obtained data is given in Fig.1). The possible explanation is that in those simulation experiments the ions have a quite wide energy distribution, 50-1500 eV with corresponding wide depth distribution of trapped deuterium. As was shown in experiments that were carried out with specially expended ion energy spectra [31, 32], in such a situation the "classical" blistering was suppressed, even when the total ion fluence significantly exceeded the threshold for blistering at any fixed ion energy.

Thus, it is not excluded that the probability for blisters to appear on the FM surfaces is small, because the energy distributions of CXA flux in ITER will be very wide [2]. However, this qualitative conclusion has to be checked by groups where the LEB phenomenon was observed [25-30]. Note, when providing such experiments, the powder metallurgy Mo and W samples should be excluded from the test, as the feature of these materials is an appearance of blisters of so called "nonradiative nature" [33] with size spread from tens to hundreds micrometers, which were observed, e.g., in [33, 34]. It should be noted in addition, that no blister formation was reported from the present-day tokamaks and it looks like the conditions in the modern fusion devices may be quite "unfavorable" for the blisters to occur. Therefore, the relevance of blister formation to the nowaday tokamak environment is yet to be proven.

5. WHAT WAS NOT MADE YET

5.1. The results of simulation experiments allowed making a motivated choice of mirror materials mostly appropriate from the viewpoint of mirror long term operation under impact of CXA flux when contamination of the mirror surface can be neglected. Several materials: single crystalline Mo and W, and Rh films of an acceptable thickness (what depends on the mirror location, i.e., on the degree of weakening the CXA flux in comparison with the flux to the first wall).

However, there was not studied yet the behavior of these materials in conditions of high fluence of neutrons (for a Mo-made retro-reflector the maximum achieved neutron flux was only $1.4 \cdot 10^{20}$ n/cm^2 [35], about two orders of magnitude less than was estimated for ITER). Such investigations probably have to be provided in a nearest time aiming the prediction of the role of transmutation on the modification of mirror material and, correspondingly, on the mirror optical characteristics. Simultaneously, the effect of volume swelling could be cleared up which was also not investigated yet.

5.2. There is no clear understanding of the mechanism of influence of mirror moderate temperature on the growth rate of a carbon-based deposit, what was observed in laboratory experiments [36] and on tokamaks DIII-D [21] and T-10 [37]. One may hypothesize two reasons for the temperature effect on the rate of C-film growth: with increasing temperature (i) the chemical erosion of the film increases [38], and (ii) the binding energy of every C atom with the surface decreases [21]. However, much more has to be done to clarify the mechanisms of cleaning and to apply them for guarantying of long-time maintenance of acceptable working characteristics of in-vessel mirrors.

The dedicated experiments are underway on different fusion devices (TEXTOR, Pilot-PSI) coupled with modeling of experimental conditions: sample temperature, parameters of ion flux, the relative concentration of carbon in the ion flux, etc. The puffing of "protective" gas in front of the mirror is another attractive option to mitigate the carbon deposition. This technique is presently under investigation.

Very little is known about the properties of Be deposits. The R&D on the mitigation and cleaning of Be-containing layers is the subject of an immediate investigations.

5.3. There is lacking clearness of the role that can play the mirror material in the C-film growth rate when mirror surface is bombarded with C and D ions. For the first time the effect of mirror material was observed on the Tore Supra tokamak when Mo, SS and Cu mirrors were exposed close each other for the whole experimental campaign (March 2003 – April 2004) [18]. Later on, the laboratory experiment was provided [39] with SS and Cu mirrors exposed to plasma of a glow discharge in D_2 with small addition of CH_4 (up to 3.5%). It was found after exposure that the SS mirror became coated with a quite thick carbon film while the Cu mirror continued to be clean and therefore became sputter eroded under impact of C and D ions.

In the TCV experiment the carbon film grew several times faster on Si sample than on Mo sample located nearby [20], this result is in a qualitative agreement with theory [24]. However, during the recent experimental campaign in TEXTOR, the substrate effect on carbon deposition on the Mo and W was not very pronounced [40] and the dedicated modeling has revealed a narrow operational window for such an effect which may significantly limit the occurrence of pronounced substrate effect. At the same time, the chemical reactivity of the surface is expected to play an important role in the deposition mitigation. Other factors, like the surface roughness are expected to play the important or even major role in the deposition efficiency.

Further experiments in this direction have to be provided approaching the conditions for some FMs in ITER where the flux of low energy C atoms and ions with relatively low energy was predicted [2] to constitute a quite high part of the flux of D and T ions and charge exchange atoms.

6. MAIN RESULTS ON MIRRORS IN FUSION DEVICES

1. The manifestation of both main mechanisms of in-vessel mirror degradation (erosion and deposition) was found out in mirror experiments carried out at fusion devices under operation. The erosion was observed for mirror samples fixed at the walls along the inner circumference in JET, and samples closely located to the plasma (LHD, Tore Supra). The main contribution into erosion is given by either working discharges (JET, LHD) or glow discharge (Tore Supra).

The deposition of contaminants (mainly carbon) was observed for mirrors fixed at the walls along the outer circumference in JET, on the samples remote from the confinement volume in LHD, and on mirrors exposed in tokamaks T-10, TEXTOR, TCV, DIII-D, TRIAM-1M, HL-2A, HT-7.

2. The rate of carbon deposition was found strongly depends on the temperature of mirror samples. The exposure of samples at moderate temperature (~150°C) significantly decreases or even completely suppresses the growth rate of deposition on the mirror surface (DIII-D, T-10).

3. The rate of carbon deposition was found to be dependent on the mirror material (TEXTOR, Tore Supra, TCV). In the Tore Supra experiment the mirrors of non carbide-forming material (copper) were strongly eroded but the mirrors of carbide-forming materials (Mo, SS) were eroded only slightly. The difference in the degree of

erosion strongly differed from what would be expected comparing the published sputtering yields of these material (SS, Cu, Mo).

REFERENCES

1. A.J.H.Donne et al. Progress in the ITER Physics Basis. Chapter 7: Diagnostics. – *Nucl. Fusion* 47, S337–S384 (2007).
2. A.Kukushkin (private communication).
3. D.V.Orlinski et al. – *Plasma Devices and Operations.* **15,** 33-75 (2007).
4. A.F. Bardamid et al. – *Surface and Coating Technology* **103-104**, 365-369 (1998).
5. A. Litnovsky et al. *Fusion Engineering and Design* 82, 123-132 (2007).
6. M. Balden et al. – *J. Nucl. Mater.* **329-333,** 1515-1519 (2004).
7. F.J.Humphreys – *J. Mater. Sci.* **36**, 3833-3854 (2001).
8. G.Hass, – J. *Opt. Soc. America* **47,** 1070-1073 (1957).
9. H.Yoshida et al., – JAERI-Research 96-062. (Data on n and k at λ=632.8nm measured by a spectroscopic ellipsometry were presented by Dr. H. Yoshida as private communication.)
10. E.T.Arakawa et al., – *Phys. Rev.* **31B** 8097-8101 (1985).
11. A. Malaquias et al. – Paper O-3.4C *at 30th EPS Conf. on Contr. Fusion and Plasma Phys.*, St. Petersburg, Russia, July 2003, ECA vol. 27A.
12. M.von Hellermann, C.Giroud, N.C.Hawkes et al. – Paper O-4.2D *at 30th EPS Conf. on Contr. Fusion and Plasma Phys.*, St. Petersburg, Russia, July 2003, ECA vol. 27A.
13. M.Mayer et al. – *Diagnostics for Experimental Thermonuclear Fusion Reactors*, 2. Edited by Stott et al., Plenum Press, New York, 1998, p. 279.
14. M.J.Rubel et al. – *Rev. Sci. Instr.* **77**, 063501 (2006).
15. V.S.Voitsenya et al. – *Plasma Device and Operations* 13, 291-300 (2005).
16. V.S.Voitsenya et al. – Paper P-5.134 *at 33rd EPS Conference on Plasma Phys.*, Rome, June 2006, ECA vol. **30I**.
17. K.Yu. Vukolov et al. – Paper P-2.76 presented at *30th EPS Conference on Contr. Fusion and Plasma Phys.*, St. Petersburg, Russia, July 2003, ECA vol. 27A.
18. M.Lipa et al. – *Fusion Engineering and Design,* **81,** 221–225 (2006).
19. P. Wienhold et al., – *J. Nucl. Mater.* **337-339,** 1116-1120 (2005).
20. G. De Temmerman et al. – Paper P-1.076 *at 32nd EPS Conference on Plasma Phys.*, Tarragona, Spain, 27 June–1 July 2005, ECA Vol. **29C** (2005).
21. A. Litnovsky et al., *Fusion Eng. Design* (2007), doi: 10.1016/j.fusengdes.2007.06.042
22. Y. Zhou et al., *Fusion Engineering and Design* **81** 2823–2826 (2006).
23. A. Litnovsky et al., *J. Nucl. Mater* **363-365**, 1395-1402 (2007).
24. D.Naujoks, W.Eckstein. – *J. Nucl. Mater.* **230,** 93-100 (1995).
25. T.Venhaus et al. – *J. Nucl. Mater.* **290-293,** 505-508 (2001).
26. D.Nishijima et al. – *Japan. J. Appl. Phys.* **44**, 1A, 380-382 (2005).
27. M.Ye et al. – *J. Nucl. Mater.* **313-316,** 72-76 (2003).
28. Sh.Takamura et al. – Presentation at ITPA-9, Daejeon, Korea, 10 – 14 October 2005.
29. D.Nishijima et al. – *J. Plasma Fusion Res.* **81**, 703-707 (2005).
30. W.M.Shu et al. – *Phys. Scr.* **T128,** 96–99 (2007).
31. I.M.Neklyudov et al. – *J. Nucl. Mater.* **115,** 134-136 (1983).
32. M.I.Guseva and Yu.V.Martynenko. *Sov. Phys. Uspekhi,* **24** (12), 996-1007 (1981).
33. M. I. Guseva et al. Atomnaya Energiya, **50** (3), 212-213 (1981) (in Russian).
34. A.F.Bardamid et al., – *Plasma Devices and Operations* **13,** 203-208 (2004).
35. T.Nishitani et al., – *Fusion Engineering and Design* **42,** 443-448 (1998).
36. W.Bohmeyer et al., – Paper P-3.184 *at 30th EPS Conference on Contr. Fusion and Plasma Phys.*, St. Petersburg, Russia, July 2003, ECA vol. **27A**.
37. K.Yu.Vukolov et al., – *Problems of Atomic Science and Technology, Ser. Thermonuclear Fusion,* 2006, issue 4, p. 21-33 (in Russian).
38. A.A. Haasz et al. – *J. Nucl. Mater.* **128 & 129,** 593-596 (1984).
39. G. De Temmerman et al. – *J. Nucl. Mater* **363-365**, 259-263 (2007).
40. A. Kreter et al. – *J. Nucl. Mater* 363-365, 179-183 (2007).

Perspectives of Use of Diagnostic Mirrors with Transparent Protection Layer in Burning Plasma Experiments

Eugene E. Mukhin*, Gennadiy T. Razdobarin*, Vladimir V. Semenov*, Sergey Yu. Tolstyakov*, Mikhail M. Kochergin*, Gleb S. Kurskiev*, Klara A. Podushnikova*, Alexandr N. Andreev†, Denis V. Davydov†, Marina G. Rastegaeva†, Yuriy P. Khimich‡, Vladimir N. Gorshkov‡, Dmitriy B. Nikitin‡, and Andrej M. Litnovsky†*

** Department of Plasma Physics, Nuclear Physics and Astrophysics, A.F.Ioffe Physico-Technical Institute, SPb, 194021, Russia*

† Division of Solid State Physics, A.F.Ioffe Physico-Technical Institute, SPb, 194021, Russia

‡ Research Institute of Optical Design, Federal Research Center Vavilov State Optical Institute, SPb, 199034, Russia

† Institute fur Plasmaphysik, Forschungszentrum Julich, Partner in the Trilateral Euregio Cluster, Ass. EURATOM-FZ Julich, D-52425 Julich, Germany*

Abstract. We evaluate using of metal mirrors over-coated with transparent protection layer for the in-vessel diagnostic systems in reactor-grade fusion devices. Ideally, these should satisfy the contradictory demands of high reflectivity and small rate degradation when being bombarded by CX atoms. The serious threat to the performance of diagnostic mirrors is surface contamination with carbon-based material eroded from carbon tiles. Via coupling the protective layer to a bulk mirror we can mitigate the deposit influence on the reflectance spectra. The regards are given to survivability in plasma environment of protected coated metallic mirrors.

Keywords: Diagnostics, Pollution, Mirror, Coating.
PACS: 50.52.70.-m

MOTIVATION TO LOOK AT THE ISSUE

The lifetime of the refractive components exposed to reactor grade plasma is to be very short. A key factor in the construction and operation of optical diagnostic in ITER is to select the material for the in-vessel mirrors. Unfortunately, the number of metals suitable for manufacturing of diagnostic mirrors is very limited. Metals resistive to sputtering (Mo, W) distinguished by small rate degradation when being bombarded by CX atoms are commonly lacking in reflectivity. Figure 1 shows that there are transparent dielectric materials with relatively low sputtering rate for the predicted mean energy of bombarded CXA less than 250eV [1].

CP988, *Burning Plasma Diagnostics*, edited by F. P. Orsitto, G. Gorini, E. Sindoni, and M. Tardocchi

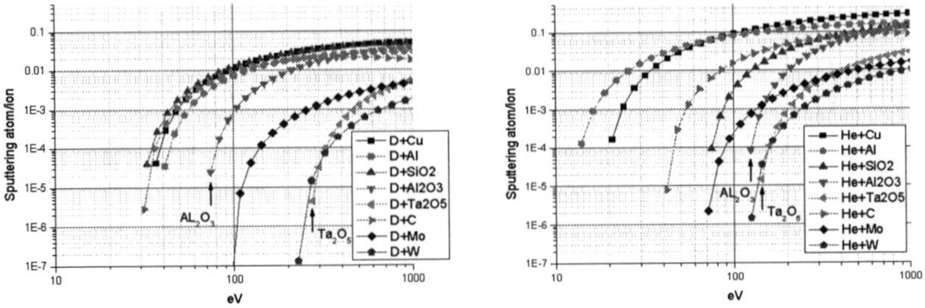

FIGURE 1. Sputtering by atoms D and He of metal mirror materials Al, Cu, Mo, W; dielectric materials SiO_2, Al_2O_3, Ta_2O_5 used for protection layer, and C as constituent part of contaminating layer (*adapted from [2]*).

Protective dielectric layers having a low sputtering rate, such as Al_2O_3 and Ta_2O_5 will reduce the risk of failure for gainful high reflective mirror materials thus increasing their lifetime prior to replacement. At present, there are some uncertainties, which prevent from getting the mirror design fully launched. Thin oxide films are typically deposited under highly non-equilibrium conditions that lead to microstructural disorder and therefore changed properties. The severe threat to the performance of diagnostic mirrors is surface contamination with carbon-based material eroded from carbon tiles by plasma. Once such a contaminating film has been deposited on a mirror, it can dramatically change the slope of reflectance spectra owing to interference of the reflected beams. The composition of contaminating film appearing on the surface of the mirror may be rather complicated depending on mirror location inside the vacuum vessel. Hydrogenated carbon films are practically transparent in the visible range in contrast to pure amorphous carbon distinguished by extinction coefficient that is never close to zero. Thus, it is natural that contaminated mirror reflectance will degrade differently in different conditions as illustrated for example in Fig.2.

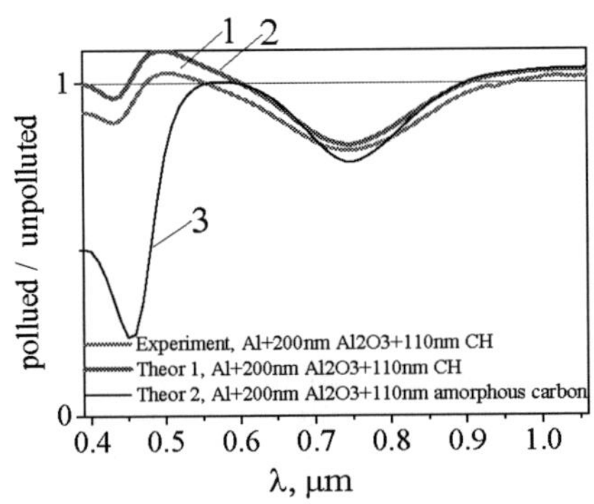

FIGURE 2. Deposition – induced spectral reflectivity normalized to clean Al mirror protected by 200nm Al_2O_3. Experimental and calculated curves 1 and 2 correspondingly show the results for hydrogenated carbon films of 110nm in thick. Curve 3 shows the calculated results for amorphous carbon.

Below the conservative assumption has been made assuming that optical properties of contaminating films attribute to amorphous carbon coating. Also interplay of Be with C, which is turning to be much severe on its consequences has not been analized in the paper because the data on the issue are still insufficient. Figure 3 shows an example of the reflectance spectra affected by surface contaminating films deposited on the mirrors. It is clearly seen in Fig.3 that deposits are of special influence for the mirrors that are lacking in reflectivity unlike for the high reflective mirrors. The reflectance spectra curves for thick protective layer ~10λ in Fig.4 are similar to those in Fig.3 except of appearing high frequency modulation of amplitude.

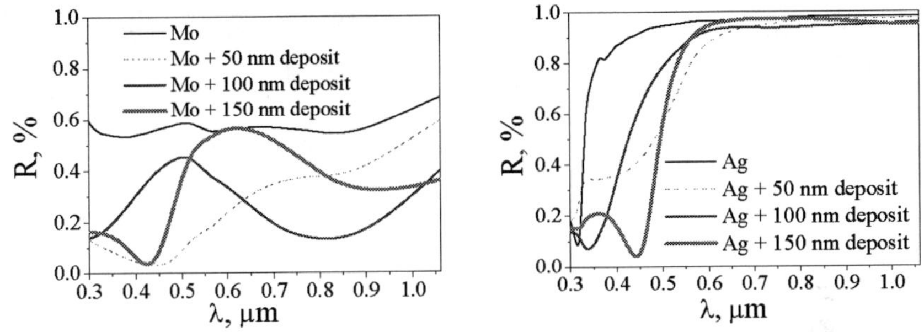

FIGURE 3. Calculated spectra for unpolarized emission reflected at 45° to normal from Mo, Ag mirrors, free of coating and contaminated with amorphous carbon layer of 50nm, 100nm, 120nm, 150nm in thick.

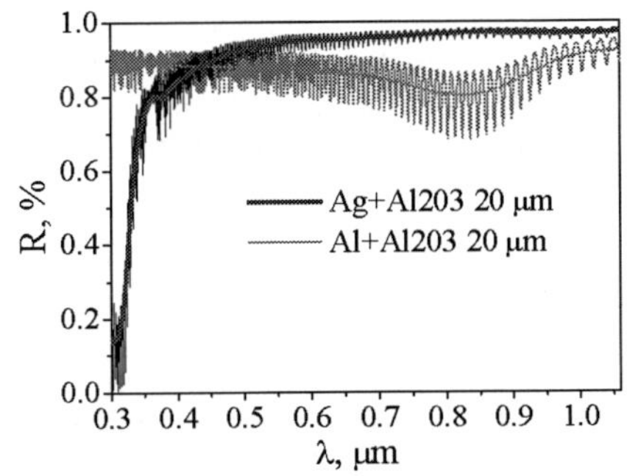

FIGURE 4. Calculated spectra for unpolarized emission reflected at 45° to normal from clean Al and Ag mirrors coated by Al_2O_3 of 20μm in thick.

Note, that the reflectance of carbon contaminated mirror can easily recover in the erosion-dominated conditions owing to the fact that sputtering threshold for the provisioned protective dielectric layer is larger than for a carbon film. Some other dielectric materials might also be used but the data on them are still insufficient [3].

TOLERANCE TO THERMAL STRESSES AND NUCLEAR IRRADIATION

In ITER, mirrors are assumed to be operating nominally at 150°C and specified to be tolerant to heating to 250°C during the bake-out campaigns [4]. Heat cycling operations are additionally introduced to ensure dimensional stability of the mirror at all stages of fabrication [5]. The problem of stresses is anticipated with regard to both thermal transients and the inbuilt residual stresses that appear at the fabrication stages of the mirror substrate. To avoid thermal mismatch for metal mirror materials and dielectric layer expansions, the sensible strategy would be to adhere a thin high reflective metal film on a substrate thermally coupled to dielectric protective layer. Such a design concept was verified in the bench test [6] showing that SiO_2 over-coated aluminum on pirex glass mirror performed well when heated up to 250 C. An approach described here can be extended addressing the materials in Table 1 perspective for mirror design.

TABLE 1. Temperature expansion coefficient of mirror materials.

Materials	**Temperature expansion coefficient α, 10^{-6} 1/K**				
Protective dielectric coating	Al_2O_3	2.04	Ta_2O_5	2.45	[7]
Candidate mirror coating material 1	Ag	18.9	Al	23.6	[8]
Candidate mirror coating material 2	Au	14.2	Cu	16.5	[8]
Mirror substrate material 1	Be	11.3	SiO_2	0.52	[8]
Mirror substrate material 2			Poly-Si	2.9	[9]

The available data on irradiation effects for fused silica -KU-1 and sapphire -Al_2O_3 draw to the conclusion that the protective dielectric coating is extremely thin to produce irradiation adverse effect on optical properties of the mirror.

SUMMARY

To satisfy the contradictory demands of high reflectivity and a small rate of sputtering when being bombarded by CX atoms the transparent layer coating on high reflective metal mirror has been evaluated. The predictive modeling has shown that the deposition induced loss of reflectivity can be minimized with use of high reflective mirror materials and would recover in the erosion-dominated conditions owing to the fact that sputtering rate for carbon is larger than for the protective dielectric layer. Metal mirrors protected by dielectric coated layer are required to survive thermal stresses and nuclear irradiation damage. Equality of thermal expansions of coated layer and bulk material is essential to maintain mirror heating stability. Thus, the solids Be, K8, Si, SiO_2 are considered as the materials of choice for the substrate of over-coated metallic reflective layer. The induced optical absorption is sooner not an issue for a thin protective layer. Much work is necessary and going on to ensure the protected mirror designs are realizable and ready to be deployed to ITER.

ACKNOWLEDGMENTS

This work is supported by Russian Federal Agency of Atomic Energy (contract # H.43.41.19.07.511).

REFERENCES

1. V. Voitsenya, A. E. Costley, V. Bardamid et al., Diagnostic first mirrors for burning plasma experiments, Rev.Sci.Instrum. **72** (2001) 475.
2. *http://www-amdis.iaea.org/cgi-bin/ALADDIN/query0.pl* Analytic fit using revised Bohdansky formula, see IPP9/82 (1993). Experimental Data in Eckstein,et.al.
3. A. Portinha, V. Teixeira, J. Carneiro, et.al., Rev.Adv.Mater.Sci. **5** (2003) 311-318.
4. R. Reichle, S. Henry, J. Miggozzi, E. Thomas, C. Walker Preliminary final report for the EFDA task TWP2002 TW2-TPDS-DIASUP-231.
5. D. B. Nikitin, D. A. Seregin, and Yu. P. Khimich, J. Opt. Technol. **72** _(2005) 926.
6. T. Hernandez, A. Moroño and E. R. Hodgson, Proceedings of the 23rd Symposium of Fusion Technology - SOFT 23, Fusion Engineering and Design **74** (2005) 793-796.
7. Chen-Chung Lee, Chuen-Lin Tien, Wean-Shyang Sheu, et al. Rev.Sci.Instr. **72** (2001) 2128.
8. David R.Lide, editor. CRC Handbook of Chemistry and Physics, volum 77th ed. CRC Press.
9. J. H. Chae, J. Y. Lee, S. W. Kong, Sensor and Actuators A: Physical **75** (1999) 222-229.

Issues And Options For Mirrors Of CXRS And H-alpha ITER Diagnostics

K. Vukolov*, A. Medvedev, I. Orlovskiy

RRC "Kurchatov Institute", Kurchatov sq.1, Moscow, 123182, Russia

The present status of investigations on the first mirror (FM) problem is discussed on the basis of the last experimental results obtained in modern tokamaks and laboratory devices. Engineering solutions that are necessary for protection of FMs for CXRS and H-alpha diagnostics are considered. Future plans for R&D that has yet to be done are presented. Results of neutron irradiation and heating tests of multilayered dielectric mirrors are given briefly.

Keywords: ITER, optical diagnostics, mirror, sputtering, deposition, radiation hardness.

INTRODUCTION

In-vessel components of optical ITER diagnostics will be exposed to significant gamma-neutron fluxes and heat loads. The "first mirror" (FM) being located in a straight view of the burning plasma will operate in most severe conditions. The FM will be subjected to erosion by charge exchange atoms (CXA) and/or deposition of contaminants depending on the diagnostic layout and local conditions (CXA fluxes and energy, mirror temperature, neutral gas composition and pressure). Both processes as well as a material transmutation under neutron irradiation can lead to degradation of the FM optical properties [1]. Special emphasis should be placed on FM integrity at plasma disruptions and other emergency cases (water cooling leak, breaking of the protective shutter and etc). The general requirements to ITER FMs are very hard: high optical reflectivity, low sputtering yield, stability under neutron fluxes and chemical stability, high melting temperature and thermal conductivity. That is why the option of mirrors for ITER diagnostics is a difficult task which demands careful engineering and wide supporting R&D program. Note that foregoing local conditions in ITER diagnostics ducts will be different in every particular case, so FM issues will be specific for every diagnostics that requires an individual approach to the FM option. Also it is necessary to provide a highest possible transmission of all collecting optics, so the multilayered dielectric mirrors with reflectivity about 100% are the best candidates for secondary mirrors of ITER optical diagnostics.

OPTICAL SYSTEM DESIGN AND INTEGRATION

The ***Charge Exchange Recombination Spectroscopy*** (CXRS) is an active spectroscopic technique that uses diagnostic neutral beam (DNB) and enables to measure the density of light impurities via quantitative spectroscopy of line intensities

CP988, *Burning Plasma Diagnostics*, edited by F. P. Orsitto, G. Gorini, E. Sindoni, and M. Tardocchi

and also to get ion temperature profiles from Doppler broadening of impurity lines at spectral range 450-660 nm [2]. As a main conception of an impurity density measurement, combination of CXRS and beam emission spectroscopy (BES) suppose to be used. It enables to deduce the ion densities without absolute calibration and beam attenuation simulation. That could simplify FM problem because in that case a spectrally uniform decreasing of mirror reflectivity does not influence on the measurement result. Transmission of CXRS optics has to be extremely high owing to low intensity of useful signals. CXRS collecting optics (see Figure 1) has a large aperture for looking at the DNB. The dimensions of its FM are 130×60 mm^2 at the input aperture is about 300×140 mm^2 for edge CXRS systems.

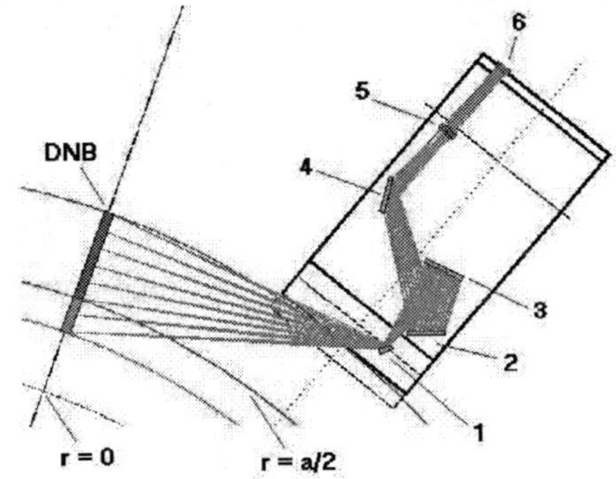

FIGURE 1. CXRS collecting optics into the equatorial port plug # 3 for measurements from r = a to r = 0.3a. 1 and 4 - flat mirrors, 2, 3 - spherical mirrors, 5 – lens, 6 - image plane.

The ***H-alpha Spectroscopy*** [3] should be solve several tasks including space resolved measurements and monitoring of hydrogen Balmer lines' intensity and isotopic ratio measurements at the plasma periphery. The diagnostics includes six independent optical paths, four of those dedicated to measurements at *Main Plasma*, and other two – for *Divertor and X-point area* observations. The input parts of all of them have a similar optical scheme well-known as endoscope (see Figure 2).

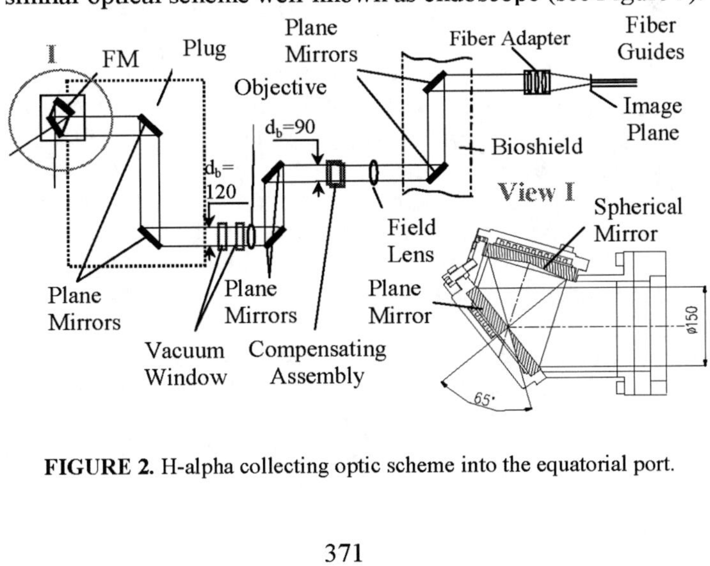

FIGURE 2. H-alpha collecting optic scheme into the equatorial port.

Such a design gives one a possibility to reduce CXA fluxes on account of small entrance pupil (diameter about of 10 mm).

The input optical systems of the both diagnostics will be placed in upper and equatorial ports. Periscope channels will be used into port plugs for decreasing of the straight neutron fluxes. Active water cooling of mirrors is foreseen for temperature maintenance at the range of 150-230°C.

OPTIONS FOR FIRST MIRRORS

The single crystal Mo is foreseen as FM material for decreasing of reflectivity degradation under CXA sputtering [4], but in that case the technological limitation about of 100 mm exists for crystal dimension. The other possibility is the using of mirrors with Mo or Rh nanocrystal coating [5, 6]. The choice of Rh as a mirror material gives advantages of higher reflectivity and chemical stability. Both materials have high melting temperature and thermal conductivity that prevents risk of melting under disruptions. In that case the deposition of contaminants could be the main reason of the FM degradation during long-time ITER operation. The semi-transparent hydrocarbon (C:H) film formation is most likely, the films can distort the spectrum of the reflected light due to interference [7]. Moreover, chemical modification of the mirror surface in hostile ITER's environments (high temperature, active radicals and plasma radiations) can result in development of thin opaque films.

Nevertheless, the experimental results of modeling of the degradation of metallic mirrors under ITER relevant conditions obtained both on tokamaks [4, 7 and 8] and on the plasma magnetron device [9] allow one to conclude that heating of the mirrors to the temperatures, at which a chemical erosion of C:H films by hydrogen atoms will prevail over deposition, prevents the C:H film deposition. In particular, at recent experiments in the DIIID divertor [8] for the mirrors exposed at the temperature of ~160°C neither carbon depositions nor degradation of optical properties was detected. That was also confirmed by our last tests on heating of Mo mirrors at depositing of atoms sputtered from the carbon cathode of the magnetron device. The grow of C:H film was stopped at the temperature about of 200°C, but at higher temperatures Mo compounds with carbon, oxygen and nitrogen arise on the mirror surface during 2 hours exposure. In result a mirror reflectivity was decreased of 10-20%.

At the moment our plan for mirror protection from C:H film deposition includes the following points: i) FM temperature should be kept near to 200°C, ii) a protective shutter has to be used during the vacuum vessel conditioning, iii) in-situ FM reflectivity monitoring should be provided. Note, that the question of Be deposition remains open up to now. Only the sputtering of deposited film by HF plasma discharges could be effective for the metallic film remove, but this way seems to be very complicated. So it is necessary to develop an engineering solution for quick FM replacement.

Technological (polishing, nanocrystal structure coating) and engineering solutions (mirror strengthening and adjustment, shutter drive, water cooling, etc) should be tested during next R&D stage. It is also necessary to perform numerical simulation of local conditions close to every FM (CXA and radiation fluxes, partial gas pressures, etc). This will make possible to model numerically and experimentally FM operation during standard ITER mode.

MULTILAYERED DIELECTRIC MIRRORS TESTS

The reflectance of multilayered dielectric mirrors can be close to 100% in a preset range of wavelengths. This makes very attractive their use as secondary mirrors for ITER optical diagnostics. Previous investigations [10] have shown that small (∅25 mm) dielectric mirrors can sustain neutron fluence of 10^{19} n/cm^2 and temperatures up to 280°C. It is supposed that heating, including radiation induced, plays the main role in damaging mirror coating at least in test conditions.

Thermal tests of large (∅100 mm) mirrors have been done recently. The samples with TiO_2/SiO_2 coating evaporated on silica glass K-8 and fused silica KU-1 substrates have been tested under continuous heat load and thermocycling. In all tests the samples have retained their coatings undamaged. The spectrum range of high reflectance was found to be shifted towards long wavelengths. The reflectance in almost all the range remained the same, except the negative peaks at which the reflectance reduced down from 90% to 80% (Figure 3).

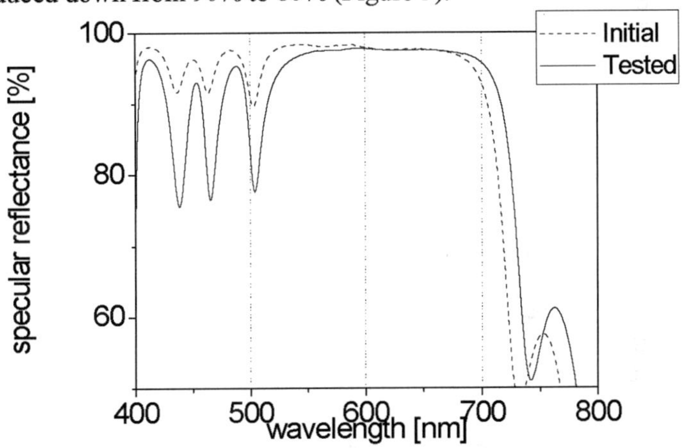

FIGURE 3. Reflectance spectrum of the large mirrors before and after the heating tests.

The results show that multilayered dielectric mirrors of large size sustain thermal loads which are supposed to be similar to ITER conditions. New samples without negative peaks in reflectance spectrum are planned to be tested soon under heating and irradiation.

ACKNOWLEDGMENTS

This work is supported by Russian Federal Agency of Atomic Energy (contract #H.43.41.19.07.511). The authors would like to express their gratitude to S. Zvonkov and T. Mukhammedzyanov for measurements on magnetron plasma device.

REFERENCES

1. V.Voitsenya, A.Costley, et al. *Rev. Sci. Instruments* **72**, No.1, 475-482 (2001).
2. S.Tugarinov, et al. *Rev. Sci. Instruments* 74, 2075 (2003).
3. K. Vukolov, A. Medvedev, in proceeding of 11th ITPA Meeting on Diagnostics, Sendai, Japan, 4–8 September 2006.
4. A. Litnovsky, G. De Temmerman, K. Vukolov, et al. *Fusion Eng. and Design* 82, 123-132 (2007).
5. A.F. Bardamid, K.Yu. Vukolov, et al. *Plasma Devices and Operations* **14**, No.2, 159-171 (2006).
6. L. Marot, G. De Temmerman, G. Covarel, A. Litnovsky and P. Oelhafen. Rhodium-coated mirrors deposited by magnetron sputtering for fusion applications, *Rev. Sci. Instruments.* Submitted.
7. K.Yu. Vukolov, M.I. Guseva, S.A. Evstigneev, A.A. Medvedev, S.N. Zvonkov, *Plasma Devices and Operations* **12**, No.3, 193 - 202 (2004).
8. A. Litnovsky et al., First tests of diagnostic mirrors in a tokamak divertor: An overview of experiments in DIII-D, to be published in *Fusion Eng. and Design* (2007).
9. K.Yu. Vukolov, A.A. Medvedev, et al., in Proc. 33rd EPS Conf. on Plasma Physics, Roma, Italy, P1.116. (19-23 June 2006) http://eps2006.frascati.enea.it/
10. K.Yu. Vukolov, I.I. Orlovskiy. *Fusion Eng. and Design* 74, Issues 1-4, 865-869 (2005).

Gamma-irradiation tests of IR optical fibres for ITER thermography – a case study

R.Reichle[1], B. Brichard[2], C. Pocheau[1], M. Jouve[1], S. van Ierschot[2], S. Martinez[2], H. Ooms[2] , F. Berghmans[2], M. Decréton[2]

[1] *Ass. EURATOM-CEA, DSM/DRFC, CEA-Cadarache, 13108 St. Paul-lez-Durance, France*
[2] *SCK-CEN, EURATOM Ass. Belgium, Boeretang 200, B-2400 Mol, Belgium*

Abstract. In the course of the development of a concept for a spectrally resolving infrared thermography diagnostic for the ITER divertor we have tested 3 types of infrared (IR) fibres in Co^{60} irradiation facilities under γ irradiation. The fibres were ZrF_4 (and HfF_4) fibres from different manufacturers, hollow fibres (silica capillaries with internal Ag/AgJ coating) and a sapphire fibre. For the IR range, only the latter fibre type encourages to go further for neutron tests in a reactor. If one restricted the interest onto the near infrared range, high purity core silica fibres could be used. This study might be seen as a typical example of the relation between diagnostic development for a nuclear environment and irradiation experiments.

Keywords: Gamma irradiation, Optical fibres, Infrared, Thermography, ITER.
PACS: 42.81.Cn, 42.88.+h, 61.80.Ed, 61.82.Ms, 07.57.Ty, 07.60.Vg, 52.70.Kz.

1. INTRODUCTION

The ITER radiation environment for diagnostics [1] requires that ceramic materials [2, 3] have to be radiation tested before considering use. In our case the question of irradiation tests has arisen in the course of the concept development of an infrared (IR) thermography diagnostic for ITER. In one of the concepts optical fibres are supposed to transport of the IR light measured, through a harsh radiation environment. Data for optical fibres under irradiation can be found mainly for silica fibres [4-9]. The transmission of these fibres is limited to the range from near ultraviolet to the near infrared (NIR). Similar data of true IR material and in particular fibres were not found in the literature. Therefore we started a dedicated irradiation campaign of a few candidate IR fibres in the context of a wider development program for this diagnostic [10]. Here we present specific results of the irradiation study and the link with the development of the diagnostic. The paper has a first part with details of the diagnostic, then the preparatory studies, the irradiation tests and finally the conclusions.

2. DETAILS CONCERNING THE DIAGNOSTIC

The diagnostic concept under investigation here is that of a spectrally resolving IR divertor thermography diagnostic using optical fibres [11]. The most interesting spectral range had been identified as being 2-6 µm. A short description of the most

CP988, *Burning Plasma Diagnostics*, edited by F. P. Orsitto, G. Gorini, E. Sindoni, and M. Tardocchi

recent version of the diagnostic and the usefulness of spectral resolution is presented elsewhere [12]. Optical fibres are considered since they have advantages over conventional optical transports with mirrors and lenses in the areas of: access to areas difficult to see and reach, number of elements to align and reflecting surfaces, number of materials to select and qualify, amount of neutron streaming and leakage for high light throughput, suitability for active measurements (using laser light as light or heat source [12]) hyper-spectral imaging (using fibre bundles and spectrally dispersive elements [13]) and immunity to vibrations. These advantages have to be weighted against disadvantages such as: lower radiation hardness than mirrors, more radiation induced absorption losses and parasitic signals from radio luminescence in the fibres than for a lens solutions using the same material, and probably lower spatial resolution. A fibre based thermography solution has been implemented at Tore Supra [14] which has a spectrally resolving detection in the wavelength range of 1-4 μm using ZrF_4 fibres for the transport of the IR light and a sapphire prism as dispersing element [13]. Some of its aspects can be seen as a prototypical for the planned divertor thermography diagnostic: Target and entire optics are mounted on the same removable cassette; the fibres are connected to the front end optics on the cassette; the mirrors are fix; the alignment of optics can be done prior to installation using eccentric elements in the fibre connectors.

3. PREPARATORY STUDIES BEFORE IRRADIATION TESTS

3.1 The harsh environment

The fibres are supposed to run nearly horizontally from the pumping hole (fig. 1 in contribution [12]) of the divertor cassette outwards to the bioshield, the total length being 8m. The radiation environment can be approximately deduced from [15]. The peak neutron flux (E > 0.1 MeV) at the divertor cassette would be about 10^{16} n/m²s (fig. 1), yielding after 0.5-0.8 years of full power operation (full ITER lifetime dose) a total neutron fluence of 2-3 $\cdot 10^{23}$ n/m². These values need to be reassessed once the components inside the port are defined.

Assuming 8 exchanges of the divertor during the lifetime of ITER, the final fluence for the fibres might come down to $4 \cdot 10^{22}$ n/m² (E> 0.1 MeV), which corresponds to about 40 MGy of ionising dose in the fusion environment [1]. The neutron fluxes may be reduced significantly if one can benefit from further shielding: 7 cm of steel will reduce the fast neutron flux by a factor of 10 [16]. In consequence the situation might be improved significantly by shifting the fibre starting point to a point shielded by the outer/upper part of the divertor cassette which contains considerable amounts of shielding steel and water.

The base temperature of the cooling water is 150°C, which rules out the use of some IR transparent material. Due to the nuclear heating, close to the divertor port, cooling of the fibres will be necessary. In the case of using silica fibres, which are very temperature resistant, it might be an interesting proposition to design a cooling system for the fibres that would maintain the fibre temperature at 300°C (see section 3.2).

When considering differences between nuclear environments of fission and fusion devices, three observations seem relevant to set up fibre irradiation experiments: 1) the ratio of nuclear heating (joint effect of neutrons and gammas) per neutron flux is higher in fission experiments than in fusion experiments. 2) The neutron energy spectra for fusion ranges up to 14 MeV whereas the fission spectra do not exceed 1 MeV. This means that some nuclear reactions may occur in fusion conditions, which are not present in fission experiments. 3) The relevant figure for the irradiation of fibres is rather the ionising dose than the neutron flux – even though it has been seen that neutrons of the same ionising dose cause somewhat more damage than gamma irradiation alone [3].

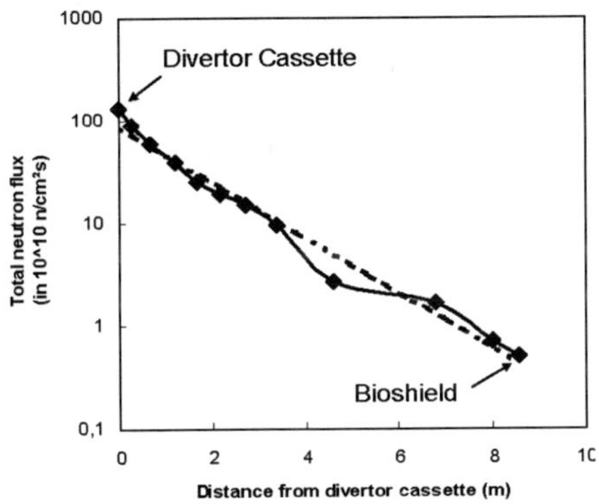

FIGURE 1: Neutron flux along the fibre path from divertor cassette to bioshield.

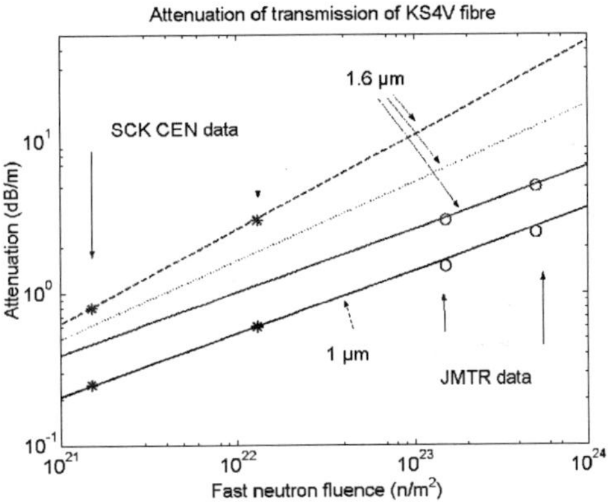

FIGURE 2: Synopsis [11] of RIA results from SCKCEN [5] and JMTR [18] for KS4V fibres.

3.2 Silica fibre data

The normal transmission losses in fibres are caused by Rayleigh scattering at the short wavelengths and IR phonon absorption at long wavelengths; fibre drawing defects and impurities intervene in between [e.g. 4]. The best silica fibre candidates in terms of NIR transmission and radiation hardness are low OH pure silica core fibres [17]. After irradiation their transmission in the range 800 – 2000 nm is much better than the transmission in the visible range. Good examples of such fibres are the KS4V fibre (FORC) [5] and the STU fibre (Heraeus) [7].

Fig. 2 [11] shows a comparison of transmission values for KS4V fibres after irradiation at 1 μm and 1.6 μm measured at the SCK·CEN [5] and at the JMTR [18] facilities. Despite significant systematic differences it seems that in the range that interests us (see section 3.1) of $4 \cdot 10^{22}$ - $3 \cdot 10^{23}$ n/m (40-300 MGy in ITER), the remaining transmission in the NIR range could be acceptable (~3db/m) and for more protected situations as mentioned previously there should be a security margin.

Radiation defects are fairly stable at room temperature. Thermal annealing usually takes place uniformly at all wavelengths. To recover optical transmission, the temperature should be raised and kept above 300°C [17].

Cerenkov radiation created in optical fibres is observed in reactor irradiation tests. It is strongest in the blue, but extends well into the NIR range [18]. However based on [6] we would expect that the Cerenkov radiation is in most plasma conditions of the divertor a smaller perturbation to the thermal measurements than the bremsstrahlung from the divertor plasma.

Radio-luminescence in silica-fibres is known to produce a clear peak at around 1270 nm [18] which is attributed to oxygen in the fibres [17]. Silica exhibits two orders of magnitude less radio-luminescence in the visible range than sapphire (which plays a role in the following) [19], but no comments have been found on radioluminescence behaviour of sapphire in the NIR and IR range.

3.3 IR fibre section – the initial choice

For IR applications Tore Supra has selected ZrF_4 fibres [14]. They are hygroscopic and more fragile but they are transparent from the visible range to 4.3 μm and work at up to 200°C in Tore Supra. Therefore we selected ZrF_4 fibres from two manufacturers (Reflex Analytical and Le Verre Fluoré) for irradiation tests. The critical glass temperature for ZrF_4 is only around 300°C, which did not raise the expectations for the radiation hardness of ZrF_4 too high.

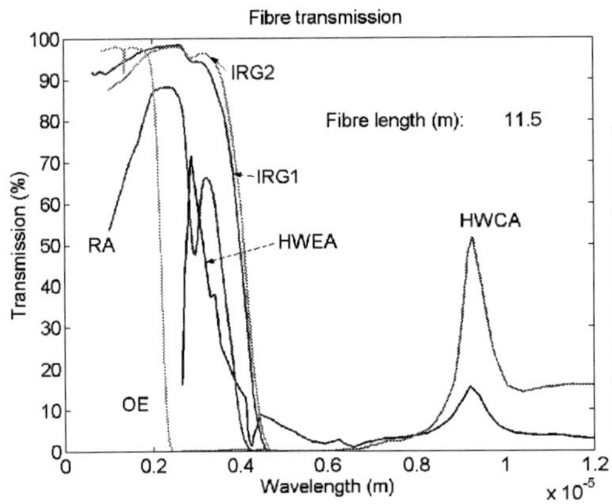

FIGURE 3: Comparison of literature values of spectral transmission values of fibres. OE: Oxford Electronics low OH silica fibre (for reference); IRG1 ZrF_4 fibre (and IRG2 HfF_4 fibre, which is irrelevant for applications in radiation environment due to Hf content) from Le Verre Fluoré, RA ZrF4 fibre from Reflex Analytical and HWEA and HWCA, hollow silica waveguides from Polymicro.

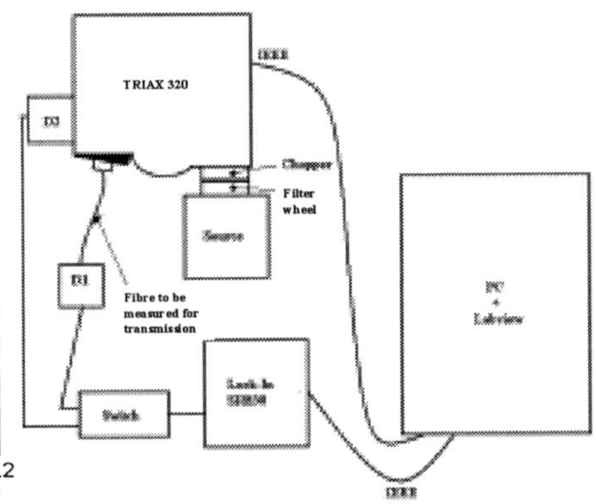

FIGURE 4: Experimental setup [10].

Two types of (hollow) silica capillary light guides (from Polymicro) were also selected. Their materials are a priori radiation hard (silica, thin inner metal and dielectric coating) and one particular patented configuration of this waveguide [20] appeared particularly interesting for photothermal measurements [11, 12]. We expected bending losses to be important for these light guides [21]. Sapphire fibres were not taken into account in the first round of tests, despite the radiation hardness credentials [22] of solid sapphire material because they are only available in short length (2m) and with only one index of refraction, which limits the application and/or increases the losses.

4. IRRADIATION TEST OF SELECTED IR FIBRES

4.1 The experimental setup

Two Co^{60} γ sources with different dose rates were available at SCK·CEN Mol for fibre irradiation: RITA with 1.5 kGy/h and BRIGITTE with 20kGy/h. It was decided that measurement in one of the reactors would only be envisaged after successful γ-testing. The CEA supplied measurement equipment (fig. 4) and fibres [10].

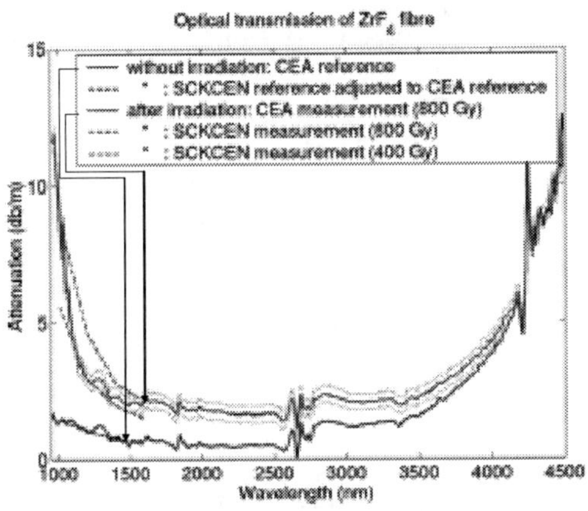

FIGURE 5: Results of preliminary off-line experiments of exposure of a ZrF_4 fibre to low dose irradiation.

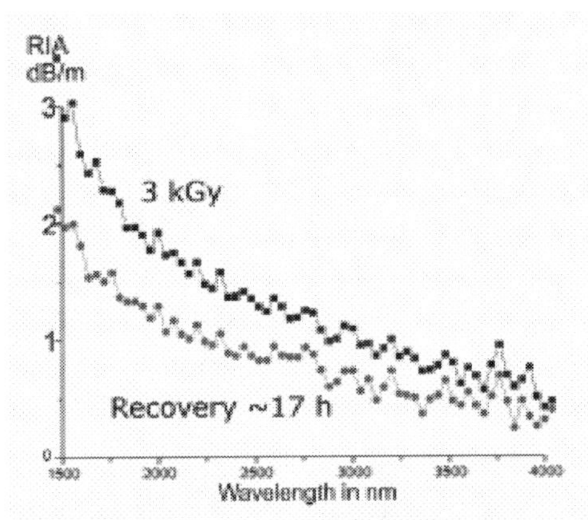

FIGURE 6: Spectrally resolved RIA for ZrF_4 [10].

4.2 Preliminary off-line measurements of ZrF_4 fibres

To prepare on-line gamma irradiation experiments, first an off-line experiment was conducted (fig. 5). A ZrF_4 fibre (from Reflex Analytical) was exposed to 400 Gy and 800 Gy of γ's and then measured in comparison with an un-exposed fibre. At wavelengths inferior of 1.5 μm there was a very noticeable Radiation Induced Absorption (RIA) - about 5 orders of magnitude larger than values known from good silica fibres. This attenuation seemed to extend at a low level also in the IR range.

4.3 Online measurements of ZrF_4 fibres

For online measurements of ZrF_4 fibres in the RITA facility, we used fibres of 2 m length to be exposed and patch-cord fibres of 11.5 m length that received very little dose themselves. Fig. 6 [10] shows the spectral distribution of the RIA in the IR range between 1.5 and 4 μm for a dose of 3 kGy immediately after irradiation and after a recovery period of 17 hours at room temperature.

Fig. 7 [10] shows for two of such irradiation/recovery cycles the RIA at 2 μm and 3.5 μm. No significant difference was observed between the two manufacturers of ZrF_4 fibres.

At elevated temperature (100°C) the recovery is more rapid than at room temperature (fig. 8) [10]. The conclusion is that the fibre is too sensitive to radiation to be used inside the bioshield of ITER. Outside use can be considered and may still be interesting for hyper-spectral imaging.

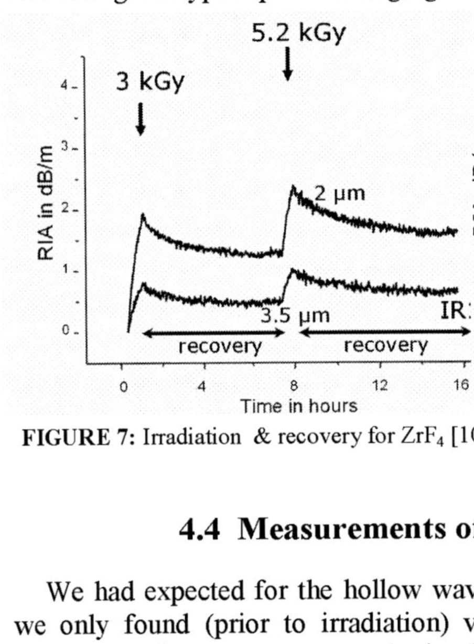

FIGURE 7: Irradiation & recovery for ZrF_4 [10].

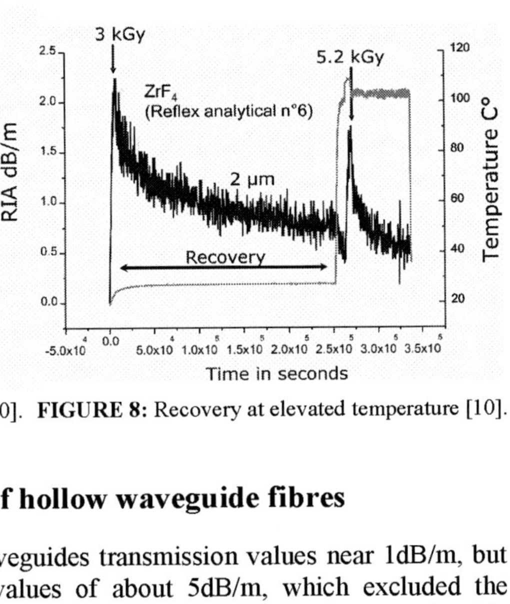

FIGURE 8: Recovery at elevated temperature [10].

4.4 Measurements of hollow waveguide fibres

We had expected for the hollow waveguides transmission values near 1dB/m, but we only found (prior to irradiation) values of about 5dB/m, which excluded the foreseen on-line measurements. An irradiation up to 27 kGy did not show a noticeable degradation in a 2 m long fibre. For an 8 m fibre a 0.4 db/m loss is possible. The conclusion on these fibres is that they are not usable due to their intrinsically low transmission. A situation that worsens still with the bending losses.

4.5 Sapphire fibre investigation

Given the difficulties with ZrF_4 fibres and hollow waveguide fibres, we decided to try a sapphire fibre. The transmission range is smaller than for ZrF_4 fibres (fig. 9).

The sapphire fibre was first briefly tested in the RITA facility. As no change in the transmission was observed, the fibre was moved to the BRIGITTE facility. Since we did not have suitably radiation hard IR patch-cord fibres these measurement were conducted in off-line mode. The fibre was exposed to a total final dose of more than

2.5 MGy (fig. 10) with little if any RIA. We think it is likely that at least part of the small reduction in transmission observed is due to losses produced by repetitive connecting and deconnecting of the fibre. This results invites us to consider now reactor testing of this fibre.

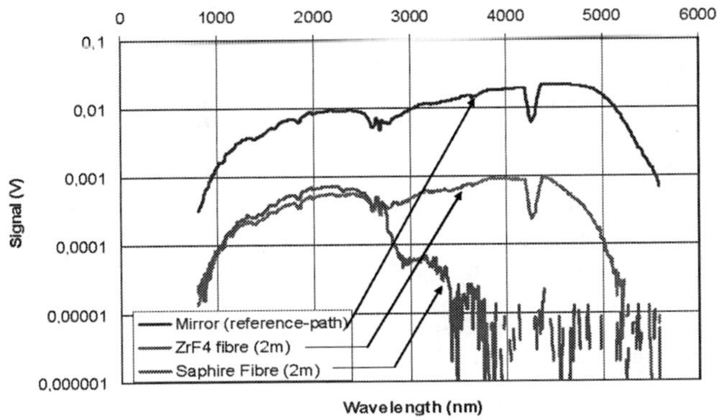

FIGURE 9: Comparison of raw data of transmission measurement taken with setup shown in fig. 4 for the reference-path to detector D2 and through different fibres (ZrF_4 and sapphire) to detector D1 [23].

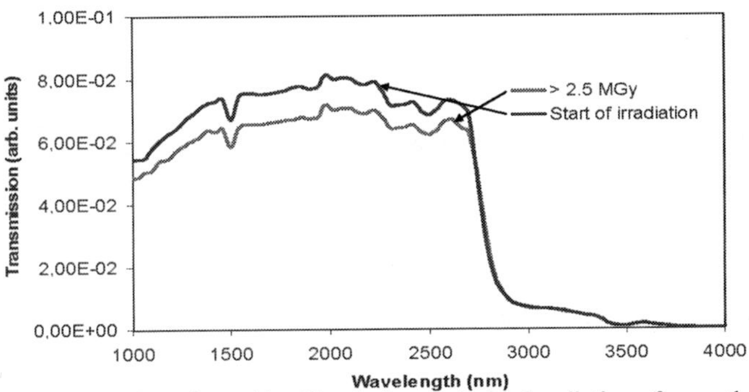

FIGURE 10: Transmission of sapphire fibre before and after irradiation of more than 2.5 MGy [23].

5. CONCLUSION, SUMMARY AND OUTLOOK

The procedure used here, was to first develop a diagnostic concept based on existing knowledge, then set up an irradiation test program with some initial off-line γ tests, followed by on-line γ tests, to lead finally to neutron tests. This takes years and it has lead to revisions of the concept of the diagnostic in question. This may be typical for other diagnostics and should be considered in terms of time estimates needed for their development. In our case we have learned that low OH pure silica fibres are suitable candidates in the NIR range for use in the whole divertor port – possibly with the only exception of the region directly at the divertor cassette itself. With provision for thermal annealing this allows to consider seriously to use such fibres for illumination purposes for active (laser-aided) thermal measurements. Cerenkov radiation and and radio-luminescence are not expected to hamper their use. The IR experiment has ruled out two of the candidates (ZrF_4 and hollow waveguides) but encourages to go further to neutron tests with the third candidate, sapphire fibres. As a potentially interesting new candidate we consider micro-structured silica fibres.

This work was partially supported by the European Communities. Views and opinions expressed do not necessarily reflect those of the European Commission.

REFERENCES

1. A.E. Costley et al., Fusion Engineering and Design 55 (2001) 331-346
2. E. R. Hodgson, this conference
3. K. Vukolov, this conference
4. B. Brichard 'Radiation effects in silica optical fibres', New Challenges for Radiation Tolerance Assessment, RADECS 2005 , Short Course Notebook, September 19-23, 2005, Cap d'Agde, France
5. B. Brichard, et al. Fusion Engineering and Design, 56-57 (2001) 917[1]
6. T Kakuta, T. Shikama, T. Nishitani, Journ. Nucl. Mat 258-263 (1998) 1893
7. H. Henschel et al., IEEE Trans. Nucl. Sci. (June 2002), 1432
8. A.T. Ramsey 'D-T radiation effects on TFTR diagnostics' , Techn. Rep. PPPL-3010, Oct 1994
9. P.D. Morgan,' Irradiation of optical fibres at JET through 14 MeV neutron production', Proc. 17th Symp. Fus. Techn., Rome 14-18 Sept 1992, 722
10. R. Reichle, B. Brichard, F. Escourbiac et al, Journ. Nucl. Mat., **363-365** (2007) 1466
11. R. Reichle et al., Proc. 32nd EPS Conf. on Plasma Phys., ECA Vol **29C** (2005) P4.083
12. R. Reichle, J.M. Badie, C. Balorin et al., this conference
13. R. Reichle, et al. Rev. Sci. Instrum. **75**,10 (2004) 4129
14. R. Reichle, D. Guilhem, R. Mitteau, J. C. Vallet. et al. Nucl. Fus. **43** (2003), 797
15. Nuclear Analysis Report, H. Ida, V. Khripunov, L. Petrizzi, NAG FDR, V 2.0 G73 DDD 2 W 0.2 (2 March 2005)[2]
16. M.M Komochkov and B.S. Sychev, 'The attenuation of a high energy neutron flux in shielding', transl. from Atomnaya Energiya, Vol 15, No4, (1963), 325
17. B. Brichard, 'Future needs for the development of radiation-resistant optical fibres for inertial and magnetic fusion machines' Contr. to Workshop on ITER-LMJ-NIF components in harsh environments, Aix-en-Provence, France, 27-29.6.2007
18. T. Kakuta, et al., J. of Nucl. Mat.,307,(2002),1277[3]
19. A. Morõno and E. R. Hodgson, J. Nucl. Mat. 258-263 (1998) 1889
20. J. A. Harington, United States Patent, Number 5,815,627 Date: Sep 29 1998
21. D. J. Gibson and J. A. Harrington, SPIE Proceedings 3849 (1999) 143
22. K. C. Baldwin et al. Journ. of Am. Ceramic Society, Jan 01 (2002)
23. R. Reichle, B. Brichard et al. 'Gamma-irradiation tests of IR optical fibres for ITER thermography' Contr. to Workshop on ITER-LMJ-NIF components in harsh environments, Aix-en-Provence, France, 27-29.6.2007

---

[1] For more details: B. Brichard, T 492 European Fusion Techn. Progr. Rep. EUR CIEMAT 93 (2001) 25 and Annexe 6

[2] For more details : L.Petrizzi, '3D Neutronic Analysis around the Cryopomp Port: some preliminary results' VHTP at ITER Garching JCT, DIU, Meeting on IVV system, Garching 18 January 2001

[3] For more details : a) T. Kakuta, T. Shikama, T. Nishitani, JAERI Research 2002-007, Irradiation effects on plasma diagnostic components (II), March 2002, 20;
b) T. Kakuta, T. Shikama, JAERI Research 2002-007, Irradiation effects on plasma diagnostic components (II), March 2002, 34
c) B. Brichard, Fusion Techn. Programme, Physics Integration-Diagnostics, EFDA TW3 - TPDC / IRRCER Del 1.0 & EFDA TW4-TPD/IRRCER Del.3 Final report part II of II, SCKCEN Mol Belgium, R4082 INSTR/507/05-01 SCKCEN Dec 2004 (2004), 49

MAGNETICS, EQUILIBRIUM AND PLASMA CONTROL

Main Features of Plasma Control

F. Crisanti*, R. Albanese† and G. Ambrosino†

*Associazione Euratom-ENEA sulla Fusione, Frascati, C.P. 65, 00044 Frascati, Italy
†Associazione EURATOM/ENEA/CREATE, Univ. Napoli Federico II, Via Claudio21, 80125 Napoli, Italy

Abstract. In the recent years Plasma Control has always increased his importance in any advanced experiment. It is now clear that ITER will not be able to operate without a quite advanced and sophisticated control apparatus. Necessarily this system will have to integrate several different aspects of the Plasma behavior. One of the most important parts of a closed loop control system is the quality of the measurement of the plasma parameters that should be controlled. Eventually, this aspect involves sophisticated and complex diagnostic apparatus. This paper presents an overview of the present status, and further studies and developments needed, in the next future, for the design and realization of an integrated plasma control system aimed at both stabilizing the plasma non-axisymmetric instabilities and controlling the most important internal plasma parameters. In particular the Edge Localized Modes (ELMs), the Neo-Classical Tearing Modes (NTM), the Resistive Wall Mode (RWM) and the Plasma Profiles control system necessities will be shortly illustrated.

Keywords: Plasma, Control, Diagnostics
PACS: 52.55.Fa, 52.70.Ds, 07.05.Dz.

INTRODUCTION

All the past years experimental evidences have clearly shown that any future reactor relevant experiment, based on the Tokamak concept, is practically impossible without a careful control of the main critic plasma characteristics. Traditionally, plasma control was mainly indicating the control of the equilibrium and stability properties and, partially, the control of some macroscopic kinetic variables (β, neutron yield…). The operations on the present large facilities [1,2] and the preparation of ITER scenarios [3] have highlighted the strong necessity of an integration of always more control parameters to fulfill the requirements of a burning plasma experiment. The simultaneous control of plasma shape, current and pressure profiles, radiation, impurity control, instabilities, etc… now appears as the most important development to provide to the plasma scenarios the necessary relevance for the next step. Just to give an idea of these problems, in Table 1 it is reported a not exhaustive list of the main Physics Quantities to be controlled, of the Sensors to be used and of the possible Actuators; in the last column the acronyms are indicating the different heating systems (NBI=Neutral Beams, LHCD=Lower Hybrid, ICRH=Ion Cyclotron, ECCD=Electron Cyclotron), and the poloidal circuit (PF). It is to bear in mind that, many times, it is the complete profile that has to be known and controlled.

CP988, *Burning Plasma Diagnostics*, edited by F. P. Orsitto, G. Gorini, E. Sindoni, and M. Tardocchi

PHYSICS ISSUES	SENSORS	ACTUATORS
Confinement & P_{Fus} : $P_{fus} \sim H(q)\ \beta_N / q^2$ $J_{boot} \sim q\ \beta_N\ (V_{loop} = 0)$	q, T_e, T_i, N_e, β_N, V_{rot}, Equilibrium, Current Profile	NBI, LHCD, ICRH, ECCD, PF
Power & Particle exhaust and fuelling Fuelling Heat Load & ELMs ^{4}He level & impurities	N_D, N_T, Heat Flux, Z_{EFF}, Radiation	NBI, Pellet & Impurity Injection, Pumping
MHD & Disruptions NTM & Sawteeth Pressure Driven, RWM Disruptions	MHD monitor, Pressure, Disruption Prediction	ECCD, Saddle Coils, Gas Jet, Pellet

Table 1. Non-Complete Overview of the Physics Issues to be Controlled and of their Integration.

This short overview will try to concentrate on the aspects of the integrated control of the plasma "internal" properties and in particular on the request to the diagnostic measurements necessary to optimize the closed loop. The "magnetic control" [4] aspect will not be presented here, but it is discussed in a separated paper [5].

Presently two plasma scenarios are mainly pursued in the attempted to optimize the fusion experiment performances. The so-called ITER standard scenario, or H mode, and the Advanced Scenario; eventually, this includes the Hybrid and the ITB (Internal Transport Barrier) scenarios. The presence of a steep edge pressure gradient is the key issue of the improved confinement in the H mode scenario, but at the same time it is the cause of strong Edge Localized Modes (ELMs) that can release very large amount of power and energy on the divertor plates. During the years several ideas have been suggested and/or tried to moderate the ELM activity (light impurities injection, launch of small fast pellets, ergodization of the magnetic edge…). All this ideas are essentially finalized to moderate the pressure (temperature and/or density) edge gradient, but, as already mentioned, this property is also the reason of the enhanced energy confinement. It is quite obvious that any attempt to moderate, in real time, the ELMs behavior must be correlated with the global plasma performances and with the divertor and first wall protection. These facts illustrate immediately the strong and integrated requested to the diagnostic apparatus. A very reliable and high spatial and time resolution of the edge density and temperature measure must be envisaged.
Another important feature of the H mode scenario is the presence of the saw-tooth activity, which can act as a trigger of Neoclassical Tearing Modes (NTM) that,

eventually, will degrade the plasma performances. Several feed-back loops based on different radiofrequency systems have been proposed and tested to locally modify the current profile and, consequently, to avoid the onset of these modes. It is immediately clear the strong importance of a good diagnostic of the modes (frequency, amplitude and localization) to have an efficient suppression of their growth.
In the so-called Advanced Scenarios the plasma current profile plays a key role for achieving the requested performances. Unluckily this quantity is one of the most difficult to measure. So far several experiments based on external current drive provided by the Neutral Beam or some Radiofrequency have been attempted, sometimes getting also satisfying results. However, in all these experiments it is clearly appeared that the weakest point of the closed loop chain is the correct identification of the real time current profile. Moreover, this quantity and the scenario performances are strongly interconnected with the pressure profile. A good real time identification of this profile is still an open point.

Eventually, these scenarios rely on the achievement of very large β values. The consequent large pressure gradient can destabilize a class of MHD instabilities named Resistive Wall Modes (RWM). In principle, these instabilities can be stabilized by external multipolar currents and/or acting on the plasma rotation.

GENERAL ASPECTS OF PLASMA CONTROL

In Fig. 1 the usual scheme of a Plasma Control system is shown. Essentially, the system is composed by two parts: the "Plant" and the "Controller". For the design of a high performance feedback control system, one of the most important pre-requisites is a good knowledge of the plant dynamics. Many times (practically very often in a fusion experiment) the system dynamics is highly non-linear, and the system knowledge is very poor.

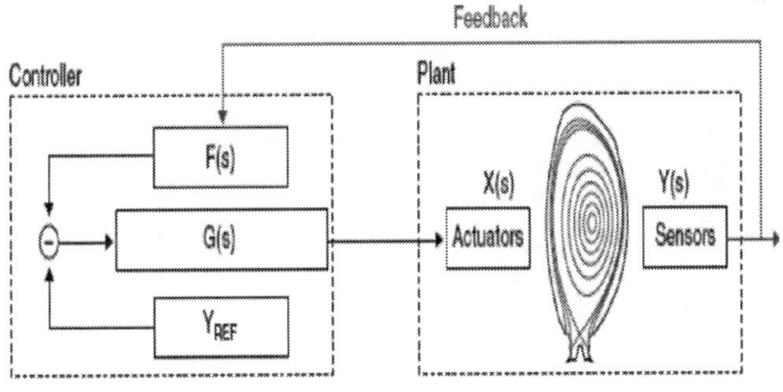

FIGURE 1. General Scheme of a Feed Back Closed Loop.

Whenever the equations ruling the system are known (see for instance the Plasma Equilibrium case [4]) a linearization of these equation is carried out and used for the design and optimization of the control law. Unlukily, many times, in the fusion experiments, the equations ruling the system evolution are not known at all, or only a very simplified version is available. Obviously, this makes quite difficult to design a high performance controller. In these cases the "Sensors" role becomes quite more important. A good quality of the controlled variables measurements can partially compensate the blindness about the system dynamic behavior. This implies that a lot of effort must be put into improving the quality of the real time diagnostics and rough data elaboration. Moreover, very often the "controlled variable" depends upon different plasma features so that a large effort must be put in the real time integration of quite different plasma diagnostics, often available with quite different levels of reliability and/or robustness.

RWM CONTROL

The achievement of high normalized beta ($\beta_N = \beta a B_T/I_p$, where Ip is the plasma current, B_T is the toroidal field, a is the minor plasma radius and β is the plasma pressure normalized to the magnetic pressure) is limited by the onset of some MHD kink instabilities, that can degrade the plasma performances up to cause a plasma disruption. The limit value of β_N largely varies depending on the presence of a superconducting ideal wall or in the absence of any wall. Of course, in a real experiment, there is always some conductive wall at fixed plasma distance and with fixed conductivity, consequently it will always exists a fixed "machine" β_N limit. But it has been theoretically and experimentally shown that this limit can be pushed towards his ideal value by "controlling" the toroidal plasma rotation [6] and/or by "magnetically controlling" the mode onset [7]. In Fig. 2a it is schematically shown the DIII-D [7] systems of magnetic probes to "individuate" the mode onset, and the external and internal sets of magnetic coils.

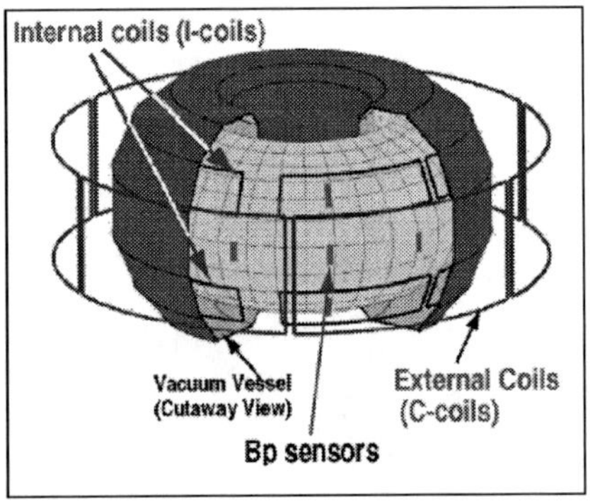

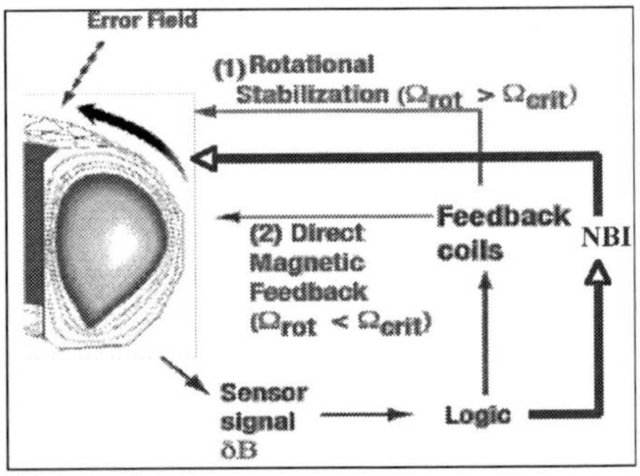

FIGURE 2. a) Schematic View of DIII-D RWM control apparatus; **b)** schematic description of an ideal RWM MIMO controller, where all the different control aspects and actuators effects are included.

The plasma rotation can be "controlled" by acting on the momentum injection provided by tangential NBI apparatus. However, the plasma rotation velocity is also affected by the presence of non-symmetric error field; consequently, a set of active magnetic coils, such the one of Fig. 2a, can also be used to indirectly modify the toroidal plasma velocity. The sensors apparatus used to actively control the mode growth is normally very simple, i.e., a set of magnetic probes; however, how and which type of information is used can play a role on the choice of what it will be controlled. For instance the use of Br probes strongly couples the sensors and the actuators, while the use of Bp pick-up coils makes, in principle, possible decouple the two ones, allowing a more straightforward control of the plasma mode. As previously mentioned, the control of the magnetic "perturbations" and plasma velocity are, in principle, coupled. Consequently an ideal control system must include this coupling in the control scheme that becomes a very complex MIMO apparatus (see Fig. 2b), where all the different actuators effect are included to control at the same time the mode onset and the plasma rotation.

NTM CONTROL

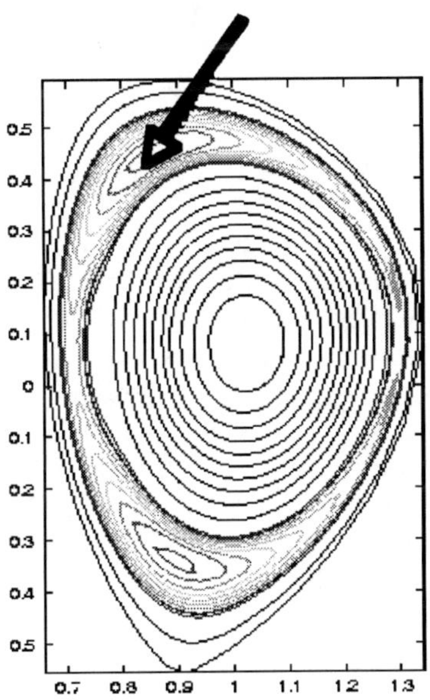

FIGURE 3. ITER-like Plasma Equilibrium, with indicated an NTM island; the arrow indicates the steering of an Electron Cyclotron beam.

Fig.3 shows generic ITER-like plasma equilibrium with the presence of magnetic islands due to Neo-Classical Tearing mode. This kind of instability appears at high β and is triggered by the saw-tooth activity. Consequently, it is likely to be present in the standard H mode scenarios, when approaching high performances. Belonging to the class of the Tearing instabilities, it can be stabilized by local modification of the current profile gradient. The most appropriate tool to have a quite localized current drive is the Electron Cyclotron. There are several different techniques that can be used to control this instability [8]. In principle the wave frequency could be varied to "follow" the island position. Unluckily, the present sources of Electron Cyclotron are not tunable in frequency. However, different methods can be applied. The toroidal field can be adjusted or the plasma column can be horizontally displaced. Both these methods, although already tested, can have some problems, when applied to large Tokamaks. The Toroidal field is not so easy to be varied on a short time scale, because of his large inductance; on the other hands, by moving the plasma position also other plasma boundary features can be affected (for instance the strike points position...). Another, more robust, way to adjust the current deposition localization is the steering of the wave beam (see Fig. 3). Eventually, the wave power can be easily tuned. This tuning can vary slowly or can be linked to the mode rotation velocity, in order to drive the current only on the O-point of the island. All the above considerations clearly show how much complex the real time NTM control problem is. Also the diagnostic system of these instabilities is vey complex. Indeed, an ideal "sensor" system must integrate several different diagnostics to be able to reconstruct the plasma equilibrium, to individuate the island location and his rotation frequency... Eventually, all this information should be elaborated and linked together to provide the necessary inputs to the control system.

ELMS CONTROL

The reference ITER scenario is the so-called H mode that relies on an edge transport barrier, and a consequent edge pressure pedestal (Fig. 4a), to have an improved global heat transport quality. However, the presence of a large edge pedestal has the drawback to drive some edge localized perturbations that can release huge amount of energy on the plasma facing materials. In the present Tokamak experiments a large typology of different ELMs [9] can occur, but the best improved confinement is obtained in the presence of quite large pressure pedestal and of the so called Type I ELMs, that have low frequency, but that can release up to several MJ of energy, per ELM, on the divertor plates. Eventually, the very hard target of an ELMs control system is the aim of controlling the pedestal pressure gradient without affecting the global confinement that is proportional to the pedestal height (see Fig. 4a).

So far there have been several different ideas in order to moderate the ELMs activity and the consequent energy release. One of the ideas is to try to modify the local magnetic topology and the consequent stability balance. This can be achieved by trying to modify the plasma boundary shape and, consequently, the local magnetic shear [9]. An ELMs control system based on the active control of the plasma boundary is, in principle, imaginable, but, in practice, it would have quite a lot of severe complications, also thinking that the plasma shaping is linked with several other plasma features.

The other basic idea relies on the matter that the energy released by any single ELM is proportional to the ELMs frequency ($\Delta W_{ELM} \sim f^{-1}_{ELMs}$) and on the fact that some small local perturbation can affect the pedestal width without degrading the pedestal height (see Fig. 4a). One of the ideas is to try to partially and locally destroy the flux surfaces topology (Fig 4a).

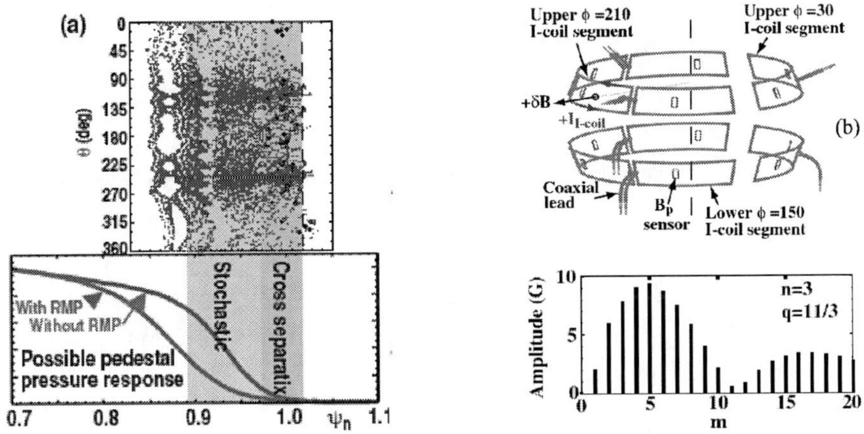

FIGURE 4. a) Pressure pedestal with and without the perturbation field and the calculated stochastic field; **b** Schematic View of DIII-D Perturbation Coils apparatus and of the produced mode spectrum

This goal has been achieved in several experiments by using some relatively low m and n number coils perturbations [9,10,11]. In Fig. 4b it is schematically shown the DIII-D [10] perturbation coil system and the produced mode spectrum. So far this technique has been used in an open loop scheme relying on the matter that the degraded region is very small, with a consequent minor reduction of confined energy. However, in an idealized experiment, it would be useful to control in real time the main features (height and width) of the pressure pedestal. Obviously this implies a quite sophisticated diagnostic apparatus, capable to measure with a high spatial and temporal resolution both the edge plasma temperature and density.

Another assessed technique, to reduce the lost ELM energy by increasing the ELMs frequency, is the attempt to determine the ELMs appearance by some local induced kinetic or magnetic perturbation. The injection [12,13] of small pellets that do not penetrate, but the plasma edge, has shown the capability to induce the onset of a small ELM for any injected pellet, as shown in Fig 5. In principle this technique is applicable in an open loop scheme, without any necessity of a Feed Back loop. However, also in this case, since what we want is to keep constant the loss of energy per ELM, it could be useful to have the possibility to change, in real time, the pellet injection frequency, whenever there could be the necessity.

Eventually the ELMs frequency can be varied/determined by using the poloidal circuits to give vertical up and down “kicks” to the plasma [14,15]. This fast and small vertical oscillation induces local magnetic perturbations that cause the onset of a single ELM per single “kick”. Obviously, this methodology has large implication with other important plasma aspects, not last the vertical stability. Consequently, although it is in principle usable in an open loop scheme, it is quite advisable to use it at least in an integrated plasma shaping and vertical stability closed loop system.

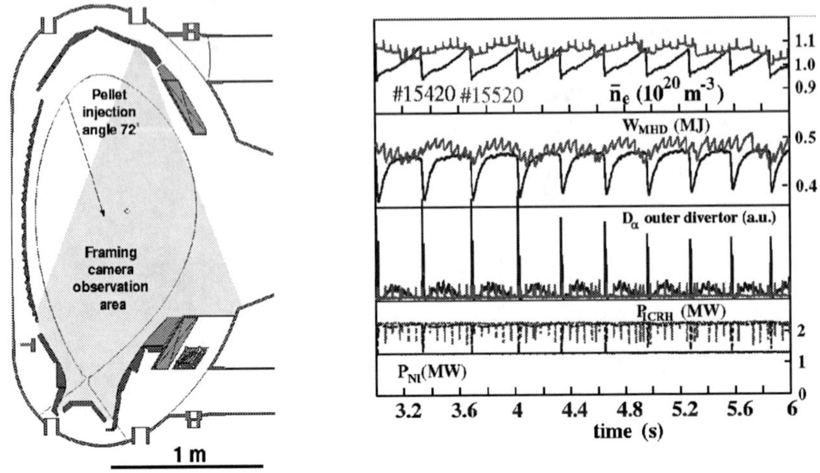

FIGURE 5. Asdex [9,13] experiment to determine the ELMs frequency by using small pellets injections.

PROFILES CONTROL

The most reliable Fusion Reactor should work in steady state. Presently the so-called Advanced Scenario [16] is trying to produce a plasma configuration with zero Loop Voltage and with a small current fraction driven non-inductively by some external tools (for instance the Lower Hybrid). This performance can be achieved if the largest amount of the plasma current is self-driven by a very high plasma pressure. Moreover, this regime can be achieved only through a careful control of the plasma current profile, but if the largest fraction of the current is self produced by the plasma

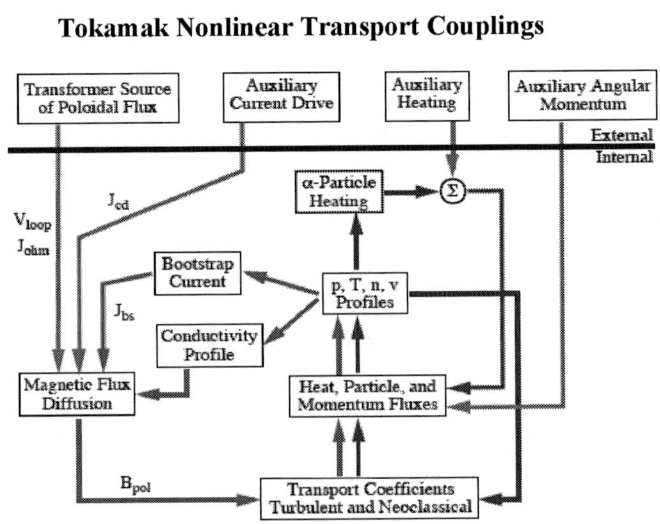

FIGURE 6. Nonlinear Couplings in Advanced Scenarios operation [18].

pressure, this, unavoidable means that both the plasma current density and pressure must be controlled at the same time and within the same loop [17].

The complexity and the high nonlinearity of such a problem is well described by Fig. 6 [18] where all the interlinks among all the “actors” are shown. Actually, in the most sophisticated cases a simultaneous control of some points of the plasma pressure and current is achieved at JET [17]. However, this is the typical case where the diagnostic capability is quite far from the control necessities. In the next future it will be mandatory to have a more robust and reliable real time reconstruction of the current and of the pressure profiles. Of course, on top of this improvement in the signals quality, it will also be necessary a strong improvement in the actuators flexibility.

CONCLUSIONS

The best way to try to summarize this short review, about the status of art in the Plasma Control, it is to show Fig. 7 (taken from [1]), where the planned attempt to integrate as much as possible control systems in JET is described.

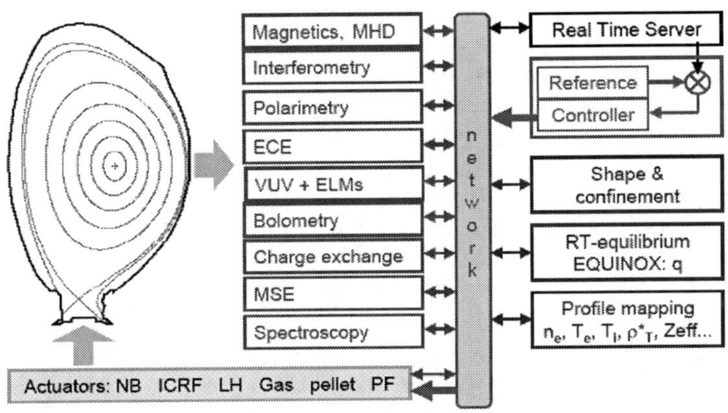

FIGURE 7. Planned Real Time Measurements and Control for Scenario integration in JET [1].

Although the figure considers only a partial fraction of the future necessities, it clearly puts in evidence the long road, that we have in front of us, to be able to fully exploit any experiment like ITER. Besides, it is straightforward that there is no chance to travel this long road without a strong improvement in many of the present diagnostic capabilities.

AKNOWLEDGEMENTS

The Authors want to thank all the people that along these years have contributed to progress on the Plasma Control. In particular the authors would like to thank all the colleagues from whom it has been "taken" the largest part of the material quoted in this review. Many of them are also quoted in the references, however a particular thank goes also to the missing ones, with whom we apologize.

REFERENCES

1. E. Joffrin et al., "Integrated scenarios in JET using real time profile control", 30th EPS, St. Petersburg, Russia, 7-10 July 2003.
2. D.A. Humphrey et al., *Nuclear Fusion*, **47**, 943-951 (2007).
3. ITER Physics Expert Group on Disruptions, Plasma Control, *Nuclear Fusion*, **39**, 2578-2625 (1999).
4. Pironti and M.L. Walker, "Fusion, tokamaks, and plasma control," *IEEE Control Syst. Mag.*, vol. 25, no. 5, pp. 30–43, Oct. 2005.
5. R. Albanese et al., "Magnetic Control of a Tokamak Plasma", Proc. "*Burning Plasma Diagnostics*", Varenna, September 2007
6. S.A. Sabbagh, et al., Nucl. Fusion **46** (2006) 635.
7. A.M. Garofalo, et al., Nucl. Fusion **47** (2007) 1121.
8. B. Lloyd, "Current Drive by Electron Cyclotron Waves", *Intern. School of Fus. Reac. Techn.*, 10th Course, Erice-Sicily: 8 –14 November 2005
9. S. Guenter, "Current Drive by Electron Cyclotron Waves", *Intern. School of Fus. Reac. Techn.*, 10th Course, Erice-Sicily: 8 –14 November 2005
10. T.E. Evans *et al* 2005 *Nucl. Fusion* **45** 595-607
11. H.R. Koslowski et al., "Type-I ELM mitigation in high triangularity and steady state regimes using low n external perturbation fields on JET", 34th EPS, Warsaw, Poland, June 2007
12. P.T. Lang et al., Nucl. Fusion **47** (2007) 754–761
13. P.T. Lang et al., Nucl. Fusion **43** (2003) 1110
14. Degeling A.W. et al 2003 Plasma Phys. Control. Fusion **45** 1637
15. P.T. Lang et al., Nucl. Fusion **45** (2005) 502–511
16. X. Litaudon et al., Nucl. Fusion **47** (2007) 1285–1292
17. D. Moreau et al., "New Dynamic-Model Approach for Simultaneous Control of Distributed Magnetic and Kinetic Parameters in the ITER-like JET Plasmas", 21th IAEA, Chengdu, China, 16-21 October 2006.
18. P.A. Politzer et al., "Stationary, High Bootstrap Fraction Plasmas in DIII-D without Inductive Current Control", 20th IAEA, Vilamoura, Porrtugal, 1-6 November 2004.

Magnetic Control of a Tokamak Plasma

R. Albanese[1], G. Ambrosino[1], F. Crisanti[2] and JET-EFDA Contributors[3]

JET-EFDA, Culham Science Centre, OX14 3DB, Abingdon, UK
[1] Assoc. EURATOM/ENEA/CREATE, Univ. Napoli Federico II, Via Claudio 21, 80125 Napoli, Italy
[2] Associazione Euratom-ENEA sulla Fusione, Frascati, C.P. 65, 00044 Frascati, Italy

Abstract. The plasma in a tokamak is magnetically confined through electromagnetic fields generated by a set of poloidal field coils, with feedforward nominal voltages or currents based on an approximate plasma model. Therefore, feedback control is necessary. A good magnetic control is essential for achieving good performance in present devices and expected to be of paramount importance in future tokamaks with a burning plasma due to the more stringent requirements in terms of plasma-wall clearance, stability margins, limited number of tolerable disruptions and so on. Magnetic fields are able to control plasma current, position and shape as well as three dimensional instabilities on the resistive time scales (resistive wall modes). They also have an influence on the control of the plasma current density profile. This paper focuses on the axisymmetric control of plasma current, position and shape, illustrating modelling assumptions and techniques, controller design procedures, assessment of controller performance and magnetic diagnostic requirements.

Keywords: Tokamaks, Magnetic diagnostics, Control systems.
PACS: 52.55.Fa, 52.70.Ds, 07.05.Dz.

INTRODUCTION

In both present-day and future tokamaks the plasma has to be stabilized and controlled in all phases of a pulse, from initiation to ramp-up, shaping, heating, current drive, stabilization, and safe termination of discharges. For high performance burning plasmas, however, the control issues will be more challenging, not only because of new problems related to the production of a considerable amount of fusion energy. As a matter of fact, to yield high efficiency, future tokamaks will operate very close to stability limits, with extreme, e.g., highly elongated configurations. For this reason an accurate regulation of plasma boundary, internal profiles, pumping, fueling, heating will be needed. In addition, due to the huge amount of kinetic and magnetic energy, more stringent requirements will have to be met in terms of plasma-wall clearance, stability margins, and number of tolerable disruptions.

Feedforward Poloidal Field (PF) coil currents for magnetic confinement are based on approximate models, therefore a feedback control is necessary. For burning plasmas there is no need to propose new concepts of magnetic control systems, but a high accuracy will be required. Therefore, conventional model-based controllers will be used. This paper focuses on the basic axisymmetric control of plasma current, position, shape, showing how most systems and techniques currently used in the JET tokamak can be applied to ITER.

---

[3] *See the Appendix of M.L. Watkins et al., Fusion Energy 2006 (Proc. 21st Int. Conf. Chengdu, 2006) IAEA, (2006)*

CP988, *Burning Plasma Diagnostics*, edited by F. P. Orsitto, G. Gorini, E. Sindoni, and M. Tardocchi
2008 American Institute of Physics 978-0-7354-0507-3/08/$23.00

MAGNETIC CONTROL IN JET AND ITER

Magnetic fields in tokamaks are mainly produced by PF coil currents, but the total magnetic field in the plasma chamber is also given by the magnetization of the ferromagnetic materials, by the eddy currents induced in the passive structures, and by the plasma itself.

Figure 1 shows the PF systems of JET [1] and ITER tokamaks [2]. JET is characterized by the presence of a magnetic circuit, copper coils and short time constants of the passive structures. In ITER the ferromagnetic materials will be limited to a number of inserts aimed at reducing the toroidal field ripple, the PF coils will be superconducting, and the time constants of the passive structures will be considerably longer (see Table 1). Table 2 reports the main functional requirements of ITER [3], for which a typical magnetic configuration is shown in Figure 2.

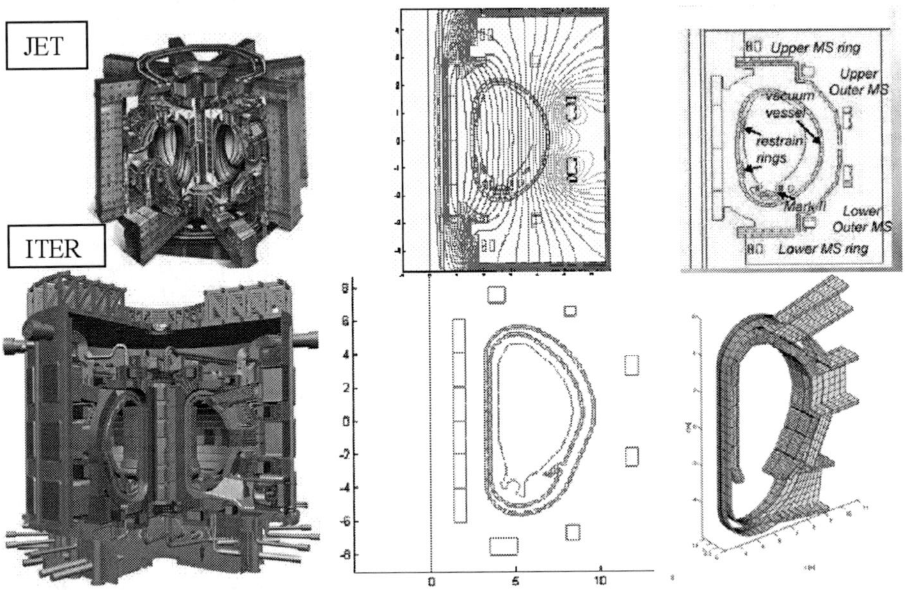

FIGURE 1. JET and ITER PF systems. On the left there are their pictorial views; in the center there are their poloidal cross sections showing PF coils (and magnetic circuit in the case of JET); on the right there are some details of the passive structures.

Model-Based Controller Design

The first control systems in tokamaks were SISO (single input single output), mainly dedicated to keep the position of the plasma, having a PID structure, designed using simple models of the system and fine-tuned during tokamak operation. The stringent plasma boundary control requirements for hot plasmas have motivated researchers to improve the modeling of plasma response as well as the design of feedback controllers. Multivariable models were used because of the high degree of coupling between the control variables, and sophisticated MIMO (multiple input multiple output) controllers were proposed so as to improve the system performance.

TABLE 1. JET vs ITER magnetic control systems.

	JET	ITER
Dimensions	R = 3 m, a = 1 m	R = 6 m, a = 2 m
Plasma current	Ip=1.5÷4.5 MA	Ip=15 MA
Growth times	τ_g = 1.5÷5 ms	τ_g = 70÷250 ms
Shape control	SC in current or gap control XSC controlling whole shape	Integrated shape (6 gaps) & Ip controller
Vertical stabilization	10 kV – 2.5 kA Dedicated FRFA circuit *being enhanced to* *ERFA (12kV 5kA)*	6 kV - 10 kA VS1 (PF2-PF5) *with alternative proposals of* *9 kV VS1 or 6 kV VS2 on CS2*

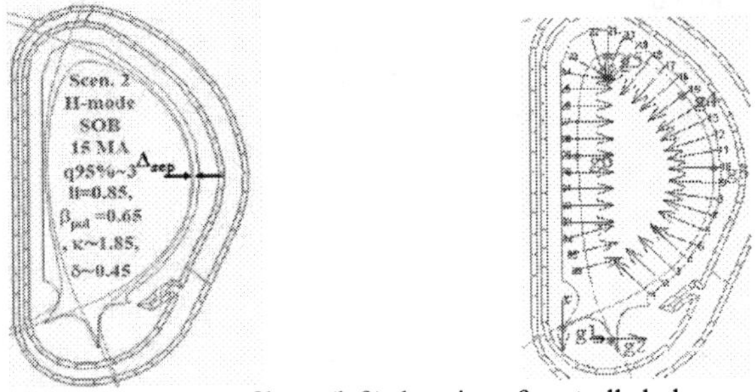

FIGURE 2. ITER: configuration at start of burn (left); location of controlled plasma-wall gaps and strike points (right).

TABLE 2. Functional requirements for ITER.

ITEM	SPECIFICATION
Disruption rate	❖ 10 % at full performance
Separatrix control at full performance	❖ Δsep > 1 cm (4 cm for reconstruction accuracy) ❖ plasma-wall gaps > 0 at all times ❖ settling time: 10 mm deviation from nominal shape after 5-20 s from disturbances (uncontrolled ELMs, ITB collapse, minor disruption, fast H-L transition) specified in terms of time behaviors of li & βp
Vertical stabilization	❖ stabilization in all phases (higher li & lower bp) ❖ recovery from above + additional disturbances
Plasma current	❖ settling time so as to keep q95% above threshold

The design of a controller typically requires the following steps. Firstly, a nominal configuration (or trajectory) of the system is selected, identifying all relevant input and output quantities. Then, a linearized model is derived, i.e., the relationships between input and output quantities are established. These relationships are in general nonlinear; however, the linear approximation is often acceptable as far as the deviations from the nominal conditions are within a certain range. The use of linear response models allows using well established techniques for controller design. The open loop relationships can be observed experimentally, but this procedure can obviously be used reliably only for configurations of machines in operation and SISO systems. For future tokamaks the relationships should be extrapolated using theoretical information, whereas the establishment of the open loop response to multiple inputs is time consuming and more likely affected by experimental errors. For these reasons, in these cases, the derivation of models based on a priori theoretical assumptions is more suitable for the design of a controller. The limits of validity of the linearized models must obviously be experimentally verified as well as the closed loop performance of the system when applying the feedback controller.

Linearized Models for Plasma Current, Position and Shape Control

The design of plasma current, position, and shape controllers is generally based on the following simplifying assumptions

- axisymmetry: the plasma-circuit system in a tokamak is assumed to be basically axisymmetric; 3D eddy current effects, if any, are taken into account via equivalent circuits;
- massless plasma: the inertial effects are neglected on the resistive time scale due to low density of the plasma, which is then assumed to evolve through equilibrium states;
- finite number of plasma dof's (degrees of freedom): the local thermo-fluid-dynamic equations can then be ignored, and the plasma profile is described in terms of few global parameters, e.g., plasma current I_p, poloidal beta β_{pol}, and internal inductance ℓ_i:

$$I_p = \int_{\Omega_p} J_\varphi d\Omega,\ \beta_{pol} = 4\int_{V_p} p\,dV \Big/ \mu_0 R_0 Ip^2,\ \ell_i = 4\int_{V_p} \left(B^2/2\mu_0\right) dV \Big/ \mu_0 R_0 Ip^2 \quad (1)$$

In the 2D plane case, completely different current profiles give the same field outside a cylindrical conductor: only the total current counts. In the 2D axisymmetric case two plasmas with the same values of I_p, β_{pol}, and ℓ_i give nearly the same magnetic field outside. This is a well known problem for current profile reconstruction, but can be exploited for modelling plasma-circuit interactions [4-7].

Using well established equilibrium codes, the magnetic flux and the plasma configuration can be determined when prescribing the vector of currents $\boldsymbol{I}$ (including circuit currents in active and passive conductors and the plasma current Ip) along with β_{pol}, and ℓ_i. The time evolution of these currents is given by:

$$d\boldsymbol{\Psi}/dt + \boldsymbol{R}\,\boldsymbol{I} = \boldsymbol{U} \quad (2)$$

$$[\boldsymbol{\Psi}, \boldsymbol{Y}]^T = \boldsymbol{\eta}\ (\boldsymbol{I},\boldsymbol{W}) \quad (3)$$

where (2) is the set of circuit equations, $\boldsymbol{\Psi}$ is the vector of magnetic fluxes linked with the circuits, $\boldsymbol{R}$ is the resistance matrix, and $\boldsymbol{U}$ is the vector of applied voltages (which are nonzero only for active circuits). The magnetic fluxes linked with the circuits, as well any other quantity of interest $\boldsymbol{Y}$, depend on the magnetic configuration and therefore are functions of $\boldsymbol{I}$, and $\boldsymbol{W}= [\beta_{pol}, \ell_i]^T$ through the Grad-Shafranov equilibrium constraint, formally indicated by $\boldsymbol{\eta}$ and practically calculated by means of numerical equilibrium codes.

Assuming $\boldsymbol{I}=\boldsymbol{I_0}+\boldsymbol{i}$, $\boldsymbol{U}=\boldsymbol{U_0}+\boldsymbol{u}$, $\boldsymbol{W}=\boldsymbol{W_0}+\boldsymbol{w}$, $\boldsymbol{Y}=\boldsymbol{Y_0}+\boldsymbol{y}$, i.e., splitting nominal and perturbed values, it is possible to get the *linearized form* of the above equations [6-7]:

$$\boldsymbol{L}^*\, d\boldsymbol{i}/dt + \boldsymbol{R}\,\boldsymbol{i} = \boldsymbol{u} - \boldsymbol{L_E}^*\, d\boldsymbol{w}/dt \quad (4)$$

$$\boldsymbol{y} = \boldsymbol{C}\,\boldsymbol{i} + \boldsymbol{F}\,\boldsymbol{w} \quad (5)$$

with $\boldsymbol{L}^* = \partial\boldsymbol{\Psi}/\partial\boldsymbol{I}$, $\boldsymbol{L_E}^* = \partial\boldsymbol{\Psi}/\partial\boldsymbol{W}$, $\boldsymbol{C} = \partial\boldsymbol{Y}/\partial\boldsymbol{I}$, $\boldsymbol{F} = \partial\boldsymbol{Y}/\partial\boldsymbol{W}$. Here $\boldsymbol{L}^*$ is the inductance matrix $\boldsymbol{L} = (\partial\boldsymbol{\Psi}/\partial\boldsymbol{I})_{plasmaless}$ modified by the presence of the plasma.

Finally, assuming $\boldsymbol{x}=\boldsymbol{i}$, $A=-(\boldsymbol{L}^*)^{-1}\boldsymbol{R}$, $B= (\boldsymbol{L}^*)^{-1}$, $E= -(\boldsymbol{L}^*)^{-1}\boldsymbol{L_E}^*)$, the *state space form* suitable for controller design can be obtained:

$$d\boldsymbol{x}/dt = \boldsymbol{A}\,\boldsymbol{x} + \boldsymbol{B}\,\boldsymbol{u} + \boldsymbol{E}\,d\boldsymbol{w}/dt \quad (6)$$

$$\boldsymbol{y} = \boldsymbol{C}\,\boldsymbol{i} + \boldsymbol{F}\,\boldsymbol{w} \quad (7)$$

This model has been validated using several codes [7-8] on various toroidal devices (e.g., JET, FTU, RFX and TCV) in different conditions, and extensively used for controller design in the above machines as well as in ITER. Figure 3 shows the comparison between linear model predictions and experimental data in JET using the CREATE-L model with a suitable schematization of the magnetic circuit and the passive structures. On the left Figure 3 shows the time behaviour of the signal of a magnetic probe during flat top; in this case, relevant for current and shape control, the eddy current effects are neglected, and it is possible to observe that the linearized model provides accurate predictions even for a the quite long time interval of 10 s. On the right is reported the simulation of a VDE; in this case the eddy current effects are fundamental for the description of the phenomenon, and the model provides a good prediction of the instability growth rate (within 10%) as well as the whole pattern of the transient magnetic field.

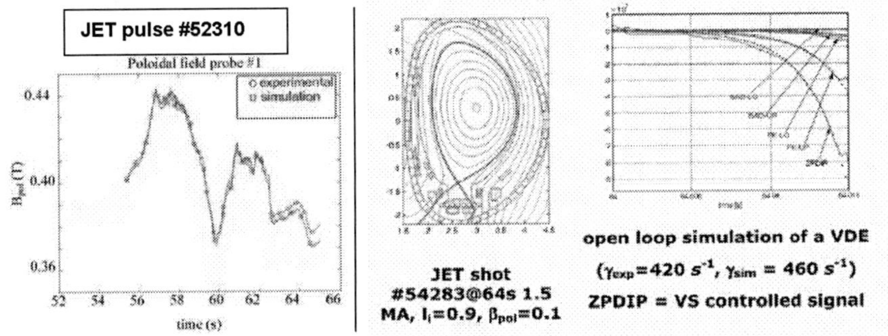

FIGURE 3. Validation of CREATE-L models in JET for shape control and vertical stabilization.

PLASMA SHAPE CONTROL

Model based shape controllers have already been implemented and tested on various devices including JET. The Extreme Shape Controller (XSC) in JET is used for extremely shaped plasmas with high triangularity and elongation. This approach seeks to control the entire plasma boundary rather than a small, fixed number of magnetic measurements (for example, magnetic fluxes) or plasma-wall gaps. The plasma is identified by a large number (up to 40) of geometrical descriptors (gaps, X-point coordinates, and strike point positions of the separatrix on the divertor targets). Finally, a singular value decomposition approach is used to optimize the control of the geometrical descriptors with the 8 available PF currents (one circuit is dedicated to plasma current control and another to vertical stabilization), even if only some of them (5 or 6) are practically able to move the plasma boundary with a reasonable effort in terms of power required. In other (non-extreme) conditions, the JET Tokamak uses the shape controller (SC) in gap or current control (usually the controlled variables are 3 gaps or strike points and 5 currents). The XSC provides better performance with large variations of β_{pol} or ℓ_i, the drawback is that a different set of controller gains has to be designed for each different nominal configuration.

A model based shape controller is proposed also for ITER. In this case there are 11 circuit currents available (some of them shared with the vertical stabilization system), shape control is integrated with the plasma current regulation, and the controlled variables are Ip and the plasma-wall gaps at 6 fixed position (see Figure 2).

Magnetic Diagnostics for Shape Control

To obtain a good feedback control scheme of a tokamak device, the main prerequisites are not only the availability of reliable models and robust control algorithms, but also the accurate real-time determination of the quantities to be controlled, requiring top high quality and reliable diagnostics. A real-time diagnostics is necessary to determine the most important plasma quantities with high accuracy and time resolution. Magnetic diagnostics, including pick-up coils, flux loops, saddle loops and Rogowski coils, diamagnetic loops, are successfully used in present tokamaks.

Most magnetic diagnostics provide the time derivative of the average field in a given region, picking up the raw signal, i.e., the voltage measured at the probe terminals and processing it by analog or digital time integration and multiplication by a suitable calibration factor.

With a sufficient number of magnetic diagnostics around the plasma boundary, it is possible to identify the main features of the plasma current distribution via a moment description:

- $m_0 = \int_\Gamma \underline{J} \cdot d\underline{n} = I_p$ plasma current
- $\left.\begin{aligned} m_1 &= \int_\Gamma z\underline{J} \cdot d\underline{n} = Z_p I_p \\ m_2 &= \int_\Gamma r^2 \underline{J} \cdot d\underline{n} = R_p^2 I_p \end{aligned}\right\}$ centroid position (8)
- . . .

These moments are given by simple integrals of the magnetic field along a closed line surrounding the plasma (e.g., the plasma current is obtained by Ampere's low), which are obviously approximated as linear combinations of the sensors available.

Plasmas with the same sets of moments have indistinguishable external magnetic configurations and hence this makes it possible to reconstruct the plasma boundary reliably using the first, more significant moments [9]. There are various reconstruction codes routinely used in present tokamaks, based on different methods, and all of them are accurate, reliable and sufficiently fast.

Both magnetic diagnostics and reconstruction codes are then adequate for ITER [10] (see Figure 4). However, due to the higher plasma magnetic and kinetic energy of ITER, the degradation of sensors caused by neutron flux should be minimized with suitable sensor design or protections. In addition, a drift of the time integration procedures is expected due to longer pulse length; hence non-magnetic diagnostics or non inductive magnetic measurements (e.g., Hall sensors) have to be added and the reconstruction codes have to be upgraded so as to take advantage of the other diagnostics to correct the drift on-line using data fusion techniques.

Figure 4 also shows the main magnetic measurements at JET [12]. Recently, the magnetic diagnostics have been enhanced to replace the various in-vessel coils no longer working properly, adding new in-vessel coils to increase the resolution of the identification tools especially in the upper and lower region [13].

Moreover, some new ex-vessel sensors (including both pick-up coils and Hall sensors) have been added to measure the magnetization state of the iron when the time integration of the signals provided by the inductive sensor starts [13]. This problem will not be crucial in ITER, which is an air core tokamak, but analogous problems are expected, due to the presence of 3D ferromagnetic inserts to reduce the toroidal field ripple and to the forementioned drift of the integrators.

A new magnetic identification tool has also been developed and successfully applied already during the commissioning phase of the sensors to check cable connections, calibration factors and so on, by comparing measured vs calculated, reconstructed or interpolated signals [14].

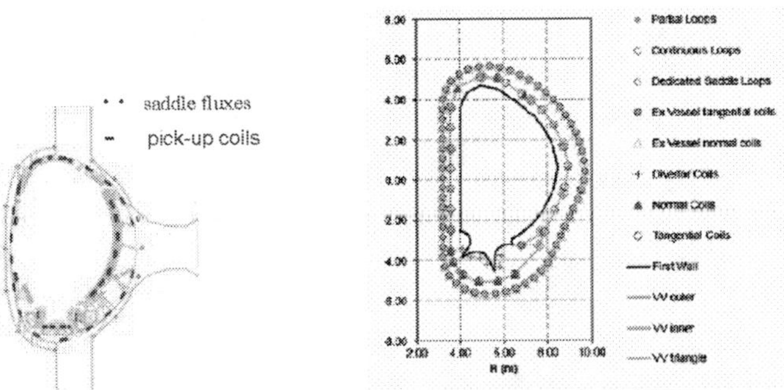

FIGURE 4. Location of main magnetic diagnostics in JET (left) and ITER (right).

VERTICAL STABILIZATION

Vertically elongated plasmas have to be used to maximize the performance-to-cost ratio (large volume and pressure) An elongated tokamak plasma is unstable with a growth rate which depends on its configuration and the surrounding conducting structures (Figure 5).

In the absence of conducting materials surrounding the plasma, the vertical instability of an elongated configuration would grow on the Alfven time scale (of the order of 10 μs for large tokamaks). Therefore the vessel, the first wall and the other conductors near the plasma should be located so as to guarantee passive stabilization. A measure of the ability of the conductors to stop the vertical instability is the stability margin m_s, initially defined as $m_s=(F_s/F_d-1)$, where F_d is the vertical destabilizing force exerted by the equilibrium field on a vertically displaced plasma (see Figure 5), and F_s the vertical stabilizing force due to the currents induced in the structures. Actually, the passive conductors only slow down the instability to the time scale of the time constant L/R. To stabilize the plasma on a longer time scale, an active feedback system is required. Several active feedback controllers have been proposed and successfully applied in various tokamaks: PD, PID, bang-bang, adaptive, etc.

Vertical instabilities were initially studied with simple rigid models, and then using the perturbed equilibrium approach of (2)-(7), with more accurate estimations of the growth rate $\gamma_g=1/\tau_g$ and the stability margin m_s (especially for triangular plasmas):

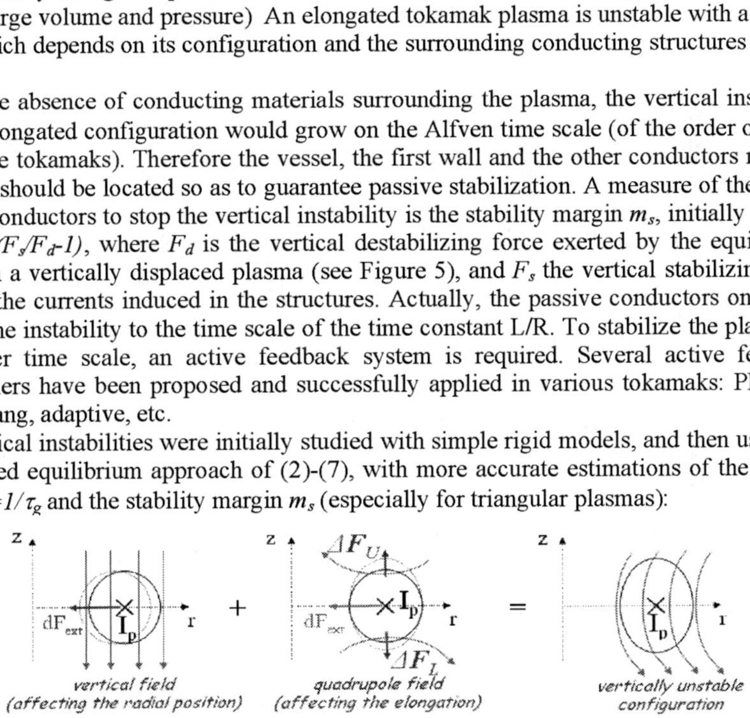

FIGURE 5. An elongated configuration: the plasma is stretched vertically by a destabilizing quadrupole field: the magnetic field pushes the plasma away in case of vertical displacement.

$$A x_u = \gamma_g x_u, \quad \gamma_g > 0 \qquad \text{(growth rate)} \tag{9}$$

$$m_s = max\{Re[\lambda(-L^* L^{-1})]\} \qquad \text{(stability margin defined in [11])} \tag{10}$$

$$m_s = -(x_u^T L^* x_u)/(x_u^T L\, x_u) \qquad \text{(alternative stability margin definition)} \tag{11}$$

where x_u is the unstable eigenvector of A and λ's are the eigenvalues of $(-L^* L^{-1})$. Other parameters of interest are the gain and phase margin of the SISO open loop control system, where the input is the applied voltage and the output the controlled variable. All these parameters depend on the plasma configuration and the passive structures.

The following procedure is currently applied for large tokamaks like JET and ITER to assess the capabilities of a VS system or determine the installed power needed (Figure 6). A set of benchmark cases (including equilibrium configurations and expected disturbances) is selected. The most critical configurations in terms of stability margin are those with high elongation and triangularity, high ℓ_i and low β_{pol}. Most disturbances can be approximated as equivalent step changes $\Delta \boldsymbol{w} = [\Delta\beta_{pol}, \Delta\ell_i]^T$, hence the initial conditions for the transient after a disturbance are given by:

$$dx/dt = \boldsymbol{A}\,\boldsymbol{x} + \boldsymbol{B}\,\boldsymbol{u} + \boldsymbol{E}\,dw/dt \Rightarrow \boldsymbol{x}(0^+) = \boldsymbol{E}\,\Delta \boldsymbol{w} \qquad (12)$$

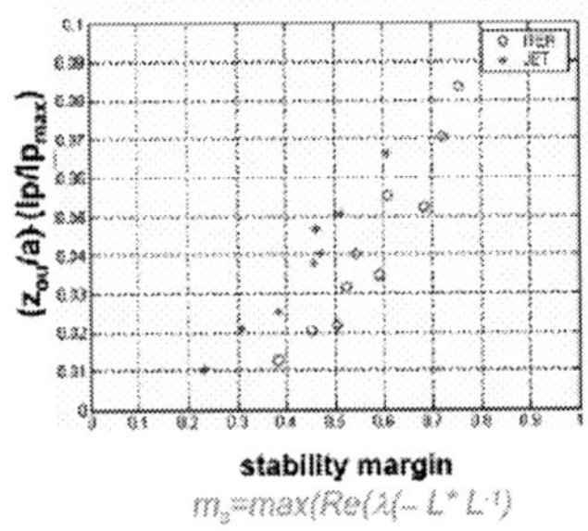

JET	ITER
10 kV – 2.5 kA Dedicated FRFA circuit	6 kV - 10 kA VS1 (PF2-5)
Ip=1.5÷4.5 MA	Ip=15 MA
τ_g = 1.5÷5 ms	τ_g = 70÷250 ms
Adaptive control on d(zpIp)/dt and I_{FRFA}	Control of dzp/dt
being enhanced to ERFA (12kV 5kA)	Evaluating alternative options 9 kV VS1 + 6 kV VS2 on CS2

FIGURE 6. Vertical stabilization capabilities of JET and ITER. The diagram on the left reports the maximum extent of a disturbance as a function of the stability margin. The tolerable disturbance is the unstable component of the vertical disturbance divided by the plasma minor radius with the plasma current scaled to full performance conditions.

The initial conditions are a linear combination of the unstable eigenvector $\boldsymbol{x}_s$ and of the stable modes:

$$\boldsymbol{x}(0^+) = a_u \boldsymbol{x}_u + \Sigma_k\, a_{sk} \boldsymbol{x}_{sk} \qquad (13)$$

All stable modes are naturally damped, hence only the unstable mode has to be stabilized. The extent of the excitation of the unstable mode can be quantified in terms of equivalent unstable component s_u of the controlled signal s:

$$s_u(0^+) = \boldsymbol{C}_s\, a_u \boldsymbol{x}_u \neq \boldsymbol{C}_s \boldsymbol{x}(0^+) \qquad (14)$$

where the controlled signal can be the plasma vertical position z, the speed dz/dt, or any other quantity excited by the growth mode, i.e., with $a_u \neq 0$; $\boldsymbol{C}_s$ is the row of the output matrix defined by (7) corresponding to the controlled signal s.

Applying a voltage step to the active circuit, it is immediate to assess whether it is able to stop the unstable motion (invert speed) before exceeding tolerable limits in terms of active current and vertical displacement. It is then possible to determine the minimum power to be installed for a given benchmark condition (or, alternatively, the maximum extent of a disturbance that can be tolerated). Of course, there are only lower bounds, because a real controller will hardly apply a full voltage step, even for a bang-bang controller like in JET as shown in the next section.

Magnetic Diagnostics for Vertical Stabilization

The controlled signal is typically related to the vertical speed of the plasma, which can in principle be obtained from the time derivatives of the measurable current moments (8) given by suitable linear combinations of flux and field derivatives:

$$I_p \dot{z}_p = \frac{d}{dt}\left(\sum_k z_k I_k\right) - z_p \dot{I}_p - \sum_c z_c \dot{I}_c \tag{15}$$

where I_k's are the currents surrounded by the set of sensors (including plasma current I_p and other currents I_c's) and z_k's their vertical positions. This algorithm requires either the absence of other currents encircled by the line of sensors or accurate current measurements and perfect knowledge of their position.

In JET (Figure 7), after the introduction of the divertor coils and the installation of MK2 divertor structure, all the measurements coming from magnetic field sensors placed on the lower part of the machine have been significantly delayed and affected by active and passive currents inside the vessel. To minimize the error due to these currents, the reconstruction is based only on flux and field measurements coming from the pick-up coils located in the upper region of the vessel. Nevertheless, the contribution of the conducting sources placed inside the vessel on the reconstructed signal (denoted as ZPDIP) is conspicuous and adds to the noise coming from the row, non-integrated magnetic signals. The ZPDIP signal is extremely sensitive to the currents flowing in the internal coils and conductors during ELMs (edge localized modes) and also in plasmaless pulses. Its behaviour during ELMs yields an equivalent undesired delay τ_d of the order of several hundreds of ms (see Figure 7), in which the unstable mode is but is amplified by a non negligible factor of $exp(\gamma_g \tau_d)$. Alternative linear combinations of signals have been proposed and tested to mitigate these problem, and promising results have been obtained [15].

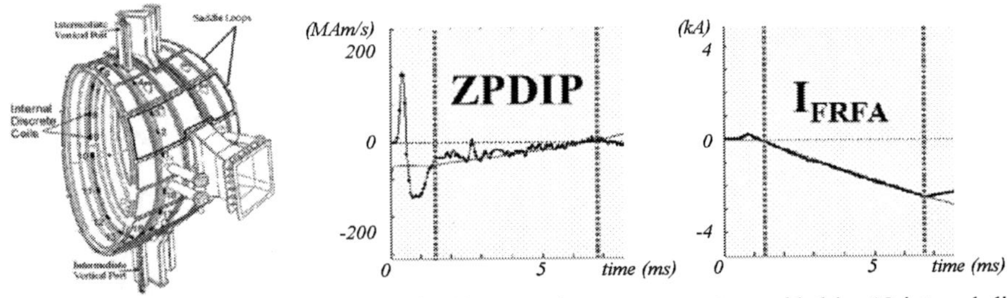

FIGURE 7. VS measurement in JET: a) the 32 magnetic measurements provided by 18 internal discrete tangential field sensors inside the vacuum vessel, and 14 saddle loops, measuring the average normal field, averaged in 4 octants to remove the effects of $n=1$ and $n=2$ modes ; b) ZPDIP signal and active current after an ELM in JET pulse #60571 ($\gamma_g \sim 280\ s^{-1}$).

In ITER similar problems are expected in terms of signal-to-noise ratio. The linear dimensions and the plasma current will be greater, the time constants of the passive currents will be longer, but the plasma instability will be slower. No continuous toroidal currents will flow inside the ITER vessel, but some magnetic sensors will be placed behind the blanket modules, hence the local transient effects of the eddy currents should be taken into account properly. The main effects of n>0 modes can be removed by taking advantage of the sensors available at various toroidal locations. The noise is not expected to be a critical problem, sophisticated VS systems are used.

CONCLUSIONS

The techniques for plasma current, position and shape control in present tokamaks are well assessed and can be used for ITER. The main problems for ITER are due to the engineering limits on voltages and currents associated with the use of superconducting coils. The issue of vertical stabilization is even more critical. VDE effects are dangerous and ITER shall operate at rather low stability margins. Therefore further studies and experimental tests on present machines needed.

As for the magnetic diagnostics for plasma current and shape control, both present sensors and accuracy of current reconstruction algorithms seem more than adequate, provided that the effects of a huge heat flux and long pulse duration can be managed. The present-day sensors and reconstruction algorithms for VS measurements seem also suitable for ITER, but careful attention has to be paid to the n>0 modes, local eddy current effects and noise.

ACKNOWLEDGMENTS

The work leading to this article was funded by the European Atomic Energy Community and is subject to the provisions of the European Fusion Development Agreement. This paper is based on a long term activity in plasma modelling, control and diagnostics carried out in cooperation with several colleagues from ENEA, CREATE, RFX, UKAEA, CEA, and EFDA. This work has been partially supported by MIUR.

REFERENCES

1. F. Crisanti et al., *Fusion Engineering and Design* , 66-68, (2003) 803-807.
2. Y. Gribov et al., *Nucl. Fusion* 47 (2007) S385–S403.
3. *ITER Physics Guidelines,* N 19 FDR 1 (01-06-07) R 0.2.
4. J. Blum, J. Le Foll, and B. Thooris, *Comput. Phys. Commun*., vol. 24, no. 3–4, pp. 235–254, Dec. 1981.
5. J.L. Luxon and B.B. Brown, *Nucl. Fusion,* vol. 22, no. 6, pp. 813–821, June 1982.
6. R. Albanese and F. Villone, *Nucl. Fusion,* vol. 38, no. 5, pp. 723–738, May 1998.
7. G. Ambrosino, R. Albanese, *IEEE Control Systems Magazine*. vol. 5, (2005) 76-92.
8. R. Albanese, G. Calabrò, M. Mattei, and F. Villone, *Fusion Eng. Des.,* vol. 66–68, (2003) 715–718.
9. L. Zabeo et al., "A New Approach to the Solution of the Vacuum Magnetic Problem in Fusion Machines", Proc. SOFT Conference, Warsaw, Poland 11th - 15th September 2006.
10. A.J.H. Donné et al., *Nucl. Fusion* 47 (2007) S337–S384.
11. A. Portone et al., Fusion Engineering and Design 74 (2005) 537–542.
12. A. Murari, et al., *Plasma Physics and Controlled Fusion,* Vol. 47, No. 3, (2005) 395-407.
13. V. Coccorese et al., *Review of Scientific Instruments*. vol. 75 No. 10, (2006) 4311-4313.
14. A. Cenedese et al., *Fusion Engineering and Design,* Vol. 74, Issues 1-4 (2005) 825-830.
15. R. Albanese et al., "Identification of Vertical Instabilities in the JET Tokamak", Proc. COMPUMAG 2007, Aachen, Germany 24th - 28th June 2007.

SOFT-X RAY DIAGNOSTICS AND TREATMENTS FOR FUTURE REAL TIME APPLICATIONS

D. Pacella [1], D. Mazon [2]

1) Associazione Euratom-ENEA sulla Fusione, Frascati, Italy
2) Euratom-CEA Ass., DSM-DRFC, Cadarache, F-13108, St Paul lez Durance, France

Abstract. This paper offers a preliminary review of the present diagnostics and data analysis techniques in the domain of Soft X-ray (SXR) emissions of fusion magnetic plasmas, including a short description of the theoretical background as well. Particular attention is devoted to the wide use of SXR tomography and impurity transport simulation. In their actual form, these techniques are not adequate to future real time applications. For this goal a step forward in the diagnosing and analysis of SXR emissions is required. The following part of the paper is therefore dedicated to the discussion of these improvements. The first one is the SXR tomography optimized for real time applications, like that one developed at Tore Supra (Cadarache, France). Fast 2-D tomographic inversions using different techniques (regularisation of Minimum Fisher, Maximum entropy, Zernicke polynomial expansions), will be optimized to be performed in a few millisecond time scale, crucial for MHD analysis, mode detection and localisation. The other goal to be pursued is the energy resolved imaging, now possible with a gas Micro Pattern Gas Detector with pixel read-out, as recently demonstrated at FTU (Frascati, Italy) and at NSTX (Princeton NJ, US), together with an ad hoc modeling of SXR emissions, compatible with future real time applications.

Keywords: Soft X-ray imaging, gas detector, magnetic plasma physics
PACS: 52.55Hc , 52.70La , 87.59-E

PRESENT STATUS OF SOFT X-RAY (SXR) DIAGNOSTICS AND CURRENT ANALYSIS

Theoretical Background

Soft X-ray emissions in magnetic fusion plasma are related to the presence of impurities in the plasma. This is made of background plasma (usually Deuterium or Hydrogen, sometimes Helium) with some impurities diluted in it, produced by the interaction of the main plasma with the vacuum chamber, limiter or divertor tiles, antennas or all the other structures inside the machine. Indeed the impurities present into the plasma depend on the nature of these elements facing the plasma: graphite (C), stainless steel (Fe, Ni, Cr..), Molybdenum (Mo), tungsten (W) and gases absorbed into the material itself (O, Cl, Ne, N…). These impurities are generally partially ionized in the plasma, depending on their atomic number and the plasma electron temperature. Light impurities are fully ionized in the plasma core, while medium and high Z elements are only partially stripped even in a many KeV plasma. We have three families of soft X-ray emissions, depending on their spectral distribution:

CP988, *Burning Plasma Diagnostics*, edited by F. P. Orsitto, G. Gorini, E. Sindoni, and M. Tardocchi

continuum emissions (W_{ff}) due to the bremsstrahlung radiation emitted by the plasma background electrons in the field of the impurity ions; semi continuum emissions (W_{fb}) due to the radiative or dielectronic recombination of the impurity ions; line transitions (W_{bb}) due to the photon emission of an electron transition during the disexcitation of the impurity ions [1]. Soft X-rays are measured as integrals B_l along the line of sight l and energy E, ε_D being the detection efficiency, Ω_D the solid angle (see the following formula). The sum in Z is extended all over the ionization states of a given impurity and over all the impurities present in the plasma.

$$B_i = \int_{l_i}\int_{E}\sum_{z}[W_{ff}^{z} + W_{fb}^{z} + W_{bb}^{z}]\Omega_D \varepsilon_D(E)dEdl$$

Since SXR emissions carry a lot of information about the plasma (electron density and temperature, impurity content, particle transport and so on), their measurement and interpretation will be crucial for all the new experiments, included ITER.

SXR Tomography

SXR Tomography is a technique widely used in all the major tokamaks. It consists of a few cameras located at different poloidal position, each one having tens of line of sight. In fig.1 the system carried out at Tore Supra [2] is described. It uses 45 viewing lines in a horizontally oriented fan and 37 in a vertical fan (see Fig. 1a), providing a poloidal and toroidal resolution of about 3 cm and 10 cm respectively. It records the integrated Soft X-ray emission between 1 to 15 keV using Silicon diodes polarized with a tension of 55V. The hardware and electronics of this diagnostics has been recently totally refurbished in order to improve the quality of the measured data. Another important particularity of this new electronics comes from the fact that it now uses VME technology for Real-Time (RT) data collection which allows slow (from 1 to 8ms) and fast (from 1μs to 1ms) RT acquisitions, the last ones being acquired on predefined time windows.

One way to address the tomography problem is to subdivide the area of the poloidal cross section where we want to reconstruct the emissivity into rectangular pixels (Fig. 1b) and to assume constant emissivity within one pixel [3].

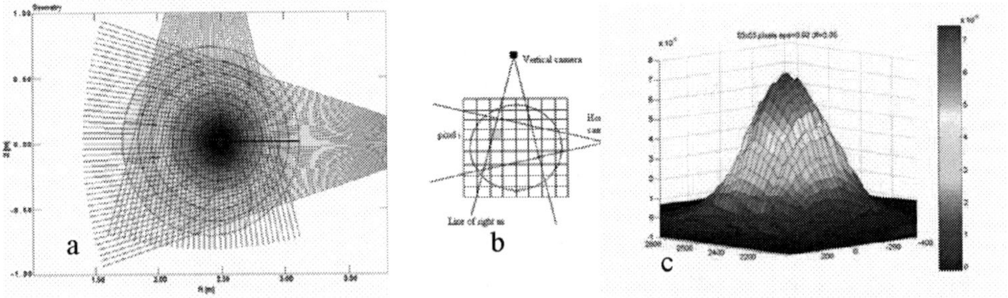

FIGURE 1. Tore Supra SXR line of sights (a). Tomographic principle (b). Example of tomographic reconstruction using a 55*55 grid pulse #37040 (c).

Then the SXR line integrated measurement f can be written in the matrix form: $f = T * g$ where g is the searched 2D local emissivity and T the geometrical matrix whose elements Tij equals the length of chord number i in pixel number j. An example of tomographic inversion, using a 55x55 pixels grid, by means of the method of Minimum Fisher Regularization is shown in fig. 1c, in a case of a Gaussian shaped emissivity profile. Other techniques of inversions have been considered (maximum entropy, polynomial decompositions and so on..) and the minimum fisher revealed as the best one for these applications.

Impurity Transport and SXR Simulation

The standard technique for modeling the SXR emissions and simulate the impurity behavior is quite complex. Starting form the first impurity, with a guess of the impurity transport coefficients (usually the convective V and the diffusion D), the ionization equilibrium for all the charge states of this impurity is calculated all over the plasma, using the experimental profiles of electron temperature and density. Many spectroscopic line transitions of the different ions are calculated and compared with the measure ones. The transport is therefore adjusted and a further iteration is made and so on until a satisfactory agreement between calculated and measured line transition intensities. This process is repeated for all the major impurities present in the discharge. Finally all the SXR emissivity profile is calculated, together with bolometry and Z effective ones. These quantities depend on the absolute values of the impurity densities, which are therefore adjusted to have an agreement with the measured ones as good as possible (fig.2)

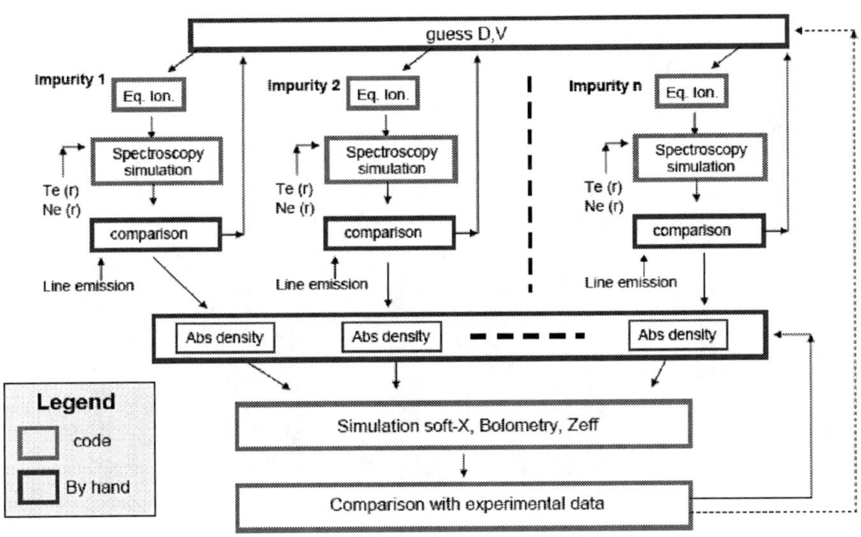

FIGURE 2. Schematic diagram for the impurity transport analysis

This technique is very detailed as description of the impurity dynamics, but requires a lot in term of X-VUV spectroscopic measurements and the simulation is unfortunately time consuming and man dependent.

PLASMA IMAGING: A NEW PROMISING APPROACH

Soft-X ray 2-D plasma imaging is still an unexplored domain in magnetic fusion experiments. The use of a new camera, based on a gas detector with a gas electron multiplier (GEM) as amplifying structure and with a two-dimensional readout with independent electronic channels per each pixel. The concept (fig. 3a) has been already tested by means of a prototype [4], with 128 pixels, carried out in Frascati in collaboration with INFN-Pisa and tested on tokamak FTU (Italy) in 2001 and NSTX (USA) in 2002-2004. Thanks to the photon counting mode, it provides 2-D imaging with high time resolution (sub millisecond), high sensitivity and signal to noise ratio. Its capability of energy discrimination [5] allows the acquisition of pictures in X-ray energy bands or to perform a spectral scan and a better contrast as well. Single pixel can be addressed and programmed.

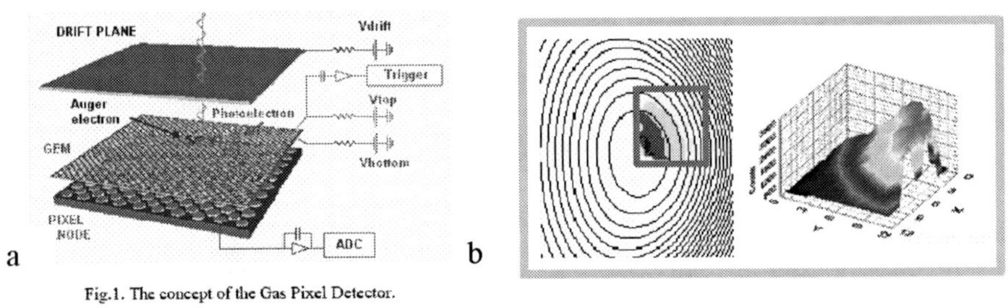

FIGURE 3. Scheme of principle of a Gas Pixel Detector with Gem as amplifier (a). Example of an X-ray image (color plot superimposed to EFIT reconstructions or in a 3D plot) obtained with the prototype PIXCS-128 at NSTX (b)

Since the detector has high and controllable intrinsic gain, it works well even at very low photon energy, ranging from 0.2 keV to 10 keV (X-VUV region).

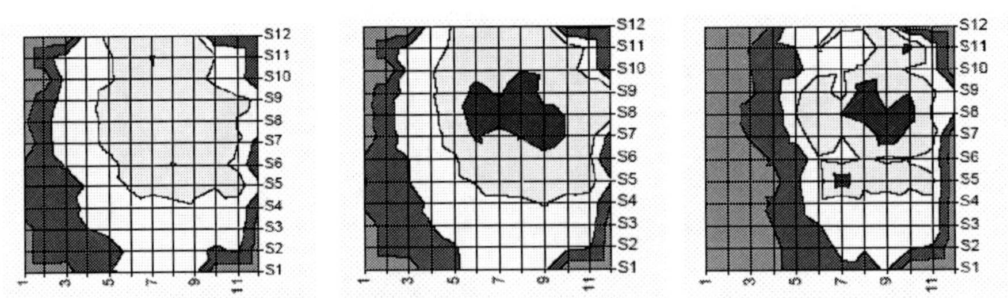

FIGURE 4. Central tangential views (40cm x 40cm), with the X-ray camera PIXCS-128 sampled at 1KHz, of the NSTX shot 107390, before an MHD mode (a), during (b) the mode and after the breaking of magnetic surfaces (c).

Preliminary experiments performed at NSTX revealed many attractive features of the direct plasma imaging [6]. Two examples are reported: plasma shaping and constraints for magnetic reconstructions in the core (fig. 3b) and direct inner MHD activity detection (fig. 4), with framing rate of 1-10 KHz, showing three different time instants, respectively before the mode, during and after the breaking of the magnetic surfaces (plasma collapse).

After this very successful prototype with 128 pixels, were the soft X-ray imaging capability at very high flux rate has been demonstrated, a way toward high pixel density gas detectors has been undertaken by the group of Gas Detectors at INFN-Pisa, where the ASIC itself becomes the pixelized anode of the Micro Pattern Gas Detector [7, 8].

A STEP FORWARD: AN IMPROVED SXR ANALYSIS IN PERSPECTIVE OF REAL TIME APPLICATIONS

It is our opinion that the future challenges in fusion experiments require a step forward in the direction of a better exploitation of the SXR emissions to diagnose and monitor the plasma. This improvement has to be done in 3 areas: upgrading of the tomographic systems and related inversion techniques, integration of tomography with energy resolved imaging and finally with an ad hoc modelling of the SXR emissions and impurity dynamics. All these work is thought in the perspective of real time applications

Tomography for real time application

As described in the first section, tomographic data are generally used in two ways: reconstruction of the 2D emissivity profiles in the cross section or by means of a time analysis of the line integrated signals (Fourier decomposition and frequency spectra), both as post process analysis (see black boxes in fig. 5).

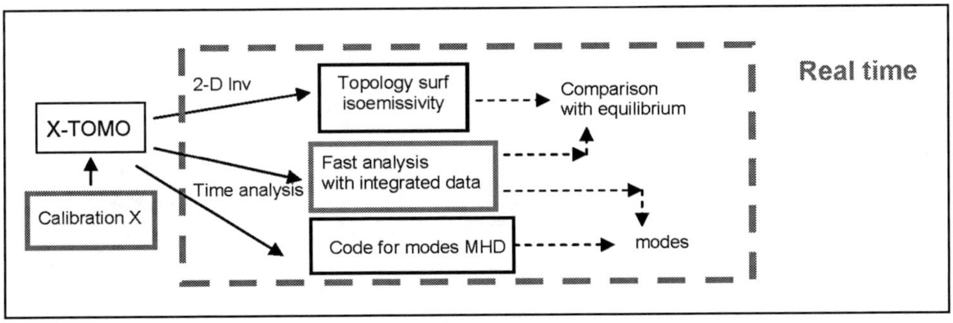

FIGURE 5. Schematic implementation of SXR tomographic data analysis in the perspective of real time applications

In the perspective of real time applications [9], the number of the lines of sight of the tomography cannot increase too much, otherwise the computing time for the data inversion is too long and not compatible with the purpose. In this case all the detectors have to be characterized with X-ray sources, measuring sensitivity, linearity, saturation, and noise. Each channel can be therefore corrected taking into account its sensitivity. The measurement of the detection efficiency as function of the X-ray photon energy is necessary as well, to simulate properly the measured SXR emissions and derive plasma parameters.

A large effort should also be provided to optimize the calculations of the emissivity profiles, looking for a good compromise between the computing time required for the reconstruction of SXR emissivity and the spatial resolution (depending on the grid pitch) enough to be useful for the magnetic reconstruction and for fast MHD instability detection.

In addition experiments should be done to study correlations between the line integrated signals of the tomography with a few parameters relevant for the magnetic reconstruction (magnetic axes) or the plasma shaping (elongation, triangularity and so on). These improvements in the direction of real time applications are summarized in a schematic way in fig. 5.

Direct imaging

Direct imaging is a powerful technique, particularly for real time applications. Preliminary experiments performed at NSTX with the prototype PIXCS-128 confirmed these expectations. Starting from the raw data of the X-ray intensity acquired with this pinhole camera looking at the central portion of the cross section, we calculated the barycentre of the X-ray emissivity. The time evolution of the barycentre is then plotted in a polar representation for the stationary case (fig. 6a, green curve) and during the instability, interpreted as a Resistive Wall Mode (red curve in fig. 6a). The sampling rate in this case is 10 Khz. With this approach MHD instabilities can de detected at their onset, in a sub millisecond time scale. The other example (fig. 6b) shows the trajectory of the barycentre of the X-ray emissivity during

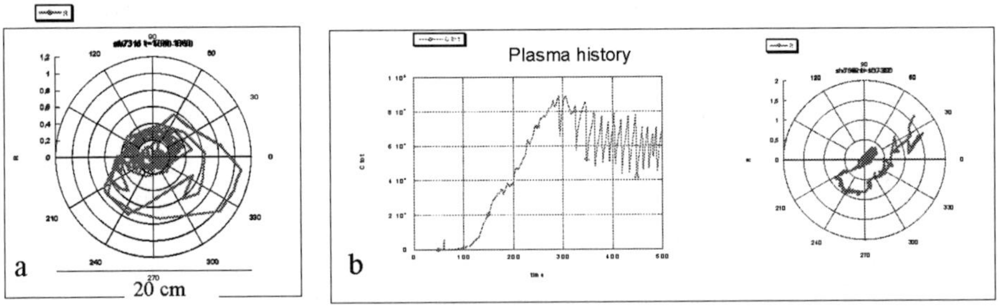

FIGURE 6. Polar plot of the position of the barycentre of the X-ray emissivity (NSTX shot 107316) in the cross section (a) in a quiescent phase (green line) and during a RWM instability (red line). Time history (b) of maximum SXR emissivity (NSTX shot 107356) and polar plot of SXR barycentre, particularly interesting during the current ramp-up phase.

the current rump up phase, opening the potential use of direct imaging to monitor the formation of the plasma column and its following evolutions [10].
Having in mind the exploitation for real time control, higher momenta of the soft X ray intensity distribution can be calculated and their correlation with the plasma parameters experimentally investigated. These preliminary results have been produced in a very favourable condition: tangential view in a spherical tokamak. Experiments in less easy optical conditions will have to be performed to validate completely the techniques.

Integration Tomography-Imaging

Tomography with silicon diodes and direct imaging with energy resolved X-ray camera are complementary diagnostics. The first one represents energy integrated measurements (typically in the range 1-20 keV), with the advantage of high time resolution (up to 1 MHz) and a stable and long lasting detector response. Due to its lower sensitivity it is used to monitor the plasma core (electron temperature greater than hundreds of eV). On the contrary energy resolved imaging with gas detector allows the acquisition in energy bands, extremely useful to increase the contrast of the region of interest or to image different parts of the plasma, characterized by different impurity radiative patterns. This second system is more sensitive, allowing the monitoring of the plasma periphery (up to a few tens of eV), very versatile in the data treatment (due to photon counting mode) and with a huge flexibility in the geometrical views (pinhole camera), both as magnification and field of view. It is clear how a step forward in the exploitation of SXR as diagnostic tool will require a strong integration of these two complementary measurements.

SXR ad hoc modelling

Last, but not the least, is the development of a simplified modelling of the plasma SXR emissivity processes, able to be optimized for real time applications. In particular a SXR simulation code should be developed, having two sets of input data. The first one is represented by the non SXR data: electron temperature and density profile, impurity related data (CXS, Zeff , a few VUV line intensities and so on…) and a few monitors of MHD modes like the ECE radiometer signals. The second set is made of SXR input data, provided by the inversion of the calibrated tomography and by the energy resolved imaging (see fig. 7). With an ad hoc simplified model of the impurity transport and the SXR emission process we could derive the impurity density profile for low, medium and high Z elements present in the plasma. The capability of inferring the impurity profile stems from the combination of tomography and energy resolved imaging. A variation of soft X-ray emissivity is generally related to particle effects (changes in electron or impurity densities) or to temperature effects (radiation pattern is strongly dependent on the electron temperature). Thanks to the energy scan that can be performed with the PIXCS camera, these two contributions can be in principle separately estimated, opening the perspective to derive and possibly control the impurity profiles in real time.

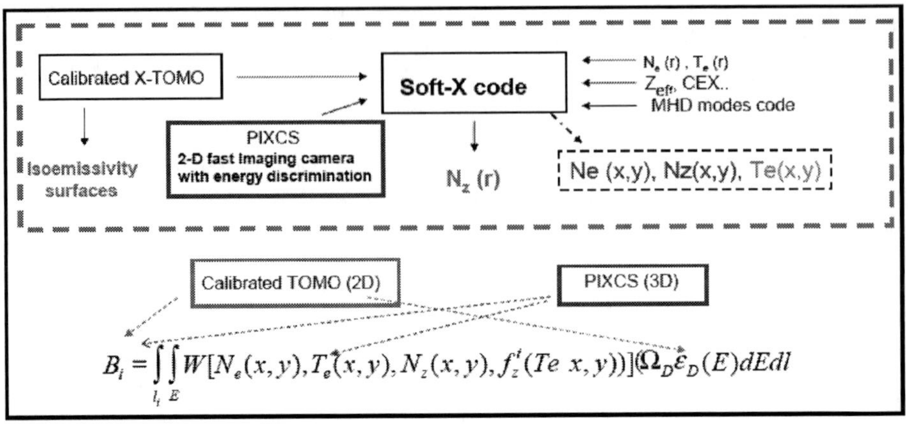

$$B_i = \int_{l_i}\int_E W[N_e(x,y), T_e(x,y), N_z(x,y), f_z^i(Te\ x,y))](\Omega_D \varepsilon_D(E) dE dl$$

FIGURE 7. Schematic diagram of a proposed simulation code, integrating tomography and energy resolved imaging, aiming to the assessment of impurity density profile, for real time exploitation

CONCLUSIONS

The package of a calibrated tomography, an energy resolved imaging, an ad hoc modelling and a simulation code capable to integrate all these data in a simplified way compatible for following real time application, can be exploited in a number of plasma physics issues. It can provide useful information for inner magnetic reconstruction and plasma shaping and position control, recognition of MHD modes, fluid effects like centrifugal forces, particle and impurity transport or monitoring different regimes, physics of pedestal of H-mode and perhaps electron temperature map too. All these subjects need to be investigated in present day machines for future real time applications in ITER.

REFERENCES

1. G. De Michelis, M. Mattioli, Nuclear Fusion, Vol. 21, No 6 (1981)
2. D. Mazon, 34th EPS Conference, Warsaw, 2007
3. M. Anton et al., Plasma Phyis. Control. Fusion, Vol 38, (1996) 1849-1878
4. D. Pacella et al., Rev Sci. Instrum., 72, No 2, (2001) 1372-1378
5. D. Pacella et al., Nucl. Instrum. Methods A 508 (2003) 414–424
6. D. Pacella et al, Plasma Physics and Controlled Fusion 46 (2004) 1075
7. R. Bellazzini et al., Nucl. Instrum. Method A 471 (2001) 41
8. E. Costa et al., Nature, 401 (2001)
9. O. Barana, D. Mazon Plasma Physics and Controlled Fusion , Vol. 49 No. 7 (2007)
10. D. Pacella et al., Journal of Instrumentation, Vol 1 , September 2006

Development in Diagnostics Application to Control Advanced Tokamak Plasma

Y. Koide

Japan Atomic Energy Agency, Naka, Ibaraki 311-0193, Japan

Abstract. For continuous operation expected in DEMO, all the plasma current must be non-inductively driven, with self-generated neoclassical bootstrap current being maximized. The control of such steady state high performance tokamak plasma (so-called 'Advanced Tokamak Plasma') is a challenge because of the strong coupling between the current density, the pressure profile and MHD stability. In considering diagnostic needs for the advanced tokamak research, diagnostics for MHD are the most fundamental, since discharges which violate the MHD stability criteria either disrupt or have significantly reduced confinement. This report deals with the development in diagnostic application to control advanced tokamak plasma, with emphasized on recent progress in active feedback control of the current profile and the pressure profile under DEMO-relevant high bootstrap-current fraction. In addition, issues in application of the present-day actuators and diagnostics for the advanced control to DEMO will be briefly addressed, where port space for the advanced control may be limited so as to keep sufficient tritium breeding ratio (TBR).

Keywords: Advanced Tokamak Scenario, Diagnostic Application, Real-time Control, bootstrap current, normalized beta, DEMO
PACS: 52.55Fa, 52.55.-s, 52.70.-m, 28.52.-s

INTRODUCTION

Recent development in diagnostic application to control advanced tokamak plasma is presented, with emphasized on the real-time control of the current profile and pressure profile under a DEMO-relevant high bootstrap-current fraction.

Although concrete design for DEMO is still under discussion, DEMO will be the last step before the first commercial plant. DEMO should convince electric supply industry that fusion reactor is attractive from business point of view. Therefore economies are one of the major requisites for DEMO.

There are two basic requirements for economically attractive steady-state tokamak reactor. The first requirement is the high bootstrap current fraction f_{BS} to reduce the power for external current driver. Since f_{BS} is proportional to poloidal beta as $f_{BS} \propto \beta_p \equiv \dfrac{\langle p \rangle}{B_p^{\,2}/2\mu_0}$, the first requirement is equivalent to high β_p.

The second requirement is high fusion power density under technologically available magnetic field strength. Here fusion power density is roughly proportional

CP988, *Burning Plasma Diagnostics*, edited by F. P. Orsitto, G. Gorini, E. Sindoni, and M. Tardocchi

to plasma pressure squared as $n_D n_T \langle \sigma v \rangle \sim n^2 T_i^2 \sim p^2$. Therefore the second requirement is equivalent to high toroidal β, which is defined as $\langle \beta_t \rangle \equiv \frac{\langle p \rangle}{B_t^2 / 2\mu_0}$.

It should be noted that, from this approximate relation, simultaneously high β_p and β_t requires high normalized beta, which is defined as $\beta_N = \beta_t \frac{I_p(MA)}{a(m)B_t(T)}$. Therefore, in order to meet the requirements of economically attractive steady-state tokamak reactor, it is necessary to operate at high β_N, that is near the pressure limit. Based on the consideration of overall plant efficiency, the target region of DEMO is set as $f_{BS} \geq 70\%$ and $\beta_N \geq 4$, which is above ideal limit without wall. In order to realize such condition, it is believed that utilizing of Internal Transport Barrier (ITB) [1, 2] in addition to H-mode is promising. However, fundamental issue in the high beta regime is MHD stability. Three major MHD instabilities are introduced in the following.

The first instability we should manage is infernal mode [3, 4], which is destabilized at low β_N under the strong linkage between j(r) and p(r) at high f_{BS}. The second instability is neoclassical tearing mode NTM [5]. NTMs are resistive tearing mode islands that are sustained by a helically perturbed bootstrap current. The resulting magnetic islands break up the magnetic surfaces, leading to the enhanced transport and can lead to disruption in a high β plasma. The third one is resistive wall mode (RWM). Ideal-MHD theory predicts that nearby perfectly conducting wall would allow tokamak operation with high plasma pressure well above the no wall limit. However, due to the finite resistivity of the first wall, these high pressure plasmas can only be sustained for short period of time before RWM is excited. In DIII-D, feedback control of n=1 coil current successfully sustained high β_N above no wall limit [6, 7].

Operating a tokamak at steady state by controlling these instabilities remains a challenging [8] but realistic prospective. DIII-D experiments have demonstrated the existence of such regimes allowing access to a high $\beta_N \sim 4$ and high $f_{BS} \geq 0.6$ with improved MHD stability [6]. In such discharges, multiple control tools with regard to j(r) control, p(r) control and RWM suppression were necessary to sustain the high β. It should be noted that it is important to mature the above proof of principle techniques into those applicable to DEMO.

RECENT PROGRESS IN REAL-TIME CONTROL OF KINETIC AND CURRENT PROFILES, AND RELEVANT DIAGNOSTICS

In this section, we discuss recent progress in real-time control of kinetic and current profiles, and relevant diagnostics.

Table 1 summarizes development of diagnostic control towards steady-state high β in present devices. In order to advance the present results to be applied to DEMO, systematic design shown in Fig. 1 is required. Key elements may consist of (1) physics understanding, (2) diagnostics development, (3) actuators development and (4) modeling and control scheme. The first three are integrated into the fourth element. Then experimental validation is carried out with subsequent feedback for improved control. During this design process, we should pay attention to whether the control scheme is applicable to DEMO.

TABLE 1. Development of diagnostic control towards steady-state high β

A route to a fully integrated control of steady-state high-β plasmas still requires a

Control		Device	Parameters to be controlled	Parameters directly controlled	Diagnostics/ code for control	Actuator	Typical period of control
ITB control	Feedback	JET	j(r) and p(r)	q(r) and ρ_{Te}/L_{Te}	EFIT(MSE), ECE	Power of ICRF, NBI and LH	10ms
		DIII-D	β_N	Stored energy	Diamagnetic loop	NB power	
			β_N and $V_t(r)$	Stored energy and $V_t(r)$	CXRS and diamagnetic loop	NB power and combination of CO/CTR NBs	
		JT-60	j(r)	$j_{LH}(r)$	MSE	LH power and $N_{//}$	10ms
			β_N	Stored energy	Diamagnetic loop	NB power	10ms
			p(r) via transport	$V_t(r)$ for enhanced transport	CXRS	Combination of CO/CTR NB power	10ms
			dp(r)/dr	dT/dr	CXRS	Combination of on/off-axis NB power	10ms
NTM control		DIII-D	j inside magnetic island		EFIT and MSE	ECCD with control of plasma position or B_t	100ms (response)
		JT-60			ECE	ECCD power with steerable mirror	10ms
RWM control		DIII-D	n=1 δBp		Magnetics	n=1 coil current	(200μs delay)
Long pulse		Tore Supra	j(r) and local heat flux to PFC	Profile of suprathermal electron emission	HXR	LH power and $N_{//}$	16ms
				LH/ICRF power	IR camera	LH/ICRF	
ELM mitigation	w/o FB	ASDEX	Energy loss per ELM	$n_e(r)$	-	Pellet	-
		DIII-D	Pedestal pressure gradient	Particle transport	-	n=3 coil current	-

sustained research effort. However remarkable progress was made in each key area, which shed light on the future direction of development.

As for NTM control, JT-60 conducted control of current inside magnetic island [9]. Here one of keys is development of ECE diagnostic with good spatial and time resolution. Another key is ECCD as an actuator with good controllability and localization. There are three steps in the procedure of the control. In the first step, coarse estimation of mode location is executed by making use of the NTM character to grow on the mode rational surfaces such as q=3/2. Here real-time plasma shape calculation gives the location of such rational surface. Then corresponding channels of ECE measurement are identified. In the second step, precise estimation of island location is executed from the profile of electron temperature fluctuation measured with ECE. Since the island structure rotates, the associated fluctuation amplitude of electron temperature takes M-shape profile, and the location of the island center is identified as the location of local minimum. The location of the island center in terms of ECE measurement is converted to the location in the electron cyclotron resonance layer, by which mirror angle of EC wave is determined. Then EC wave homes on the island center by the steering mirror. These calculations take 8ms, and mirror angle is controlled every 10ms, which is shorter than the island growth time. NTM was completely suppressed when EC wave was injected [9].

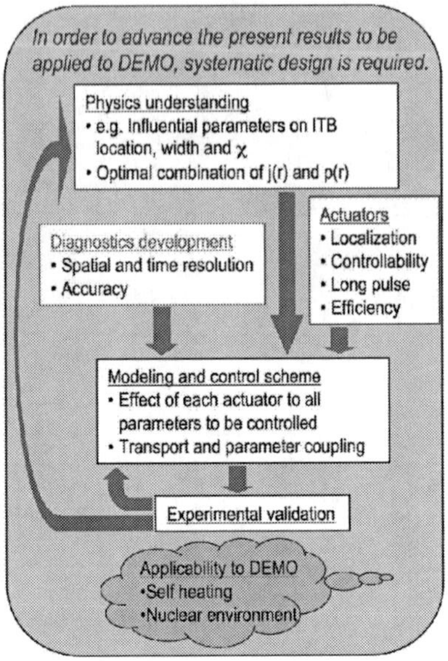

FIGURE 1. Design process of diagnostic control towards steady-state high β

Another example of real-time NTM suppression is DIII-D result [10]. Here real-time equilibrium reconstruction is used to identify the location of rational surface in real space. The location in R and Z of the ECCD is calculated using the TORAY-GA code for an earlier discharge with similar properties. The change needed for toroidal field strength B_t in order to bring the peak ECCD into coincidence with the q=2 surface at the same Z is calculated and controlled in real-time. As a result, 2/1 NTM was successfully suppressed and high β_N of 3.3 was achieved, which is near no wall ideal stability limit.

Physics understanding of NTM suppression is also progressed. It was found that experimental evolution of the 2/1 island width for the various ECCD locations is reproduced by the theoretical prediction and that well aligned ECCD is required for full suppression [11]. This kind of information should be reflected to modeling for improved control.

Next topic is real-time control of ∇p at high f_{BS} in JT-60. Here control of transport by changing toroidal rotation velocity $V_t(r)$ has been tried, which approach is important in burning plasmas dominated with self-heating. We consider that possible role of reduced CTR (anti-parallel to the plasma current) momentum input for avoiding instability is as follows [12]: reduced CTR momentum input reduces rotation shear near q_{min}; then transport rate is enhanced possibly by the reduced E_r shear which may enhances turbulence transport; as a result, ∇p is reduced below the stability limit. Since it is difficult to forecast the timing of q_{min} to pass across 4, the following real-time control is developed. In the first step, q-profile is evaluated from local pitch angle of magnetic field measured with MSE. In the second step, the estimated q_{min} is compared with upper and lower values set around integer value such as 4. If q_{min} enters the region defined by these values, CTR-NB injection is stopped for the set interval Δt. When q_{min} escapes from the region, CTR-NB is re-injected. Injection of CTR beam is controlled every 10ms. As a result of the successful CTR-NB control, plasmas with f_{BS}~70% was sustained for 8s, and q(r) reached almost steady state [12]. It was found that response of q_{min} to ∇p control becomes sensitive at high f_{BS} as shown in Fig. 2 [12]. The dependence should be reflected to the control scheme. It should also be pointed out that too sensitive response of q_{min} may lead to MHD instability at low ∇p. Therefore simultaneous j(r) control with ∇p control is required under such strong linkage.

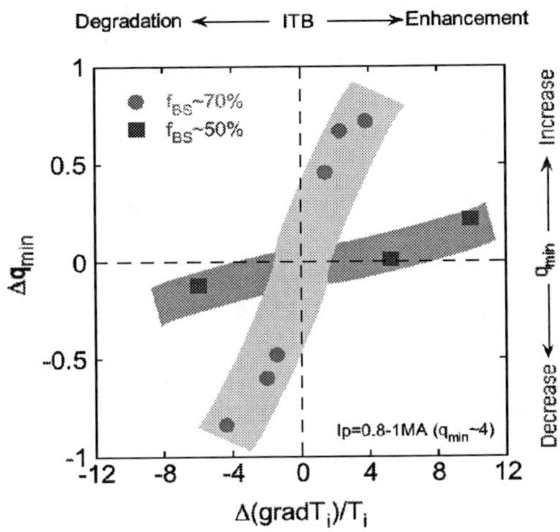

FIGURE 2. Sensitive response of q_{min} to ∇p control at high f_{BS} [12]

Combined control of j(r) and p(r) has been in progress in JET [13, 14, 15], which is essential aspect in the development of reliable high performance operation. Specifically, multi-input-multi-actuator feedback control scheme for the multi-parameter control is used, and real-time control of weak ITB in reversed shear discharges was conducted. Here diagnostics used are MSE and ECE and actuators used are ICRF, NBI and LH. Control scheme and model used are characterized as follows: real-time detection of $T_e(r)$ from ECE and real-time q(r) from Grad-Shafranov solver are used; ∇T_e and reversed shear simultaneously controlled by requests from controller to the three actuators every 10ms; p(r) and j(r) response models are being based on step or modulation experiments of the actuators. The criterion of $\rho_{T_e}^* \geq 0.014$ for the ITB strength and position is used according to empirical database analysis [16, 17] where $\rho_{T_e}^* \equiv -\frac{1}{Z_i}\sqrt{\frac{A_i m_p}{q_e}}\frac{1}{B_t}\sqrt{T_e}\frac{1}{T_e}\frac{\partial T_e}{\partial R}$. The controller succeeds in both counteracting the current diffusion and moving the barrier into the more external region of the plasma (β_N=1.8, f_{BS}=50%, f_{NBCD}~20%, f_{LHCD}~25% [18]). Confirmation of robustness of this approach in much longer time scale than resistive time and control of strong ITB, including ITB in ion system, will be carried out in the future.

As for RMM control, DIII-D conducted control of n=1 magnetic field component, where diagnostic used are magnetics and actuators used is n=1 coil current [6, 7].

As for long pulse operation, Tore Supra conducted control of j(r) and local heat flux to plasma facing component, where diagnostic used are hard X-ray detectors and IR camera, respectively, and actuators used are LH power and its $N_{//}$ and ICRF power, respectively [19].

As for ELM mitigation, ASDEX controlled energy loss per ELM, where actuators used is pellet injector [20].

SOME COMMENTS ON DEVELOPMENT OF ADVANCED CONTROL TOWARDS DEMO

The first comment is on the spatial limitation for diagnostics in DEMO from the viewpoint of consistency with the sufficient tritium-breeding ratio (TBR). Discussion here is based on the SlimCS design [21] as an example of DEMO. In this design, net TBR above 1.05 is required. On the assumption of the local TBR of 1.35 and moderate frame thickness of blanket module of 15mm, area of 7.5m^2 is remained for the heating, current drive and diagnostics systems. In the SlimCS design, 60MW is assigned for the CD power. In the case of NBCD of 60MW, area of 6m^2 out of the 7.5m^2 is devoted to NBI, based on the power flux of 10MW/m^2. As a result, only 1.5m^2 is left for diagnostics, which is in contrast to the experimental reactor ITER. Therefore effective usage of limited port space is required for diagnostics system to keep sufficient net TBR.

Second comment is the importance of study of determinant factor of key control parameters. In the previous section, pressure gradient control by using NB torque was discussed. However neutral beams have basically poor spatial controllability with regard to $V_t(r)$ control. In order to develop alternative tool, it is important to identify the determinant factor of plasma rotation, which is now an active research area.

Third comment is on the beam diagnostics. CXRS and MSE are indispensable diagnostics for experimental tokamak devices. These beam probe diagnostics may require dedicated diagnostic beams in DEMO. This is because the measurement should be made irrespective of the injection of NB units for the plasma control, and because the energy of heating beam may be too high for the measurements. It seems not be obvious that these beam diagnostics are consistent with DEMO from the viewpoint of available space.

I hope these issues will be resolved in JT-60SA and ITER.

SUMMARY

Recent progress in real-time control of kinetic and current profiles, and relevant diagnostics were discussed.

Control of major instabilities in steady-state high β plasmas (infernal mode, NTM, RWM) has been tackled in many tokamaks. Although a route to a fully integrated control still requires a sustained research effort, the information and experience gathered of real-time control development provides a good basis for the perspective of DEMO. Especially, effectiveness of actuator with good controllability and localization was proved by combining diagnostic with good spatial and time resolutions (e.g. ECCD with ECE as a NTM killer). Value of q_{min} sensitively responds to the ∇p control under high f_{BS}, thereby q_{min} critically approaches MHD unstable region. Therefore simultaneous j(r) control with ∇p control is required. In order to meet the requirement, development of multi-input-multi-actuator feedback control scheme for the multi-parameter control is in progress. Effective usage of limited port space is required for diagnostics system, including diagnostic beam, to keep sufficient net TBR. In order to develop effective actuators, identification of determinant factor of key control parameters is important.

ACKNOWLEDGMENTS

The author would like to express hearty thanks to Drs. Gohil, Luce, Wade, La Haye, Reimerdes, Tala, Mazon, Eich, Challis, Sylvain, Litaudon, Hobirk, Isayama, Sakamoto, Suzuki, Kamada, Hatae, Sugie and Kusama for providing information. Many thanks to Drs. Tobita and Nishio for fruitful discussion.

REFERENCES

1. Y. Koide, M. Kikuchi, M. Mori et al., *Phys. Rev. Lett.* **72**, 3662-3665 (1994).
2. T. Fujita, T. Fujita, S. Ide et al., *Phys. Rev. Lett.* **78**, 2377-2380 (1997).
3. T. Ozeki, M. Azumi, S. Tokuda, S. Ishida, *Nucl. Fusion* **33**, 1025-1035 (1993).
4. J. Manickam et al., *Nucl. Fusion* **27**, 1461-1472 (1987).
5. R. J. La Haye, *Phys. Plasmas* **13**, 055501 (2006).
6. A. M. Garofalo, E. J. Doyle, J. R. Ferron et al., *Phys. Plasmas* **13**, 056110 (2006).
7. M. Okabayashi, J. Bialek, A. Bondeson et al., *Nucl. Fusion* **45**, 1715–1731 (2005).
8. J. B. Lister, *Nucl. Fusion* **40**, 1167-1181 (2000).
9. A. Isayama and the JT-60 Team, *Nucl. Fusion* **43**, 1272-1278 (2003).
10. R. Prater, R. J. La Haye, T. C. Luce et al., *Nucl. Fusion* **47**, 371–377 (2007).
11. A. Isayama, N. Oyama, H. Urano et al., *Nucl. Fusion* **47**, 773–782 (2007).
12. Y. Sakamoto, H. Takenaga, T. Fujita et al., *Nucl. Fusion* **47**, 1506-1511 (2007).
13. D. Mazon, X. Litaudon, D. Moreau et al., *Plasma Physics and Controlled Fusion* **45**, L47–L54 (2003).
14. A. Murari, E. Joffrin, R. Felton et al., *Plasma Phys. Control. Fusion* **47**, 395–407 (2005).
15. T. Tala, L. Laborde, D. Mazon et al., *Nucl. Fusion* **45**, 1027–1038 (2005).
16. G. Tresset et al., 2001 *Controlled Fusion and Plasma Physics, Proc. 28th Eur. Conf. (Madeira, 2001)* vol 25A (Geneva: European Physical Society) p 533
17. G. Tresset, X. Litaudon, D. Moreau and X. Garbet, *Nucl. Fusion* **42**, 520-526 (2002).
18. D. Mazon (private communications).
19. E. Joffrin, J.F. Artaud, O. Barana et al., *Nucl. Fusion* **47**, 1664-1672 (2007).
20. P. T. Lang et al., 2003 Europhysics Conference Abstracts:Proc. 30th EPS Conf. on Controlled Fusion and Plasma Physics (St Petersburg, 2003) (Geneva: European Physical Society) vol 27A (ECA) paper P-1.129.
21. K. Tobita and S. Nishio (private communication).

REAL-TIME EQUILIBRIUM RECONSTRUCTION IN A TOKAMAK

Jacques Blum, Cédric Boulbe and Blaise Faugeras

Laboratoire J.-A. Dieudonné (UMR 6621)
Université de Nice Sophia-Antipolis, CNRS
Parc Valrose
06108 Nice Cedex 02
France
email: jblum or boulbe or faugeras @unice.fr

Abstract. This paper deals with the numerical reconstruction of the plasma current density in a Tokamak and of its equilibrium. The problem consists in the identification of a non-linear source in the 2D Grad-Shafranov equation, which governs the axisymmetric equilibrium of a plasma in a Tokamak. The experimental measurements that enable this identification are the magnetics on the vacuum vessel, but also polarimetric and interferometric measures on several chords, as well as motional Stark effect or pressure measurements. The reconstruction can be obtained in real-time using a finite element method, a non-linear fixed-point algorithm and a least-square optimization procedure.

Keywords: Tokamak, Grad-Shafranov, inverse problem, real-time, plasma equilibrium
PACS: 02.30.Jr, 02.30.Zz, 02.60.Pn, 02.70.Dh, 41.20.Gz

INTRODUCTION

The problem of the equilibrium of a plasma in a Tokamak is a free boundary problem in which the plasma boundary is defined either by its contact with a limiter or as being a magnetic separatrix. Inside the plasma, the equilibrium equation in an axisymmetric configuration is called Grad-Shafranov equation [1, 2, 3]. The right-hand side of this equation is a non-linear source which represents the toroidal component of the plasma current density.

An important problem is the identification of this non-linearity [4, 5, 6]. The aim of this paper is to present a method for real-time identification from experimental measurements, such as magnetic measurements on the vacuum vessel, polarimetric measurements (integrals of the magnetic field over several chords), MSE (Motional Stark Effect) and pressure measurements. The pressure is supposed to be isotropic. For the anisotropic pressure case, one can refer to [7].

The next section is devoted to the mathematical modelling of the equilibrium problem in axisymmetric configurations. The inverse reconstruction problem is adressed in the last section.

CP988, *Burning Plasma Diagnostics*, edited by F. P. Orsitto, G. Gorini, E. Sindoni, and M. Tardocchi

MATHEMATICAL MODELLING OF AXISYMMETRIC EQUILIBRIUM OF THE PLASMA IN A TOKAMAK

The equations which govern the equilibrium of a plasma in the presence of a magnetic field are on the one hand Maxwell's equations and on the other hand the equilibrium equations for the plasma itself.

The magnetostatic Maxwell's equations as follows are satisfied in the whole of space (including the plasma):

$$\begin{cases} \nabla \cdot B &= 0 \\ \nabla \times (\dfrac{B}{\mu}) &= j \end{cases} \tag{1}$$

where B represents the magnetic field, μ is the magnetic permeability and j is the current density. The first relation of (1) is the equation of conservation of magnetic induction and the second one is Ampere's Theorem.

The momentum equation for a plasma is

$$\rho \frac{du}{dt} = j \times B - \nabla p \tag{2}$$

where u represents the mean velocity of particles and ρ the mass density. At the resistive time-scale the first term can be neglected [8] and the equilibrium equation for the plasma is

$$\nabla p = j \times B \tag{3}$$

This equation (3) means that the plasma is in equilibrium when the force ∇p due the kinetic pressure p is equal to the Lorentz force of the magnetic pressure $j \times B$. We deduce immediately from (3) that

$$B \cdot \nabla p = 0 \tag{4}$$

$$j \cdot \nabla p = 0 \tag{5}$$

Thus for a plasma in equilibrium the field lines and the current lines lie on isobaric surfaces ($p = const.$); these surfaces, generated by the field lines, are called magnetic surfaces. In order for them to remain within a bounded volume of space it is necessary that they have a toroidal topology. These surfaces form a family of nested tori. The innermost torus degenerates into a curve which is called the magnetic axis.

In a cylindrical coordinate system (r, z, ϕ) (where $r = 0$ is the major axis of the torus) the hypothesis of axial symmetry consists in assuming that the magnetic field B is independent of the toroidal angle ϕ. The magnetic field can be decomposed as $B = B_p + B_\phi$, where $B_p = (B_r, B_z)$ is the poloidal component and B_ϕ is the toroidal component. From equation (1) one can define the poloidal flux $\psi(r,z)$ such that

$$\begin{cases} B_r &= -\dfrac{1}{r}\dfrac{\partial \psi}{\partial z} \\ B_z &= \dfrac{1}{r}\dfrac{\partial \psi}{\partial r} \end{cases} \tag{6}$$

Concerning the toroidal component B_ϕ we define f by

$$B_\phi = \frac{f}{r} e_\phi \tag{7}$$

where e_ϕ is the unit vector in the toroidal direction, and f is the diamagnetic function. The magnetic field can be written as:

$$\begin{cases} B & = & B_p + B_\phi \\ B_p & = & \frac{1}{r}[\nabla\psi \times e_\phi] \\ B_\phi & = & \frac{f}{r} e_\phi \end{cases} \tag{8}$$

According to (8), in an axisymmetric configuration the magnetic surfaces are generated by the rotation of the flux lines $\psi = const.$ around the axis $r = 0$ of the torus.

From (8) and the second relation of (1) we obtain the following expression for j:

$$\begin{cases} j & = & j_p + j_\phi \\ j_p & = & \frac{1}{r}[\nabla(\frac{f}{\mu}) \times e_\phi] \\ j_\phi & = & (-\Delta^*\psi) e_\phi \end{cases} \tag{9}$$

where j_p and j_ϕ are the poloidal and toroidal components respectively of j, and the operator Δ^* is defined by

$$\Delta^*. = \frac{\partial}{\partial r}(\frac{1}{\mu r}\frac{\partial .}{\partial r}) + \frac{\partial}{\partial z}(\frac{1}{\mu r}\frac{\partial .}{\partial z}) \tag{10}$$

Expressions (8) and (9) for B and j are valid in the whole of space since they involve only Maxwell's equations and the hypothesis of axisymmetry.

In the plasma region, relation (4) implies that ∇p and $\nabla\psi$ are colinear, and therefore p is constant on each magnetic surface. This can be denoted by

$$p = p(\psi) \tag{11}$$

Relation (5) combined with the expression (9) implies that ∇f and ∇p are colinear, and therefore f is likewise constant on each magnetic surface

$$f = f(\psi) \tag{12}$$

The equilibrium relation (3) combined with the expression (8) and (9) for B and j implies that:

$$\nabla p = -\frac{\Delta^*\psi}{r}\nabla\psi - \frac{f}{\mu_0 r^2}\nabla f \tag{13}$$

which leads to the so-called Grad-Shafranov equilibrium equation:

$$-\Delta^* \psi = r p'(\psi) + \frac{1}{\mu_0 r}(f f')(\psi) \tag{14}$$

where Δ^* is the linear elliptic operator given by (10) in which μ is equal to the magnetic permeability μ_0 of the vacuum.

From (9) it is clear that right-hand side of (14) represents the toroidal component of the plasma current density. It involves functions $p(\psi)$ and $f(\psi)$ which are not directly measured inside the plasma.

In the vacuum, the magnetic flux ψ satisfies

$$-\Delta^* \psi = 0 \tag{15}$$

The equilibrium of a plasma in a domain Ω representing the vacuum region is a free boundary problem. The plasma free boundary is defined either by its contact with a limiter D (outermost flux line inside the limiter) or as being a magnetic separatrix (hyperbolic line with an X-point, X). The region $\Omega_p \subset \Omega$ containing the plasma is defined as

$$\Omega_p = \{x \in \Omega,\ \psi(x) \geq \psi_b\} \tag{16}$$

where either $\psi_b = \max_D \psi$ in the limiter configuration or $\psi_b = \psi(X)$ in the X-point configuration (see Fig. 1)

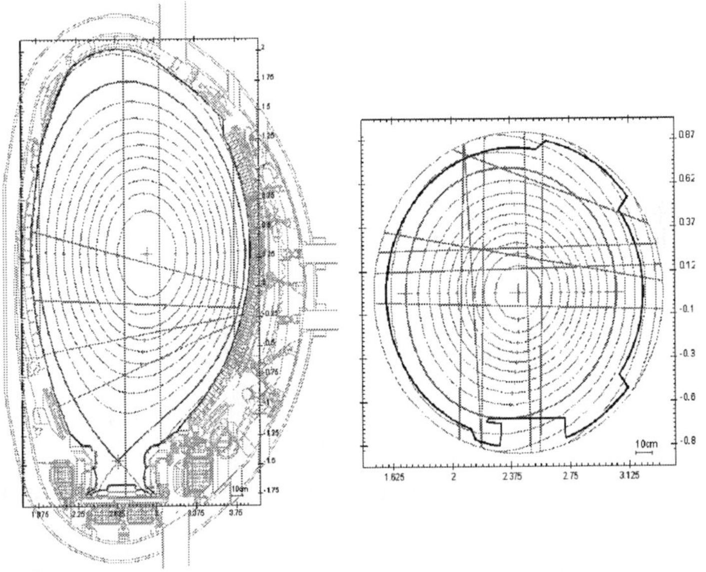

FIGURE 1. Definition of the plasma boundary (thick blue line). Left, JET (Joint European Torus) example, X-point configuration. Right, TORE SUPRA (the CEA-EURATOM Tokamak at Cadarache) example, limiter configuration (the limiter is represented by the black line). The thin blue lines represent iso-contours of ψ.

Assuming Dirichlet boundary conditions, h, are given on $\Gamma = \partial\Omega$ which is the poloidal cross-section of the vacuum vessel, the final equations governing the behaviour of $\psi(r,z)$ inside the vacuum vessel, are:

$$\begin{cases} -\Delta^*\psi = [rA(\bar{\psi}) + \frac{1}{r}B(\bar{\psi})]\chi_{\Omega_p} & \text{in}\,\Omega \\ \psi = h & \text{on}\,\Gamma \end{cases} \tag{17}$$

with $A(\bar{\psi}) = p'(\bar{\psi})$ and $B(\bar{\psi}) = \frac{1}{\mu_0}(ff')(\bar{\psi})$, $\bar{\psi} = \dfrac{\psi - \max_{\Omega}\psi}{\psi_b - \max_{\Omega}\psi} \in [0,1]$ in Ω_p (this normalized flux is introduced so that A and B are defined on the fixed interval $[0,1]$), χ_{Ω_p} is the characteristic function of Ω_p.

The aim of the following section of this paper is to provide a method for the real-time identification of the plasma current i.e. the non-linear functions A and B in the elliptic equation (17).

THE INVERSE PROBLEM

Experimental measurements

The given experimental measurements are:

- the magnetic measurements
  - $\psi(M_i) = h_i$ on Γ, given by the flux loops (see Fig. 2). Thanks to an interpolation between points M_i these measurements provide the Dirichlet boundary condition h.
  - $\frac{1}{r}\frac{\partial\psi}{\partial n}(N_i) = g_i$ on Γ, which corresponds to the component of the magnetic poloidal field, measured by the magnetic probes (see Fig. 2), which is tangent to the vacuum vessel. Indeed from Eq. (8) the tangential component of B_p is equal to the normal component $\frac{1}{r}\frac{\partial\psi}{\partial n}$ of $\frac{1}{r}\nabla\psi$.
- the polarimetric measurements which give the Faraday rotation of the angle of infrared radiation crossing the section of the plasma along several chords C_i:

$$\int_{C_i} n_e(\bar{\psi})B_{\parallel}dl = \int_{C_i} \frac{n_e(\bar{\psi})}{r}\frac{\partial\psi}{\partial n}dl = \alpha_i$$

  where n_e represents the electronic density which is approximately constant on each flux line, $B_{\parallel}$ is the component of the poloidal field tangent to C_i and $\frac{\partial}{\partial n}$ represents the normal derivative of ψ with respect to C_i.
- the interferometric measurements which give the density integrals over the chords C_i

$$\int_{C_i} n_e(\bar{\psi})dl = \beta_i$$

- the kinetic pressure measurements obtained from density and temperature measurements, for instance in the equatorial plane:

$$p(r,0) = p_d(r)$$

- and MSE (Motional Stark Effect) angle measurements taken at different points $x_i = (r_i, z_i)$:

$$m(B_r, B_z, B_\phi)_i = \gamma_i$$

with

$$\tan(m(B_r, B_z, B_\phi)) = \frac{a_1 B_r + a_2 B_z + a_3 B_\phi}{a_4 B_r + a_5 B_z + a_6 B_\phi}$$

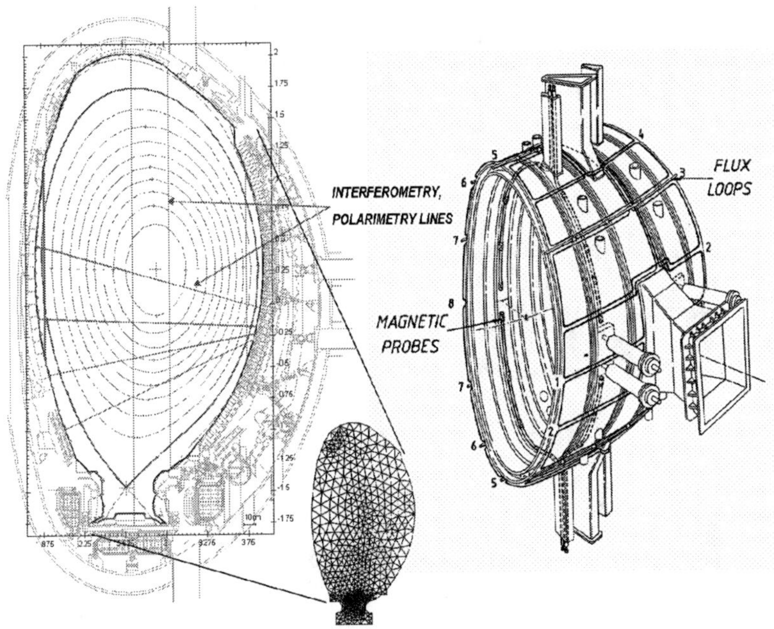

FIGURE 2. Left: the straight green lines represents the chords used for polarimetry and interferometry measurements. Right: part of the vacuum vessel. At the bottom middle an example of finite element mesh used for numerical simulations (see next Section).

Statement of the inverse problem

The numerical identification problem is formulated as a least-square minimization with a Tikhonov regularization. The cost function is defined as:

$$J(A, B, n_e) = J_0 + K_1 J_1 + K_2 J_2 + K_3 J_3 + K_4 J_4 + J_\varepsilon \tag{18}$$

with

$$J_0 = \sum_i \left(\frac{1}{r}\frac{\partial \psi}{\partial n}(N_i) - g_i\right)^2$$

$$J_1 = \sum_i \left(\int_{C_i} \frac{n_e}{r}\frac{\partial \psi}{\partial n} dl - \alpha_i\right)^2$$

$$J_2 = \sum_i \left(\int_{C_i} n_e dl - \beta_i\right)^2$$

$$J_3 = \int_{R_{min}}^{R_{max}} (p(r,0) - p_d(r))^2 dr$$

$$J_4 = \sum_i (m(B_r, B_z, B_\phi)_i - \gamma_i)^2$$

and K_1, K_2, K_3 and K_4 are weighting parameters enabling to give more or less importance to the corresponding experimental measurements [5].

The inverse problem of the determination of A and B is ill-posed. Hence a regularization procedure can be used to transform it into a well-posed one [9]. The Tikhonov regularization term J_ε constrains the function A, B and n_e to be smooth enough and reads:

$$J_\varepsilon = \varepsilon_1 \int_0^1 [A''(x)]^2 dx + \varepsilon_2 \int_0^1 [B''(x)]^2 dx + \varepsilon_3 \int_0^1 [n_e''(x)]^2 dx$$

where ε_1, ε_2 and ε_3 are the regularizing parameters.

It should be noticed that the electronic density n_e does not intervene in Eq. (17). However as soon as we want to use the polarimetric measurements it is necessary to include n_e (and hence interferometry) in the identification procedure. The inverse problem can finally be formulated as,

$$\begin{cases} \text{Find } A^*,\ B^*,\ n_e^* \text{ such that :} \\ J(A^*, B^*, n_e^*) = \inf J(A, B, n_e) \end{cases} \tag{19}$$

Numerical identification

Problem (17) is solved using a finite element method [10]. Let $H^1(\Omega)$ and $V = H_0^1(\Omega)$ denote the usual Sobolev spaces. The finite element approximation is based on the following weak formulation:

$$\begin{cases} \text{Find } \psi \in H^1(\Omega), \text{ such that } \psi = h \text{ on } \Gamma, \text{ and} \\ \displaystyle\int_\Omega \frac{1}{\mu_0 r} \nabla \psi \cdot \nabla v dx = \int_{\Omega_p} [rA(\bar{\psi}) + \frac{1}{r} B(\bar{\psi})] v dx \quad \forall v \in V \end{cases} \tag{20}$$

Classically Ω is approximated using triangles by a polygonal domain Ω_h, the space V is approximated by a space V_h of finite dimension n. A $P1$ finite element method is used, in which functions of V_h are affine over each triangle and continuous on the whole domain.

Let K denote the finite element stiffness matrix. Let us also (abusively) denote by $\psi \in \mathbf{R}^n$ the components of the magnetic flux fonction approximated in V_h.

The unknown functions A, B and n_e are approximated by a decomposition in a reduced basis $(\phi_i)_{i=1,...m}$

$$A(x) = \sum_i a_i \phi_i(x)$$
$$B(x) = \sum_i b_i \phi_i(x)$$
$$n_e(x) = \sum_i c_i \phi_i(x)$$

This basis can be made of different types of functions (polynomials, splines, wavelets, etc ...) [6]. Let u be the vector of $\mathbf{R}^{3m}$ defined by $u = (a_1, \ldots, a_m, b_1, \ldots, b_m, c_1, \ldots, c_m)$. With these notations the discretization of problem (20) can be written as follows:

$$\begin{cases} \text{Given } u \in \mathbf{R}^{3m}, \text{ solve the fixed} - \text{point equation} \\ \tilde{K}\psi = D(\psi)u + h \end{cases} \tag{21}$$

Where $D(\psi)$ denotes the $n \times 3m$ "plasma current matrix", and $\tilde{K}$ is the stiffness matrix modified in order to impose the Dirichlet boundary condition represented by h.

The discrete inverse optimization problem is:

$$\begin{cases} \text{Find } u \text{ minimizing :} \\ J(u) = ||C(\psi)\psi - k||^2 + u^T \Lambda u \\ \text{with } \psi \text{ satisfying (21)} \end{cases} \tag{22}$$

where $C(\psi)$ is the observation operator. The quantity $C(\psi)\psi$ represents the outputs of the model corresponding to the experimental measurements, given in a previous subsection, denoted by k. The matrix Λ represents the regularization terms. The first term of J in Eq. (22) corresponds to $J_0 + K_1J_1 + K_2J_2 + K_3J_3 + K_4J_4$ and the second to J_ε.

In order to solve this problem we use an iterative algorithm based on fixed-point iterations for Eq. (21) and the normal equation of Eq. (22).

Algorithm

At the n-th iteration, ψ_n and u_n are given. The non-linear mapping $u \mapsto \psi(u)$ is approximated by the affine relation

$$\psi = \tilde{K}^{-1}[D(\psi_n)u + h]$$

and the cost function to be minimized by

$$\begin{aligned} J(u) &= ||C(\psi_n)\psi - k||^2 + u^T \Lambda u \\ &= ||C(\psi_n)\tilde{K}^{-1}D(\psi_n)u + (C(\psi_n)\tilde{K}^{-1}h - k)||^2 + u^T \Lambda u \\ &= ||E_n u + F_n||^2 + u^T \Lambda u \end{aligned}$$

with obvious notations. The normal equation

$$(E_n^T E_n + \Lambda)u = -E_n^T F_n$$

is solved to update u_n to u_{n+1}. Then a fixed-point iteration for Eq. (21) enables the update of ψ_n to ψ_{n+1}

$$\psi_{n+1} = \tilde{K}^{-1}[D(\psi_n)u_{n+1} + h].$$

Since the algorithm is usually initialized with the equilibrium at a previous time step, two or three fixed-point iterations are usually enough to ensure convergence.

Equinox software

Based on the algorithm presented above, a C++ software, called EQUINOX [11, 12, 13] has been developed in collaboration with the Fusion Department at Cadarache, and has been implemented for JET (Joint European Torus) and for TORE SUPRA (the CEA-EURATOM Tokamak at Cadarache). Figure 3 shows a graphical output of Equinox. With all these techniques it is possible to follow the quasi-static evolution of the plasma equilibrium, either in TORE SUPRA or JET configurations, with free boundaries defined either by limiter contact or with an X-point. It is also possible to simulate ITER configurations.

CONCLUSION

We have presented an algorithm for the identification of the current density profile in Grad-Shafranov equation from experimental measurements.
The decomposition of the unknown functions $p'(\psi)$ and $ff'(\psi)$ in a reduced basis makes it possible to do the reconstruction in real-time.
The choice of this reduced basis must still be improved and optimized (robustness, precision, ...).
Real-time reconstruction makes possible future real-time control of the current profile [14].

AKNOWLEDGEMENTS

The authors are very thankful to K. Bosak for his enormous work in developing the real-time EQUINOX software and to E. Joffrin and S. Bremond for their contributions into the domain of real-time identification and control.

REFERENCES

1. H. Grad, and H. Rubin, "Hydromagnetic Equilibria and Force-free Fields," in *2nd U.N. Conference on the Peaceful uses of Atomic Energy*, Geneva, 1958, vol. 31, pp. 190–197.
2. V. Shafranov, *Soviet Physics JETP* **6**, 1013 (1958).

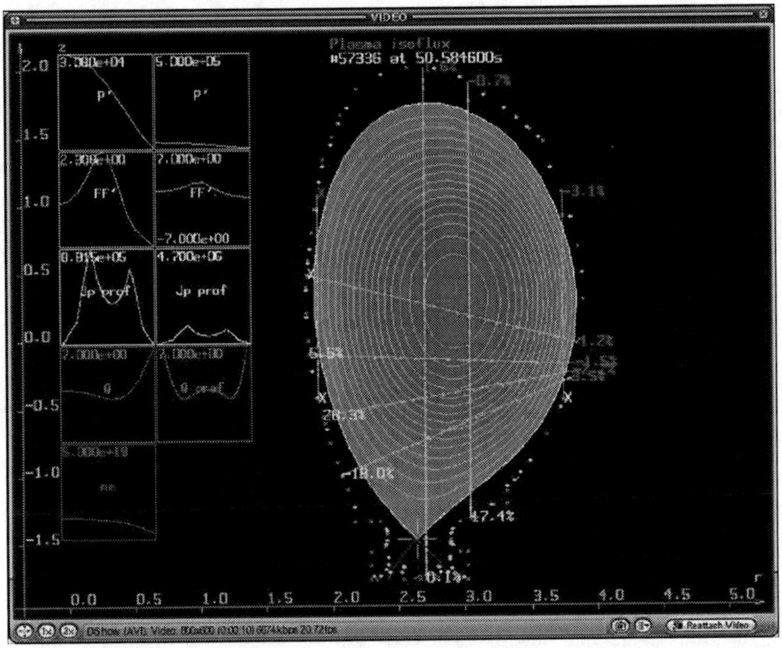

FIGURE 3. An output of EQUINOX. The plasma is in an X-point configuration. On the left column the identified p', ff' and n_e functions as well as the toroidal current density j and the safety factor q are displayed in terms of ψ and of r (in the equatorial plane).

3. C. Mercier, *The MHD approach to the problem of plasma confinement in closed magnetic configurations*, Lectures in Plasma Physics, Commission of the European Communities, Luxembourg, 1974.
4. L. Lao, J. Ferron, R. Geoebner, W. Howl, H. St. John, E. Strait, and T. Taylor, *Nuclear Fusion* **30**, 1035 (1990).
5. J. Blum, J. Lazzaro, J. O'Rourke, B. Keegan, and Y. Stefan, *Nuclear Fusion* **30**, 1475 (1990).
6. J. Blum, and H. Buvat, "An inverse problem in plasma physics : The identification of the current density profile in a Tokamak," in *IMA Volumes in Mathematics and its Applications, Large Scale Optimization with applications, Part 1: Optimization in inverse problems and design*, edited by Biegler, Coleman, Conn, and Santosa, 1997, vol. 92, pp. 17–36.
7. W. Zwingmann, and P. Stubberfield, "Equilibrium Reconstruction of Tokamaks Discharges with Anistropic Pressure," in *27th EPS Conference on Contr. Fusion and Plasma Phys.*, Budapest, 2000, URL `http://epsppd.epfl.ch/Buda/pdf/p3_062.pdf`.
8. H. Grad, and J. Hogan, *Physical review letters* **24**, 1337–1340 (1970).
9. A. Tikhonov, and V. Arsenin, *Solutions of Ill-posed problems*, Winston, Washington, D.C., 1977.
10. P. Ciarlet, *The Finite Element Method For Elliptic Problems*, North-Holland, 1980.
11. K. Bosak, *Real-time numerical identification of plasma in tokamak fusion reactor*, Master's thesis, University of Wroclaw, Poland (2001), URL `http://panoramix.ift.uni.wroc.pl/~bosy/mgr/mgr.pdf`.
12. J. Blum, K. Bosak, and E. Joffrin, "New applications of Equinox code for real-time plasma equilibrium and profile reconstruction for tokamaks," in *12th ICPP International Congress on plasma physics*, Nice, 2004, URL `http://fr.arxiv.org/abs/physics/0411181`.
13. K. Bosak, J. Blum, E. Joffrin, and F. Sartori, "Real-time plasma magnetic equilibrium reconstruction for tokamaks," in *EPS Conference on plasma physics*, Saint-Petersbourg, 2003.
14. E. Joffrin et al., *Plasma Physics and Controlled Fusion* **45**, A367–A383 (2003).

Real-Time Profile Control for Advanced Tokamak Operation

D. Mazon[1], D. Moreau[1], M. Ariola[2], G De Tommasi[2], J. Brzozowski[3], R. Felton[4], I. Nunes[5], F. Piccolo[4], F. Sartori[4], L. Zabeo[4] and JET EFDA contributors *

JET-EFDA Culham Science Centre, OX14 3DB, Abingdon, UK
[1] *Euratom-CEA Association, DSM-DRFC, Cadarache, 13108, St Paul lez Durance, France*
[2] *Euratom-ENEA-CREATE Association, Univ. Napoli Federico II, I-80125 Napoli, Italy*
[3] *Euratom-VR Association, Fusion Plasma Physics, EES, KTH, Sweden*
[4] *Euratom-UKAEA Association, Culham Science Centre, Abingdon, OX14 3DB, U. K.*
[5] *Euratom-IST Association, CFN/IST, Lisbon, Portugal*
* *See M.L.Watkins et al., Fusion Energy 2006 (Proc. 21st int. Conf. Chengdu IAEA, (2006)*

Abstract. Simultaneous control of the plasma shape, the magnetic and kinetic plasma profiles (such as the safety factor, q(x), and gyro-normalized temperature gradient, ρ_{Te}*(x), respectively) and the boundary flux is being investigated on JET, and has potential applications in the operation of ITER steady state advanced tokamak discharges. The control of radially distributed parameters was achieved for the first time on JET in 2004 [1-4]. The controller was based on the static plasma response only. The approach newly implemented on JET aims to use a dynamical plasma model, all the available heating and current drive (H&CD) systems, and the poloidal field (PF) system in an optimal way to achieve a set of requested magnetic and kinetic profiles. This paper describes the new model-based optimal profile controller which has been tested during the last 2007 experimental campaign. The controller aims to use the combination of heating and current drive systems - and optionally the PF system. First experimental results of current profile control obtained during the last 2007 JET campaign are presented.

Keywords: Tokamak, Internal transport Barrier, Real-time profile control, Model based controller.
PACS: 52-55.Fa, 02-30.Yy

STATE-SPACE PLASMA RESPONSE MODEL IDENTIFICATION

The structure of the model stems from a set of transport equations,

$$\mu_0 \frac{\partial j}{\partial t} = -\nabla \times \nabla \times E, \qquad \frac{\partial n}{\partial t} = -\nabla.\Gamma + S_n, \qquad \frac{3}{2}\frac{\partial (nT)}{\partial t} = -\nabla.Q + S_T \tag{1}$$

in which couplings are retained with no loss of generality. The system (Eq. 1) is linearized around an equilibrium reference state so that it can ultimately be cast in the generic form of a state space model. In doing so, the state variables appear naturally to be the variations of the internal poloidal magnetic flux, Ψ, and of the temperature, T, and the state space model reads:

CP988, *Burning Plasma Diagnostics*, edited by F. P. Orsitto, G. Gorini, E. Sindoni, and M. Tardocchi
2008 American Institute of Physics 978-0-7354-0507-3/08/$23.00

$$\partial\Psi/\partial t = A_{11}\Psi(t) + A_{12}T(t) + B_{11}P(t) + B_{12}n(t) + U.V_{ext}(t) \quad (2)$$

$$\varepsilon\,\partial T/\partial t = A_{21}\Psi(t) + A_{22}T(t) + B_{21}P(t) + B_{22}n(t) \quad (3)$$

with inputs $P = [P_{LH}, P_{NBI}, P_{ICRH}]$, the heating and current drive input powers, and V_{ext}, the surface loop voltage. The distributed variables $\Psi(x)$ and $T(x)$, where x is a radial coordinate, are projected onto a finite set of trial functions using a Galerkin scheme so that the original partial differential system of equations reduces to an ordinary linear differential system where U is known and $A_{i,j}$, $B_{i,j}$ are matrices of appropriate dimensions which are to be identified from experimental data. The small (constant) parameter, ε, represents the ratio between the energy confinement time and the characteristic resistive diffusion time ($\varepsilon \ll 1$).

TWO-TIME-SCALE APPROXIMATION

To take advantage of the small parameter ($\varepsilon \approx 0.05$ in JET), the control technique proposed here is based upon the theory of singularly perturbed systems and multiple-time-scale expansions [5]. We therefore seek two models of reduced orders, a slow model,

$$\partial\Psi/\partial t = A_s\Psi + B_s u_s \qquad \text{together with} \qquad T_s = C_s\Psi + D_s u_s \quad (4)$$

and a fast model ($\tau = t/\varepsilon$),

$$\partial T_f/\partial\tau = A_f T_f + B_f u_f \quad (5)$$

where $T=T_s+T_f$, and where u_s and u_f are the slow and fast components, respectively, of a vector, $u=u_s+u_f$, containing all the inputs (P, n and V_{ext}).

Having identified a set of relevant state variables, it can prove practical to apply the control to some output parameters which are more directly linked with MHD stability or internal transport barrier physics. The inverse of safety factor, $\iota(x)$, and gyro-normalized temperature gradient, $\rho_{Te}^*(x)$ [6], have been chosen and are thus introduced into the state-space model. As for $\Psi(x)$ and $T(x)$, a Galerkin approximation is used and in the following, the notations Ψ, T, μ and ρ will refer to the coefficients of the $\Psi(x)$, $T(x)$, $\iota(x)$ and $\rho_{Te}^*(x)$ expansions, respectively. Noticing that $\iota(x) \propto \nabla\Psi(x)$ and $\rho_{Te}^*(x) \propto \nabla Te(x)/\ Te(x)$, linearizing these expressions, differentiating the basis functions and assuming that the time variations of factors such as the toroidal magnetic field and toroidal magnetic flux are not essential and do not depend intrinsically on the power inputs, it appears relevant to seek a model with direct matrix relations between the Galerkin coefficients Ψ and μ, on one hand, and between T and ρ on the other hand. Within the two-time-scale approximation, this yields:

$$\mu = C_{\mu,\psi}.\Psi \quad \text{and} \quad \rho_s = C_{\rho,\psi}\Psi + D_{\rho,\psi}u_s \quad (\text{or} \quad \rho_s = C_{\rho\mu}\mu + D_{\rho\mu}u_s) \quad (6)$$

$$\text{and} \quad \rho_f = C_{\rho,T}.T_f \quad (7)$$

which complete the system (Eq. 4-5). The identified two-time-scale model is then used to construct and design a controller which can respond faster to rapid plasma events, while converging slowly towards the requested high performance plasma state (on the resistive time scale) [8].

Experimental results

The next step was to apply this controller to a real plasma discharge. In order to validate the controller, and as an intermediate step, we concentrated during the 2007 campaign on the control of the q-profile only. The main actuators were the LH, ICRH, NBI powers and the loop voltage. Open loop modulations of those actuators were performed around a reference state in order to identify the plasma model. A highly triangular (δ=0.45) configuration was chosen to perform both open and closed loop experiments with a toroidal magnetic field of 3T, a plasma current of 1.5MA and an average density of about 3.5×10^{19} m-^{3}.

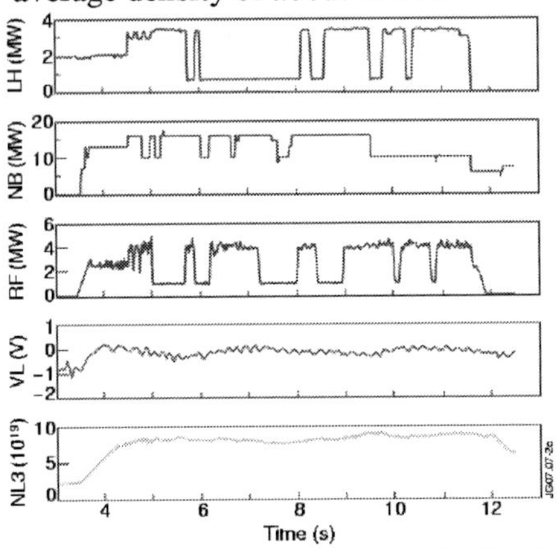

FIGURE 1. Input powers for modulated pulse #67876, loop voltage and line integrated density.

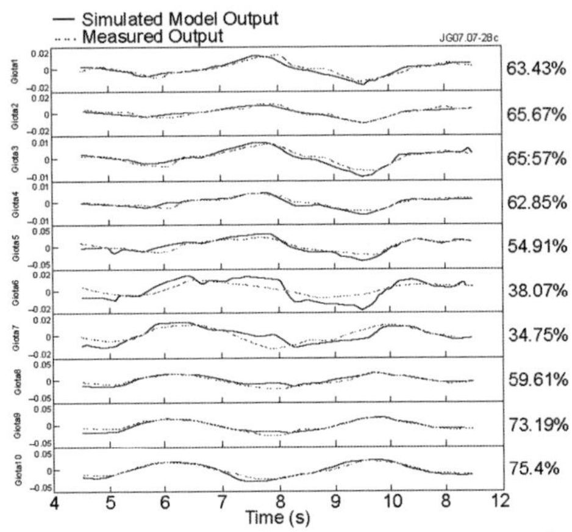

FIGURE 2 Comparison between reconstructed and measured iota profile for pulse #67876 at 10 fixed normalized radii.

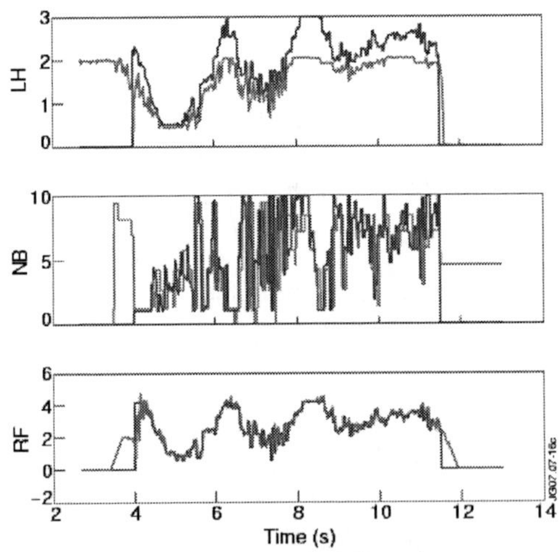

FIGURE 3. Requested and delivered powers (LHCD, NBI, ICRH) for the controlled pulse #70395

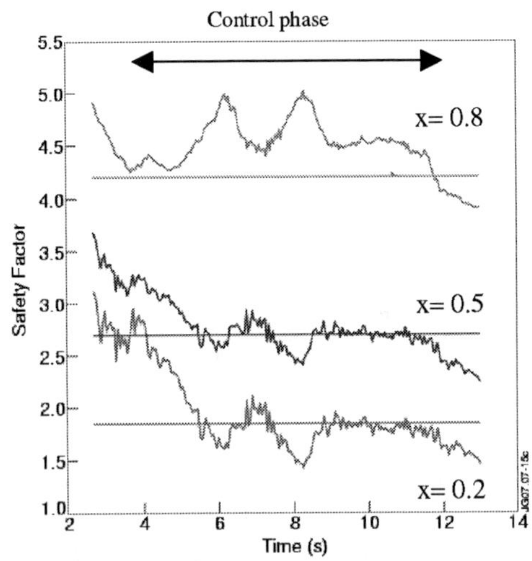

FIGURE 4.. Control of the safety profile at 3 normalised position using the 3 H&CD actuators.#70395. Vloop constant (32 mV/rad).

Examples of such modulation of the heating and current drive actuators are given in Fig. 1. As the q profile evolves on the resistive time scale, those modulations were relatively slow. In a same way, control of the loop voltage was possible due to the development of the JET Extreme shape controller [7, 9]. Comparison between the measured data and the reconstructed data using the identified model can be seen in Fig. 2. Good agreement was found and feedback control experiments of the current profile were then performed. Control of q profile was performed over a period lasting more than 7s using the 3 H&CD systems while keeping the loop voltage at a constant value (see Figs. 3 & 4), and for a shorter period with the loop voltage included in the loop (Fig. 5). An important step has been made in demonstrating the validity of this new controller and the potential for controlling the safety factor profile.

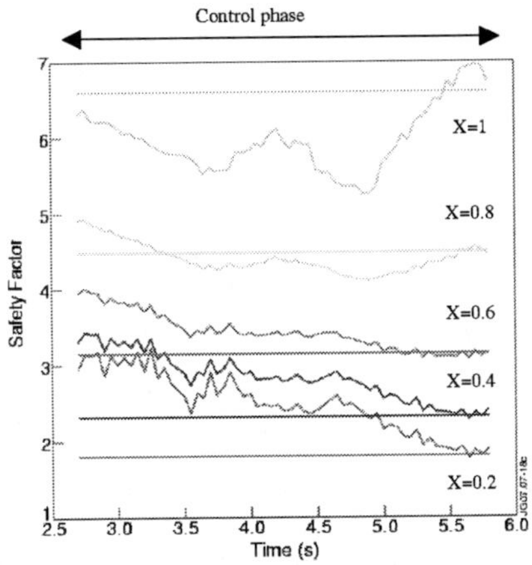

FIGURE 5. Control of the safety profile at 5 normalised position, x=1 (green), x=0.8 (cyan), x=0.6 (magenta), x=0.4 (blue), x=0.2 (red) using the 3 H&CD actuators.#70395.

ACKNOWLEDGEMENT

The work leading to this article was funded by the European Atomic Energy Community and is subject to the provisions of the European Fusion Development Agreement.

REFERENCES

1. Moreau. D., et al., "Development of Integrated Real-Time Control of Internal Transport Barriers in Advanced Operation Scenarios on JET", Fusion Energy 2004 (Proc. 20th Int. Conf. Vilamoura, 2004), IAEA-CSP-25/CD, IAEA, Vienna (2005), CD-ROM file EX_P2_5,
2. Laborde, L., et al., "A Model-Based Technique for Integrated Real-Time Profile Control in the JET Tokamak", Plasma Phys. Contr. Fus. **47** (2005) 155.
3. Mazon, D., et al., "Active Control of the Current Density Profile in JET", Plasma Phys. Contr. Fus. **45** (2003) L47.
4. Moreau, D., et al., "Identification of the Dynamic Plasma Response for Integrated Profile Control in Advanced Scenarios on JET", Proc. 33rd EPS Conf. on Plasma Physics, Roma (2006), paper P1.069.
5. Kokotovitch P.V., Singular Perturbation Methods in Control: Analysis and Design, Academic Press, London (1986).
6. Tresset, G., et al., "A Dimensionless Criterion for Characterising Internal Transport Barriers in JET", Nucl. Fusion **42** (2002) 520.
7. Ambrosino, G., et al., "A New Shape Controller for Extremely Shaped Plasmas in JET", Fusion Engineering and Design **66-68** (2003) 797.
8. Moreau. D., et al., "New Dynamic Model Approach for simultaneous profile control", 21st IAEA Fusion Energy Conference, Chengdu China (2006).
9. Ariola .M., et al., "Plasma Shape and Boundary Flux control at JET", 34nd EPS Conf. on Plasma Physics, Warsaw (2007), poster P5.140M.

Magnetic and X-ray Diagnostics for Plasma Position Measurements in the Ignition Experiment Ignitor

F. Bombarda[1], F. Alladio[1], G. Pizzicaroli[1], A. Licciulli[2], M. Fersini[2], D. Diso[3], E. Paulicelli[4]

[1]*Associazione Euratom-ENEA sulla Fusione, Frascati, Italy,*
[2]*Università di Lecce, Italy,*
[3]*SALENTEC, Lecce, Italy,*
[4]*Università di Bari, Italy*

Abstract. The ignition experiment Ignitor will produce high performance plasma regimes with neutron fluxes at the first wall comparable to those expected in power generating reactors (10^{15} n/cm^2/s at the first wall). An appreciable degradation of the inorganic insulators surrounding the conductors is expected and the measurement of basic plasma parameters, such as plasma current and position, by electromagnetic diagnostics can become problematic. Thus an R&D program, aimed at the development of effective and affordable devices for electromagnetic diagnostics with higher damage threshold, is being carried out. Prototype diagnostic coils of different shapes have been manufactured, in collaboration with SALENTEC and Università di Lecce. At the same time, an alternative plasma position control method is being explored, based on the diffraction and detection of the soft X-ray radiation emitted, mostly by Molybdenum impurities, near the top or bottom of the plasma column, where the distance of the plasma last closed magnetic surface from the wall is only few millimeters. The underlying principles and general layout of the new system is essentially an adaptation of a 2-dimensional curved crystal spectrometer.

Keywords: Ignitor, Magnetic Diagnostics, Plasma Position Control.
PACS: 52.70.-m

ELECTROMAGNETIC DIAGNOSTICS

The design of the full set of electromagnetic diagnostics for the Ignitor experiment [1] and their integration with the plasma chamber has been completed. All the in-vessel pick-up coils for B_p and B_T fluctuation measurements, saddle coils (for magnetic flux), Rogowsky's (for plasma current), and diamagnetic loops (for stored energy) will be attached to the vacuum chamber behind the Molybdenum first wall tiles. Burning plasma experiments pose demanding conditions on the insulation materials used for magnetic measurements: high neutron and γ radiation fluxes increase the conductivity, and can cause electromotive forces and thermoelectrical sensitivity; good insulation has to be maintained at the kV level, while the vacuum vessel can reach temperatures up to 470 K during baking operations; electromechanical stresses caused by disruption events have to be sustained. In the special case of the Ignitor experiment, which is the only device aiming at reaching ignition conditions, the coil design is challenged also by space requirements, with thin conductors (200μm) bent to small curvature radii (<6

CP988, *Burning Plasma Diagnostics,* edited by F. P. Orsitto, G. Gorini, E. Sindoni, and M. Tardocchi mm) and reduced overall sizes. Commercially available pre-insulated ceramic wires (e.g. Ceramawire Ni) are not able to meet all these requirements; nonetheless, they provide a good starting material. We have aimed at improving the insulation quality by submerging the coil in a bath of magnesium oxide compound at elevated refractoriness, chemically inert, and neutron radiation resistant. Prototype coils have been manufactured, of different design. In one case, the pre-insulated nickel coil is completely encased in a plastic alumina container, which can be cast in any shape, also curved ones. A suspension based on Magnesium Oxide (MgO > 87.8%) and inorganic bindings is poured into the container, where the coil is placed; the suspension is kept drying for 24 hours in air at room temperature, and a sintering procedure is carried out at 1100°C in inert atmosphere of N_2 or Ar (treatment in air causes an increased brittleness of the Ni wire due to oxidation). The cover is sealed on the alumina container by means of glazes with high vacuum tightness. To guarantee an optimal sealing, the glaze is also baked at 1100°C. For a second prototype, the Ni wire was initially wound on one of the NSTX original rectangular Macor support, which proved unsuitable to sustain the subsequent high temperature treatments. After impregnation with the liquid magnesia suspension, a thick layer of Mg oxide/fiber was applied, and baked at 1100 °C (Fig. 1). This material can then be machined and finally a vitreous coating is applied for vacuum tightness.

FIGURE 1. Rectangular coil prototype (NSTX design): initial winding, MgO wrapping, and detail of the 2D reinforcing material in Alumina Nextel 610.

New supports have been produced (see Fig. 2), in this case by modeling a block of sintered alumina by means of a numerically controlled machine.
The difficulties relative to the MgO impregnation are well illustrated by Fig. 3, which shows a roughly wound, double layer coil partially covered with MgO: the powdery consistency of the dried compound tends to easily detach from the wires. Finding proper manufacturing solutions is the subject of ongoing research activities.

FIGURE 2. Newly machined alumina support.

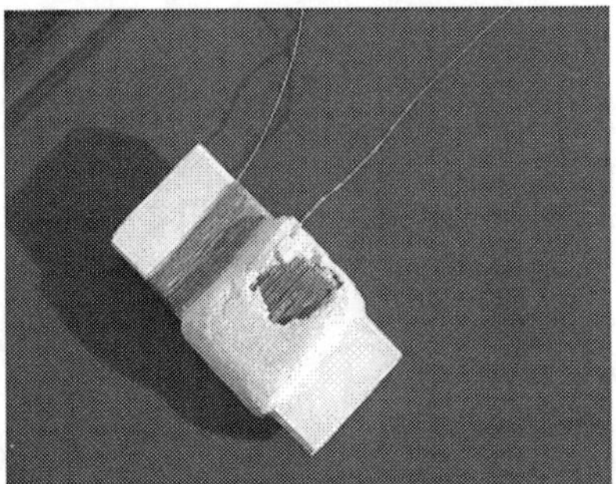

FIGURE 3. MgO over a double layer coil.

X-RAY DIAGNOSTICS FOR PLASMA POSITION CONTROL

The control of the plasma position is crucial in Ignitor as for every elongated configuration. The capability of the Poloidal Field Coils system to provide an effective vertical stabilization of the plasma column has been investigated using the CREATE_L response model [2], also taking into account the possible failure of the relevant electromagnetic diagnostics. An additional means for monitoring the (vertical) position of the plasma column may be provided by the detection of the soft X-ray radiation emitted at the edge. This is a combination of continuous bremsstrahlung and line radiation that, in the case of Ignitor, is emitted mostly from Molybdenum. The Ignitor geometry allows the lower or upper sections of the plasma to be viewed from the horizontal ports. For the purpose of plasma position control, the system needs to be sufficiently fast (>1 kHz) and possibly provide an output signal to the control system without additional inputs from other diagnostics.

A cylindrical Multilayer Mirror (MLM) placed at a suitable location inside the port can selectively diffract the radiation and focus it on a space resolving Gas Electron Multiplier (GEM) detector outside the vessel, not in direct view of the plasma, where the front-end electronics can be properly protected (Fig. 4). The high counting rates allowed by GEMs allow the possibility to detect any plasma movement with sufficient

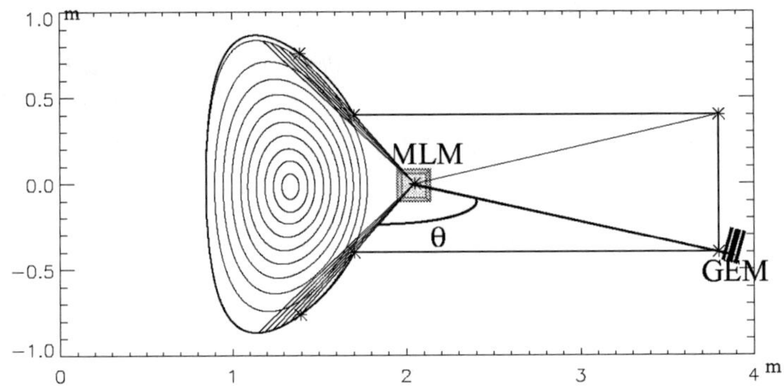

FIGURE 4. Cross section of the Ignitor horizontal port. A diffracting element (MLM) can be placed inside the port to observe the radiation emitted at the top (and/or bottom of the machine and focus it onto detectors outside the vacuum vessel.

time resolution to be used for real-time feedback control of the vertical plasma position. For radiation wavelengths of 20-50 Å thin Beryllium windows can still be used to separate the machine high vacuum from the rough vacuum in the detection arm attached to the external part of the port.

The cylindrical Johann mounting is self-focussing in the (meridian) diffraction plane. In order to obtain a space resolved image of the plasma in the poloidal direction, a horizontal slit needs to be located between the MLM and the detector. The use of spherical mirrors is prevented, in the case of Ignitor, by the specific port geometry, but toric ones could be considered in successive optimization of the system design. In this first step we have defined the essential optical parameters of the spectrometer. Two possible types of measurement are considered: the first adopts two symmetrical spectrometers, looking at the upper and lower regions of the plasma, respectively. The shift in vertical position is then deduced as a difference between the two signal intensities at each time. The second method relies on the time evolution of the signal from a single system pointed either to the top or the bottom. This will require a more sophisticated algorithm to distinguish signal changes associated with position shifts from variations of the background plasma parameters, but it offers the advantage of a simpler construction. The X-ray emission for Ignitor plasmas at ignition, both in the nominal extended limiter configuration and for a shifted plasma limited at the bottom of the first wall, but otherwise same plasma parameters, have been simulated and line brightness profiles of the radiation impinging on the MLM are shown in Fig. 5. The variation in brightness for the electric dipole transition of Mo^{14+} at 50.444 Å appears to be as large as a factor of two, thanks to the very high sensitivity of the line emission to the temperature profile along the line-of-sight. The full optical path to the detector remains to be completed in order to estimate the spectrometer throughput, nevertheless these first results indicate that X-ray measurements may provide a complementary method for plasma position control, and possibly replace the electromagnetic diagnostics in case of radiation-induced failure of the latter.

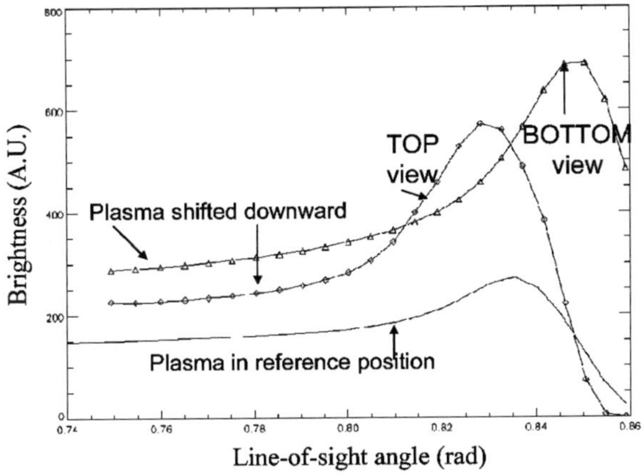

FIGURE 5. Brightness profiles for the Mo^{14+} dipole line at 50.444 Å, for the reference plasma position (solid line), and for the plasma column shifted to the bottom. The triangles refer to the lower view, the diamonds to the upper one.

ACKNOWLEDGMENTS

This work is supported in part by ENEA and Università di Bari of Italy, and by the US DOE

REFERENCES

1. B. Coppi, A. Airoldi, F. Bombarda, et al., *Nucl. Fusion* **41(9)**, 1253 (2001).
2. G. Ramogida, R. Albanese, F. Alladio, et al., *Proceed. of 24th SOFT Conference,* Warsaw (Poland), 2006, Paper P2-C-233.

MHD Activity in FTU Discharges With a Liquid Lithium Limiter

A. Botrugno[a], M.L. Apicella, P. Buratti, G. Mazzitelli, P. Smeulders

Associazione Euratom-ENEA sulla Fusione, Centro Ricerche di Frascati
Via Enrico Fermi 45, 00044 Frascati, Italy
a) E-mail address: antonio.botrugno@frascati.enea.it

Abstract. Experiments on plasma configurations with Liquid Lithium Limiter (LLL) have been performed on FTU to investigate liquid lithium facing the plasma. The liquid lithium surface with capillary porous system (CPS) has been inserted in the scrape off layer (SOL) of the FTU vacuum chamber in various plasma discharges. MHD activity in discharges with and without LLL has been investigated and compared. The main MHD effect observed with LLL is a reduction of the amplitude of tearing modes with poloidal (toroidal) number m=2 (n=1). The magnetic activities in FTU are analysed by means of a set of poloidal field pick-up coils. A total of 57 coils are installed at various toroidal and poloidal positions. In this paper we will describe in details the diagnostic apparatus and experimental results.

Keywords: MHD, magnetic diagnostics, lithium limiter.
PACS: 52.55.Tn ; 52.70.Ds

THE LIQUID LITHIUM LIMITER

The interest on liquid lithium as plasma facing material for fusion devices is growing up in the last years. Two main scientific issues are under investigation: a) the use of liquid lithium as first wall material b) the effects of a Li coated wall (lithization) on plasma performances. The DIII-D experiments with a lithium sample exposed to the lower divertor put in evidence the critical role of J x B forces on tearing the molten lithium off the container under high power exposure [1]. Subsequent T-11M experiments pointed out the possibility to overcome this problem, by using a lithium confinement based on capillary porous system (CPS) [2].

Recently a liquid lithium limiter (LLL) based on CPS has been installed on FTU in the framework of the joint program between ENEA and TRINITI & Red Star aimed to investigate liquid material facing the plasma. The LLL is showed in Fig 1. It is composed by three similar units based on a mesh of capillaries of stainless steel 304, with pore radius 15 μm and wires diameter 30 μm. The dimensions of each module of the LLL are 100 mm and 34 mm in poloidal and toroidal direction, respectively. Lithization is performed during the discharge by inserting the LLL about 2 cm away from the last closed magnetic surface (LCMS) in the shadow of the main TZM toroidal limiter. Li atoms are produced on LLL during the discharge by physical sputtering and evaporation and then spread on the vacuum chamber walls by the discharge itself.

CP988, *Burning Plasma Diagnostics*, edited by F. P. Orsitto, G. Gorini, E. Sindoni, and M. Tardocchi

Several experiments have been performed with LLL including various ohmic plasma discharges characterized by the following parameters: toroidal magnetic field $B_T \sim 6$ T; plasma current $I_p \sim 0.5 - 0.7$ MA; average electron density from 0.2 up to 2.6 10^{20} m^{-3}. The physical effects of lithization on plasma characteristics are very similar or even better than those of boronization. With lithium, the effective ion charge (Z_{eff}), the radiation losses and the wall particle recycling are reduced leading to improved plasma operations [3]. In this paper, we will focus on differences in MHD activity between plasma discharges with pure metallic walls and lithized walls. In the third section an overview of magnetic diagnostics used to detect MHD activity is presented.

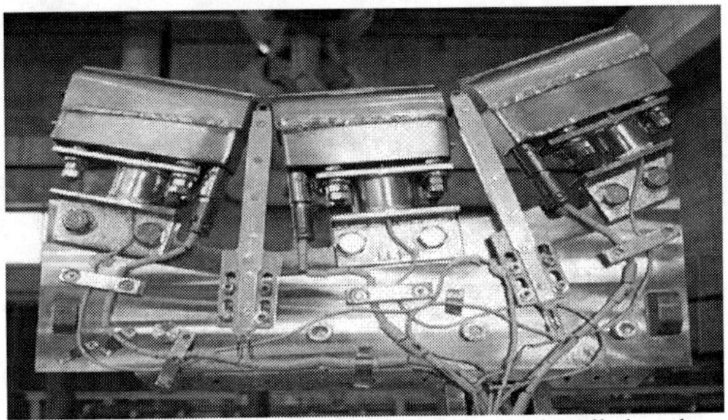

FIGURE 1. Photograph of three units of lithium limiter installed on the support used for introducing it inside FTU.

ANALYSIS OF MHD ACTIVITY WITH LLL

Comparison between discharges after lithization and discharges with lithium-free vessel evidences a stabilizing effect of lithium at low electron density ($\langle n_e \rangle \sim 0.5$ 10^{20} m^{-3}). Further analysis and higher statistic would be necessary to assess if the higher electron temperature, generally observed at plasma boundary with LLL, could improve the mode stability and which would be the acting mechanisms.

An example of this behaviour is represented by shots number 28474 and 28544, which are characterized by the same $B_T \sim 6$ T, $I_p \sim 0.5$ MA, and very similar electron density ($\langle n_e \rangle \sim 0.6$ 10^{20} m^{-3}) and temperature ($T_{e0} \sim 2$ keV) as function of time. Fig. 2 shows the results of Fourier analysis of signals detected by pick-up coils. The amplitude of oscillation is displayed in scale of grey, from white (low amplitude) to black (high amplitude) as a function of frequency and time. The Fourier analysis shows a tearing mode with frequency between 6 and 7 KHz in the shot with lithium-free vessel (n. 28474) and a very low MHD activity in the shot with lithium (n. 28544). Moreover, the mode analysis pointed out that the tearing mode has poloidal (toroidal) number m=2 (n=1).

Another example is given in Fig. 3, which shows the comparison between shots number 28482 and 28556 which have the same $B_T \sim 6$ T, but higher current ($I_p \sim 0.75$ MA) than previous shots. They have very similar electron density ($\langle n_e \rangle \sim 0.6$ 10^{-20} cm^{-3}) and temperature ($T_{e0} \sim 2.6$ keV), too. The Fourier analysis shows an evident

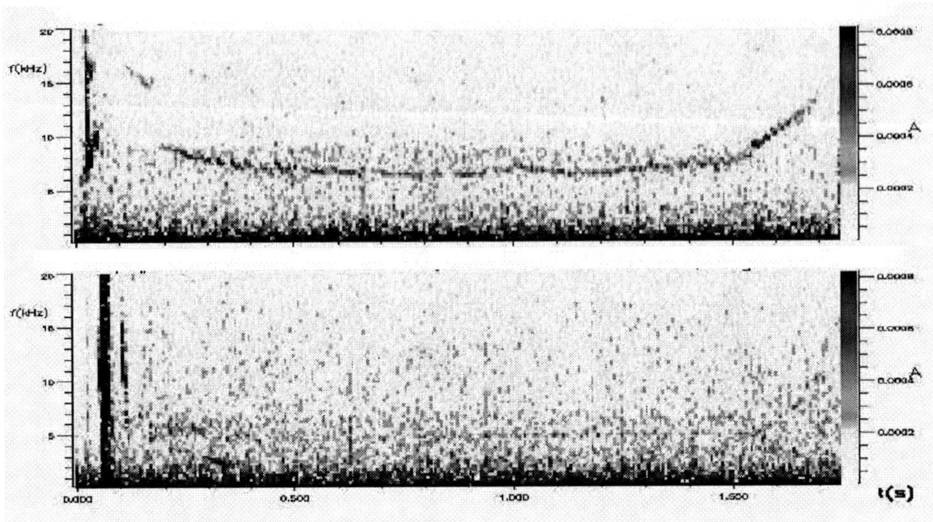

FIGURE 2. Spectrograms of a magnetic pick-up coils for shots n. 28474 (up) and n. 28544 (down).

tearing mode with poloidal (toroidal) number m=2 (n=1) in shot without lithium (n. 28482), while the same tearing instability is strongly reduced in shot with lithium (n. 28556).

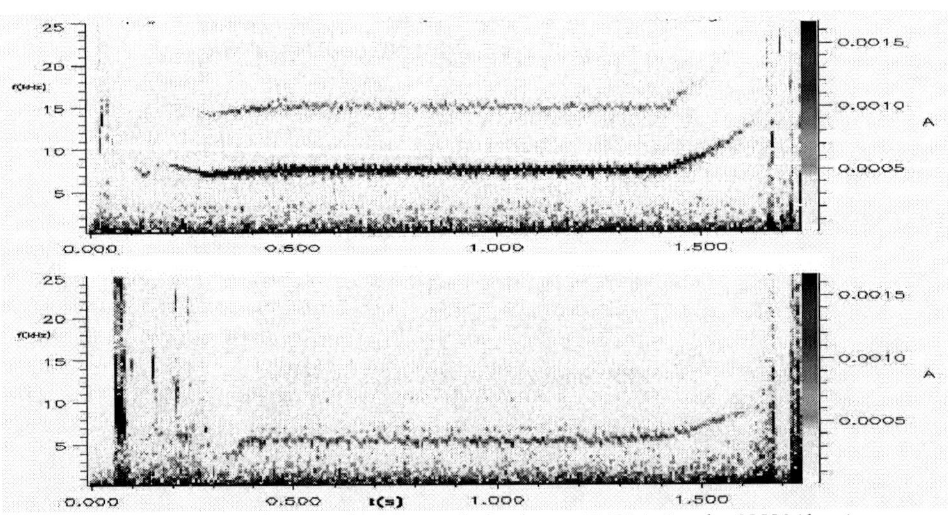

FIGURE 3. Spectrograms of a magnetic pick-up coils for shots n. 28482 (up) and n. 28556 (down).

Unfortunately, this kind of analysis is not easily extendible to discharges with medium and high density because of poor number of discharges with similar characteristics. As an indicative example at high density, the shots number 28847 and 29919, which have $B_T \sim 6$ T and $I_p \sim 0.7$ MA, can be compared in the temporal range

0.5–1.0 sec. only. Even in this case, the tearing instability at 6 KHz with poloidal (toroidal) number m=2 (n=1) in shot without lithium (n. 28847) is strongly reduced by the presence of lithium (n. 29919).

The intensity of the instabilities is reported in Tab.1 with other characteristics of the discharges. The ability of lithium to reduce the instabilities grows at high toroidal current; on the contrary, there are no evident correlations between reduction of instability and core electron temperature or electron density. However, we need a richer statistics to determine a real tendency of the discharges.

The stabilizing effects of the lithium can be related to the modification of the plasma edge temperature, this effect is now under investigation.

TABLE 1. Intensity of Instability.

Toroidal Magnetic Field (T)	Toroidal Current (MA)	Electron Density (10^{20} m^{-3})	Electron Temperature (KeV)	Intensity of instability (arbitrary units) (Shot without lithium)	Intensity of instability (arbitrary units) (Shot with lithium)
6.0	0.50	0.6	1.9 - 2.2	1.2 (28474)	0.6 (28544)
6.0	0.75	0.6	2.6 - 2.8	16 (28482)	3 (28556)
6.0	0.70	2.4 - 2.6	1.0	8 (28847)	2 (29919)

MAGNETIC DIAGNOSTICS ON FTU

The magnetic activities in FTU are analysed by means of a set of pick-up coils. A total of 57 poloidal field pick up coils are installed at 6 different toroidal and 34 different poloidal positions inside the ports behind the toroidal limiter structure and the poloidal limiter. The coils are electrically isolated from the electronics and the data acquisition through special designed air-core transformers with a tertiary loop that compensates for stray field pickup. These air-core transformers cut off frequencies below 1 KHz, which has an additional advantage that the growth of large MHD modes, which frequency slows down in time, does not so easily lead to saturated signals in the ADC's. 16 channels are recorded at present by the fast data acquisition system of National Instruments over 1.8 s at 500 KHz and filtered at 300 KHz for anti-aliasing. This system has allowed to find a new type of plasma mode between 40 and 80 KHz [4] and also broad-band plasma turbulence during additional heating. A large number of coils have been installed at one toroidal position but at various poloidal positions in order that the deformation of the pitch of the field line on which the mode resides can be measured. This deformation is caused by the combination of the external vertical magnetic field and of the Shafranov shift and is usually called the theta-star effect. The purpose of this particular coil configuration is to obtain this information of the local Shafranov shift of the rational q-surface and hence to be able to extract additional information on the current and pressure profiles with the aid of equilibrium codes. This is in addition of course to the standard information on mode behaviour.

REFERENCES

1. J. N. Brooks et al., *Fusion Sci. Technol* **47**, 2005, p. 669.
2. V. A. Evtikhin, I.E. Lyublinski, et al., *Plasma Phys. Cont. Fusion* **44**, 2002, p. 955.
3. M. L. Apicella, G. Mazzitelli, et al., *Journal of Nuclear Materials* **363-365,** 2007, p. 1346.
4. P. Buratti, P. Smeulders et al, *Nucl. Fus.*, **45**, 2005, p. 1346

Development and Testing of in-Vessel Equilibrium Pick-up Coils for ITER

S. Peruzzo, G. Chitarin, S. Dal Bello, R. S. Delogu, L. Lotto, A. Sottocornola, C. Taccon

Consorzio RFX, Associazione EURATOM-ENEA sulla fusione, Corso Stati Uniti 4, 35127 Padova, Italy

Abstract. A new concept of magnetic pick-up coil has been developed, aimed at reducing radiation induced thermoelectric effects which could affect the quality of the signals of the magnetic diagnostic for ITER. A winding made of a ceramic-coated conductor and impregnated with ceramic filler is proposed as an alternative to the Mineral Insulated Cable, presently indicated in ITER reference design. Some prototypes of the new in-vessel pick-up coils have been designed, built and tested, to assess the thermoelectric characteristics of the coils.

Keywords: magnetic diagnostics; pick-up coils.
PACS: 28.52.Lf; 52.70.Ds

INTRODUCTION

The magnetic diagnostics for ITER plasma control must be installed in a hostile environment characterized by severe neutron irradiation and plasma heat loads. The heat generation due to neutron irradiation is the cause of thermoelectric phenomena such as RITES and TIEMF [1, 2]; these effects can produce steady state spurious signals, which could be the main and dominant cause of errors for the magnetic measurements of ITER, up to estimated values larger than 30%. New concepts for the construction of pick-up coils are therefore essential to reduce the thermal gradient in the coil and consequently the spurious voltage. During the design review carried out during 2005 [3], a preliminary design based on flexible ceramic-insulated conductor was proposed in order to keep the temperature difference within the coil below 10°C, which induces a measurement error of the order of 1%, acceptable for a standard operation of pick-up coils. Consequently a R&D activity on pick-up coil prototyping has been launched in the framework of EFDA Technology Workprogramme.

DESIGN AND MANUFACTURE OF PROTOTYPE COILS

The design of the pick-up coil prototypes and the relative impregnation system started from the conceptual idea to develop a process similar to the manufacture of coils insulated with fiberglass and impregnated with resin. Figure 1a shows the design concept of the pick-up coil, with the following components:

- 0.5 mm diameter flexible ceramic-coated conductors (thin inter-turn insulation);
- fluid ceramic filling ("potted") for improved thermal conductivity between turns and to the external Case;

CP988, *Burning Plasma Diagnostics*, edited by F. P. Orsitto, G. Gorini, E. Sindoni, and M. Tardocchi

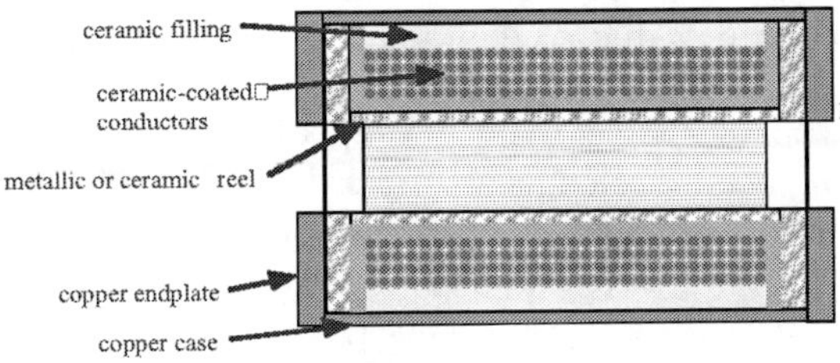

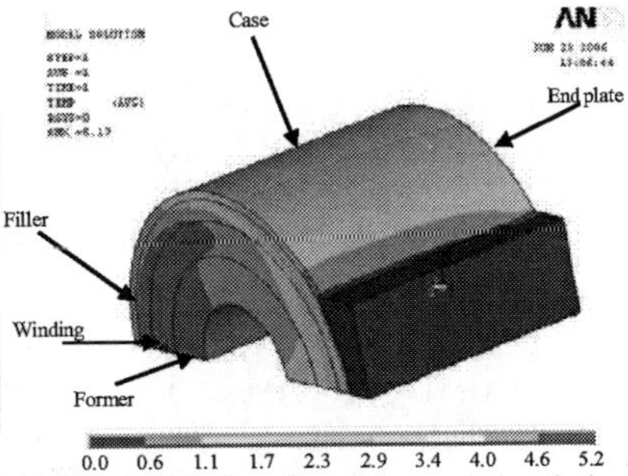

FIGURE 1a. Design concept of the in-vessel pick-up coil. **FIGURE 1b** Results of thermal simulation.

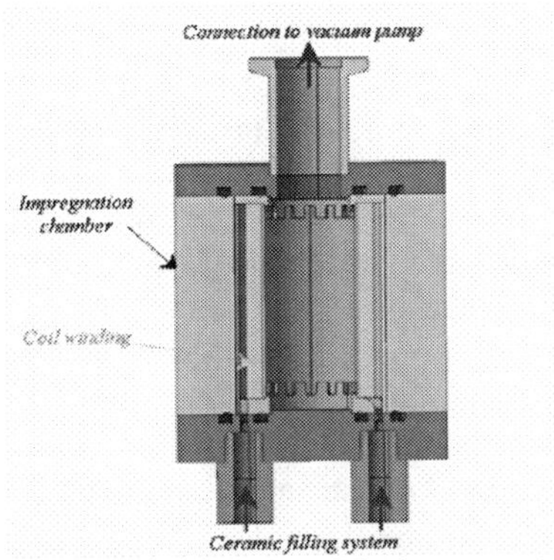

FIGURE 2a. Scheme of the impregnation system. **FIGURE 2b** Impregnation chamber

- thin stainless-steel (or machinable ceramic) Reel;
- copper external Case and End-plate providing good thermal anchor to the vessel (properly slotted to avoid electromagnetic shielding effect);
- ground insulation only if necessary (plasma sprayed or fiberglass tape).

Figure 1b reports the results of thermal analysis obtained with a 3D finite element model of the sensor, with a similar approach described in [3]. With respect to the previous analysis, a different geometry of the Case was implemented in order to obtain a better heat transfer from the sensor to the vessel. The best solution obtainable with this modification, and a proper choice of materials, provides a maximum temperature drop within the coil of 5°C.

The basic idea for the impregnation system was based on the use of low viscosity ceramic material to be inserted through coil winding in vacuum conditions, to avoid air entrapment, and to be cured in a proper tank, to guarantee good adhesion of ceramic to wire and structures. Figure 2a shows the CAD model of the impregnation system, based on an impregnation chamber, connected to vacuum and filling auxiliary systems, tight fitting the coil. This impregnation process requires a semi-permeable former, stiff enough to sustain the winding, able to allow vacuum pumping from the inner side, but not permeable to ceramic flowing from the outer side.

Figure 2b shows the impregnation chamber manufactured, made of Plexiglass in order to allow an easy visual inspection during the impregnation process. The top flange was connected to a rotary pump able to create a vacuum of approximately

10^{-2} mbar. The filling system was based on a simple nozzle manually operated. The design of the impregnation chamber, in particular the number of nipples in the bottom flange connected to the filling system, has been adjusted during the tests to guarantee a uniform insertion of the ceramic.

The coil Reel was made of a slotted stainless steel cylinder, to guarantee sufficient stiffness for the winding, with a thin metallic cloth spot welded on the external surface, to assure uniform penetration of the ceramic.

A set of 10 prototypes were realized with different wires (bare steel wires and ceramic coated wires), different winding procedures (manual or automatic) and different impregnation procedures (during or only after winding).

TEST OF PROTOTYPES

Preliminary tests on filler material and ceramic-coated conductors

Several ceramic-based fluid materials are commercially available for high temperature and vacuum compatible applications. A sub-set of materials already used for similar applications for RFX and JET [4, 5] was subjected to preliminary tests aimed at studying the capability of liquid ceramic (at different dilution) to penetrate through a metallic net (of different wire thickness and openings), simulating the coil winding and former structure. Ceramic adhesive Ceramabond™ 503, produced by Aremco Products Inc. was finally selected for prototype construction, due to the satisfactory fluidity and possibility to vary the dilution of the material.

The choice of ceramic coated conductors was marked by the limited number of suppliers. Two batches of conductor were procured by different suppliers. Unfortunately during preliminary insulation resistance tests, performed to assess the reliability of the insulation, both conductors evidenced poor electrical characteristics, with breakdown voltage to ground around 100 V and insulation resistance of 1 m long samples below 100 MΩ, which are unacceptable for the proposed applications (and not in conformity with respect to the specifications). The conductors were anyway used for the manufacture of prototypes to assess the feasibility of the impregnation process, as described in the following Section. Nevertheless the possibility to obtain a conductor with adequate and reliable electrical properties is still an open issue that should be demonstrated with further tests.

Filler penetration test

During the R&D several prototypes have been realized and tested modifying the Reel geometry, the wire type, winding and impregnation method. Once defined the best filler material and procedure, some prototypes were put through out-gassing tests and micrographic analysis of coil section, to test the uniformity and reliability of impregnation. The following conclusions can be drawn from test results:

- The obtained uniformity and density of ceramic in the outer region of the winding pack is acceptable (Fig. 3a); this should guarantee a satisfactory and repeatable thermal path of the winding towards the external Case which represents the heat-sink of the probe;

- The presence of the filler material between turns and between layers is not clearly quantifiable and not repeatable (Fig. 3b); nevertheless, the obtained winding packing seems to be adequate to provide an acceptable thermal contact between turns. This assumption shall be confirmed by thermal tests that could be performed with conductors having adequate electrical properties;
- During the curing process of the ceramic material (up to 370 °C, following the supplier instructions) some superficial longitudinal cracks appeared (Fig. 3a), probably as an effect of differential thermal deformation during solidification between the ceramic material and the metallic Reel. This problem should be overcome with a proper tuning of the curing procedure.

FIGURE 3a. Coil after curing process.

FIGURE 3b micrographic inspection of a coil section

The poor electrical properties of the ceramic-coated conductors prevented the thorough assessment of the new concept of magnetic pick-up coil. Nevertheless the tests performed gave confidence on the feasibility of the impregnation process, even though this should be significantly improved together with the reliability of the ceramic coating of the conductors. Further tests with this aim are presently in progress.

ACKNOWLEDGMENTS

This work, supported by the European Communities under the contract of Association between EURATOM/ENEA, was carried out within the framework the European Fusion Development Agreement. The views and opinions expressed herein do not necessarily reflect those of the European Commission. The authors would like to acknowledge A. Barzon, F. Degli Agostini, M. Fincato and L. Ranzato (RFX Team) for the technical support.

REFERENCES

1. G. Vayakis, at al. "Radiation-induced thermoelectric sensitivity (RITES) in ITER prototype magnetic sensors" Review of Scientific Instruments Volume 75, Number 10, October 2004
2. L. Vermeeren, M. Wéber, "Induced voltages and currents in copper and stainless steel core mineral insulated cables due to radiation and thermal gradients" , *Fusion Engin. and Design* (2007) in press.
3. G. Chitarin, L. Grando, S. Peruzzo, C. Taccon, "Design developments for the ITER in-vessel equilibrium pick-up coils and halo current sensors", *Fusion Engineering and Design* (2007) in press.
4. G. Serianni, et al., "Development, tests, and data acquisition of the integrated system of internal sensors for RFX", *Rev. Sci. Instrum.* Vol 75, No 10 (2004), 4338-4340.
5. S. Peruzzo, et al., "Status of design and manufacture of the Upper Coils and Outer Poloidal Limiter Coils subsystems for the JET-EP magnetic diagnostic", *Fus. Engin. and Design* 74 (2005) 733-738

Real-time Magnetic Field Pitch Angle Estimation With a Motional Stark Effect Diagnostic Using Kalman Filtering

R. Coelho* and D. Alves* and JET EFDA contributors[1]

JET-EFDA Culham Science Centre OX14 3DB, Abingdon, UK
**Associação Euratom/IST, Centro de Fusão Nuclear, 1049-001 Lisboa, Portugal*

Abstract. The real-time amplitude estimation of selective harmonics from an Avalanche Photo Diode (APD) signal of a Motion Stark Effect diagnostic is addressed using a Kalman filter. The proposed technique is shown to be much more robust and provide less noisy estimates than a lock-in amplifier scheme. In addition, the negative impact of Edge Localised Modes (ELMs) is minimized, reducing significantly the biasing in the amplitude estimation and ultimately allowing for the pitch angle estimation in the vicinity of the ELM. The inherent biasing in the amplitude estimation due to the 50Hz modulation in the NBI power grid is also easily circumvented with such a technique, rendering dispensable any further filtering of the data.

Keywords: Stark Effect, real-time filtering, magnetic field pitch angle.
PACS: 52.70Kz, 52.25Os, 52.55Fa

INTRODUCTION

The real-time control of a burning plasma confined within a Tokamak device such as ITER remains a primary goal in fusion research. In particular, the characterisation of the safety factor q(R) profile is of great concern since it is a valuable indicator of the operating scenario and of potentially disruptive plasma instabilities. The Motional Stark Effect (MSE) diagnostic plays a leading role in the determination of the poloidal magnetic field in neutral beam heated Tokamak discharges, by measuring the polarisation angle of the broadened D_α emitted light (σ or π components) due to the local $\mathbf{v}_b \times \mathbf{B}$ electric field [1]. Using two photo-elastic modulators (PEMs) operating at frequencies ω_1 and ω_2, a linear polariser and an avalanche photo diode (APD), polarisation angle modulations are converted into amplitude modulations of electric signals. At JET, the MSE system operates with 20kHz and 23kHz modulators and the magnetic pitch angle (γ), obtained from the $+\pi$ component, is calculated from all 25 spatial channels, covering the full outboard minor radius, according to

$$\tan(2\gamma(t)) = \frac{C_{21}A_{DC}(t) + C_{22}A_{23}(t) + C_{23}A_{46}(t) + C_{24}A_{40}(t)}{C_{11}A_{DC}(t) + C_{12}A_{23}(t) + C_{13}A_{46}(t) + C_{14}A_{40}(t)} \qquad (1)$$

[1] See the Appendix of M.L.Watkins et al., Fusion Energy 2006 (Proc. 21st Int. Conf. Chengdu, 2006) IAEA, (2006)

CP988, *Burning Plasma Diagnostics*, edited by F. P. Orsitto, G. Gorini, E. Sindoni, and M. Tardocchi
2008 American Institute of Physics 978-0-7354-0507-3/08/$23.00

where $A_{ij}(t)$ denotes the time dependent amplitude of a particular harmonic of the APD signal and C_{ij} are diagnostic calibration coefficients [2]. The determination of the relevant amplitudes is conventionally carried out with software/hardware lock-in amplifier schemes [3]. Although it is a good compromise for offline analysis, such a scheme may strive to overcome some of the typical obstacles found in a real-time analysis, namely, the impact of large ELMs (typically type-I), an imperfect phase compensation in the phase lock-in scheme and eventually the contamination by a 50Hz component arising from the power supplies. To circumvent such obstacles, in this paper, an approach founded on Kalman filtering techniques is adopted for the real-time estimation of the APD harmonics involved in the determination of the magnetic field pitch-angle.

REAL-TIME FILTERING MODEL

The standard Kalman filter is a predictor-corrector, optimal least square linear estimator for the time advance of some state variable x_k ($x_k \in \mathrm{IR}^n$ with k indexing time samples) [4]. Given an estimate $\hat{x}_{k-1}$, the *a priori* estimate at sample k (denoted by $\hat{x}_k^-$) is derived from the linear process model that is assumed for the state variable. Since a lock-in amplifier is intended to estimate the amplitude of periodic signals immersed in noise, the process model that needs to be considered is a periodic signal generator [5], modeled through a rotation matrix, i.e.

$$\hat{x}_k^- = F\hat{x}_{k-1} \tag{2}$$

$$F = \begin{bmatrix} F_1 & & & \\ & F_2 & & \\ & & \ddots & \\ & & & F_N \end{bmatrix} \tag{3}$$

$$F_i = \begin{bmatrix} \cos(\Delta\theta_i) & -\sin(\Delta\theta_i) \\ \sin(\Delta\theta_i) & \cos(\Delta\theta_i) \end{bmatrix} \tag{4}$$

where F_i is a 2×2 rotation matrix with angle given by $\Delta\theta_i = 2\pi\omega_i/\omega_s$ and $\omega_s/2\pi$ is the sampling frequency. The block matrix F encompasses all the frequencies that have a relevant amplitude in the APD signal. For JET data, this includes, besides the DC, 23, 40 and 46kHz components, several beating combinations resulting from the crystal retardance and PEM modulation process [1,2].

The *a posteriori* estimate $\hat{x}_k$ is a correction to the *a priori* estimate $\hat{x}_k^-$, taking into account the difference between the real (z_k is the APD signal at time sample k) and the *a priori* estimate ($H\hat{x}_k^-$ where H is a transfer matrix) of the measurements, weighted by a *filter gain* K_k:

$$\hat{x}_k = \hat{x}_k^- + K_k\left(z_k - H\hat{x}_k^-\right) \tag{5}$$

where the *filter gain* K_k minimizes the *a posteriori* error covariance matrix, denoted by $\hat{P}_k = E\left[(x_k - \hat{x}_k)(x_k - \hat{x}_k)^T\right]$, yielding

$$K_k = P_k^- H^T \left(H P_k^- H^T + R\right)^{-1} \tag{6}$$

The *a priori* and *a posteriori* estimated covariance matrices P_k^- and P_k are given, respectively, by

$$P_k^- = F P_{k-1} F^T + Q \tag{7}$$

$$P_k = \left(I - K_k H\right) P_k^- \tag{8}$$

with the process and measurement error covariance matrices are given by Q and R, respectively. For the harmonic generator process, the gain (K) is a function of the ratio r/q only, assuming R=rI and Q=qI [5]. Since the above algorithm provides the estimation for both the in-phase and quadrature components of the individual harmonics, their amplitude at a given time sample is readily estimated as

$$A_i = \sqrt{\hat{x}_{2i-1}^2 + \hat{x}_{2i}^2} \tag{9}$$

EXPERIMENTAL RESULTS

The above scheme can be suitably adapted to cope with some of the particular demands of MSE experimental data. Firstly, sign changes in the amplitudes, relevant for channels in the vicinity of the magnetic axis, are resolved by appropriately mixing the inphase and quadrature estimates of both the relevant APD harmonics and associated reference waveforms. Secondly, grid induced 50Hz oscillations in the estimated harmonic amplitudes can be removed by applying a successive filter estimating DC and 50Hz, without recurring to any type of box car filtering or data smoothing. The biasing of ELMs (type-I) on the amplitudes' estimate, particularly evident on single-phase lock-in amplifier schemes, as shown in Figure 1 (a), can be mitigated by changing the filter gain K. In fact, reducing the filter gain by several orders of magnitude, much less weight is given to the measurements, corrupted by the ELMs. The filters' role, during an ELM, is thus to assure that the estimated APD components remain with approximately the same amplitude. A logarithmic derivative controller is used to identify the ELM onset from the data whereas a baseline controller is used to mark the ELMs' end and remove the artificial decrease in the gain. This is shown in Figure 1 (b), where the Kalman filtered results are shown with (in grey) and without (in black) the ELM mitigation. The D_α signal at the outer divertor leg evidencing the ELMs is also shown on the bottom of Figure 1 for clarification. The removal of the 50Hz oscillation is also evident. A 250 point box car averaging and decimation, used in the lock-in amplifier scheme, is also used for the Kalman filter to place both methods at equal grounds.

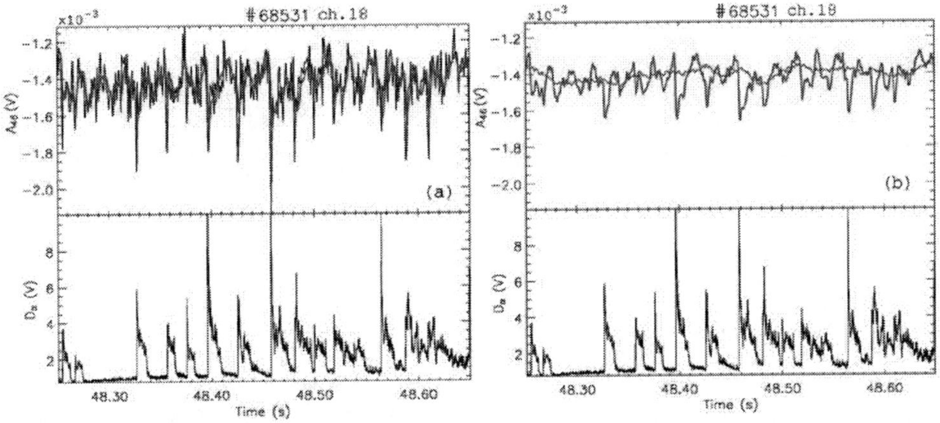

FIGURE 1. Amplitude of the 46kHz component for channel 18 of shot #68531 using a lock-in amplifier and the Kalman filter algorithm with no ELM/50Hz removal (a). When removing both (b) the ratio r/q, set to 1×10^5 in between ELMs, is reduced by 3 order of magnitude during the ELMs.

CONCLUSIONS

An efficient and flexible Kalman filtering model for estimating the amplitudes of the periodic APD components that are relevant for the magnetic pitch angle real-time calculation was presented. Besides providing a significant improvement over the elementary lock-in amplifier scheme regarding noise attenuation, biasing from ELMs and other external forced oscillations can be effectively mitigated from the estimated amplitudes, contributing to a higher quality MSE data to be used subsequently in the equilibrium q-profile reconstruction.

ACKNOWLEDGMENTS

This work, supported by the European Atomic Energy Community and "Instituto Superior Técnico", has been carried out within the Contract of Association between EURATOM and IST and is subject to the provisions of the European Fusion Development Agreement. Financial support was also received from "Fundação para a Ciência e Tecnologia" in the frame of the Contract of Associated Laboratory . The views and opinions expressed herein do not necessarily reflect those of the European Commission, IST and FCT.

REFERENCES

1. F. M. Levington et al, *Rev. Sci. Instrum.* **61**, 2914 (1990).
2. N. C. Hawkes et al, *Rev. Sci. Instrum.* **70**, 894 (1999).
3. D. Alves, *Fusion Eng. And Design* **71**, 175 (2004).
4. R. E. Kalman, *Transactions of the ASME–Journal of Basic Engineering*, **82** (Series D) 35-45 (1960).
5. S. Bittanti and S. M. Savaresi, *IEEE TRANSACTIONS ON AUTOMATIC CONTROL*, **45**, 1718 (2000).

Algorithms for the Control of NTM by Localized ECRH. Principles and Requirements of the Real Time Diagnostic and Control System

G. D'Antona[1], S. Cirant[2], D. Farina[2], F. Gandini[2], E. Lazzaro[2], W. Treuterer[3] and A. Manini[3]

[1]Politecnico di Milano - Dip. di Elettrotecnica - P.za L. da Vinci,32 – 20133 Milano – Italy
[2]Istituto di Fisica del Plasma – Ass. CNR-ENEA-EURATOM – Via Cozzi 53 – 20125 Milano – Italy
3Max Plack Institut für Plasmaphysik – Boltzmannstraße 2 - D-85748 Garching, Germany

Abstract. The diagnostics requirements for the control of NTM instabilities is outlined stressing the importance of correctly managing the estimate uncertainty by the control system. A methodology for the Bayesian assimilation of model predictions and observations is outlined together with an example of application.

Keywords: Bayesian estimation, *NTM* control.
PACS: 52.70.-m

INTRODUCTION

In this paper the diagnostic requirements for the control strategy and the suppression of *NTM* instabilities in a tokamak will be outlined. The topic will be assessed as a decision making problem under risk and uncertainty. In its most essential formulation the decision problem can be stated as the way to maneuver the *ECRH* antenna angles, subject to some static and dynamic actuator constrains, given the uncertain knowledge concerning the amplitude and position of the *NTM* instabilities and about the locations of the *ECRH* power deposition in the plasma.

The actuator control variables in this case are the antenna steering angle $\boldsymbol{\alpha}$ for each *ECRH* line and the power modulation $\mathbf{P}_{ECH}$ for each gyrotron. As aforementioned the information available for the control/decision task are the amplitude W_{NTM} and radial position r_{NTM} of the NTM instability and the deposition radiuses $\mathbf{r}_{DEP}$ together with $\mathbf{P}_{ECH}$ and $\boldsymbol{\alpha}$. In the following we will suppose for simplicity that there is only one instability subject to control, but the approach outlined can be extended to the case of more instabilities.

If the information concerning W_{NTM}, r_{NTM}, $\mathbf{r}_{DEP}$, $\mathbf{P}_{ECH}$ and $\boldsymbol{\alpha}$ were perfectly known, i.e. with negligible uncertainty, together with a dynamical model $M(\cdot)$ of the instability growth:

$$\begin{bmatrix} \dot{W}_{NTM}(t) \\ \dot{r}_{NTM}(t) \end{bmatrix} = M\left(W_{NTM}, r_{NTM}, \mathbf{r}_{DEP}, \mathbf{P}_{ECH}, \boldsymbol{\alpha}, t\right), \quad \begin{bmatrix} W_{NTM}(t_0) \\ r_{NTM}(t_0) \end{bmatrix} = \begin{bmatrix} W_0 \\ r_0 \end{bmatrix} \tag{1}$$

CP988, *Burning Plasma Diagnostics*, edited by F. P. Orsitto, G. Gorini, E. Sindoni, and M. Tardocchi

classical control theory would allow the design of an open loop control law with the objective of minimizing W_{NTM} given the actuators constraints. If the uncertainty of the model (1) was too large a feedback loop control strategy would be necessary. In this case the success of the control objective would be attained only if little uncertainty was affecting the knowledge of the quantities W_{NTM}, r_{NTM}, $\mathbf{r}_{DEP}$, α and $\mathbf{P}_{ECH}$ (actually for *NTM* stabilization, since part of the control problem is tracking r_{NTM} by $\mathbf{r}_{DEP}$, it would be sufficient to have a low uncertainty on the tracking error $\mathbf{e}=r_{NTM}-\mathbf{r}_{DEP}$ only, even with large uncertainties on r_{NTM} and $\mathbf{r}_{DEP}$).

In practice the uncertainty affecting the knowledge of W_{NTM}, r_{NTM}, $\mathbf{r}_{DEP}$ and α is far to be negligible, and thus the likely of a wrong decision taken by the control system is large. For this reason it is mandatory to explicitly process together with the estimates of the quantities involved in the decision procedure also their uncertainties. For example figure 1 shows the instability tracking with two different levels of r_{NTM} uncertainty. When the uncertainty is high the *ECRH* deposition radiuses are far from each other and the *ECRH* power is switch off. When the r_{NTM} uncertainty drops the various $\mathbf{r}_{DEP}$ converge r_{NTM} and the *ECRH* power is switched on.

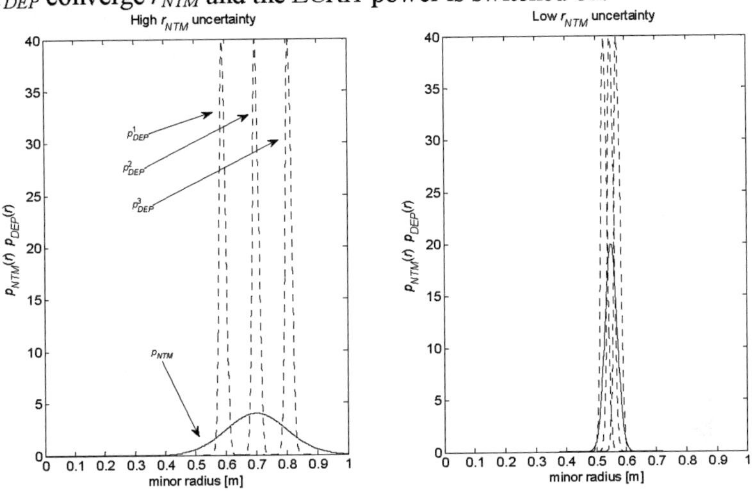

FIGURE 1. Steering of $\mathbf{r}_{DEP}$ for the case of high and low r_{NTM} uncertainty

Formally the information about all the quantities whose knowledge is affected by uncertainty will be described by random vector variables characterized by a proper joint probability density function (*pdf*). From this point of view there is no reason to discriminate among uncertain information coming from either direct measurements or physical models: it is always a proper *pdf*. When both these two sources of information are available their respective information must be assimilated together. For *NTM* stabilization we can count on the plasma equilibrium code for predicting r_{NTM} and on a ray tracing code for $\mathbf{r}_{DEP}$; figure 1 shows the block diagram of the diagnostic and control system including two assimilation algorithms.

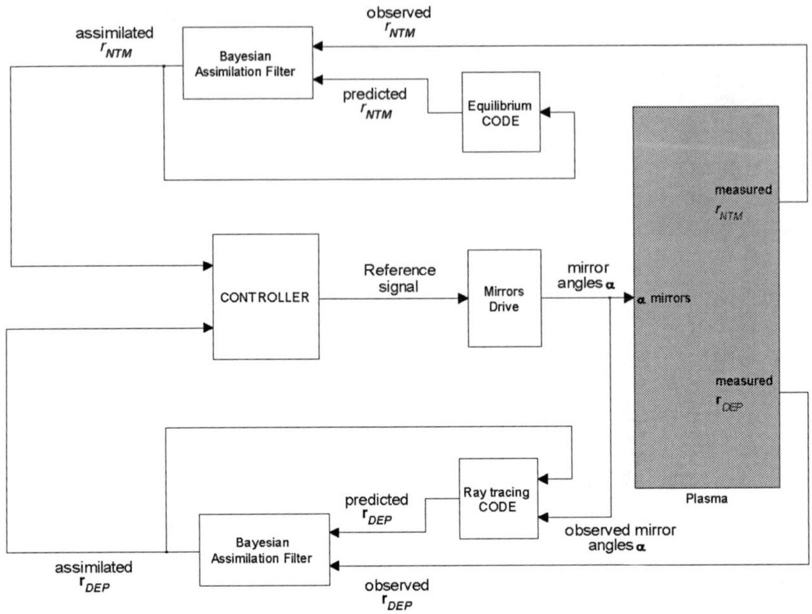

FIGURE 2. Block diagram of the *NTM* control system with two model-data assimilation filters for $\mathbf{r}_{DEP}$ and r_{NTM}

BAYESIAN MODEL-DATA ASSIMILATION PARADIGM

The assimilation algorithm is named Bayesian filter in figure 2 since it is based on the Bayes' formula:

$$p(r \mid \mathbf{d}) = \frac{L(\mathbf{d} \mid r) \cdot p(r)}{p(\mathbf{d})} \propto L(d \mid r) \cdot p(r) \tag{2}$$

In (2) the likelihood function $L(\mathbf{d}|r)$ is the stochastic model of the diagnostic system generating the data vector **d**, the a-priori *pdf* $p(r)$ is the stochastic model predicting the position r either of the *NTM* instability ($r=r_{NTM}$) or of the *ECRH* deposition radius ($r=r_{DEP}$), while the a-posteriori *pdf* $p(r!\mathbf{d})$ is the estimate of r given the observation data vector d. The number $p(\mathbf{d})$ at denominator of (2) is the evidence *pdf* representing the agreement among the observation vector **d** and the a-priori model of the quantity r which have generated the data **d**. The evidence can be evaluated as:

$$p(\mathbf{d}) = \int_{IR} L(\mathbf{d} \mid r) p(r) dr \tag{3}$$

The evidence (3) is also used by the control system since action is taken, i.e. *ECRH* power is switch on, only if there is a high agreement among physical models and observations.

Figure 3 shows an example of application of equations (2) and (3) for the estimation of r_{DEP} for one *ECRH* line using a rough time invariant a-priori *pdf* $p(r)$ and the correlation among the *ECRH* power modulation and the *ECE* channels response as the data vector **d** for shot No. 17107 in ASDEX Upgrade. Figure 4 shows the estimate (mean and standard deviation) and the its evidence versus time.

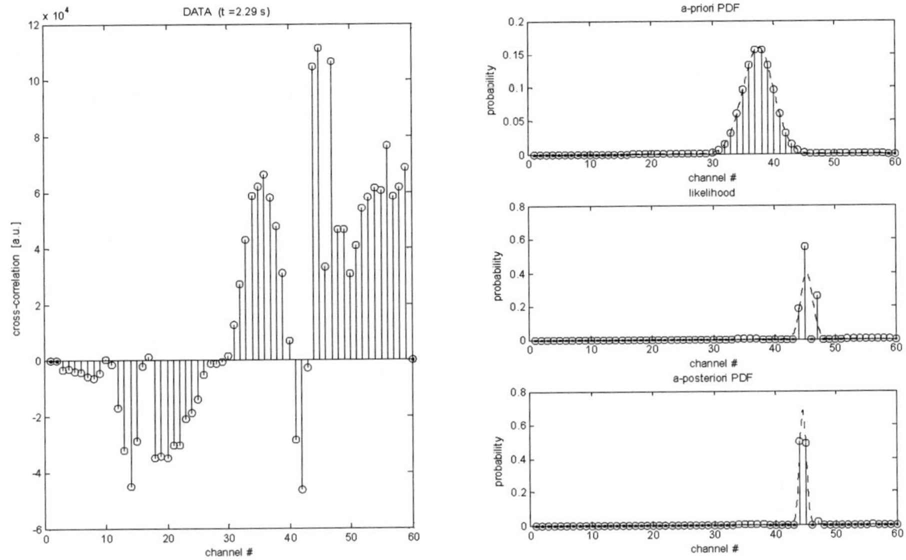

FIGURE 3. r_{DEP} estimate for shot No. 17107 in ASDEX Upgrade by PECH/ECE cross-correlation at time 2.89 s

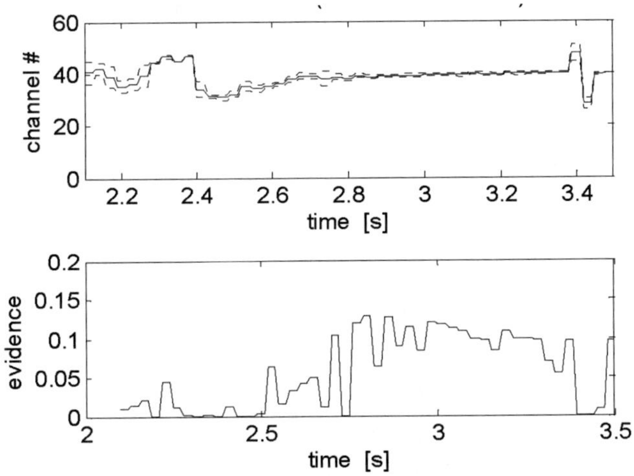

FIGURE 4. r_{DEP} estimate (mean and standard deviation) and its evidence for shot No. 17107 in ASDEX Upgrade by PECH/ECE cross-correlation

REFERENCES

1. S. Cirant et alt., *Journal of Phys.: Conference series 25*, 223-233 (2005).
2. D. Simon, *Optimal State Estimation,* Hoboken: John Wiley & Sons, 2006.

INTEGRATED DATA ANALYSIS

How to Extract Information and Knowledge from Fusion Massive Databases

A.Murari [1)], J.Vega[2)], J.A.Alonso[2)], E.De La Luna[2)], J.Farthing[3)], C.Hidalgo[2)], G.A.Rattá[2)], J.Svensson[4)], G.Vagliasindi[5)], and JET-EFDA Contributors*

JET-EFDA, Culham Science Centre, OX14 3DB, Abingdon, UK
1) Consorzio RFX-Associazione EURATOM ENEA per la Fusione, I-35127 Padova, Italy.
2) Asociación EURATOM-CIEMAT para Fusión, CIEMAT, Madrid, Spain
3) EURATOM/UKAEA Fusion Association, Culham Science Centre, Abingdon, UK
4) Max-Planck-Institut für Plasmaphysik, Teilinstitut Greifswald, EURATOM Association, Wendelsteinstr.1, 17491 Greifswald, Germany
5) Dipartimento di Ingegneria Elettrica Elettronica e dei Sistemi-Università degli Studi di Catania, 95125 Catania, Italy
**See the Appendix of M.L.Watkins et al., Fusion Energy 2006 (Proc. 21st Int. Conf. Chengdu, 2006) IAEA, (2006*

Abstract. The need to understand and control the dynamics of reactor grade fusion plasmas requires the analysis of increasing amounts of data, which at JET can reach easily the level of several GBytes per shot. Therefore a series of new approaches are being pursued to store the data and to retrieve the required information. They range from loss less data compression techniques, to wavelets and Structural Pattern Recognition methods. Since the information available is very often affected by high level of uncertainties and the phenomena to be studied are complex and nonlinear, the inference problems in this field of plasma physics are particularly delicate. Even in this perspective innovative solutions are under development. In particular a range of Soft Computing approaches have already been implemented at JET. The most successful are Bayesian statistics for the integration of diagnostic measurements, Data Mining techniques to study the nonlinear correlation of various variables and Fuzzy Logic to include the knowledge of the experts even if formulated in linguistics terms. Specific methodologies are being investigated for real time control, since it poses some specific issues. In this field, a combination of hardware and software solutions, like Cellular Nonlinear Networks, are often necessary to satisfy the needs of both speed and reliability.

Keywords: Data mining, information retrieval, nuclear fusion, structural pattern recognition.
PACS: 42.30.Sy.

1. INTRODUCTION

Fusion plasmas are nonlinear, complex systems, which have to be controlled to a high degree of reliability in order to achieve the final goal of thermonuclear fusion for energy generation. They are also quite delicate physical entities, very difficult to access and therefore, in many cases, the required information must be derived indirectly on the basis of the natural emission of the plasma, like electromagnetic radiation or particles. The objectives of data processing are also quite different

CP988, *Burning Plasma Diagnostics*, edited by F. P. Orsitto, G. Gorini, E. Sindoni, and M. Tardocchi
2008 American Institute of Physics 978-0-7354-0507-3/08/$23.00

compared to other traditional fields of physical research, like particle physics. In high energy physics, very often the main task of the analysis consists of isolating the products of a specific reaction, among a myriad of particles generated by competing and unwanted phenomena. This is not the case in Magnetic Confinement Nuclear Fusion (MCNF), where, at least in theory, one should be able to make sense of all the data collected in each single discharge. In this respect, the similarities are higher with the health sciences, in which a holistic view of the patient is required to optimize the result of therapies or operations. Such a comprehensive view of fusion plasmas will be also desirable in the final reactor, to find the best compromise between the often contrasting objectives of maximizing the efficiency of the machine and minimizing the risk of incidents. On the other hand, this global understanding of all the available data is not an easy task in modern machines, since the amount of information, derived from each plasma shot, keeps increasing. At JET, for example, already more that 10 GBytes of data have already been acquired in some specific discharges and in normal experiments several Gbytes of data are the typical output. Since the first discharge at JET, the increase in the amount of collected data has followed the Moore law (doubling every two years) and this trend is expected to continue in the future. The data output of JET is also expected to keep increasing and indeed in the next years JET diagnostics are estimated to be able to produce an addition amount of about 50 GBytes of data per shot at full performance. In ITER, the next step device, the situation will be more critical in this respect, because an even greater amount of data will be necessary to assure the safe operation of the device, the understanding of the physics and the proper control of the phenomena in preparation for the prototype reactor.

The management of such an amount of information presents of course a series of issues. The most evident are related to the data transmission and storage. In particular various forms of first data processing and data compression are under study to optimize the data management over JET network, with particular attention to the implications for the real time control of complex phenomena in context of scenario development for ITER (see section 2). In the last years, the problem of information retrieval has also become a very sensitive issue. In addition to producing more data, JET has also become a much more international experiments, with hundred of scientists visiting from all the countries of the EU and also from outside Europe. New tools, based mainly on Structural Pattern Recognition techniques, have therefore been developed to help the scientist in the first data screening (see section 3). These methods have been recently extended also to bidimensional signals since the use of cameras, both visible and infrared, have become much more widespread in the last years, following a general trend in science and industry.

In addition to the issue of information retrieval, fusion plasmas of reactor relevance pose also a formidable inference problem. Since they are very delicate physical objects, they cannot be perturbed too much in order to perform the necessary measurements. Therefore information about their internal status must be derived indirectly, sometimes with a significant level of uncertainties. These difficulties, combined with the fact that very often non linear complex phenomena dominate their behaviour, have motivated the community to develop more robust statistical inference methods to both validate and combine the data of various diagnostics and to falsify theories. The most successful approach tested so far is based on Bayesian theory, which has been applied, among other things, to derive the magnetic topology of JET discharges (see section 4).

Even if progress in magnetic confinement fusion has been significant in the last decades, several issues remain obscure and comprehension of many aspects far from satisfactory. The correlation between many variables, together with various sources of uncertainties, increases the difficulty of many problems. Since many experiments have already been carried out, more data driven theory could be a solution to improve the rate of progress in the next years. The first steps in this direction have already been moved with the adoption of various data mining techniques to explore in particular the non linear correlation between the many variables which can cause disruptions (see section 5). Other Soft Computing methods, other than the statistical ones already mentioned, are also receiving increased attention. From Fuzzy Logic to Neural computing, the successful applications increase in number and ambition (see section 6). Since real time applications very often required ad hoc solutions to guarantee the required performance in terms of reliability and speed, some specific technologies are being explored in this field, with particular attention to image processing, since the use of visual diagnostics is expected to increase significantly in the near future (see section 7).

2. DATA STORAGE AND FIRST SIGNAL PROCESSING

As already mentioned, JET has demonstrated that its diagnostic can produce more than 10 Gbytes of data per shot and this amount of information is expected to reach more than 50 Gbytes with the installation of the new measuring systems. Since during a typical two shifts operational day JET can have between 30 and 40 good shots, the quantity of information to be handled is clearly already enormous and it is going to increase. In ITER the data to be acquired will certainly be 1-10 Tbytes per shot. On the other hand, as in any typical experiment, it is difficult to tell which signals will be the most relevant for the understanding of the physics and the progress of the scientific programme. Any loss of information is therefore to be avoided because tit could affect the interpretation of the experiments. Therefore in order to manage the vast amount of data produced only loss less compression techniques are an option in a fusion device. In JET and other facilities, like TJ-II, the so called delta methods have been successfully adopted. They are based on the observation that magnetic confinement fusion plasmas are relatively stable physical entities, whose parameters tend to change quite slowly in time. Therefore, the difference between two subsequent time slices of the same signal has a probability distribution which peaks at zero and decreases rapidly as the value of the difference increases. The strategy therefore consists of storing only the difference between subsequent time slices, assigning the shortest symbol to the value of zero and progressively the longer symbols to higher incremental values, which have a much lower probability of occurring. The detailed description of the solutions tested on TJ-II and JET are reported in [1]. With this technique, the average compression rate on TJ-II waveforms (almost 3 million signals) is 69.87%, where the compression factor is defined as

$f = (1 - \frac{compressed\ storage}{no\ compressed\ storage})*100$, *i.e.* a 70% of compression means that only 30 bytes out of 100 bytes are written. Application to JET waveforms (62567 time-series data) gives f = 49.59%. It is worth pointing out that this approach can be applied very successfully also to bidimensional signals and its effectiveness has already been verified. Table I shows a comparison of this technique with well-known compaction procedures. All of them were applied to typical video frames of JET infrared and visible cameras. In addition to the higher compression rate, it should be mentioned that the above delta method can be used under real-time requirements whereas the other procedures are delayed techniques (compression only can be carried out having available all the data). The real-time software compression on frames of 256x256 pixels takes 5.5 ms (< 200 frames/s) in a Pentium IV computer with Windows XP. The use of dedicated hardware, for instance Field Programmable Gate Arrays (FPGA), will allow a significant reduction of this time.

TABLE 1. Compression rates for images.

	(%)
Delta approach	93.05
WinZip	87.30
Compress() UNIX command	85.52
pack() UNIX command	36.78

3. INFORMATION RETRIEVAL

Even once the data has been efficiently and safely stored, the problem of retrieving the desired information remains a huge task. In JET, this has become particularly evident under the new diffuse organization, with scientists from all over Europe participating in the experimental campaigns and the following analysis of the measurements. Only the preliminary task of identifying the most relevant discharges, to concentrate on them the subsequent analysis efforts, has become an incredibly time consuming and sometimes frustrating work. In order to alleviate this problem and render the first data screening a much more efficient process, completely new way to organise the data bases is under intensive investigation. In the new paradigm adopted, signals are not organized according to the traditional scheme of signal name, shot number and time slice. On the contrary, the objective consists of giving the scientists the opportunity to select, on a simple visual interface, the part of a signal they are interested in and then it is the task of the software system to extract the shot numbers and time intervals in which the same signal shape is present. This new approach, which is meant to complement not to substitute the old database structure, is illustrated in fig 1.

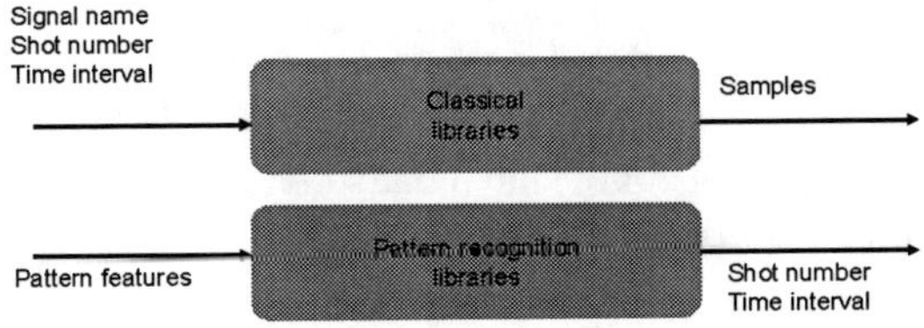

FIGURE 1. Comparison between the traditional structure of fusion data (top) and the new one.

The methods to obtain these objectives are based on pattern recognition, which is a mathematical formulation of the more general problem of object identification. The general task of recognising objects is a characteristic goal of all living beings and pervades their daily life, since it is essential in basic activities like search for food or navigation, identification of friends and enemies etc. Pattern recognition is the discipline which studies the mathematical aspects of these activities, so that they can be carried out automatically by computers. In fusion the most promising techniques for the first data screening are the ones of structural pattern recognition. They are based on the general structural shape of the signals more than on their detailed mathematical properties (unlike correlation or coherence estimators). This choice is based on the fact that in plasma physics many phenomena can indeed be identified by the form of the signals they generate in the diagnostic measurements. And these shapes in the signals are indeed what the physicists normally use to identify the various phenomena of interest during the visual inspection of the measurements, a preliminary activity to all type of subsequent data analysis. The main idea therefore consists of developing methodologies that can perform a first screening of the data extracting the signal shapes which are similar to a prototype defined by the user. The first step in achieving this goal is the extraction of the relevant features from the signals. On the basis of these features, criteria of similarity have to be defined, together with a suitable distance, which would allow assessing automatically the degree of similarity between different signals.

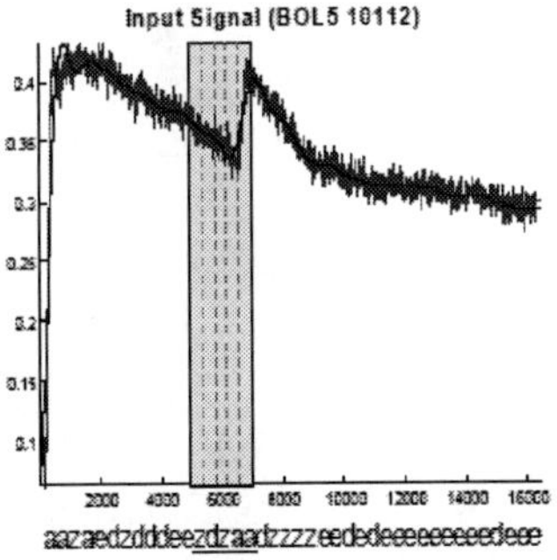

FIGURE 2. Pattern recognition of patterns inside signals using primitives based on the gradient.

To exemplify these concepts the approach based on primitives is particularly instructive, in addition to being very effective in finding patterns within time traces.

The version, already implemented in JET, is based on the division of time traces in a series of short intervals, inside which the gradient of the signal is calculated. The gradient levels are codified in a suitable number of groups, depending on their value, and a letter is attributed to each group. The signals can therefore be represented as a sequence of letters, codifying the amplitude of the signals gradient in each time interval (see fig.2). Finding a certain signal pattern is therefore reduced to a string comparison problem, which can be carried out very efficiently. The method provides also a lot of flexibility because differences from the original letter sequence can be allowed, relaxing the constraints on the search degree of similarity. Also filters can be applied to impose the search for signal with the same time base and amplitude or more generally simply signals parts with the same shape can be looked for (independently from amplitude and time duration). Alternative approaches have also been tested. For example, to retrieve entire time traces and not only parts of them, an efficient method is base on the wavelet transforms, which play the role of sophisticated filters [2]. Then the search is performed in the space of the transforms on the appropriate parameters. Some pilot software, implementing these solutions for monodimensional signals, is already operational at JET. Figure 3 illustrates the typical performance for the case of a generic time trace. Both the time needed for the search and the addition overhead in terms of memory occupation are very reasonable even for cases in which many hundreds of signals are screened. This methodology has also been recently extended to bidimensional images. In this case, again a wavelet transform is applied to the image and then the original shape is retrieved.

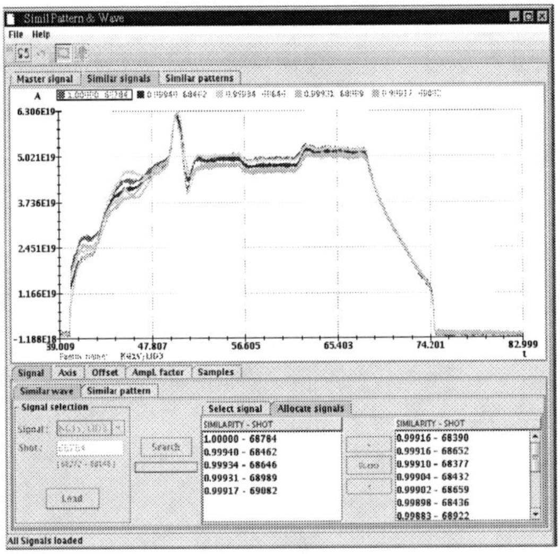

FIGURE 3. Pattern recognition of entire waveforms. The essential features are extracted by applying a wavelet transform. The signals are then grouped in similarity classes using a tree structure to expedite the search by avoiding the need to traverse the whole database.

4. DATA MINING

As described in the previous sections, techniques for storing big amounts of data and for retrieving information efficiently from massive databases have been introduced. These methods have already been implemented at JET and other devices and they have proven to be very useful. On the other hand, plasmas of fusion relevance are very complex and nonlinear systems and therefore very important information can remain hidden in the signals, without the scientists being aware of it. Another important issue could be the tendency to oversimplify physical issues, which is always a danger when very complex phenomena have to be studied and solutions are dispersed in large amounts of information. The methods used to address these problems fall in the category of "data mining", which is defined as the collection of techniques used to extract hidden information from large amounts of data. Data mining techniques have been originally developed by private companies for various purposes, ranging from product development to market targeting. These approaches allow, for example, automated prediction of trends and behaviour and the automated discovery of unknown patterns (whereas the pattern recognition methods described in the previous section of course assume that the signal shape to be retrieved has already been identified).

In the context of MCNF, a particularly reach application field of data mining techniques is expected to be the quest for nonlinear correlations among signals, which is also a scientific issue of wider application than simply plasma physics. The example described in the following uses classification and regression trees to determine the relevance of various quantities for the prediction of disruptions. Disruptions are sudden losses of confinement which lead to the fast and unforeseen termination of a plasma discharge. They are the result of instabilities that go out of control to the point of causing complete collapse of the discharge. Disruptions are particularly troublesome events not so much for the interruption of the plasma but for the potential danger to the machine integrity. Indeed they can cause very high heat loads on the plasma facing components and strong electromagnetic forces on the vacuum vessel and surrounding structures. The instabilities, which can evolve to the point of causing disruptions, have very complex dynamics and they can derive their energy from many different sources. It has therefore resulted impossible so far to reach a satisfactory theoretical understanding of the disruption phenomenology. As a consequence, no satisfactory model is available to predict the occurrence of disruptions and in the last years significant efforts have been devoted to devising non algorithmic predictors. In order to better understand the interplay between the various causes and to formulate better predictors, an analysis of the importance of various signals for the prediction of disruptions has been undertaken using Classification and Regression Trees (CART) methods. CART is a non-parametric and fully nonlinear technique to solve classification ad regression problems. In our case it is used to describe on output variable, the fact that a discharge is disruptive or not, as a function of different interpretative variables. Given the value of the output variable and the values of the signals which are believed to be of relevance for the interpretation of the phenomenon, the method builds automatically a classification tree. This is achieved by traversing the whole database of the input variables and trying to find the signal which better divide the discharges into classes, one with only disruptive and one with only non-disruptive discharges. This approach is said to try to maximize the purity of the child nodes, in the sense to split the classes in the most coherent way. In general, of course, no variable is enough to obtain two pure child nodes so the process is repeated or the child nodes using other variables until a pure classification is obtained (or the list of variables is exhausted). The hierarchy of the nodes in the final tree allows defining the relative importance of the various variables for the classification of the phenomenon under study, the disruptions in our case.

Table II summarises the results of an application of the CART method to a JET database of disruptions. The various columns refer to different time intervals before the occurrence of the disruption. It appears very clearly how the relative importance of the various signals changes significantly depending on the considered time. This information has been included in the design of a Fuzzy Logic predictor to improve its performance as described in section 7. This technique and other classification methods, based on automatic clustering and unsupervised learning, are being investigated further and this field of research is expected to gain significant momentum in the next years.

TABLE 2. The importance of various variables for the prediction of disruptions as derived by application of the CART software. The various columns refer to different time intervals before the disruption.

[td – 440, td - 100]		[td – 200, td - 100]		[td – 320, td - 220]		[td – 440, td - 340]	
Variable	Importance	Variable	Import.	Variable	Import.	Variable	Import.
dW_{dia}/dt	100.0	dW_{dia}/dt	100.0	dl_i/dt	100.0	Ipla	100.0
dl_i/dt	82.88	dl_i/dt	50.46	dW_{dia}/dt	99.20	dl_i/dt	69.42
Ipla	70.88	$d\beta_p/dt$	37.01	Ipla	79.40	P_{net}	68.23
P_{net}	67.96	P_{net}	35.03	l_i	70.47	l_i	54.78
dq_{95}/dt	54.90	q_{95}	27.54	q_{95}	62.06	dq_{95}/dt	53.92
Dens	53.24	β_p	24.80	P_{net}	59.00	P_{inp}	49.58
$d\beta_p/dt$	52.96	l_i	24.37	β_p	57.05	dW_{dia}/dt	38.78
l_i	52.36	Loca	22.74	dq_{95}/dt	56.19	q_{95}	37.97
Loca	46.75	P_{inp}	20.49	$d\beta_p/dt$	38.68	Loca	37.39
P_{inp}	45.26	Ipla	16.36	P_{inp}	38.41	$d\beta_p/dt$	36.57
β_p	43.47	Dens	13.34	Dens	37.91	Dens	35.55
q_{95}	33.55	dq_{95}/dt	9.87	Loca	30.13	β_p	28.77

5. THE INFERENCE PROBLEM AND SOFT COMPUTING METHODS

As already mentioned in the introduction, fusion plasmas present a formidable inference problem. The reasons are manifold. First of all, the purpose of the analysis is holistic and more similar to the health sciences than to other big physics experiments like high energy accelerators. In principle in plasma physics one should be able to make sense of all the available data and show that they fall in a coherent picture, including all the aspects of the experiment. Even if only one of the measurements is anomalous, compared to the expectations of the model, this discrepancy could mean that there something profoundly wrong with the interpretation of the discharge (provided of course the diagnostic data are themselves correct). On the other hand, fusion plasmas are complex systems, which depend on many quantities. The interaction between the various quantities is also very often profoundly non linear and therefore impossible to model correctly, at least with the present day computational resources. Moreover, today measurements are affected by many causes of uncertainties. This is due, on the one hand, to the inherent architecture of Tokamak machines, whose main objective is the production of energy for commercial purposes and which therefore have to be designed according to various constraints, limiting the flexibility of diagnostics. On the other hand, the plasmas themselves are very delicate entities, which would be altered by too intrusive measuring systems, again an analogy with diagnosis of the human body in medicine. Therefore many important properties of fusion plasmas are derived indirectly by the emissions of the plasmas, i.e. radiation, neutral and charged particles, neutrons etc. For all these reasons, algorithmic methods and traditional inference techniques have shown several limitations in the field of plasma physics. In the last years therefore much attention has been devoted to the so called "soft computing" disciplines. These new methodologies have explicitly been developed to address complex inference problems about systems with high uncertainties. Since human reasoning can handle high level of uncertainty and complexity in several situations, some aspects of the human mind have had a strong influence on these approaches. Some of the most widely applied "Soft computing" approaches are probabilistic reasoning, neural computing and Fuzzy Logic.

In the field of probabilistic inference, the main method actively pursued at JET is Bayesian Statistics, which provides a coherent framework to integrate the data of various measuring systems to obtain a specific plasma quantity and to integrated prior information in the process (see section 6). Neural networks have been a field of research for many years to address issues for which logarithmic solutions are lacking and expected to be particularly difficult to develop. Since a broad literature covers this aspect, Neural Networks are not discussed any further in this paper. On the contrary, Fuzz Logic is a quite new entry in the field of Nuclear Fusion. Its ability to include expert knowledge and its transparent character, have allowed to make several progress in which other methods had exhausted their capability (see section 7).

6. BAYESIAN STATISTICS

For the reasons described in the previous section, large nuclear fusion experiments present a remarkable interpretation problem and require complex processes of inference, which sometimes have to rely on sophisticated physics models and complex inversion algorithms, to provide a good estimate of the physical quantities of interest from the available external measurements. Due to these difficulties, the extraction of physical information from measurements and its validation is undergoing a fundamental methodological revision and more systematic, statistical techniques are being implemented. The general approach of 'Bayesian Probability Theory' is proving very suited to the difficulties inherent in associating error bars to the diagnostic signals and in deriving better estimates of the physical parameters needed to analyse the experiments. Based on Bayes theorem, it constitutes now a solid body of work that can be applied to a series of different problems, ranging from measurement interpretation to theory falsification.

In JET Bayesian statistic is being considered now to treat in a coherent way the uncertainties in the measurements, both random and systematic. This is expected to provide a better evaluation of errors bars and the determination of various parameters, for which independent measurements exist, with higher accuracy. More ambitiously, it constitutes the backbone of the diagnostic integration programme, which aims at deriving the plasma state from a unique Bayesian estimator, capable of including a vast majority and possibly all the diagnostic information available. This methodology is based on the observation that all the element of the inference process, from the model parameters to the measurements and prior information, should be represented as probability distribution functions. Bayesian statistic is an already developed tool to combine all these probability distribution functions, the inputs, to obtain the probability distribution functions of the required physical quantities, the output of the model [3].

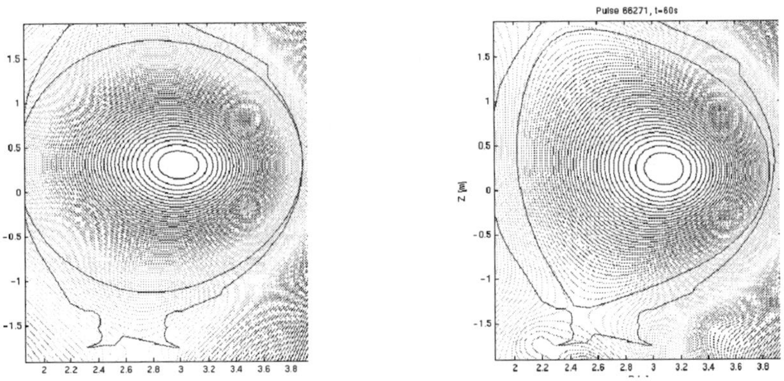

FIGURE 4. Magnetic topology for limiter (top) and x-point bottom JET discharge #66271 obtained with a Bayesian estimator without any hypothesis on the equilibrium.

On the route of this ambitious programme, the first step is the derivation of the magnetic topology of the plasma. In addition to providing the basic element of confinement in a Tokamak, the magnetic fields represent also an essential ingredient to interpret the measurements of practically all the diagnostics. The topology of the magnetic fields indeed are the most natural choice for the reference frame into which combine all the available measurements. The main problem is that in the past the magnetic filed geometry was determined with the help of complex equilibrium codes, which rely on various hypotheses and do not allow an easy determination of the error bars to be associated with the final results. It was therefore decided in JET to develop a current tomography i.e. a Bayesian derivation of the magnetic fields of the plasma on the basis of the pure measurements and without any hypothesis on the equilibrium.

To this end, JET plasmas and the surrounding conducting structures have been modelled with a series of current beams. The values of the currents in these beams are optimised on the basis of the prior information in order to reproduce best the available measurements of the pick-up coils in a statistical sense. The results are very positive as shown in fig.4. The error bars to be associated to essential parameters of the magnetic topology, like the magnetic axis or X-point position, can be easily derived as shown in fig.5 [4]. The next step in this research programme will be the systematic inclusion in the inference process, leading to the magnetic topology, of the internal magnetic measurements, like the Motional Stark Effect and Polarimetry.

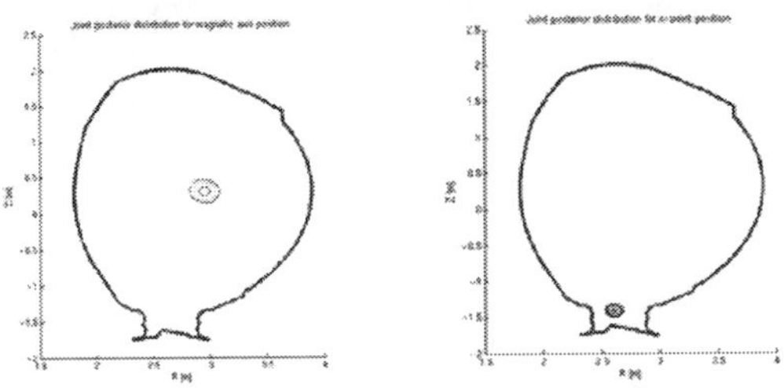

FIGURE 5. Error bars for the magnetic axis position (left: 6-8 cm) and X-point position (right: 2-3 cm).

7. FUZZY LOGIC

In addition to the complexity of the plasmas and the uncertainties in the systems and the measurements, and sometimes because of these facts, some specialist knowledge of difficult phenomena is often patrimony of experts, who do no have the background or the opportunity to formulate it in mathematical terms. Sometimes therefore there is a relevant amount of information, which is difficult to express in a way useful for the wide community because it remains at linguistic or in any case at an informal level. A case in point, to use an example already mentioned, is the one of disruptions. These sudden events, which produce a complete loss of confinement and can be very dangerous for the integrity of the machines, are induced by many different causes, which can interact nonlinearly in complicated way. Moreover, for a series of reasons, the boundary between the safety and dangerous regions of the operational space are very often blurred and difficult to pinpoint exactly. All these aspects have suggested the investigation of "Fuzzy Logic" as a possible tool for both understanding and prediction of disruptions. This recent form of logic is indeed based not on crisp sets, like Boolean logic, but on fuzzy sets, whose membership is defined in completely different way from the traditional one, which hinges on the so-called "Law of the Excluded Middle", which states that an element X must either be in set A or in set not-A. On the contrary, in the scheme of Fuzzy Logic the membership of an element to a set can take any value in the whole interval [0,1] and not only the Boolean discrete values 0 and 1. In more detail, a fuzzy set F defined on U (called the universe of discourse) is given by $F = \{(u, \mu_F(u)) \mid u \in U\}$ where $\mu_F(u)$ is the membership function, a curve that specifies how each element of U is mapped to the real interval [0, 1], that is $\mu_F(u):U \to [0,1]$. In other terms, for each $u \in U$, the membership function defines the degree of membership of u to the fuzzy set F, which can vary continuously from zero (no membership) to one (full membership) according to the particular properties of the fuzzy set. It can be demonstrated that fuzzy logic is in reality a superset of traditional Boolean logic. Therefore, also the traditional logic operations, like AND, OR etc, can be generalised to operate on fuzzy sets. On this basis, entire fuzzy systems can be defined, which consist of a series of fuzzy rules and an inference mechanism. In practice they allow to operate on fuzzy sets and perform inferences to derive conclusions again expressed in terms of fuzzy sets. If numerical values are required as the result of the computation, specific rules exist to perform weighted averages over the fuzzy sets to obtain such numerical outputs.

Pioneering work on the application of these fuzzy methods to disruption prediction has been carried out at JET [5]. Various predictors, with about ten signals as inputs and almost forty fuzzy rules, have been devised to provide the probability of disruption for any phase of the discharge. Their results are very competitive with other alternatives but their main advantage has proven to be their transparency. Indeed, using the information obtained from the correlation analysis performed with the CART approach and described in section 4, it has been possible to optimise the predictors for the various phases of the discharge. This work culminated in the definition of a series of predictors, each one giving the probability of disruption for a certain time in the future [6]. The outputs of these predictors, being probabilities of disruptions about future times, can be integrated in proper controller, which can evaluate the risk and decide what safety actions to undertake accordingly. Indeed, on the basis of the probability given by the predictor and the potential danger of the discharge, a coherent evaluation of the risk can be performed. Moreover, since these fuzzy predictors provide the probablity for various times in the future a disruption is to be expected, the intervention strategy can be optmised taking into account the tools available to mitigate the disruption and their time response.

8. REAL-TIME AND IMAGE PROCESSING

Real time control of fusion plasmas is a field of great development and it will play an essential role on the route to a viable solution for a nuclear reactor. Certainly with the increase in the dimensions and performance of Tokamak plasmas, more quantities will have to be controlled in real time to guarantee safety of operation and stability of the configuration. More ambitiously, real time control could also play a direct role in pushing the boundaries of present scenarios and help in improving performance. In JET for example, the real time system has increased significantly in the last years and it involves now more than 30 diagnostics for a total of about 500 acquired signals.

In this general context, data processing for real time purposes certainly presents some specific issues which have to be carefully taken into account both in the design phase of measuring systems and during the execution of the experiments. In the case of feedback control, velocity of computation and reliability of the results are of course at a premium, whereas the constraint on accuracy can very often be relaxed at least partially. Therefore specific solutions are sometimes required. Other aspects of real time applications of diagnostics, which very often require ad hoc solutions, are the needs of portability and easy maintainability. These are the main reasons why, in the past, at JET several problems have been addressed with the help of Artificial Neural Networks. From the calculation of the total radiated power to the determination of the plasma internal inductance, ANNs have provided very efficient solutions. Nowadays, ANNs are also investigated for even more ambitious tasks, like the identification in real time of specific atomic lines for the control of the plasma wall interactions [7].

A very recent trend, motivated among other things by rapid developments in electronic technologies, is the convergence of hardware and software for the solution of complex tasks in real time. A general example, of course, is the widespread use of Field Programmable Gate Arrays, which can be reconfigured in real time depending on the needs of the experiment. Another drive behind the adoption of new technologies for control is also the increased use of cameras, in both the visible and the infrared parts of the spectrum, to address a series of issues, ranging from surveillance and safety to physical investigation of fast phenomena. Since images are particularly demanding in terms of computational resources, new solutions are being investigated to achieve the necessary speed and accuracy. For these reasons, at JET the new approach of Cellular Neural Networks is being actively pursued. Cellular Nonlinear Networks are originally mathematical objects, whose abstract definition was introduced in 1988 [8]. They consist of 1, 2 or 3 dimensional infinite arrays of non linear dynamic units called cells. These cells are located at the nodes of the array and interact locally, via weighted connections, within a finite radius r, which is called the neighbourhood of the cell and can be varied depending on the application. Various mathematical models have been developed to compute the core cell value on the basis of its input and the status of the neighbourhood cells and nowadays libraries of functions, called templates, are available for implementing the basic operations on the cells. These templates allow performing typical fundamental steps of image manipulation, like thresholding or interpolation. The mathematical paradigm of the CNNs has found many computational applications, ranging from differential equation solutions to modelling of complex biological systems. With regard to the various aspects of image processing, the local connectivity of CNNs is particularly advantageous, because it allows easy implementation with well developed VSLI technologies. Moreover, the topological architecture of CNNs is well suited to parallel processing of the cells, with the ensuing reduction of the computational time. The hardware implementation adopted for the applications at JET is the ACE16K [9], the third generation of CNN chips, which consists of two layers of CMOS components. The first is an array of 128X 128 identical analog cells, designed to act as photo-receptors. The second layer contains the analog processing unit (implementing the CNN processing paradigm) and a series of digital processing units (for storage, programming and other functionalities). The chip can perform up to 330 GOPS (Gyga operations per second) and can therefore compute linear convolutions of 3X3 neighbourhoods in less than 3 μs.

This technological solution has already been adopted at JET to determine the position of the plasma strike points, the region in the divertor where the plasma intercepts the material boundaries [9]. At the moment it is being successfully used for surveillance purposes, in particular to detect the so called hot spots, small parts of the vacuum vessel where the plasma wall interactions become too intense and cause excessive increase in the local temperature. The algorithms to identify the hot spots using JET wide angle infrared camera have already been tested and the performance is reported in table III.

TABLE 3. The performance achieved with the CNNs for the detection of the hot spots

ALGORITHM	Time (ms)
Static detection (full frame)	~ 110
Dynamic detection (full frame)	~ 100
Static detection (128x128)	~ 19-20
Dynamic detection (128x128)	~ 20-21

ACKNOWLEDGEMENT

The work leading to this article was funded by the European Atomic Energy Community and is subject to the provisions of the European Fusion Development Agreement.

REFERENCES

1. J. Vega et al. "Encoding technique for high data compaction in databases of fusion devices". Rev. Sci. Ins. 67, 12 (1996) pp. 4154-4160.
2. J. Vega et al. "Data mining technique for fast retrieval of similar waveforms in Fusion Massive databases". Accepted in Fus. Eng. Des.
3. J. Svensson et al. "Integrating Diagnostic data analysis for W7-AS using Bayesian graphical models". Rev. Sci. Ins. 75, 10 (2004).
4. J. Svensson, A. Wemer. "Current Tomography for Axisymmetric Plasmas". Submitted to Nuclear Fusion.
5. G. Vagliasindi, A. Murari, L. Fortuna, P. Arena. "Fuzzy logic approach to disruption prediction at JET". Submitted to IEEE Transactions on Plasma Science
6. A. Murari, G. Vagliasindi, P. Arena, L. Fortuna, O. Barana, M. Johnson. "Prototype of an adaptive disruption predictor for JET based on fuzzy logic and regression trees". Submitted to Nuclear Fusion.
7. O. Barana, A. Murari, I. Coffey. "Artificial neural networks for real time diagnostic of high Z impurities in reactor relevant plasmas". To be published in IEEE Transactions on Instrumentation and Measurements.
8. L. O. Chua, L. Yang. "Cellular Neural Networks: Theory and Applications". IEEE Trans. Circuits and Systems. Part I, 35, pp. 1257-1290, 1988.
9. G. Vagliasindi et al. "Application of Cellular Neural Network Methods to Real Time Image Analysis in Plasma Fusion". To be published in IEEE Transactions on Instrumentation and Measurements.

Integrated Data Analysis for Fusion: A Bayesian Tutorial for Fusion Diagnosticians

Andreas Dinklage[1], Heiko Dreier[1], Rainer Fischer[2], Silvio Gori[2], Roland Preuss[1] and Udo von Toussaint[2]

Max-Planck-Institut für Plasmaphysik
EURATOM Association
[1]Wendelsteinstr. 1, 17491 Greifswald, Germany
[2]Boltzmannstr. 2, 85748 Garching, Germany

Abstract. Integrated Data Analysis (IDA) offers a unified way of combining information relevant to fusion experiments. Thereby, IDA meets with typical issues arising in fusion data analysis. In IDA, all information is consistently formulated as probability density functions quantifying uncertainties in the analysis within the Bayesian probability theory. For a single diagnostic, IDA allows the identification of faulty measurements and improvements in the set-up. For a set of diagnostics, IDA gives joint error distributions allowing the comparison and integration of different diagnostics results. Validation of physics models can be performed by model comparison techniques. Typical data analysis applications benefit from IDA capabilities of nonlinear error propagation, the inclusion of systematic effects and the comparison of different physics models. Applications range from outlier detection, background discrimination, model assessment and design of diagnostics. In order to cope with next step fusion device requirements, appropriate techniques are explored for fast analysis applications.

Keywords: magnetic confinement fusion, data analysis, data fusion, integrated data analysis, Bayesian probability theory
PACS: 02.50.Cw, 07.05.Fb, 07.05.Mh, 52.70.-me

DATA ANALYSIS FOR MAGNETIC CONFINEMENT FUSION DEVICES

Next step fusion devices define several new requirements for data analysis. Since the demonstration of fusion reactor capabilities aims at steady-state operation, the data processing paradigm needs to be shifted from a *shoot-and-collect* to a *view-and-react* philosophy. Therefore, fast and reliable physics information is mandatory to use the experimental devices most efficiently and off-line analysis schemes as in pulsed devices will restrict operational capabilities. Long-time scales due to plasma-wall effects and magnetic field relaxation will require intelligent control schemes which employ physical models in multi-variate adaptive controllers.

The major challenge for data analysis in magnetic fusion experiments is to link many different heterogeneous information sources [1]. The reason for the heterogeneity lies in the variety of methods in plasma diagnostics which is required to

CP988, *Burning Plasma Diagnostics*, edited by F. P. Orsitto, G. Gorini, E. Sindoni, and M. Tardocchi

determine the parameters for the assessment of device performance. In engineering, e.g. robotics, this link of different signals is called *data fusion* [2] which aims at reduction of uncertainties by a synergistic combination of multiple measurements. The task for fusion data analysis goes even one step ahead since the link of data requires extensively the use of physical models, e.g. equilibrium calculations or transport considerations.

Data analysis in fusion poses several system specific issues. The large number of intricate dependencies makes magnetic confinement fusion devices (MCF) a complex, large system. The complexity is even enhanced by the fact that the information suffers from heterogeneity not only in space and time but particularly in quality. In consequence, the validation of data is a demanding and time consuming task. Any progress towards an accelerated data analysis process allows one to increase both the number of analyzed fusion experiments and the quality of analyses. A vision is a framework aiming at automated procedures for analyses, validation and the detection of new features (e.g. pattern recognition techniques or *data mining*). In addition to an increased efficiency for an off-line analysis, the potential benefit of advanced data analysis methods is to allow a direct feed-back both to device operation (see e.g. [3]) and the planning of experiments.

Present data analysis methods fail to meet with these requirements for large fusion devices in general. For burning plasma experiments, new data analysis issues resulting from deterioration of signals in harsh environments may even complicate the analysis task. The key issue why *data fusion* in MCF devices is troublesome is the lack of *standardization of uncertainty measures*. Uncertainties are the key for a quantitative assessment of different data – the error is the information which allows to put decisions on agreement of data or to indicate disagreements and failures.

A concept to meet with typical fusion data analysis issues and new data analysis requirements is the Integrated Data Analysis (IDA). From an information science point of view, *data fusion* in IDA employs Bayesian *Probability Theory* (BPT). The strict formulation of the data analysis problem within BPT is the required data analysis standard. Moreover, the first-principle based calculus of BPT allows a couple of further applications typically occurring in fusion science ranging from outlier detection, background estimation to model selection for confinement studies. Although the thorough assessment of uncertainties and the technical treatment of high dimensional integrals lead to substantial effort in the data analysis process, techniques for fast analysis are available as well: Bayesian Neural Networks can be used for fast implementations of complex analysis tasks since the Bayesian pruning algorithm leads to efficient networks. Moreover, the proposed data analysis recipe allows for tools for the design of diagnostics and the planning of experiments.

This paper introduces the IDA concept after a brief revision of some data analysis requirements. Examples for applications are introduced.

Survey of Data Analysis Requirements

Issues, resulting requirements and possible IDA techniques to cope with the requirement are summarized in Table 1, particularly including issues from steady-state devices and harsh environments. The purpose of this list is to give a survey on the present requirement discussion and to arrive at a prioritization of tasks.

Table 1: Data Analysis Requirements for Fusion Devices.

Issue	**Requirement**	**Integrated Data Analysis Technique**
Data Analysis Standard	**Framework for standardized data analysis**	**Standardized Bayesian error treatment**
Global data consistency	- standardized error analysis - coherent linkage of different data analysis procedures - linkage to modeling	Bayesian formulation (sum and product rule for pdfs)
Systematic errors	- error propagation for model parameters	hyper parameter, marginalization
Amount of data	Automated analyses	Bayesian formulation (standard) Bayesian Neural Networks
Reactor like operation	On-line physics analysis	Laplace approximations, BNN, IDA
Data mining	Novelty detections Data warehousing	BNN, probability density functions, standard formulation
Error propagation with interdependencies/parameter correlations	- generalized error propagation - treatment of non-Gaussian errors	Bayesian formulation, Integrated Data Analysis
Sensitivity optimization	Super-Fit (joint analysis)	Integrated Data Analysis
Model choice	Technique for model comparison	Bayesian model comparison
Direct model assessment (e.g. data in transport models)	- inference of derived quantities (e.g. gradients)	Bayesian formulation (marginalization, forward model)
Signal identification	Signal – background separation	Mixture models
Radiation induced effects	outlier identification	Mixture models, Robust fitting (non-Gaussian likelihood)
Use of physics data for device control	Fast analyses for real-time use	Simplified/adapted data analysis models Bayesian Neural Networks
Identification of transient effects	Combination of data for in-situ calibration	Bayesian formulation
Planning of experiments	Optimized set of control parameters for experimental settings	Bayesian Experimental Design
Diagnostic Design	Optimization of settings	Bayesian Experimental Design
Robust/redundant set of diagnostics in harsh environments	Multi-tasking diagnostics, Use of information in case of failures	Meta-Diagnostics Integrated Analysis
Identification of failures & mismeasurements	validation tools	Probability density function Interdependencies (IDA)

BASIC CONCEPTS

Bayesian Probability Theory is introduced in several textbooks, see e.g. [4,5]. Here, we give a summary introducing the basic concepts and notions.

Data analysis is regarded to be a learning process: knowledge (or ignorance) on a parameter of interest is updated with knowledge from measurements. Mathematically, the updated knowledge on a parameter q due to data d can be phrased in a conditional probability

$$p(q \mid d, I) \tag{1}$$

which reads to be the conditional probability p to find the parameter q given the data d. By explicitly mentioning the background information I, it is acknowledged that information is always context sensitive. Both the quantity and the data can be vectors. In the Bayesian concept p encodes the uncertainty on the quantity q.

The calculus of conditional probabilities is based on the sum and the product rule for probabilities. Bayes theorem results from product rule:

$$p(q \mid d, I) = \frac{p(d \mid q, I) \times p(q \mid I)}{p(d \mid I)} \tag{2}$$

where I is the background of the prior information $p(q \mid I)$ on the quantity q, i.e., the knowledge on q without data d. IDA uses the prior to encode physics or technical knowledge, e.g. positivity of temperatures or detection limits. In order to describe the measurement, the likelihood function $p(d \mid q, I)$ describes the probability to find data d given a quantity q under background information I. The likelihood function employs the forward calculation and the error statistics of the measurement. The evidence $p(d \mid I)$ ensures that the sum of all possible outcome of q is normalized. The actual result $p(q \mid d, I)$ is called the posterior probability. It represents the update from $p(q \mid I)$ with data d. Hence, Bayes theorem (2) provides the prescription for learning from data, i.e. the transformation from a less informative state to a more informative one.

THE BAYESIAN RECIPE

In the Bayesian formulation the course of actions in data analysis (parameter estimation) is:

- a clear statement of the data analysis problem, i.e. the forward calculation from the quantity of interest to the data and the background (context) information.

- formulation of a data model (likelihood) with a quantification of all uncertainties (provided by experimental assessments or by physically justified assumptions, e.g. densities have to be positive)

- quantification of the prior information

- the inference step: the application of Bayes theorem

- the focusing step: marginalization onto the quantity of interest.

The final steps are an application of the product and sum rules. The marginalization is a technique to derive estimates of quantities from the full posterior: the posterior contains all parameters which enter the forward model like, e.g., calibration factors c_i as well as quantities to be inferred, e.g. densities n ($q = \{c_i, n\}$). For the estimation of the specific quantity of interest, one to apply the marginalization rule to extract the desired information from the full posterior, in this example:

$$p(n \mid d,I) = \int p(n, c_i \mid d,I)\, dc_i \tag{3}$$

The result of Eq. (3) is the marginal posterior of quantity n. Since Eq. (3) incorporates the uncertainty of c_i in the estimation of n, the marginalization is a generalization of Gaussian error propagation laws.

These two basic rules can be applied both for the analysis of a single diagnostics and for a set of diagnostics: independent probabilities need to be simply multiplied. Therefore, the rules (2) and (3) are the framework of an integrated data analysis.

Parameter estimation, as discussed, is one of the basic problems in data analysis. A second basic problem in data analysis is the question if a model or hypothesis should be preferred to a different one. This problem is called model comparison. In the Bayesian framework this can be treated by regarding the 'quantity of interest' to be a hypothesis or a model. A typical example is the question if a polynomial of nth degree fits data better than one of (n+1)th degree. In usual approaches, one has to put a decision on the trade-off between model complexity (which generally leads to lower discrepancies between data and fits) and model simplicity (which is the question if a lower complex model already explains the data within the error margins of the experiment). Bayesian model comparison allows one to compare the probability of different models M_i which is quantified by the odds ratio O:

$$O = \frac{p(M_i \mid d,I)}{p(M_j \mid d,I)}. \tag{4}$$

If models are not to be preferred a priori, the odds ratio can be rephrased in the ratio of global likelihoods [4] which give the probability to find the data if a model is given.

Since this quantity includes integration over the model parameter space, a higher complexity is punished by weighting the enlarged parameter space. This principle is called Occam's razor which states that a simpler model is to be preferred when it explains the data. It is clear, that a measure for model consistency must incorporate errors (as done in Eq. (4)), since those allow one to judge if deviances are significant.

A third basic problem is the design of measurements, i.e. which settings are to be chosen to optimize the outcome of an experiment. In the Bayesian experimental design [6] it is proposed to employ a utility function. A utility function based on information theory is the Kullback-Leibler distance [7] which measures the information gain of a measurement by comparing the posterior probability with the probability without a measurement:

$$U_{KL} = \int p(\theta \mid d,I) ld\left(p(\theta \mid d,I) \middle/ p(\theta \mid I) \right) d\theta \qquad (5)$$

Marginalizing the utility function over an expected data space yields an expected utility EU which depends on the settings η only:

$$EU(\eta) = \int p(D \mid \eta, I)\, U\, dD \qquad (6)$$

The expected utility can be maximized with respect to η in order to arrive at the requested optimum setting.

IDA OF SINGLE DIAGNOSTICS: FROM ONE ...

An example for the application of IDA is the analysis of Thomson scattering data on Wendelstein 7-AS [8]. As usual for Thomson scattering, the purpose of the measurement is the determination of electron densities and temperatures. Background information is given by the details of the experimental set-up, calibration measurements and so on. The forward function is given by the data descriptive equation of the measurement:

$$D_{Th}^{(i,j)} = c_{geom}^{(i)} P_i n_e^{(i)} \sigma_{Th} \int \tau^{(i,j)}(\lambda) S\left(\lambda, T_e^{(i)}, \Theta^{(i)}\right) d\lambda \qquad (7)$$

The data model considers background radiation, amplifications and all respective noise sources. For the specific formulation of the likelihood and prior probability density functions one needs to assess all parameters $\vec{\eta}$ entering the model, a good guidance is given by the full data model which transfers the physics origins of the measurement to the data as measured. The likelihood for data d measured with an error σ is given by the normal distribution according to the principle of maximum entropy, i.e., the most uninformative likelihood function for given data and errors:

$$p(d \mid n_e, T_e, \vec{\eta}, \sigma, I) = \frac{1}{\Pi_1^n \sqrt{2\pi}\sigma_k} \exp\left(-\frac{1}{2} \sum \left(\frac{d_k - D_k(n_e, T_e, \vec{\eta})}{\sigma_k} \right)^2 \right) \qquad (8)$$

Here, the likelihood function resembles the χ^2 misfit statistics in accordance with the maximum entropy principle [5]. For the priors on density and temperature, flat functions cut at high values were chosen (e.g. 25 meV < T_e < 10 keV).

It should be noted that the largest effort in the application of the Bayesian recipe lies in the specification of all parameters and errors. The formalized treatment requires numerical techniques for high-dimensional integration.

The Bayesian analysis was found to have a couple of important benefits meeting with fusion specific requirements discussed in the introduction. Moreover, the Integrated Data Analysis formulation can be proposed to serve as a data analysis standard. Then, a reliable comparison of results can be performed. Second, the result contains all (non-linear) correlations introduced by the forward model. In the case of Thomson scattering, a correlation of electron temperature and density is introduced by Eq. (7). As will be discussed later, these correlations are informative and are barely used in traditional data analysis approaches. Third, the sensitivity of the Thomson scattering analysis in W7-AS was enhanced. This was of particular interest for formerly unexpected high density H-mode regimes with very high densities. Fourth, systematic effects can be included in the Bayesian approach by additional model parameters. These model parameters are assigned with an appropriate prior distribution [4, 5]. In the case of the Thomson scattering analysis alignment factors were introduced describing differences in the optical path in laboratory calibrations and at the plasma vessel. The technique to treat these systematic effects is marginalization. Fifth, the data model can be used to assess the impact of different errors (e.g. systematic alignment errors or statistical noise due to electronics). This assessment is valuable for the design and maintenance of diagnostics.

… TO MANY: INTEGRATED ANALYSIS OF DIAGNOSTICS SETS

The step to an integrated data analysis is conceptually straight-forward once diagnostic models are available. Provided the data are taken independently – a trivial prerequisite for separate measurements – the linked result is proportional to the joint product of likelihoods and prior probabilities. IDA infers the quantities inductively (calculation of the probability to find a quantity, cf. Fig. 1(b)), differently to traditional approaches which perform a deductive analysis (Fig. 1(a). Fig. 1 shows the concept for a combination of Thomson scattering and ECE. The traditional approach requires iterative validation steps, each of which individually affected by additional information. This leads to complicated validation steps which require large effort for documentation. The IDA standardized formulation is compact and transparent. The joint posterior is the result of the integrated analysis.

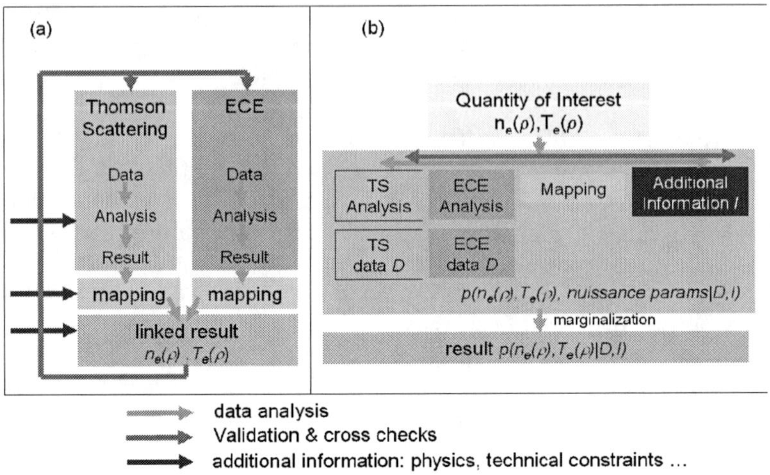

FIGURE 1. Traditional (a) vs. IDA (b) data fusion (combination of Thomson scattering and ECE).

The feasibility of this approach was demonstrated in Ref. [9]. The results show the power of the integrated approach with respect to the reduction of uncertainties of combined analyses: the joint (integrated) probability density function can be regarded as a meta-diagnostic which takes its (heterogeneous) data from different diagnostic units. This approach allows one to combine the measuring capabilities of different diagnostics to make use of their redundancies. An application of this idea is relevant to diagnostics in harsh environments where set-ups might suffer from radiation effects or from a lack of accessibility for maintenance.

A technical representation of IDA is given by probabilistic graphical models [10]. IDA can be used to evaluate uncertainties of formerly not accessible quantities, e.g. the error of the equilibrium (mapping). IDA also allows one to use physics constraints. Fig. 2(a) shows a flow diagram of already analyzed data, fits and model quantities; the link of which is given by the arrows representing probabilistic or deterministic relationships. In the analysis, the data are n_e, T_e, T_i and E_r measurements fitted to smooth model functions. Those are used as input for equilibrium calculations and for transport analyses. The latter provides a link of density and temperature profiles to the radial electric field. Moreover, the particle Γ and heat fluxes Q are outcome of this analysis and can be compared with results from separate modeling results (solid curves in Fig. 2 (b)). The joint posterior can be regarded as a *super-fit* to all data which is shown in Fig. 2(b) [11]. The uncertainty of the fit is indicated by the grey shaded histograms in Fig. 2(b).

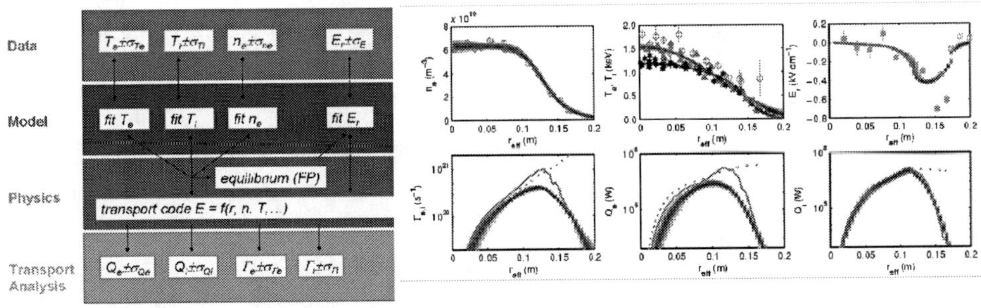

FIGURE 2. The left panel shows information sources and the data fusion scheme for a super-fit including equilibrium calculations and transport equations to link, e.g. data for the radial electric field. The right panel shows the result; note the errors in the quantities from the transport analysis.

MORE APPLICATIONS OF BAYESIAN PROBABILITY THEORY IN FUSION

In addition to conceptual improvements of the data fusion issue for fusion, Bayesian probability gives several solutions to fusion typical problems. The following selection gives some examples.

The idea to model systematic effects by additional model parameters can be extended to data descriptions consisting of two mutual exclusive hypotheses. An example is a typical spectroscopic issue: does a data point belong to a peak or to background? Mixture models consist of a sum representing the exclusive propositions and the occurrence of which are assigned to their probability: the probability to find a datum consist of the probability to find the datum to be a spectral line and the probability, that the datum is not a spectral line. In the introduced terminology, this probability is a hyper-parameter which can be marginalized. This technique was successfully applied to background estimation [12] and outlier estimation [13].

Rather than inferring parameters of the data model, the Bayesian framework may also be employed for the design of fusion measurements [14, 15]. The design is considered to provide experimental settings to measure given quantities of interest optimally as suggested in the previous chapter. So far, optimization studies of interferometer chords and Thomson scattering spectral filters were performed. Physically motivated optimization targets can be defined and cost-vs.-efficacy considerations can be performed on the basis of information gain measures. The outcome indicates the impact of errors on the optimum configuration. An application of the idea to planning of experiments is underway.

The model comparison technique can be used to test different basic plasma models against confinement data. The idea consists in the application of the Kadomtsev/Connor-Taylor invariance principle: If a basic equation is invariant against scaling transformation, then any derived relationship is invariant again. An example is basic plasma equations (e.g. Fokker-Planck kinetic equation) and energy confinement scaling laws. Model differences were found, e.g., in different β regimes of stellarator confinement data [16].

Artificial neural networks have several applications. A key problem is the so-called pruning, i.e. to reduce the neural network to relevant parts. Here, Bayesian neural networks (BNN) employ model comparison techniques to identify the relevant parts of the network. Again, errors play a key role in this consideration. BNNs also allow the identification of novelties since confidence measures for the reconstructed quantities are part of the BNN result [17].

SUMMARY AND OUTLOOK

The concept of integrated data analysis is a framework meeting with many data analysis issues of contemporary fusion devices. IDA also faces new challenges for next step fusion machines. The applications of the IDA concept indicate feasibility and the data analysis techniques were implemented in routine operation and applied to on-line data analyses. IDA offers a frame-work for standardized data-analysis.

ACKNOWLEDGMENTS

This paper is dedicated to Prof. V. Dose who introduced the author to the Bayesian variant. This work was partially supported by Deutsche Forschungsgemeinschaft through Sfb TR24.

REFERENCES

1. Dinklage, A., Fischer, R., and Svensson J., *Fusion Sci. Technol.* **46**, 355-364 (2004).
2. Wald, L., *IEEE Transactions on Geoscience and Remote Sensing* **37**, 1190-1193 (1999).
3. D' Antona, G., Cirant, S. Farina, D. et al., "Algorithms for the Control of NTM by Localized ECRH. Principles and Requirements of the Real Time Diagnostics and Control System" this conference, AIP Conference Proceedings , 2008
4. Gregory, P., *Bayesian Logical Data Analysis for the Physical Sciences*, Cambridge: Cambridge Univ. Press, 2005.
5. Sivia, D. S., and Skilling, J., *Data Analysis: A Bayesian Tutorial*, Oxford: Oxford Science Publications, 2nd Edition, 2006.
6. Shannon, C.E., *Bell Syst. Techn. J.* **27**, 379 (1948).
7. Challoner, K., Verdinelli, I., *Stat. Sci.* **10**, 273 (1995).
8. Fischer, R., Wendland, C., Dinklage, A., et al., *Plasma Phys. Contr. Fusion* **44**, 1501-1519 (2002).
9. Fischer, R., Dinklage, A., and Pasch, E., *Plasma Phys. Contr. Fusion* **45**, 1095-1111 (2003).
10. Svensson, J., Dinklage, A., Geiger, J., et al., *Rev. Sci. Instrum.* **75**, 4219-4422 (2004).
11. Dinklage, A., Beidler, C.D., Fischer, R., et al. "Integrated Interpretative Transport Modelling" in *31st EPS Conference on Plasma Physics,* edited by P. Norreys and H. Hutchinson, Geneva: European Physical Society, 2004, pp. O.2.03.
12. Fischer, R. et al. *Phys. Rev. E* **61**, 1152 (2000).
13. Dose, V. et al. *Nucl. Fusion* **41**, 1671 (2001).
14. Fischer, R., "Bayesian Experimental Design" in *Bayesian Inference and Maximum Entropy Methods in Science and Engineering,* edited by S. C. Schmidt et al., AIP Conference Proceedings 735, New York: American Institute of Physics, 2004, pp. 43-61.
15. Dreier, H., Dinklage, A., Fischer, R., et al., *Rev. Sci. Instrum.* 77, 10F128-10F131 (2006).
16. Dinklage, A., Maaßberg, H., Preuss R., et al Nucl. Fusion **47**, 1265-1273 (2007).
17. v. Toussaint, U, Gori, S., Dose, V., Neural Networks **19** 1550-1557 (2006).

Structural Pattern Recognition Techniques for Data Retrieval in Massive Fusion Databases

J. Vega[1], A. Murari[2], G. A. Rattá[1], P. Castro[1], A.Pereira[1], A. Portas[1] and JET-EFDA Contributors*

JET-EFDA, Culham Science Center, OX14 3DB, Abingdon, UK
[1]Asociación EURATOM/CIEMAT para Fusión. Avda. Complutense, 22. 28040 Madrid, Spain
[2]Consorzio RFX-Associazione EURATOM ENEA per la Fusione. I-35127 Padua, Italy
**See the Appendix of M.L.Watkins et al., Fusion Energy 2006 (Proc. 21st Int. Conf. Chengdu, 2006) IAEA, (2006)*

Abstract. Diagnostics of present day reactor class fusion experiments, like the Joint European Torus (JET), generate thousands of signals (time series and video images) in each discharge. There is a direct correspondence between the physical phenomena taking place in the plasma and the set of structural shapes (patterns) that they form in the signals: bumps, unexpected amplitude changes, abrupt peaks, periodic components, high intensity zones or specific edge contours. A major difficulty related to data analysis is the identification, in a rapid and automated way, of a set of discharges with comparable behavior, *i.e.* discharges with "similar" patterns. Pattern recognition techniques are efficient tools to search for similar structural forms within the database in a fast an intelligent way. To this end, classification systems must be developed to be used as indexation methods to directly fetch the more similar patterns.

Keywords: Data mining, information retrieval, nuclear fusion, structural pattern recognition.
PACS: 42.30.Sy

INTRODUCTION

So far, data retrieval in fusion databases has been based on a classical model of shot number for input and signal samples for output. However, high level tools to look for data according to scientific and technical criteria are a main challenge for present (JET) and future fusion devices (W7X and ITER). A new paradigm for data access has been proposed not based on shot number but on pattern recognition [1]. The basis for this rests on the fact that diagnostics translate physical properties into patterns with a direct correspondence between the physical behavior of the plasma and the structural shapes that are generated in the signals. The new paradigm means to evolve the input/output data retrieval mechanisms towards a new archetype in which the input is a pattern and the outputs are the shot numbers and the locations (temporal and/or spatial) where the pattern appears.

Pattern recognition is the scientific discipline whose aim is the classification of objects into a number of categories or classes. Key elements to implement pattern recognition systems are the signal representations (feature vectors contain the object attributes of distinctive nature), the classification systems (grouping similar objects into clusters) and the similarity measures (to compare how similar two objects are).

The crucial element to achieve efficiency when searching for similar signals is the classification method. It allows the reduction of the searching space just to the most

CP988, *Burning Plasma Diagnostics*, edited by F. P. Orsitto, G. Gorini, E. Sindoni, and M. Tardocchi
2008 American Institute of Physics 978-0-7354-0507-3/08/$23.00

probable signals of containing a similar pattern. Each cluster represents a reduced searching space in which the objects more likely to be similar can be found.

STRUCTURAL PATTERN RECOGNITION IN FUSION

Different pattern recognition systems can be developed for diverse signal collections. A signal collection is the complete set of recorded signals (individual waveforms or images) for all the discharges of interest. Two very general approaches to search for data are considered: the entire signal search (ESS) and the "pattern in signal" search (PISS). An entire signal is a complete image or waveform. ESS allows the searching of similar entire signals within the database. PISS permits seeking for specific patterns inside signals through the whole database.

Feature extraction processes, classification systems and similarity measures are developed for general purpose data retrieval methods. Given a target signal, the searching process of similar waveforms begins with the classification of the initial signal into one of the clusters of the classification system. Then, the similarity measure is only computed between the signals in the cluster. In this way, the waveforms can be ordered according to the similarity value.

Entire Signal Search

Temporal Evolution Signals

Due to the fact that temporal evolution signals are the dominant waveform type in fusion, attention was focused on this kind of signals.

- Feature extraction
  - Haar wavelet transform: $\mathbf{u} = (u_1, u_2, \ldots u_M) \Leftrightarrow \mathbf{h_u} = (h_1, h_2, \ldots h_N), N << M$
    - Retains most relevant time/frequency information
    - Strong dimensionality reduction
  - Feature vector: $\mathbf{h_u} = (h_1, h_2, \ldots h_N)$
- Classification system: multilayer classification system (fig. 1)
- Similarity measure
  - Normalized inner product: $S_{\mathbf{uv}} = \left|\cos\alpha\right| = \dfrac{\left|\mathbf{h_u} \cdot \mathbf{h_v}\right|}{\left\|\mathbf{h_u}\right\| \cdot \left\|\mathbf{h_v}\right\|}, \quad 0 \le S_{\mathbf{uv}} \le 1$

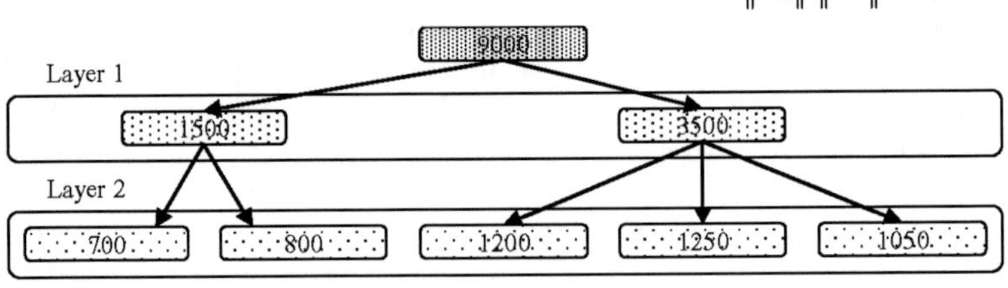

FIGURE 1. Different clustering criteria can be defined between layers. Clusters are split into smaller ones through the different layers. Numbers represent the number of signals in the clusters. Properties of the multi-tier scheme are described in [2].

A fully description of the ESS approach to the TJ-II stellarator database can be found in [2]. Applications to the JET database were shown in [3].

Images

- Feature extraction
  - Thresholding + 2D Haar wavelet transform
    - Thresholding: intensity high-pass filter
      - $\mathbf{U} = (u_{ij}), i = 1,...,M, j = 1,...M \, (u_{ij} = 0 \; if \; u_{ij} < threshold)$
    - 2D Haar: $\mathbf{U} = (u_{ij}) \Leftrightarrow \mathbf{H}_{\mathrm{U}} = (h_{ij}), i = 1,...,N, j = 1,...N, N < M$
  - Feature vector: $\mathbf{H}_{\mathrm{U}} = (h_{ij}), i = 1,...,N, j = 1,...N$
- Classification system
  - 2^{NxN} clusters defined by pixels with simultaneously intensity > 0
- Similarity measure
  - Euclidean distance: $d(\mathbf{U},\mathbf{V}) = \left\| \mathbf{H}_{\mathrm{U}} - \mathbf{H}_{\mathrm{V}} \right\|_{Euclidean}$

Results with this technique were given in [4].

"Pattern In Signal" Search

Temporal Evolution Signals

- Feature extraction
  - Haar wavelet transform + delta calculation + primitive computation
    - Haar: $\mathbf{u} = (u_1, u_2, ..u_M) \Leftrightarrow \mathbf{h}_{\mathbf{u}} = (h_1, h_2, ...h_N), N << M$
    - Delta calculation: $\delta_i = h_{i+1} - h_i, i = 1...N-1$
    - Primitive: label assignment according to (δ_i / Δ) (fig. 2)
  - Feature vector: string of characters $\mathbf{v} = "eeabed"$
- Classification system
  - Relational database
    - Recognition problem => pattern-matching problem
- Similarity measure
  - Euclidean distance: $d(\mathbf{v},\mathbf{w}) = \sqrt{\sum_{i=1}^{patternlength} \left(\delta_{\mathbf{v}i} - \delta_{\mathbf{w}i}\right)^2}$

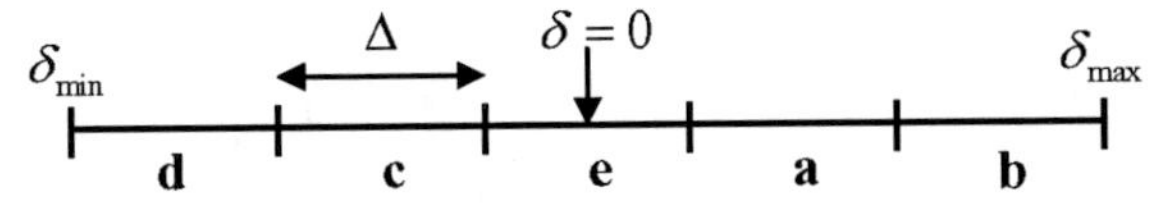

FIGURE 2. Labels (five values in this example: d, c, e, a, b) are assigned depending on the delta value between consecutive points of the Haar transform.

Similar techniques not based on the delta transformation but on the slope of a linear fitting of segmented waveforms (both equal length segments [5] and variable length segments [3]) were previously developed.

Images

- Feature extraction
    - Thresholding + 2D Haar wavelet transform + primitive computation
        - Thresholding: intensity high-pass filter
            - $\mathbf{U} = (u_{ij}), i = 1,...,M, j = 1,...M (u_{ij} = 0 \, if \, u_{ij} < threshold)$
        - 2D Haar: $\mathbf{U} = (u_{ij}) \Leftrightarrow \mathbf{H_U} = (h_{ij}), i = 1,...,N, j = 1,...N, N < M$
        - Primitive: label assignment according to $\left(h_{ij} / \Delta h\right)$ (fig. 3)
    - Feature vector: 2D string of characters $\mathbf{V} = (v_{ij}), i = 1,...,N, j = 1,...N$
- Classification system
    - Relational database
        - Recognition problem => pattern-matching problem
- Similarity measure
    - Euclidean distance: $d(\mathbf{V},\mathbf{W}) = \left[\|\mathbf{H_V} - \mathbf{H_W}\|_{Euclidean} \right]_{pattern\ row\ and\ columns}$

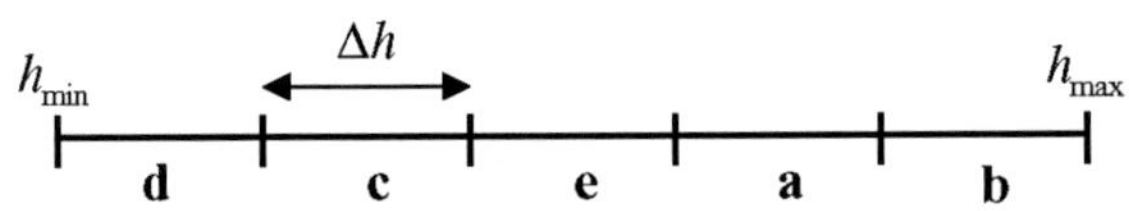

FIGURE 3. Labels (five values in this example: d, c, e, a, b) in the pixels are assigned depending on the values of the Haar transform.

This technique is going to be applied to visible fast cameras in JET.

ACKNOWLEDGEMENTS

The work leading to this article was funded by the European Atomic Energy Community and is subject to the provisions of the European Fusion Development Agreement.

REFERENCES

1. J. Vega and JET EFDA Contributors. "Intelligent methods for data retrieval in fusion databases". 6th IAEA Technical Meeting on Control, Data Acquisition and Remote Participation for Fusion Research. 4-8 June 2007. Inuyama (Japan) (http://tm2007.nifs.ac.jp/). (Sent to Fusion Engineering and Design).
2. J. Vega, A. Pereira, A. Portas, S. Dormido-Canto, G. Farias, R. Dormido et al. "Data mining technique for fast retrieval of similar waveforms in Fusion massive databases". Sent to Fusion Engineering and Design.
3. G. A. Rattá, J. Vega, A. Pereira, A. Portas, E. De la Luna, S. Dormido-Canto et al. "First applications of structural pattern recognition methods to the investigation of specific physical phenomena at JET". 6th IAEA Technical Meeting on Control, Data Acquisition and Remote Participation for Fusion Research. 4-8 June 2007. Inuyama (Japan) (http://tm2007.nifs.ac.jp/). (Sent to Fusion Engineering and Design).
4. D. Raju, J. Vega, P. Castro, G. A. Rattá, A. Murari, G. Vagliasindi and JET EFDA Contributors. "Structural pattern recognition for image processing in fusion plasmas". 6th IAEA Technical Meeting on Control, Data Acquisition and Remote Participation for Fusion Research. 4-8 June 2007. Inuyama (Japan) (http://tm2007.nifs.ac.jp/).
5. S. Dormido-Canto, G. Farias, J. Vega, R. Dormido, J. Sánchez, M. Santos et al. "Search and retrieval of plasma waveforms: structural pattern recognition approach". Rev. Sci. Ins. 77 (2006) 10F514.

Performance analysis of ITER tomographic systems

S. Kálvin* and L.C. Ingesson†

*MTA KFKI–Research Institute for Particle and Nuclear Physics, EURATOM Association, P.O.Box 49, H–1525 Budapest–114, HUNGARY
†EFDA Close Support Unit - Garching, Boltzmannstr. 2, D-85748 Garching, Germany

Abstract. A method based on the Bayesian statistical analysis is proposed to measure the performance of ITER tomographic systems. This method can quantify the performance of the systems and helps in the diagnostic design.

Keywords: Tomography, ITER, Bayesian statistical analysis
PACS: 28.52.Av

INTRODUCTION

Performance analysis to predict the likely capability of the diagnostics to meet the ITER measurement requirements is an important aspect of the design process. Tomographic measurement, Bolometry, soft X-ray and neutron diagnostic, extract information on the 2D emissivity distributions of the plasma from multi–channel spatially non localized measurement. That is the data analysis involves the derivation of physical plasma parameters from the measurements with tomographic reconstruction. Finding the solution of the tomographic problem is ill-posed: the measurements themselves do not describe a unique solution and the solution is sensitive to noise. Therefore to find adequate tomographic method for performance assessment is an important duty in order not to underestimate the performance and in consequence not to overestimate the requirements for the measuring system.

The performance assessment presented here is based on the data evaluation of the tomographic measurement in order to extract physically meaningful quantities. The performance of the tomographic measurement can be given to be objective and quantitative giving the estimation uncertainty of some scalar physically meaningful quantities. With the help of Bayesian statistical analysis the expected utility, that is the gain of information from the measurement can be determined. The tomography and performance assessment for ITER neutron diagnostic is presented here, but similar method is being applied for ITER SXR and bolometer system.

TOMOGRAPHIC RECONSTRUCTION METHOD

The performance assessment is based on tomography method which must have the flexibility to deal with the following characteristics in the field of plasma diagnostic: limited number of lines of sight and viewing directions, irregular coverage, finite beam

CP988, *Burning Plasma Diagnostics*, edited by F. P. Orsitto, G. Gorini, E. Sindoni, and M. Tardocchi

width of the detection, noisy and uncertain data and additional background on the measurement. Series expansion constrained optimization tomographic method copes with the above characteristics. The lack of the information given by the measurement can be replaced by *a priori* information on the emissivity profile such as the assumption that the emissivity profile follows the magnetic flux surfaces [1].

The ITER neutron tomography based on multichannel collimator measurement will be used for several physics study, although detection of the neutrons burn in a fusion plasma is a complicated process [2] . Neutrons can reach the detector flying through the collimator without collision, scattering in the collimator tube and after many scattering in the wall and the shielding component. While the first component can be directly used to determine the plasma neutron source profile but if the other two contributions are significant the simple tomographic reconstruction is failed. It was found that all of the contributions can be taken into account properly and the plasma neutron emissivity profile can be determined with reasonable accuracy using appropriate geometric function. The geometric function which is the relation between the neutron source profile and the measured signal can be extracted from detailed MCMP calculation. In *Fig 1* an example can be seen demonstrating the possibility of the tomographic reconstruction in the case of high scattered contribution.

PERFORMANCE ASSESSMENT

Planing diagnostic leads to the requirement to optimize the outcome from the measurement. The analysis of the outcome of the diagnostic for several technically possible alternative designs can help to decide between alternatives. In some cases, the optimization is much restricted by technical constrains, but even in those cases the performance of the chosen design has to be investigated. Estimation of physically meaningful quantities from the measurement is a possible optimization target. This kind of diagnostic design is successfully applied in the field of plasma diagnostic [3]. Part of the investigation is the virtual diagnostic. This describe the future experiment including its error statistic producing artificial measurement data. The virtual diagnostic is fed with physical quantities of interest and enable us to compare the outcome with its input.

The outcome is quantified with the utility function $U(D)$ depending on the data D. The data represent the expected measurement depending on the expected range of quantities of interest (ϕ), the diagnostic $D = f(\phi)$ and the error statistic of the measurement. Please note here that the diagnostic contains not only the measurement hardware but the data analysis (tomographic method) as well. A possible utility function is the Kullback-Leiber distance U_{KL} which measures the information gain from the ignorance on a quantity ϕ before a measurement to the knowledge after data D are taken [4],

$$U_{KL}(D) = \int d\phi P(\phi|D) log(P(\phi|D)/P(\phi)) \tag{1}$$

An integration(averaging) over the range of expected data (D), where the evidence of the data is represented by the probability density function P(D), yields the expected utility,

$$EU = \int dDP(D)U_{KL}(D) \tag{2}$$

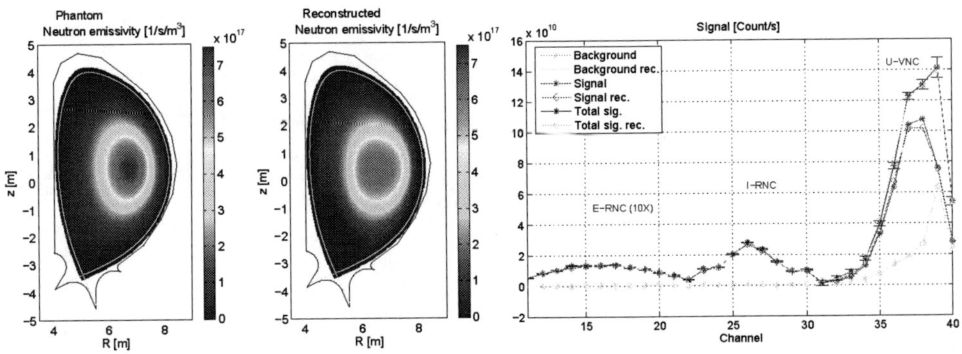

FIGURE 1. Result of a reconstruction of the neutron source profile from the measurement of radial neutron cameras and upper vertical camera. Left: emissivity profile of the phantom and the reconstruction. Right: Different component of the simulated and back-calculated measurement.Signal: direct and collimator scattered neutrons. Background: scattered from the wall. The Total sig. is the input of the tomographic reconstruction.

$P(\phi)$ is the prior probability function describing the prior knowledge of the parameter of interest. The posterior distribution $P(\phi|D)$ describes the updating of the knowledge in the light of new data. Bayesian probability theory is applied:

$$P(\phi|D) = P(D|\phi)P(\phi)/P(D) \tag{3}$$

The likelihood $P(D|\phi)$ is the probability for getting the observed data supposing given parameter ϕ. The likelihood can be determined by a series of numerical measurement systematically varying the ϕ parameter and adding random noise to the simulated measurement representing the unceartinity of the measurement. In the example given, the determination of the horizontal shift of the center of the emissivity profile relative to the magnetic axis (ϕ) is analyzed. Measuring the shift the horizontal position of the center of the gravity of the reconstructed 2D emissivity distribution is used (D). The result of the series of numerical simulations can be seen in the left side plot of *Fig 2*. The red dots are the data while the blue diamonds with error bars represent its averaged values and standard deviation. The blue and red curves represent functions fitted to the averaged value ($\hat{D}(\phi)$) and to the standard deviation($\hat{\sigma}(\phi)$), respectively. The likelihood can be constructed as:

$$P(D|\phi) \approx exp(-(D - \hat{D}(\phi))^2/2\hat{\sigma}^2(\phi)) \tag{4}$$

Supposing flat prior in the interval of $\phi = [-0.5, 0.5]$ the expected utility can be calculated using *Eq 2*. The EU, that is the information gain, is connected with the estimation uncertainty. As the EU is an average quantity for all possible experimental scenarios(ϕ), therefore we can not convert the EU directly into estimation uncertainty.

For any particular data the expected value and the standard deviation of the parameter estimation can be calculated from the posterior probability density function.

$$E_\phi(D) = E[P(\phi|D)] \quad \Sigma_\phi(D) = \sqrt{VAR[P(\phi|D)]} \tag{5}$$

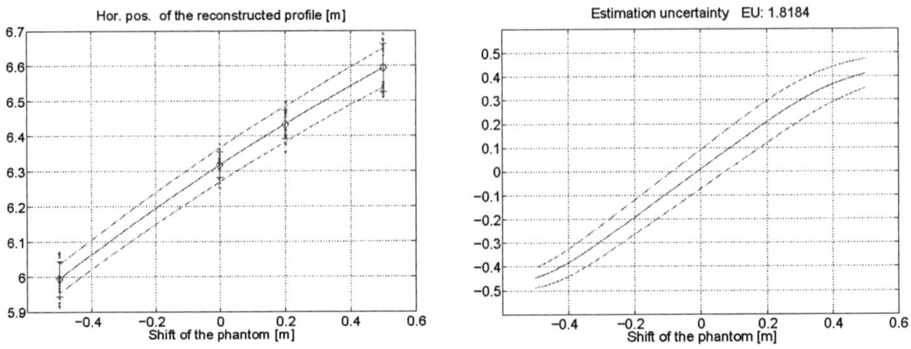

FIGURE 2. Left: The results of the numerical simulations for determination of the likelihood. Right: The estimation uncertainty of the measurement.

For any parameter values, averaging the above expected value and standard deviation taking into account the probability of the data ($P(D|\phi)$

$$< E_{\phi} > (\phi) = \int dD E_{\phi}(D) P(D|\phi) \quad < \Sigma_{\phi} > (\phi) = \int dD \Sigma_{\phi}(D) P(D|\phi) \qquad (6)$$

gives the expressions depending on the parameter. The averaged standard deviation can be considered as the estimation uncertainty for different the experimental scenarios considered (for different ϕ value). The result of this calculation can be seen on the right side plot of *Fig 2.* The blue curve is the averaged expected value, while the red curves represent the averaged standard deviation.

The consequence is that the assessed performance of the measurement depends not only on the hardware capability of the system (LOS arrangement, S/N ratio, etc.) but depends on the data analysis method (the tomographic method) and the expected range of the emissivity profile as well.

ACKNOWLEDGMENTS

This work, supported by the European Communities under the contract of Association between EURATOM and Euratom Association HAS, was carried out within the framework of the European Fusion Development Agreement under EFDA Contract EFDA/03-1116,EFDA/05-1320 and EFDA/06-1447. The views and opinions expressed herein do not necessarily reflect those of the European Commission.

REFERENCES

1. L.C. Ingesson et al., *Nucl. Fusion* 38, (1998) pp.1675
2. A.V. Krasilnikov, at al., *Instruments and Experimental Techniques,* Vol. 47, No. 2, (2004), pp. 139-143
3. H. Dreier at al., *Review of Scientific Instruments,* 77 10F323 (2006)
4. D.V. Lindley, *Bayesian Statistics, A Review,*SIAM Philadelphia, 1972

Cellular Neural Network for Real Time Image Processing

G. Vagliasindi[a], P. Arena[a], L. Fortuna[a], G. Mazzitelli[b], A. Murari[c], and JET-EFDA Contributors*

JET-EFDA, Culham Science Center, OX14 3DB, Abingdon, UK
[a]Dipartimento di Ingegneria Elettrica Elettronica e dei Sistemi - Università degli Studi di Catania, I-95125 Catania, Italy
[b]ENEA-Gestione Grandi Impianti Sperimentali, via E. Fermi 45, I-00044 Frascati, Rome, Italy
[c]Consorzio RFX-Associazione EURATOM ENEA per la Fusione, I-35127 Padova, Italy

Abstract. Since their introduction in 1988, Cellular Nonlinear Networks (CNNs) have found a key role as image processing instruments. Thanks to their structure they are able of processing individual pixels in a parallel way providing fast image processing capabilities that has been applied to a wide range of field among which nuclear fusion. In the last years, indeed, visible and infrared video cameras have become more and more important in tokamak fusion experiments for the twofold aim of understanding the physics and monitoring the safety of the operation. Examining the output of these cameras in real-time can provide significant information for plasma control and safety of the machines. The potentiality of CNNs can be exploited to this aim. To demonstrate the feasibility of the approach, CNN image processing has been applied to several tasks both at the Frascati Tokamak Upgrade (FTU) and the Joint European Torus (JET).

Keywords: cellular nonlinear networks, image processing, tokamaks, nuclear fusion
PACS: 07.05.Pj, 87.57.N-, 52.55.Fa

INTRODUCTION

CNNs are array of simple, identical, locally interconnected nonlinear dynamic circuits called cells, each of which interacts, via weighted connections, with the cells in the neighborhood of a limited radius. Due to their structure, they are particularly suited for image processing tasks.

Moreover, thanks to their local interaction on a limited radius, CNNs can be easily implemented as analog and mixed-mode circuits. This kind of implementation, leading to an individual circuit for each cell, permits to process each pixel in a parallel way. CNNs can, therefore, represent a valid alternative to traditional methods as far as real time image processing is concerned.

To demonstrate the feasibility of the CNN approach to real time image processing in nuclear fusion, several applications of CNN-based hardware and software techniques were developed in the last years at FTU and JET, exploiting the output

* See the Appendix of M.L.Watkins et al., Fusion Energy 2006 (Proc. 21st Int. Conf. Chengdu, 2006) IAEA, (2006)

CP988, *Burning Plasma Diagnostics*, edited by F. P. Orsitto, G. Gorini, E. Sindoni, and M. Tardocchi
2008 American Institute of Physics 978-0-7354-0507-3/08/$23.00

from video cameras installed in the two devices. Problems like real time detection of striking points position, hot spots and MARFE have been successfully addressed.

An overview of these developed techniques will be reported, together with a brief introduction on CNN technology.

CELLULAR NONLINEAR NETWORKS

The concept of CNNs was introduced in 1988 by L.O. Chua [1]. The architecture of the CNN is made up of a basic circuit called cell, containing linear and nonlinear circuit elements. Each cell in a CNN is connected to its local neighboring cells, so a direct interaction occurs only among adjacent cells.

The neighborhood of the cell on the i-th row and j-th column, denoted by C(i, j), has the following definition:

$$N_r(i,j)=\{C(k,l)\ |\max\{\ |k-i|,|l-j|\ \}\le r\},\ 1\le k\le M, 1\le l\le N \qquad (1)$$

where r is a positive integer number, which fixes the dimension of the neighborhood.

A CNN is entirely characterized by a set of nonlinear differential equations associated with the cells in the circuit. Several mathematical model for the state equation of the single cell has been proposed since their introduction. The model as implemented in the CNN Universal chip family [2], also called Full Signal Range (FSR), is given by the following set of relations:

$$C_x\frac{dx_{ij}(t)}{dt}=-\frac{1}{R_x}g(x_{ij}(t))+\sum_{C(k,l)\in N_r(i,j)}A(i,j;k,l)y_{kl}(t)+\sum_{C(k,l)\in N_r(i,j)}B(i,j;k,l)u_{kl}(t)+I \qquad (2)$$

and the output equation is:

$$y_{ij}(t)=x_{ij}(t) \qquad (3)$$

where u, x, and y denote the input, state, and output of the cell, respectively; R_x and C_x are the values of the linear resistor and linear capacitor, that determine the time constant of the circuit; $A(i, j; k, l)$ and $B(i, j; k, l)$ are the feedback and control templates respectively; I is the bias term, that is constant for all the CNN cells; $g(x)$ is the nonlinear function in the state equation (2), described in [2].

One of the main applications of CNNs is image processing. In this case the input image is mapped on the CNN in such a way that each image pixel is associated with the input or initial state of a particular cell. The CNN evolution implies a transformation of the input image into the corresponding output image obtained directly by equation (2). In this contest, the template operators work like the instructions in a programming code. A huge amount of templates and template algorithms for a variety of tasks is already available in the literature [3].

The hardware prototype system used in this application is based on the CNN Universal Chip prototype, which is a 128x128 mixed-signal CNN chip [4], developed following the concept of Single Instruction Multiple Data (SIMD) architectures. It is named ACE16K, where ACE is the acronym of Analogic Cellular Engine to underline the mixed signal nature of the chip (analog and logic).

ACE16K can be described basically as an array of 128x128 identical, locally interacting, analog processing units designed for high speed image processing tasks requiring moderate accuracy (around 8bits). The system contains a set of on-chip peripheral circuitries that, on one hand, allow a completely digital interface with the host, and on the other provide high algorithmic capability by means of conventional programming memories where the algorithms are stored. Implemented in this way, CNNs have the same computational power as Universal Turing machines and therefore they can be applied to a variety of tasks [5].

STRIKE POINT DETECTION AT JET

In Tokamak plasmas, the divertor is the region of the vacuum vessel explicitly designed to handle power losses. The intersection of the separatrix, the last closed flux surface, with the divertor target plate represents a strike point. In JET the position of the strike points is mainly derived from magnetic measurements obtained through loops and pick-up coils located around the vacuum vessel. The code used at JET to determine the plasma shape and therefore also the position of the strike points is XLOC [6]. Another very useful diagnostic to derive information about the power deposition in the JET divertor is represented by infrared imaging. At the time of the experiments, two IR cameras framing the divertor region were available at JET, measuring the infrared radiation in the interval 3-5 μm with a resolution of 128x128 pixels [7]. A specific procedure was developed to derive the position of the strike points from the data of the infrared cameras exploiting the capabilities of CNN technology. A detailed description of the procedure is given in [8]. It is capable of supplying the position of both the inner and the outer strike points within 20 ms. In particular, the time to identify the strike points in the image is comprised between 13 and 19 ms, depending on the brightness condition of the starting frame.

HOT SPOTS DETECTION IN JET VACUUM VESSEL

One of the future ITER relevant enhancements at JET will be the installation of a Beryllium wall. Since Be is much more vulnerable than stainless steel (the present JET wall material) preserving the integrity of the plasma facing components will be one of the main issues in future JET experiments. Detecting on time the presence of hot spots, i.e. regions of the first wall where the temperature approaches dangerous levels, is considered crucial in the safety strategy. CNN technology was applied to the real time identification of hot spots.

Two different algorithms for the identification of the hot spots were developed. The first one, for the so called static detection, performs the analysis of a single frame at the time. It is more suited to the monitoring of the fixed parts of the machines, like the plasma facing components (limiters and divertor). A second approach, called dynamic detection, is based on the difference between subsequent frames. The latter algorithm is more complicated but it allows to follow the growth of the hot spots and their movements inside the field of view. A detailed description of both the algorithms is reported in [9].

Both algorithms were tested using frames acquired by JET new wide angle infrared camera [10]. The execution time is in the order of 60 ms to process a 384x384 frame while the time required to process a 128x128 subframe is about of 10 ms. This higher time resolution could be used to monitor particularly delicate parts of the machine, like the RF antennas, with a higher time resolution.

EARLY MARFE DETECTION AT FTU

The MARFE [11] is a radiation instability which appears in tokamaks as a toroidally symmetric ring of enhanced radiation. It usually occurs on the inner side of the torus. An uncontrolled MARFE could limit the maximum density achievable during a discharge or evolve to a disruption. In consequence of this real time control of MARFEs would be a useful mean to extend the plasma operation parameters and to avoid dangerous disruptions. Camera observations of a wide poloidal portion of the plasma edge represent a good candidate for early detection of the MARFE onset and the development of a feedback control in order to mitigate its effects.

An algorithm has been devised in order to monitor the incipiency of a MARFE [12]. The whole algorithm has been implemented on the CNN based chip ACE16K. The data from the video monitoring system installed at FTU which provide a frame every 20 ms [13] was used in this work. The time taken by the algorithm to process two consecutive frames is about 16 ms, which is below the interframe rate. This real time detection system can help in carrying out a safe termination of the experiment when the probability of the occurrence of a disruption is high, thus preventing the tokamak from mechanical and thermal stresses.

ACKNOWLEDGMENTS

The work was partially supported by the Project "Real-time visual feature extraction from plasma experiments for real time control," funded by ENEA-EURATOM. The work leading to this article was funded by the European Atomic Energy Community and is subject to the provisions of the European Fusion Development Agreement.

REFERENCES

1. L. O. Chua, L. Yang, *IEEE Transactions on Circuits and Systems - I*, vol. 35, pp. 1257-1290, 1988.
2. S. Espejo et al., *International Journal of Circuit Theory and Applications*, vol. 24, pp 341-356, May-June 1996.
3. T. Roska et al., *CNN Software Library (Template and Algorithms) Version 7.3*, Hungarian Academy of sciences ed. Budapest, Hungary: Computer and Automation Institute, 1999.
4. A. Rodríguez-Vázquez, el al., *IEEE Transactions on Circuits and Systems-I*, vol.51, no.5, May 2004
5. L.O. Chua, K.R. Crounse, *IEEE Transactions On Circuits And Systems - I*, vol. 53, no.4, April 1996.
6. F. Sartori, A. Cenedese and F. Milani, *Fusion Eng. Des.*, vol. 66-68, pp. 735-739, 2003.
7. V Riccardo et al, *Plasma Phys. Control. Fusion*, vol. 44, pp. 905-929, 2002.
8. P. Arena et al., *Rev. Sci. Instrum.* 76, Issue 11, Article 113503 (November 2005).
9. G. Vagliasindi et al., "First Application of Cellular Neural Network Methods to the real time identification of Hot Spots in JET", submitted to *IEEE Transactions on Plasma Science*.
10. E. Gauthier et al., *24th Symposium on Fusion Technology (SOFT 2006)*, 11-15 September 2006, Warsaw, Poland, to be published in Fusion Engineering and Design.
11. B. Lipschultz, *Nuclear Fusion*, vol. 24, n. 8, Aug. 1984, pp. 977-989.
12. P. Arena et al, *IEEE Transactions on Plasma Science*, Vol. 33, Issue 3, June 2005, pp. 1-9.
13. R. De Angelis et al., *Rev. Sci. Instrum.* 75, Issue 10, pp. 4082-4084, October 2004.

Author Index

A

B

C

D

E

F

G

H

I

J

K

L

M

T

V

W

Y

Z